Marine Conservation Ecology

Marine Conservation Ecology

John Roff and Mark Zacharias

with early contributions from Jon Day

publishing for a sustainable future

London • Washington, DC

First published in 2011 by Earthscan

Earthscan Ltd, Dunstan House, 14a St Cross Street, London EC1N 8XA, UK
Earthscan LLC, 1616 P Street, NW, Washington, DC 20036, USA
Earthscan publishes in association with the International Institute for Environment and Development

For more information on Earthscan publications, see www.earthscan.co.uk or write to earthinfo@earthscan.co.uk.

ISBN: 978-1-84407-883-7 hardback
ISBN: 978-1-84407-884-4 paperback

Typeset by JS Typesetting Ltd, Porthcawl, Mid Glamorgan
Cover design by Susanne Harris

A catalogue record for this book is available from the British Library.

Library of Congress Cataloging-in-Publication Data

Roff, J. C., 1942-
Marine conservation ecology / John Roff and Mark Zacharias.
p. cm.
Includes bibliographical references and index.
ISBN 978-1-84407-883-7 (hardback) – ISBN 978-1-84407-884-4 (pbk.) 1. Marine resources conservation. 2. Marine ecosystem health. 3. Marine ecology. I. Zacharias, Mark, 1967- II. Title.
GC1018.R64 2011
577.7–dc22

2010049767

At Earthscan we strive to minimize our environmental impacts and carbon footprint through reducing waste, recycling and offsetting our CO_2 emissions, including those created through publication of this book. For more details of our environmental policy, see www.earthscan.co.uk.

Printed and bound in the UK by MPG Books, an ISO14001 accredited company. The paper used is FSC certified.

For Elizabeth and Karen

CONTENTS

Figures, Tables and Boxes

Figures

Tables

Boxes

Preface

Marine Conservation Ecology: Concepts and Frameworks

The oceans have traditionally been conceived of as boundless and beyond the realm of significant human impacts; the great fisheries of the world were once considered as essentially limitless. By 1883, even though economic pressures on fisheries were already being felt, T. H. Huxley boldly announced that 'Any tendency to over-fishing will meet with its natural check'. We now know that these conceptions are false. Worldwide, fisheries are in decline – perhaps irreversibly in some cases – and marine habitats have become extensively degraded. Marine conservation is no longer an option – it is a pressing necessity. However, despite many well intentioned – but generally disjointed – efforts, marine conservation is not yet firmly based on ecological foundations. Nevertheless, we can recognize general principles, concepts and even paradigms, and use these as an experimental basis for planning frameworks and decision making. This in turn requires ecological classifications. The history of science (whatever the discipline) shows how '...the development of comprehensive theoretical systems seems to be possible only after a preliminary classification has been achieved' (Nagel, 1961). This is what this book is all about, but there is little here that is canonical. It has often been said that biology has only one theory – evolution by natural selection. If this is so, then conservation has no real theories at all, it is a pragmatic science. We, the authors, simply hope to present a practical set of approaches to marine conservation that can be used as the basis of frameworks for planning.

We cannot protect the whole of the marine environment – some parts of it will always be open to exploitation and disturbance due to resource use by humans. Therefore, just as in terrestrial conservation, choices need to be made as to what will be protected. But how, on what basis, according to what criteria, do we select those pieces of the ocean that we should protect so as to maximally preserve (or at least efficiently protect) the greatest proportion of the components of marine biodiversity? This is the prime thrust and interest of this book.

We argue that places to be protected (which we shall generally refer to generically as MPAs – marine protected areas) need to be selected systematically on the basis of our knowledge of the ecology of the seas. By 'ecology' we mean simply 'the science of relationships between organisms and their environments'. Such ecological information is inevitably incomplete. This book therefore deals with the principles and criteria underlying the selection of MPAs, rather than the specifics of where conservation is required. However, we shall include specific examples of the process of selection of protected areas. In this book we cannot hope to address all the problems of marine conservation. However, through an examination of ecological principles, and a study of the nature of the marine environment, we can present the issues that need to be addressed in a systematic way. What we are attempting then is a 'codification' of the undertaking of marine conservation. By this we mean that we shall examine the various problems that arise in attempting to deal systematically with the practicalities of marine conservation, to show what they are and how they can be dealt with from available or obtainable data. The main aim of this book therefore is to present the major ecological concepts of and approaches to marine conservation, with an emphasis on protected areas – MPAs.

Books currently available on 'marine conservation' tend to fall into several categories. There are those that emphasize the conservation of individual species, especially rare and endangered species, typically marine mammals; those whose emphasis is on resource conservation, typically commercially exploited species – especially fisheries; others that issue 'wake-up calls' or 'calls-to-action' concerning marine biodiversity and the need for marine conservation; and those that deal with the identity and operation of MPAs based on socio-economic principles. We applaud all these efforts and approaches to marine conservation, and the need for another text on marine conservation may not be readily apparent. However, none of the available texts on marine conservation primarily emphasizes the overall ecological foundations of the discipline. Several of them are not scientifically rigorous, and contain little in the way of hard scientific information. In general they do not consider or integrate science-based approaches to marine conservation, or the ecological logic for planning and establishment of MPAs.

The purpose of this book then is to present the science of marine biodiversity and marine conservation. Specifically, we deal with marine biodiversity and the importance of establishing MPAs to offset threats to it. This in turn requires an understanding of the structure and function of marine environments and the ecological foundations of approaches to and options for marine conservation. We show how conservation initiatives, primarily those based on the establishment of MPAs, can be firmly based on accepted ecological principles, and can be integrated into comprehensive regional and international frameworks. Such a systematic approach, starting with the codification of the components of marine biodiversity, can ensure that we both rationalize the roles of existing MPAs and identify needs for further ones. Only such coordinated regional, national and international planning can ensure that global marine biodiversity is adequately evaluated, represented and protected.

This book explores the theory and practice of marine conservation, and importantly the process of establishment of MPAs, from the perspective of ecological principles. There is, in our opinion, a clear need for a foundational text that emphasizes marine conservation from an ecological approach. Planning for conservation also necessarily involves: social, political, economic, legal and ecological issues. However, of all these subjects, ecological issues have perhaps been most weakly addressed in terms of marine conservation. It is this important gap in marine conservation knowledge and practice that we seek to fill. In an era of change from uni-disciplinary through multi-disciplinary to inter- and trans-disciplinary studies, and with an enterprise as complex as marine conservation, it may seem anachronistic to base a text solely on the discipline of ecology. In developing conservation plans, it has frequently been argued that we cannot view ecological systems in isolation: human socio-economic activities must also be accounted for. We completely agree. However, all conservation plans must be firmly grounded in ecological principles if they are to systematically address the requirement for conservation of the components of biodiversity. Conservation without the reality of the human dimension is incomplete; conservation without the ecological dimension is unreal! We should perhaps recall the etymology of the word 'ecology' (Gk. Oikos – Home). If we cannot understand our own home then we cannot be of much assistance elsewhere beyond it whether in space, time or discipline.

Because our focus in this book is on the ecological basis and principles for marine conservation, we shall be less concerned with other aspects of marine conservation such as: international conventions, marine management, policy, legislation, enforcement, socio-economics and the human dimensions in general. These subjects will inevitably impinge upon our material, but other recent texts deal with them in much greater depth. Thus this book is not so much about human regulatory efforts or human impacts on the oceans (though they will enter into it), but rather it is about the steps we can take to protect the components of marine biodiversity by systematically reserving areas, and how such actions can be founded in ecological principles. We well recognize the significance of other disciplines and the need to integrate them in the process of planning for marine conservation as a whole, but our overarching theme is the ecology of marine biodiversity and its conservation. Nor have we emphasized subjects well covered elsewhere in standard texts – such as population biology and ecology, and fisheries biology. A primary interest is on MPAs – not because the establishment of MPAs is the only thing we should do, but because it has been repeatedly shown that MPAs are effective in protecting 'pieces' of the marine environment and their species. MPAs can be thought of as a necessary but not a sufficient contribution to integrated marine conservation.

Our emphases are primarily on: the identity and components of marine biodiversity, potential approaches to the conservation of marine biodiversity, its relationships to environmental structure and heterogeneity, and the planning of practical marine conservation strategies at the regional and national levels. The main rationale for this approach is that: although concerns for the preservation of marine biodiversity are truly global and international, nevertheless most planning and practical initiatives to conserve marine biodiversity will be undertaken at the national and regional levels. The intention is not to ignore or gloss over other aspects of – or disciplines in – marine conservation; rather it is to emphasize the fundamental importance of ecological knowledge and planning.

An ecological approach to marine conservation requires fundamental knowledge of: the environment, the organisms that inhabit those environments and their habitats, and the biological and physical interrelationships between them. In short, we need to know about the structure and function (processes) of the marine environment in physical and biological terms. It is these environmental and ecological foundations, and the principles that we can derive from them, that are vital in developing our strategies for marine conservation, yet unfortunately this is precisely what is missing from much of the marine conservation text book literature. Even in the primary research literature, such subjects are scattered and unsystematically treated.

A single text that would systematically and comprehensively cover marine environments, the range and diversity of their habitats and communities, biological and physical interrelationships structure and function, and conservation strategies and protected areas planning would be an overwhelming treatise. We do not propose such a treatise. The reader must perforce be referred to other sources for more comprehensive and systematic treatments of each of the major component disciplines and their techniques on which marine conservation planning must call. However, this text is an attempt to organize systematically an approach and overall frameworks, and – at least at a foundational level – to present the various disciplines and indicate their place and role in marine conservation.

There will inevitably be some repetition in this book because there are so many interactions among the subjects to be considered. But the flow of ideas and subjects in the book is as follows.

First, we treat the fundamental issues of what marine biodiversity actually is, and why we should be concerned about it – because of its significance and the various threats to it. Next we describe the basic structure of the marine environment – its major divisions, physico-chemical properties, ecology and biological communities. We continue by examining measures to address the threats to marine biodiversity, by means of various approaches to marine conservation – including strategies for designating MPAs. Here we consider the benefits of marine conservation and the need for a systematic approach and scientific knowledge of the oceans in the decision-making process. We argue for and rationalize the need for a systematic set of approaches, based on hierarchical 'structural' and 'functional' attributes at the genetic, population, community and ecosystem levels of organization. Because most conservation initiatives have been in terrestrial environments, and are based on terrestrial ecological principles, we examine how and why marine systems are different from terrestrial ones, and why they must be treated differently.

We then separately consider the approaches that could be taken towards marine conservation at each 'level' of ecological hierarchies from global to local, and from 'ecosystems' to genes. We consider global biogeographic classification schemes, and what is meant by ecosystem-level approaches to marine conservation. Regional 'representation' at the habitat or ecosystem level and the significance of geophysical attributes of marine environments as surrogates for marine community types is then dealt with, including an examination of relationships between habitats and community properties. Next, crossing the boundaries of the ecological hierarchy, we examine relationships between individual species and ecosystem processes in distinctive areas, including seasonally migrant species that exploit them, and show how they can be recognized and defined. Continuing at the species level, the distribution of species diversity in global and regional 'hotspots' is considered, along with underlying causative factors. Still at the species level, we examine 'focal species' and why some species may deserve more attention than others. Finally, we come to the genetic level and show its growing significance to understanding processes at all the other levels of the ecological hierarchy. The marine coastal zone and deep seas and high seas are each treated separately for a variety of reasons that are explained and rationalized. Fisheries management and its implications and impacts on marine biodiversity as a whole are then considered, and the emerging and vital process of integrating conservation of fisheries and biodiversity is explored.

The next task is to integrate the different approaches to marine conservation based on all the separate levels of the ecological hierarchy, in order to define potential 'sets' of candidate protected areas. Such 'sets' of MPAs should – collectively and ideally – afford protection to 'all' the elements of marine biodiversity. Here we show how the number, size and boundaries of MPAs can be defined, and argue how the proportion of a region to be protected can be established. Following this we examine the concept and meaning of 'value' itself, and suggest how to assess and evaluate conservation efforts. We consider the criteria for selecting 'sets' of candidate MPAs, and for the establishment of 'networks' of MPAs, based on patterns of connectivity from analysis of oceanographic and genetic data. The important but often neglected process of monitoring of conservation efforts is next considered. Finally, we indicate some of the many remaining problems for marine conservation.

Inventories of the global set of MPAs have now been assembled; however, we still have an incomplete idea of what level of protection these areas actually afford to their marine inhabitants or visitors. Most importantly, we still have little idea of what, either individually or collectively, these areas contribute to the protection of the world's components of marine biodiversity. This is surely the next step – to codify the contributions of the world's 'set' of MPAs in terms of their roles in biodiversity protection, and their roles as members of regional, national and international members of networks of MPAs. Following this, a global gap analysis programme (GAP) will indicate what we are missing and where. However, without an overall framework such as presented in this book, it is far from clear how this task would be accomplished.

Optimistically, human populations will eventually stabilize, and our environmental impacts on the globe will become sustainable. Whether this will happen or not is not really in question. The question is simply whether the transition to population stability and environmental sustainability is achieved gently or catastrophically. This book is concerned with the interim – between now and a sustainable future. It is concerned with no less than the fundamental ecological and environmental principles of how to go about planning to conserve the full array of marine biodiversity assets of our planet. How do we plan for a network of new-age Noah's Arks (or perhaps better – Noah's submarines!) to carry over our assets for the future? Ultimately MPAs should become redundant – once humans have learned to live in harmony with their environment and to treat it with respect.

The intended audience for our book consists primarily of senior undergraduate and graduate students of marine biodiversity and conservation, government and non-government agencies and their planners, managers and practitioners who are responsible for the implementation of national, regional or local strategies. However, we hope the text will also appeal to a broader audience with interests in marine ecology and marine conservation, and that it will help them to place their own discipline and actual or potential role into perspective.

Reference

Nagel, E. (1961) *The Structure of Science*, Hackett, Cambridge

Acknowledgements

A book such as this is not written solely by its authors. Experiences, personal contacts and conversations, presentations at scientific meetings – all build to the integration of ideas and concepts. If we were to thank all those friends, colleagues and students (and those whose students we were ourselves) who have influenced and guided our thinking, research and planning it would constitute a very long list indeed; we learn and teach each other. However, in one way or another we are primarily indebted to the following:

JCR (john.roff@acadiau.ca) thanks: Andrew Lewin, Michelle Greenlaw, Susan Evans, Shannon O'Connor, Tanya Bryan, Joerg Tews, Jennifer Smith, Hussein Alidina, Sabine Jessen, Susan Evans, Colleen Mercer-Clarke, Hans Hermann, Bob Rangely, John Baxter, Magda Vincx, Karim Erzini, Cheri Reccia, Josh Laughren, Mark Taylor, Gordon Fader, Jon Day and Vincent Lyne. Special thanks go to the graduate students of the Erasmus Mundus programme, classes of 2009 and 2010, at University of the Algarve in Portugal and University of Gent in Belgium. For help with specific chapters I thank: Chapter 6, Tanya Bryan and David Connor; Chapter 8, Maria Buzeta Innes and Michelle Greenlaw; Chapter 11, Michelle Greenlaw and Shannon O'Connor; Chapter 12, Vera Agostini, Salvatore Arico, Elva Escobar Briones, Malcolm Clark, Ian Cresswell, Kristina Gjerde, Susie Grant, Deborah Niewijk, Arianna Polacheck, Jake Rice, Kathryn Scanlon, Craig Smith, Mark Spalding, Ellyn Tong, Marjo Vierros and Les Watling; Chapter 14, Susan Evans, John Baxter, Michelle Greenlaw, Shannon O'Connor and Andrew Lewin; Chapter 15, Sofie Derous, Tundi Agardy, Hans Hillewaert, Kris Hostens, Glen Jamieson, Louise Lieberknecht, Jan Mees, Ine Moulaert, Sergej Olenin, Desire Paelinckx, Marijn Rabaut, Eike Rachor, Eric Willem, Maria Stienen, Jan Tjalling van der Wal, Vera van Lancker, Els Verfaillie, Magda Vincx, Jan Marcin Węsławski and Steven Degraer ; Chapter 16, John Crawford, Robert Rangeley, Jennifer Smith, Sarah Clark Stuart, Ken Larade, Hussein Alidina, Martin King, Rosamonde Cook, Priscilla Brooks and Josh Laughren; Chapter 17, Tanya Bryan and Joerg Tews.

MZ (Mark.Zacharias@gov.bc.ca) thanks: Hussein Alidina, Jeff Ardron, Rosaline Canessa, Chris Cogan, Phil Dearden, Jon Day, Dave Duffus, Zach Ferdana, Leah Gerber, Ed Gregr, Ellen Hines, Don Howes, David Hyrenbach, Sabine Jessen, David Kushner, Nancy Liesch, Olaf Niemann, Carol Ogborne, Mary Morris, Charlie Short, Mark Taylor and Nancy Wright, and extends apologies to anyone he has forgotten.

As we planned and started to organize this book and its subject matter, we have been significantly influenced by Jon Day and his earlier work with us. In recognition of Jon's earlier contributions with us, and his continued marine conservation activities at the Great Barrier Reef Marine Park Authority, we acknowledge his contribution on the title page of this book.

Over many years, both authors have worked extensively with several national and international conservation organizations, with special interests in marine conservation. Prime among these organizations has been WWF Canada, a tireless champion of applied marine ecology and marine conservation. It has been a privilege for us to work with many colleagues in this organization (see acknowledgements). In recognition of the international efforts of WWF (World Wide Fund for Nature) in marine conservation, the authors have donated their royalties to WWF to be used in the furtherance of marine conservation.

Acronyms and Abbreviations

AEC	adaptive evolutionary conservation
AFLP	amplified fragment length polymorphism
ACM	Atlantic continental margin
BVM	biological valuation map
CA	correspondence analysis
CBCRM	community-based coastal resource management
CBD	Convention on Biological Diversity
CCAMLR	Convention for the Conservation of Antarctic Marine Living Resources
CETAP	Cetacean and Turtle Assessment Programme
CoML	Census of Marine Life
CPUE	catch per unit effort
CRM	coastal resource management
CZM	coastal zone management
dB	decibel
DCA	detrended correspondence analysis
DCCA	de-trended canonical correspondence analysis
DEFRA	Department for Environment, Food and Rural Affairs
DEM	digital elevation model
DFO	Department of Fisheries and Oceans Canada
DPS	distinct population segment
EAF	ecosystem approach to fisheries
EAM	ecosystem approaches to management
EBM	ecosystem based management
EBSA	ecologically and biologically significant area
EcoQO	ecological quality objectives
EEZ	exclusive economic zone
ENSO	El Nino Southern Oscillation
ESA	US Endangered Species Act
ESU	evolutionary significant unit
FAO	Food and Agriculture Organization
GAP	gap analysis programme
GBRMP	Great Barrier Reef Marine Park
GHNMCA	Gwaii Haanas National Marine Conservation Area
GIS	geographic information system
GOOS	global ocean observing system
HSI	habitat suitability indices
ICM	integrated coastal management
ICZM	integrated coastal zone management
IDH	intermediate disturbance hypothesis
IFQ	individual fishing quota
IGOS	Integrated Global Observing Strategy
IOC	Intergovernmental Oceanographic Commission
IUCN	International Union for the Conservation of Nature
IVQ	individual vessel quota
JNCC	Joint Nature Conservation Committee
LME	large marine ecosystem
MARPOL	marine pollution (short form of International Convention for the Prevention of Pollution from Ships)
MEF	modified effective fetch
MEoW	marine ecoregions of the world
mi	mile
MIP	molecular imprinting
MNCR	Marine Nature Conservation Review

MPA	marine protected area
MRU	marine representative unit
MSPR	mortality and spawning potential ratio
MSY	maximum sustained yield
mtCOI	mitochondrial cytochrome oxidase I
mtDNA	mitochondrial DNA
MU	management unit
MVP	minimum viable population
NAO	North Atlantic Oscillation
NAEWC	North Atlantic Right Whale Consortium
Nc	census population size
Ne	effective population size
NER	non-extractive reserve
NMFS	National Marine Fisheries Service
NIS	non-indigenous species
nmi	nautical mile
NMS	non-metric multidimensional scaling
NOAA	National Oceanic and Atmospheric Administration
NRC	National Research Council
OBIS	ocean biogeographic information system
OCP	organochlorine pesticide
OOI	Ocean Observatories Initiative
OSPAR	Oslo and Paris Convention
OSY	optimum sustained yield
OTN	Ocean Tracking Network
PAH	polycyclic aromatic hydrocarbon
PCA/PAC	priority conservation area
PCB	polychlorinated biphenyl
PCDD	polychlorinated dibenzodioxin
PCDF	polychlorinated dibenzofuran
PCM	Pacific continental margin
PcoA	principal components analysis
PCM	possibilistic C-means
PCR	polymerase chain reaction
ppm	parts per million
ppt	parts per thousand
PSAMP	Puget Sound Ambient Monitoring Programme
PSR	pressure-state-response
psu	practical salinity unit
PVA	population viability analysis
RAP	Representative Areas Programme
RAPD	random amplified polymorphic DNA
RMP	Revised Management Procedure
ROI	return on investment
RFLP	restriction fragment length polymorphism
SEK	scientific ecological knowledge
SFG	scope for growth
SMEH	seamount endemicity hypothesis
SPUE	sightings per unit of effort
SSB	spawning stock biomass
SSBR	spawning stock biomass per recruit
STR	short tandem repeat
TAC	total allowable catch
TEK	traditional ecological knowledge
TIE	toxicity identification and evaluation
UKMMAS	UK Marine Monitoring and Assessment Strategy
UNCBD	United Nations Convention on Biodiversity
UNCED	United Nations Conference on Environment and Development
UNCLOS	United Nations Convention on the Law of the Sea
UNEP	United Nations Environment Programme
UNICPOLOS	United Nations Informal Consultative Process on Oceans and the Law of the Sea
UNTBB	unified neutral theory of biodiversity and biogeography
NODC	US National Oceanographic Data Center
VRM	vector ruggedness measure
WMO	World Meteorological Organisation
WWF	World Wide Fund for Nature

1

Introduction: Why Marine Conservation is Necessary

Significance, threats and management of the oceans and biodiversity

We set sail on this new sea because there is knowledge to be gained.

John F. Kennedy (1917–1963)

Fundamental significance of the oceans

Homo sapiens has a very biased view of planet Earth; its proper name should be Oceanus or Water. The oceans are the dominant feature of our planet, covering nearly 71 per cent of its surface. Indeed, a view of the Pacific Ocean of our 'Earth' from space shows hardly any land at all (Colour Plate 1a). Although most of us now live in cities, removed from direct interaction with natural environments, as a terrestrial species humans are nevertheless familiar with the 'structures' of the land – the physiography of its mountains and valleys and landscapes. The plants and animals of the land comprise our food and natural environments, and we also daily encounter the terrestrial 'processes' such as radiation from the sun, rain and winds.

We have no such inherent perceptions for the oceans. Their structures and physiography – canyons, seamounts, depths and plains – are hidden from us. The character of seawater and ocean 'climate', and oceanic processes including the myriad types of water motions are not appreciated. The wind waves we see as we travel the surface of the oceans are largely irrelevant to its biota. Apart from an occasional meal of fish, the plants and animals of the oceans are alien to us – indeed we would need a microscope to see the most common among them. This perceived remoteness of the oceans was probably responsible for the predominant interest in terrestrial conservation at the expense of conservation of the oceans (see Irish and Norse, 1996).

The oceans contain a unique molecular substance – water – whose anomalous properties would not be predicted from comparisons to other related compounds (see e.g. Franks, 1972). Life on Earth (hereafter 'earth') originated in the oceans and is only possible because of the unique physico-chemical properties of water. Together, the thermal, colligative and dielectric properties of water circumscribe both the characteristics of life on earth and its physical limits and distribution. Life on earth can exist from the summits of mountains to the depths of the oceans. With a few minor exceptions (including mercury and oils) water is the only naturally occurring liquid on earth. It is THE essential ingredient of – and

Box 1.1 Most of the properties of the oceans depend on the properties of water itself

Property	Comparison to other liquids	Importance
Heat capacity	Highest except for NH_3	Planetary thermostasis and heat transfer
Latent heat of fusion	Highest except for NH_3	Thermostatic effects
Latent heat of evaporation	Highest of all liquids	Thermostasis and heat transfer
Thermal expansion	Temperature of maximum density	Controls circulation of the oceans
Surface tension	Highest of all liquids	Cell physiology and ecology
Dissolving power	Highest of all liquids	Major implications for physical and biological processes
Dielectric constant	Highest of all liquids	Enables high chemical dissociation
Transparency	Relatively high	For photosynthesis, predation
Heat conduction	Highest of all liquids	Outweighed by eddy processes

Source: Adapted from Sverdrup et al (1942)

for – all life as we know it. In the oceans it provides not only habitats for an enormous diversity of life forms, but also buoyancy for the largest organisms the world has ever-known – the great whales. Although they are air-breathing animals like humans, they cannot support their own mass on land.

The oceans are responsible for the regulatory control of conditions on earth, including climate in both the oceans and on land; the oceans modulate and moderate the terrestrial climate. It is no exaggeration to state that life in the oceans could continue perfectly well in the absence of any land on our planet at all. However, life on land without both the climate control and water reservoir of the oceans is unthinkable. In the South Pacific Ocean, the El Nino Southern Oscillation (ENSO) drives global climates, regionally modified by variations in other oceans such as the North Atlantic Oscillation (NAO). Sea temperatures partly determine the generation and intensity of destructive typhoons and hurricanes.

Marshall McLuhan (1962) in his seminal works first defined the concept of 'the global village'. With the subsequent rise of environmental movements and expansion of global trade and communications, the significance of the oceans to us – a terrestrial species – has finally dawned. Human civilization has now reached a point where its actions can cause changes at the planetary level. Global issues, including climate change, rising levels of CO_2 and global warming, now dominate our environmental concerns. But it is the homeostatic effects of the oceans – their

Box 1.2 Importance of the oceans

Globally, the oceans are the:

- main reservoir of water: 71 per cent of the earth's surface is covered by oceans; less than 0.5 per cent is freshwater;
- main place for organisms to live; they comprise over 99 per cent of the inhabitable volume of the 'earth';
- main planetary reservoir of O_2;
- possible main planetary producer of O_2 from phytoplankton;
- planetary thermal reservoir and regulator;
- medium for longitudinal heat transfer and circulation;
- major reservoir of CO_2 especially in HCO_3^-, $CO_3^=$ forms;
- habitat for enormous diversity of living organisms, from bacteria to whales;
- reservoirs of enormous resource potential, both renewable and non-renewable, oil, minerals, etc.; also, about 50 per cent of global carbon fixation occurs in the sea.

productive and regulatory capacity – that have in large part mitigated our adverse environmental effects, and prevented things from being much worse than they presently are.

The primary producers of the oceans provide about one half of our atmospheric oxygen, and the deep oceans provide a major sink for the sequestration of atmospheric CO_2. Perhaps the most frightening scenario of potential environmental disaster is the possibility that deep ocean circulation may again cease (as it has in past geological periods), but this time with 'run-away' global warming. In more immediate human terms, the oceans are a major source of protein from fisheries, and the major trade routes among nations. Coastal zones provide an abundance of natural resources and nursery and recruitment areas for exploited species. The list goes on!

The present state of marine systems

The oceans are in a parlous state. For centuries, the oceans were thought be immutable and immune to human activities. Fish were plentiful and the capacity for the oceans to absorb human waste was believed to be unlimited. In 1605, Hugo Grotius – a Dutch jurist – laid the foundations for the International Law of the Sea by formulating the new principle '*Mare liberum*' that the sea was international territory and all nations were free to use it for trade. Apart from a narrow coastal fringe that could be protected by land-based cannon, the seas had become a 'commons' – open to all to use and abuse. Predictably, and historically, two things happened: the 'tragedy of the commons' (Hardin, 1968) and progressive protection of coastal seas (as exclusive economic zones (EEZs)). The commons is progressively being 'fenced-in', but the tragedy continues.

Although the fact that humans have the capacity for massive disturbance in marine environments has been known at least since the extinction of the Steller's sea cow in 1868, marine conservation did not become an international issue until the appeals in the 1950s and 60s by authors such as Rachel Carson (1962) and Jacques Cousteau's prolific output of books, films and television series, and organizations such as Greenpeace. As a result of these appeals and rising public awareness and concern, conservation efforts in the marine environment began in earnest with international conventions and programmes such as the London Dumping Convention, the 1973 MARPOL (International Convention for the Prevention of Pollution from Ships), the United Nations Convention on the Law of the Sea (UNCLOS) (1982) and the International Whaling Commission (1946).

However, despite these early conventions, the state of marine environments has continued to deteriorate significantly. Stocks of once globally abundant fishes such as cod, herring and tuna have in many instances become ecologically and commercially extinct. Over one million whales were harvested in a 100 year period, and only the eastern Pacific grey whale has recovered to near pre-exploitation levels. Elevated levels of pollutants are found in most marine species, even those living in the Arctic and Antarctic regions. Restaurants in California that serve tuna and certain other fish are required by law to post warnings to customers about the high levels of heavy metals in fish.

Tens of thousands of kilometres of coral reefs have bleached in recent years as a result of increased ocean temperatures, which may be aggravated by the addition of greenhouse gasses from the combustion of fossil fuels. Important breeding, feeding, mating and resting areas for migratory species have been affected by human activities. This is merely a brief summary of the continuing degradation of marine systems. Those interested in detailed accounts of the effect of human activities on marine environments should read the comprehensive works by Norse (1993), Thorne-Miller and Catena (1991) and the National Research Council (1995).

Unfortunately, as time goes by and new generations of people interact with the oceans, our human memories and expectations of the 'natural state' of the oceans also undergo progressive change. This generational change of perception of the state of the oceans has been captured in two memorable aphorisms from Daniel Pauly – 'The shifting baseline syndrome' (a term coined in 1995) and 'Fishing down marine food webs'

(Pauly et al, 1998). The first of these sayings captures the idea that although the oceans are progressively being degraded, each human generation comes to accept the degraded state as the norm. Nevertheless, whatever we currently have is still the majority of what the oceans have (or likely ever had – see below) and merits our determined conservation efforts. The second saying reflects the reality that fisheries resources of the oceans are returning smaller and smaller organisms; smaller members of species once dominated by larger populations, and smaller species once ignored or undervalued by fishers. The changing history of our views of the oceans and especially of the history of fishing fleets have been documented by Roberts (2007). We are surely and ever more rapidly reducing the biodiversity of our oceans by reducing the number of species, having an impact on habitats and their communities, and indeed destroying whole ecosystems.

What has been done to address the problems?

Humanity's response to our deteriorating marine environment has been predictably slow, reactive and piecemeal. Delays in responding to these environmental crises are exacerbated due to the fact that most marine environments are still viewed as a global commons resource, where there is little incentive to any one nation to address these issues, as problems must be solved at an international level. Early efforts at marine conservation were based on either the management of a single overexploited species (broadly referred to as single-species management) or the focus of attention on a particular environmental threat (e.g. a type of pollutant).

The discipline termed 'fisheries management' was developed to address the over-exploitation of single-species fish stocks. Fisheries management was initially based on the principals of maximum sustained yield (MSY), borrowed from forest management, which led to continued unsustainable harvest rates due to an inadequate understanding of the life histories of fish stocks and causes of variability in their populations. Recently, the traditional emphasis on management of single-species fish stocks has been changed to 'ecosystem-based management'. This has come with the realization that exploitation of single species has ecological and environmental impacts and implications well beyond the populations of the exploited species themselves, and with a renewed interest and appreciation of the structures and processes of the oceans themselves.

In nearshore areas, a similar holistic approach to management, termed 'coastal zone management', was initiated to try to integrate human activities with the goal of management and conservation of ecological systems. Coastal zone management reflected the realization that the abiotic and biotic components of marine systems were linked across spatial and temporal scales, and that any environmental change may have consequences throughout the food web.

More recently, another integrative approach, based on the conservation of defined spaces – marine protected areas (MPAs) – has been advocated as a way to protect the ecological functions of a community within a specified area such that the benefits of preserving an area may 'spill over' into adjacent areas. This book will attempt to deal with all three of these approaches to marine conservation, but with considerable bias towards the last.

How will this book address these problems?

This book is not about the litany of environmental problems in the oceans, nor is it primarily about management options and techniques. It is a book about marine biodiversity, marine conservation and ways to find solutions based on an understanding of the natural ecological hierarchies of the oceans. The purpose of this book is not to examine any one specific management construct – there are several other texts that address these topics – but to examine the various approaches to conserving marine biodiversity in light of the ecological structures and functions (processes) of marine environments. This book will provide the reader with a comprehensive canon of conservation frameworks that can be applied in all marine systems.

It is our belief that those responsible for the management and conservation of marine environments often overlook ways to conserve and manage marine environments, as they do not always fit within the traditional management systems they are familiar with. We centre this book on the conservation of marine biodiversity and its components, across the ecological hierarchy, rather than focus on any particular population, community, habitat or ecosystem. This is done because we feel that the practice of marine conservation, based on ecological principles, should be applicable from the global to the local level, and from ecosystems, through habitats and communities to individual cases of separate species and their populations.

The foundation of this book is therefore ecological in character, respecting the natural organization of the environment and biota of our planet. As Dobzhansky (1973) said of Charles Darwin: 'Nothing in biology makes sense except in the light of evolution.' To paraphrase this sentiment we could say that: 'Nothing in biodiversity conservation makes sense, except from the perspectives of ecology and the environment.'

What is biological diversity?

Biodiversity (biological diversity as coined by E. O. Wilson, 1988) is, put simply, the richness and variety of life in the natural world. The international *Convention on Biological Diversity* (United Nations Environment Programme (UNEP), 1992) defines biodiversity as 'the variability among living organisms from all sources, including … terrestrial, marine and other aquatic ecosystems and the ecological complexes of which they are a part; this includes diversity within species, between species and of ecosystems'. The term 'biodiversity' therefore includes biological diversity and ecological diversity across the organizational hierarchy through genetic, species and ecosystem levels (see e.g. Gray, 1997).

The concept of biological diversity is widely misunderstood and variously interpreted, even within the scientific community. In its narrowest sense the term biodiversity is often used synonymously with species diversity – but it is far more than this. As defined above, the term 'biodiversity' includes the diversity of genes, species and their populations, communities and ecosystems, as well as the dynamic processes that change them and their environments. The rationale for such a broad definition is based on the realization of the basic hierarchical organization of nature, and that no level of the hierarchy can exist without the support and interactions of all the other levels. For example: species cannot thrive without suitable habitats within which to live, habitats cannot exhibit any constancy of conditions without the ecosystem level processes that maintain them, and so on across the whole hierarchy of ecological and environmental interactions. In the broadest sense, we should perhaps speak of ecological diversity and/or environmental diversity.

It has been suggested that the concept of biodiversity is too 'all-encompassing' because it represents the sum total of all living things and their planetary life-support systems. In other words the term has become so general and all-encompassing (because it includes everything) that it has become meaningless; the currency has become debased. We do not agree, and we shall use the term in its broadest sense. Perhaps the most significant aspect of the concept of biodiversity is that it allows its components to be identified and analysed from spatial, temporal and ecological perspectives. This permits a hierarchical context and an approach to environmental and ecological problems that we might well not otherwise appreciate. The full significance of this should become apparent in subsequent chapters.

Why should marine biodiversity be protected?

Biodiversity is really the value of our biosphere and its environment, but because biodiversity encompasses 'everything' its value and benefits are not easily defined or categorized. However, now that Marshall McLuhan's 'global village' has become a reality, we need to categorize the components of biodiversity and approach their conservation in a systematic and responsible way. We can then recognize 'who should do what – and why'.

The reasons for protecting biological diversity are complex and encompass environmental,

Box 1.3 A brief history of marine biodiversity conservation

While many indigenous cultures, particularly Pacific island cultures, efficiently managed marine resources using many of the same approaches (e.g. closed areas, catch and size limits) used today, modern marine conservation is relatively new and has traditionally lagged behind terrestrial conservation in nearly every aspect. This is primarily due to the difficulty in understanding and measuring human impacts to marine systems relative to the terrestrial realm combined with the reality that humans generally have less of a connection with marine environments and therefore are more difficult to engage on marine conservation issues. This situation has led to the following perspectives/circumstances over the past several centuries:

- 1800s: Marine resources are thought to be inexhaustible.
- 1900s: Key fisheries are thought to be inexhaustible.
- 1960s: Major fish populations decline; traditional fishing communities break down; ecosystems deteriorate.
- Current: Ocean governance is fragmented; diverse impacts are not managed in a coordinated manner; human-induced ecosystem shifts have occurred; oceans are managed independent of the terrestrial environment.

However, marine biodiversity conservation is beginning to catch up to terrestrial conservation efforts. While many conservation efforts have been led by individual nations (e.g. Australia and the Great Barrier Reef) a number of key international laws and conventions have begun to recognize the importance of marine conservation and management and include the following:

- 1972: Stockholm Declaration commits signatories to conservation of biodiversity, sustainable use of marine environments and the use of the maximum sustained yield concept.
- 1982: UN Convention on the Law of the Sea commits signatories to conserving fish stocks, preventing introductions of alien species, and considering species interactions in management.
- 1992: Rio Declaration commits signatories to applying the precautionary approach, marine protected areas, and use of traditional knowledge in decision-making.
- 1992: Agenda 21 commits signatories to the conservation of fish stocks, application of integrated coastal zone management, consideration of climate change, and financial incentives to conserve.
- 1995: UN Agreement on Straddling Stocks commits signatories to the more cooperative management of migratory fish stocks and stocks with broad geographic ranges.
- 1995: FAO Code of Conduct commits signatories to end destructive fishing practices, adopt selective fishing gear, consider local marine users/communities in decision-making and support fisheries research.
- 2001: FAO Reykjavik Declaration commits signatories to applying the ecosystem approach to fisheries management.
- 2002: World Summit on Sustainable Development commits signatories to honour previous agreements as well as to coordinate and better cooperate on marine agreements.

Significant progress has been made with respect to conserving marine environments. Currently, nearly every commercial species of significance has some type of science-informed management plan developed across jurisdictions that is often based on the ecosystem approach to fisheries (Chapter 13). Whether these plans are adhered to by fishers and political institutions, and whether they are sufficient to prevent over exploitation of fish stocks, is another matter. In addition, nearly 1 per cent of the ocean's surface is captured by some type of protected area designation; however, many of these protected areas continue to allow extractive activities. Many national and international legislative tools are now in place to assist with marine conservation and management efforts; however, without the political will and public pressure to implement these tools, the condition of many marine habitats and their communities will continue to decline.

(Christensen et al, 2007; Guerry, 2005).

economic and social benefits (e.g. Beaumont et al, 2007), though it is fair to say that stronger rationale, whether scientific, socio-economic or ethical should be developed (see e.g. Duarte, 2000). The rationale for protecting biodiversity falls into several major categories which can be summarized as: Intrinsic Value, Anthropocentric Value (ecological goods and services for humans) and Ethical Value. Some of these reasons are summarized here.

Intrinsic Value This is a rather contentious issue – that is, that the components of biodiversity, species and natural systems have their own worth independent of human needs or considerations. For most ecologists and environmentalists, this has become largely a philosophical issue unless related to the concept of ecological functioning and how ecosystems 'work'.

Intrinsic Value reconsidered We shall explicitly consider the concept of the 'value' of marine biodiversity and how marine environments can be ranked for conservation purposes, in such terms in Chapter 15.

Anthropocentric Value – vulnerability Loss of diversity generally weakens entire natural systems; every species can be considered to play a role in maintaining healthy ecosystems upon which humans ultimately depend. When simplified by the loss of diversity, ecosystems become more susceptible to natural and artificial perturbations, or may change completely in state (e.g. on the Scotian Shelf, Canada, removal of key predators combined with effects of bottom trawling have completely changed the character of an entire ecosystem (Frank et al, 2005)). Species listed as endangered and threatened include several marine mammals (Hoyt, 2005); many of them are key to ecosystem functioning and are also valuable from an economic, ethical or aesthetic perspective.

Anthropocentric Value – renewable resources Biological diversity represents one of our greatest untapped natural resources and future potential. Our marine areas contain innumerable raw materials that could provide new sources of food, fibre and medicines, and new discoveries continually contribute to scientific and industrial innovations. The pharmaceutical potential of thousands of yet-to-be-discovered marine products to provide life-saving or commonly used drugs is an example of the almost untapped potential of our oceans for sustainable economic use. Nature has repeatedly proved to be a much better chemist than mere mortals – over 60 per cent of all anti-tumour agents and anti-infective agents introduced worldwide over the last 15 years have had a natural product structure in their background (Newman and Cragg, 2007). Only over the last couple of decades has the immense potential of the marine environment as a source of undiscovered chemical structures begun to emerge. For example, recent research indicates the synthesis of a protein produced by mussels (which in nature helps the shellfish stick to rocks) may be useful to close wounds that would otherwise require stitches. Given that we know so little about our marine resources, the potential for life-saving or beneficial pharmaceuticals is enormous and is expanding every year.

Anthropocentric Value – non-renewable resources The socio-economic value of non-renewable resources has historically over-ridden concerns for the natural environment. Hopefully with changing environmental values, and the application of the concept of ecosystem-based management, we shall progressively see a reconciliation of biodiversity and resource values.

Anthropocentric Value – ecosystem goods and services Humans benefit from natural areas and depend on healthy ecosystems. The natural world supplies our air, water and food, and supports human economic activity. Much of the world's protein comes from marine sources, and MPAs are an important mechanism to enhance commercial fisheries species. Marine fisheries around the world are clearly heavily overexploited and not sustainable. The establishment of protected areas is one of the few positive steps taken that reverse this trend. The growing discipline of 'natural capital valuation' is beginning to document the 'goods and services' provided to humans by the components of biodiversity of the natural world, including the attributes that lead to aesthetic and recreational values.

Anthropocentric Value – 'insurance' From an ecological stance this is perhaps the most fundamental and important concept. We still have much to learn about the oceans. Their depths are literally as unknown to us as the far side of the moon. Ultimately, we do not know the full environmental and ecological significance of the components of marine biodiversity or how they function in concert. Protection of the oceans can therefore be regarded as a sort of insurance policy; as we destroy its components we cannot predict the consequences. The Precautionary Principle (see Chapter 13) advocates a willingness to accept credible threat in advance of hard scientific proof. Although the principle has been widely adopted, first by the European Union, it has not been generally or seriously implemented in the oceans – even in the face of hard scientific evidence.

Three fundamental things are clear. First, a planet containing the oceans without land is perfectly viable; but a planet consisting of land without the oceans is not viable; the oceans regulate the homeostatic mechanisms of our world. Second, as we progressively degrade natural communities and ecosystems the world reverts to its more primitive microbial dominated systems; humans are unlikely to vandalize a world to the extent that it does not support life, but we could see a world that does not support a wealth of species – including our own. Third, we simply do not know how far we can degrade natural habitats and their communities before effects are irreversible; in some cases parts of the oceans seem already to have reached 'alternative stable – and undesirable states'. Conservation ecologists frequently invoke the concept of 'ecological integrity', although in the oceans its real meaning is awkward and not clearly understood (see Chapters 16 and 17).

Ethical Value – nature and responsibility The argument is made that humans are simply a part of nature and that we should not endanger our own environment. We humans are the only species on the planet that has the capability of driving many others to extinction. Environmental ethicists also stress that we have a moral responsibility to protect the environment and the other species on our planet. The concept of environmental stewardship for example is a fundamental part of Judeo-Christian religions.

Variation in biodiversity over geological and historical time

The taxonomic diversity on our planet has varied greatly in the past (see Signor, 1994; Sepkoski, 1997), but may presently be at an all-time high (see Figure 1.1), possible because of the greater spatial separation of the land masses and the abundance of shallow seas and consequent high habitat diversity (as a function of spatial heterogeneity – see Chapter 8). However, the rate of species extinctions is probably also at an all-time high (with the exception of mass extinction events) due to human environmental disturbances. We therefore seem to be living in paradoxical times, currently experiencing a bounty of the greatest species richness but also annihilating them at the greatest rate!

The process of evolution occurs in both the biological and inorganic world, and in the terrestrial and aquatic environments. The lesson of evolution is that all forms of life on earth change as the environment itself changes. Individual species (including humans) must either adapt to these changes or become extinct. As far as we know, the human species is the only one to have caused the extinction of many other species.

As a terrestrial species, our major pre-historical and historical environmental effects have been on land and in freshwaters. However, human effects on the oceans are now substantial and growing; they range from local to global in scale. No part of the oceans is now removed from human influence. Efforts to conserve our oceans are now vital, not just for the benefit of local human environmental and socio-economic health, but because of global concerns.

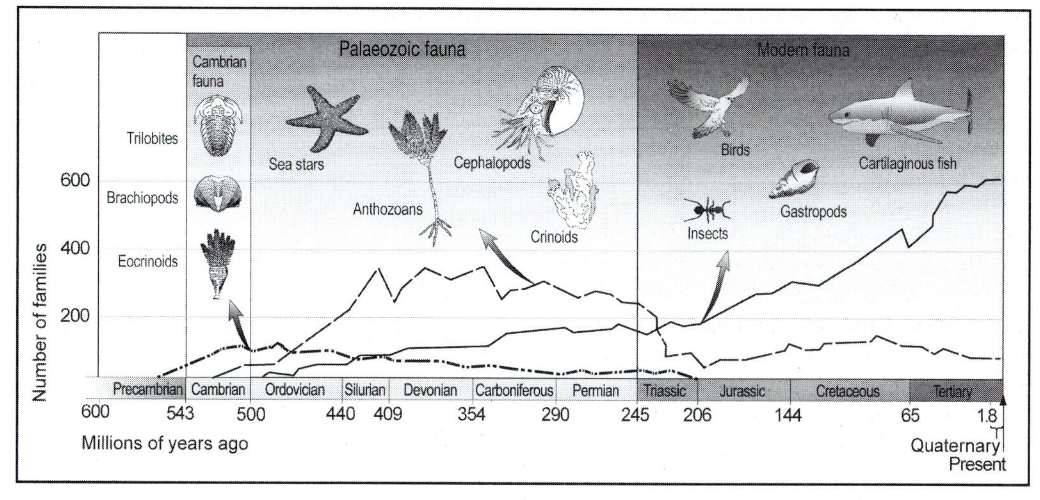

Source: Redrawn from Sadava et al, 2006

Figure 1.1 Variations in taxonomic diversity on earth over geological times

The components of marine biodiversity

Human beings classify things – either explicitly or implicitly. Classification of things seems to be an inherent human characteristic, necessary to codify the world around us – its features, its changes, its dangers, its resources. This propensity to classification permeates society as well as science. In science it leads to the classification of sub-atomic particles and the periodic table of elements. In biological sciences it led to Linnaeus's system of naming and classifying plants and animals (the *Systema Naturae*, 1758) and to Darwin's theory of evolution by natural selection. It is impossible to image modern biology without these foundational methods of organizing and categorizing the relatedness of life forms on the planet.

In a concept as large as biodiversity, similar organizational frameworks are also required. We should capitalize on the human requirement for classifications in order to understand the nature of biodiversity. A matrix of the components of marine biodiversity (see Table 1.1) has been proposed by Zacharias and Roff (2000), wherein the compositional levels of the hierarchy and corresponding structural and functional components are indicated. This framework is a marine adaptation of a hierarchical ecological framework based on the work of Noss (1990) that separates biodiversity into compositional, structural and functional attributes at the genetic, population, community and ecosystem levels of organization. Classifications of components can in fact be made at any level within the ecological hierarchy and in various ways. Such classifications help us to develop perspectives, for example on what exists, what has been degraded or lost (and to what extent or in what proportion), where priorities should be assigned, what is more sensitive, vulnerable and so on.

In Table 1.2 we expand on the matrix of Table 1.1 to show some more specific components of marine biodiversity. It is important to realize here that the Structural components are time-independent – that is, they have formal physical 'dimensions' L^3 or M (mass) only. The Functional or Process components of biodiversity are changing over time and therefore have dimensions of $M.T^{-1}$ or just T^{-1}. This sort of tabulation is not meant to be an exhaustive or inclusive listing of components, but rather it is an indication of what the components of marine biodiversity actually comprise at each of the levels of organization of the ecological hierarchy. It is a checklist

Table 1.1 Compositional, structural and functional attributes of biodiversity proposed by Zacharias and Roff (2000) for marine environments, contrasted to the terrestrial framework of Noss (1990)

Compositional		Structural		Functional	
Noss (1990)	Zacharias and Roff (2000)	Noss (1990)	Zacharias and Roff (2000)	Noss (1990)	Zacharias and Roff (2000)
Genes	Genes	Genetic structure	Genetic structure	Genetic processes	Genetic processes
Species, populations	Species, populations	Population structure	Population structure	Demographic processes, life histories	Demographic processes, life histories
Communities, ecosystems	Communities	Physiognomy, habitat structure	Community composition	Interspecific interactions, ecosystem processes	Organism/habitat relationships
Landscape types	Ecosystems	Landscape patterns	Ecosystem structure	Landscape processes and disturbances, land-use trends	Physical and chemical processes

of what biodiversity is that we can use as a reference for conservation studies and planning. In later chapters we shall expand on the contents of this table, and separately consider the relevance of each of these levels of the marine biodiversity hierarchy and their components.

Some preliminary definitions and remarks are in order here. Within the ecological hierarchy we proceed from genes, through species and their populations to the community level (Table 1.1); all these levels describe strictly biological components. Communities are indicative of particular kinds of habitats (which are defined in terms of their abiotic environmental characteristics), and sets of habitats and their communities comprise ecosystems. Habitats are thus abiotic entities, while ecosystems are compositely biotic and abiotic in nature. Note that the terms habitat and ecosystem are sometimes used interchangeably. Our preference is to use them in their original and classical meanings. In summary, by using the ecological hierarchy we are already considering a hybrid system of classification employing both biotic and abiotic (geophysical) features. Note also that other levels of a hierarchy could be interpolated, such as habitats or landscapes.

The landscape level of ecology (often inserted between habitats and ecosystems) can also be described either in abiotic or biotic terms, or in some combination of both. The equivalent term in marine conservation 'seascapes' is slowly coming into general use, sometimes as the synonym 'marine landscapes' to describe coastal features. A terrestrial landscape typically comprises some set of habitats of variable type, within a recognizable landform (valley, hill etc.). Although the term seascape should be analogous, in fact we shall generally use it in a more restrictive sense to mean an array of a particular kind of habitat type (see Chapter 5).

Marine landscapes – seemingly at first an odd or compromise term, is in fact quite appropriate where it is defined primarily by local or regional topography (landform). Where described primarily by the characteristics of the water column the term seascape is more appropriate. The marine conservation community has not settled on this level of terminology and we shall use both.

The marine environment may also be defined and described in spatial terms. Here the range is from global biogeography to genes. Terms such as marine provinces, ecoregions, geomorphic units, representative areas and so forth are used, and their meanings and usage will become clear in later chapters. In order to define these units of the biosphere we make use of some combinations of biotic and abiotic data,

Table 1.2 An expansion of the marine biodiversity framework from Zacharias and Roff (2000), showing attributes of structures (statics) and processes (function or dynamics) arranged under the population, community and ecosystem levels of organization

Genetic		Species/ Population		Community		Ecosystem	
Structures	Processes	Structures	Processes	Structures	Processes	Structures	Processes
1. Genetic structure	1. Mutation	1. Population structure	1. Migrations	1. Community structure	1. Succession	1. Water masses	1. Ocean currents
2. Genotypes	2. Genotype differentiation	2. Population abundance	2. Dispersion	2. Species diversity	2. Predation	2. Temperature	2. Tidal currents
3. Fitness	3. Genetic drift	3. Distribution	3. Retention	3. Species richness	3. Competition	3. Salinity	3. Physical disturbances
4. Genetic diversity	4. Gene flow	4. Focal species	4. Migration/ counter drift	4. Species evenness	4. Parasitism	4. Water properties	4. Gyres
5. Stock discrimination	5. Natural selection	5. Keystone species	5. Growth/ production	5. Species abundance	5. Mutualism	5. Boundaries	5. Retention mechanisms
	6. Inbreeding	6A. Indicator species – condition	6. Reproduction	6A. Representative communities	6. Disease	6. Depth/pressure	6. Pelagic–benthic coupling
	7. Non-random mating/ sexual selection	6B. Indicator species – composition	7. Recruitment	6B. Distinctive communities	7. Production	7. Light intensity	7. Entrainment
	8. Directional selection	7. Umbrella species		7. Biome types	8. Decomposition	8. Stratification	8. Biogeochemical cycles (inc. nutrient dynamics/ energy flow)
	9. Stabilizing selection	8. Charismatic species		8. Biocoenoses		9. Bottom topography	9. Seasonal cycles (physical and biological)
	10. Disruptive selection	9. Vulnerable species		9. Species–area relationships		10. Substrate type	10. Productivity
	11. Micro-evolution	10. Economic species		10. Transition areas		11. Geophysical anomalies (inc. frontal systems)	11. Hydrosphere–atmosphere equilibria
	12. Genetic erosion	11. Phenotypes		11. Functional groups		12. Wave exposure	12. Hydrosphere–lithosphere equilibria
	13. Speciation	12. Population fragmentation		12. Heterogeneity		13. Patchiness	13. Eddy diffusion/ turbulence/internal waves
	14. Macro-evolution	13. Meta-populations		13. Endemism		14. Nutrients	14. Mixing/stabilization
				14. Alternate stable states		15. Dissolved gasses	15. Upwelling/ convergences
				15. Symbioses		16. Anoxic regions	16. Divergences
				16. Biomass			17. Ecological integrity
							18. Erosion/ sedimentation
							19. Desiccation

Box 1.4 Definitions of some terms used throughout the book

A **structure** is any measurable quantity whether biotic or abiotic. Structures have no dimension of time, but they change over time and space as the result of physical and biological processes.

A **process** is any quantity, physical or biological, that varies over time, causing changes in structures. All processes have dimensions of time.

By **environment** we specify the sum total of all external influences (physical, chemical and other biological) on living organisms.

By **ecology** we mean the study of organisms in relation to their environment.

We shall use the term 'ecological hierarchy' to mean the array of biological entities, both structures (instantaneously observable quantities) and processes (or functions – the rates at which observable quantities change). This hierarchy spans biological and physico-chemical environmental structures and processes from the genetic to the ecosystems level.

The terms **genetic, species and population** are clear and we shall use these in conventional ways.

The term **community** is used in many senses. Most biologists accept that it is a vague term, and want to keep it this way. Originally it meant a group of species that interacted (either actually or potentially) in some ways. It still means this, although in practice, except in the very simplest communities, the specific ways in which members of communities actually interact at any given time is not known. A more neutral term is an 'assemblage' of species. Here we simply imply that a set of species are generally found to co-occur.

A **habitat** is a physically defined region of the environment, usually accepted as housing a defined community type. It is usually more straightforward to recognize and map habitat types than to define the communities associated with them.

Landscapes is a term used in terrestrial ecology to define regions of the earth that contain sets of habitats, generally of somewhat different types. Landscapes are therefore analogous to geomorphological features (see Chapter 5).

The corresponding term in marine ecology would be **seascapes**. This is a relatively new term that is not yet widely used. It is used here to define a particular set of habitats of similar type. Seascapes are therefore NOT equivalent to geomorphological features.

The term **ecosystem** is a useful one, but difficult to define. In its original sense, as applied for example to a lake, the intent of the term was clear. It was used to describe a 'chunk' of the environment that contained several natural communities of organisms, and that was more or less clearly circumscribed. In the marine environment, although the term 'ecosystem' is frequently used, because of the continuous nature of the medium (as opposed to the discontinuous nature of lakes for example), ecosystems are not readily defined (except somewhat arbitrarily). We shall, however, use the terms 'ecosystem structures' or 'ecosystem level processes' as more or less synonymous with habitat structures and processes.

including data on the physiography and oceanography of the marine environment. Direct information on the distribution of marine biota is often sparse. We therefore must have recourse to other abiotic data, which act as surrogates or indicators of expected or predicted distributions of the biota themselves. These characteristics may be described as either 'enduring' (predominantly physiographic) or as 'recurrent' (predominantly oceanographic).

Genetic level

The genetic variation in a population, both within individuals and among the members of a population, ensures that the vicissitudes of the environment can be met by at least some of its members. It ensures that Natural Selection can operate on the inherent variation within and among species; if a population consisted entirely of a single clone with no genetic variation, then that entire population is in danger of mass extinction in the face of some environmental change to which it cannot adapt. The genetic level of organization in fact contains vast amounts of yet untapped information that is becoming of fundamental importance to marine conservation, as we shall examine in Chapter 10.

Species level

Of all the species known to science, about 80 per cent are terrestrial, but there are more orders and phyla in the sea. In fact, all phyla of animals are found in the sea, a majority of these in benthic environments, and one third of all the phyla are exclusively marine. If plants and protists are also considered, then at least 80 per cent of all phyla include marine species. In addition, the relative abundance of marine species may be greater than presently considered, since more marine species are unknown (Thorne-Miller and Catena, 1991). The distribution of species diversity among taxa is very uneven – some taxa such as the arthropods are very species-rich, whereas others contain few species. Biologists debate why this should be so in terms of the 'adaptability and success' of fundamental body plans.

At the species level, we frequently seem to make the implicit assumption that some species and their habitats are more valuable than others. This presumably was – and may still be – the rationale for conservation efforts directed at individual species. But how can we or should we make such decisions? Should we recall the famous Orwellian dictum: 'All animals are created equal – but some are more equal than others'? In fact (in Chapter 9) we will show that there are several ways in which we can rationalize that we should pay disproportionate attention to selected species.

Communities and habitats level

There are no clear answers as to why species are distributed as they are among higher taxa, but what should we expect from species diversity distribution among communities and their habitats? Why should some habitats and their communities be richer in species than others? We shall make a preliminary exploration of some of these questions in Chapter 3, for example in comparisons of plankton and benthos. In fact reasons for higher or lower species diversity within communities are not well known though theories and explanations abound. Species diversity in communities is related to a variety of factors that will be explored, primarily in Chapters 6 and 8.

Ecosystem level

At the habitat or ecosystem level and above, we encounter an array of considerations that need to be disentangled for conservation purposes. At the level above (or including) that of ecosystems lies the discipline of global biogeography. This is a discipline historically older than biological conservation, where interest has typically centred on describing distributions of particular individual taxonomic groups such as echinoderms, molluscs or fish. In this book our interest is less in the particular taxonomic groups themselves, and more in other directions. Specifically we shall look at ecological boundaries and how they relate to changes in the species composition of communities irrespective of taxonomic groups – that is, how we can classify and define the distributions of whole arrays of biota. Secondly, from global to regional and local scales, we shall examine patterns and factors underlying the distribution of the complement of species – species richness.

The overall intention then is to classify marine environments from the global to the local level so as to recognize and define their distributions and patterns, and to facilitate analysis of their biodiversity components so that marine conservation initiatives can be undertaken in a coordinated fashion and according to ecological principles.

Some marine ecosystem types have been grossly degraded by human activities, especially in the coastal zone. Other ecosystem level processes – especially at the global scale, for example ocean circulation – have traditionally been seen as independent of human effects. We are finally realizing that even this is not true as the effects of global warming are felt on ocean circulation!

Threats to marine biodiversity

There are potentially many ways to collate and discuss threats to marine biodiversity, but threats can be broadly categorized as a result of overharvesting, pollution, habitat loss, introduced species and global climate change (see e.g. NRC, 1995; Gray, 1997). The following sections provide a brief discussion of these impacts on the marine environment and indicate how this book intends to address these threats. Many marine areas have a range of biota rivalling or exceeding that of tropical forests. However, the diversity of life in our oceans is now being dramatically altered by rapidly increasing and irreversible human activities. Although there are differing views of present and potential threats to coastal and marine biodiversity, those shown in Table 1.3 are among the most important.

Overharvesting

The unsustainable harvest of marine populations is perhaps the most serious threat to marine environments worldwide. Overharvesting is not a new phenomenon in the oceans. Many traditional cultures either removed the available species from their local marine environment and moved onto harvesting other areas, or had to develop some methods of regulating the timing and amount of harvest from certain areas in order to avoid overexploitation of populations. The advent of the industrial revolution resulted in the increasing mechanization of fish harvesting so that species such as large whales and offshore pelagic fish – that were previously difficult to catch – were now accessible in a commons environment which was owned by no one. Over one million whales have been harvested, and most species have been reduced to levels where they are considered endangered or threatened. Most populations of palatable fish stocks have been seriously depleted, and currently there is evidence that humans have fished down food webs and will continue to do so (e.g. Pauly et al, 1998). This book is not primarily about fisheries. But the vital issue of how to link fisheries management with broader marine conservation objectives through ecosystem-based approaches (e.g. Gislason et al, 2000; Hughes et al, 2005), is considered in Chapter 13.

Table 1.3 Examples of threats to marine biodiversity

Risk or speed of degradation of biodiversity	Threatening process
High ↑	Physical habitat destruction (e.g. reclamation, dredging)
	Blast fishing using explosives, *meting*[1] (either can annihilate a coral reef)
	Toxic pollution (e.g. chemical spills)
	Chemical fishing (e.g. cyanide)
	Introduction of exotic organisms
	Loss of genetic variability
	Biological invasions
	Overexploitation/overfishing
	Bioaccumulation of noxious materials (e.g. heavy metals)
	Indirect pollution (pesticides, herbicides in runoff)
	Disease/parasite infection
	Depletion of spawning sites
	Sea dumping of dredge spoil
	Incidental take/by-catch
	Destruction of adjoining watersheds
	Impacts of adjacent land-use practices (e.g. aquaculture)
	Effluent discharge (sewage, pulp/paper)
	Natural events (cyclones, tsunamis)
	Direct marine pollution, ocean dumping
	Downstream impacts from dams, dykes, etc
	Net/debris entanglement
	Siltation
	Noise pollution
	Toxic blooms/red tides
	Thermal pollution
	Climatic change – rising sea temperatures
	Sea-level rise
↓	Salinity changes
Low[2]	Indigenous take

1 *Meting* is an emerging threat that involves the indiscriminant removal of all organisms from reefs using metal crow bars to rip away coral cover to harvest species such as abalone and clams.
2 The 'high–low' scale on the left side of this table is approximate only; it seeks only to indicate that some threatening processes have a higher risk and/or speed of impact on marine diversity than others. Moreover the relative order of the various threatening processes on this scale is open to conjecture.

Pollution

There is no question that pollution from a variety of sources has affected every marine system on earth. Indigenous human populations in

Arctic areas are the most contaminated people on earth, as a result of ingesting marine fish and mammals which bioaccumulate toxins due to their high trophic levels. It was generally assumed that pollutants reached the oceans primarily in runoff from rivers. In some coastal areas this will indeed be the case, but overall transport of pollutants through the atmosphere to the oceans is more important. The types and lists of pollutants appear endless, including artificial radionuclides, petroleum hydrocarbons, chlorinated hydrocarbons, metals, carcinogens, mutagens, pesticides, excess nutrients causing nuisance and toxic algal blooms, endocrine disrupters, physical debris and so on. The persistence and longevity of pollutants in the marine environment, and their ecosystem-level effects on marine biota and ultimately humans, are of growing concern. Nevertheless, again, this book is not primarily about pollutants; it is about the components of biodiversity that may be affected by various kinds of pollutants. This is the backdrop against which their impacts can be judged.

Habitat loss

Habitat loss is probably the most serious threat to biodiversity in terrestrial environments due to the removal of larger vascular vegetation on which many species depend for food and shelter. Loss of marine habitat is primarily a concern in coastal nearshore and intertidal marine environments. Increasing pressure in coastal systems has come from a combination of: shipping – with attendant infrastructure and transportation; other construction and modification of natural coastlines; fishing; recreational activities; and increased land runoff – including nutrients and suspended solids. The types of habitats in these areas which can be 'lost' include marine macrophytes (kelp), mangroves, sea grasses, corals and other biotic communities (e.g. sponges, sea pens, sea fans, aphotic corals) as well as abiotic habitats, such as intertidal and estuarine mud flats and other areas which are dredged or subject to dumping. Habitat loss in deeper marine environments and the pelagic ocean is a more vague construct as these habitats are primarily composed of either oceanographic (e.g. currents, gyres, fronts) or physiographic (e.g. seabed composition) structures and processes which are more resilient to human activities – or less immediately impacted. Loss of marine habitat is significant not only from an ecological perspective, but also increasingly from a socio-economic perspective. The interaction of human effects and natural marine processes is most evident in coastal waters, where strategies to prevent habitat loss or mitigate effects and restore habitats are encompassed in (integrated) coastal zone management initiatives (see Chapter 11).

Introduced species

Species introductions (also termed invasive, exotic and non-native species) have probably been occurring for as long as humans have used the oceans for exploration and trade. There is evidence that many species we believed to be native are now thought to have been introduced through marine transportation prior to the industrialized era. Transport in the ballast water of ships appears to be the main mode of travel, and impacts are generally observed mainly in coastal waters and estuaries. While the introduction of larger species such as the green crab (*Carcinus maenas*), the alga *Calaupera taxifolia* and the comb 'jellyfish' *Mnemiopsis leidyi* has been well publicized, most species introductions are less obvious, and are found in the phytoplankton and zooplankton. A single 24-hour study in Washington State, USA found over 110 non-native species. Some of these invasive species can have dramatic local socio-economic effects, with different species of jellyfish having major impacts on fisheries and even coastal human recreation. Outbreaks of jellyfish have now been reported from locations around the world, probably caused by a combination of species invasions, overfishing leading to food web disruption and local water temperature increases.

Global climate change

There is no doubt that the earth's climate changes over time and that these cyclical changes occurred long before humans became the dominant species on the plant. Global climate changes have been responsible for mass extinctions in the past and the earth's climate will continue to vary, resulting in future mass extinctions. Sea levels have been known to deviate up to 85m during the Quaternary period, which inhibits the evolution of established marine communities in coastal and shelf environments. Humans have also been shown to have an impact on global temperatures, and since the 1980s there has been considerable debate on separating out the natural and anthropogenic contributions to climate change. Human activities that affect climate change include the release of CO_2 through the burning of fossil fuels, and large-scale deforestation – which lowers the total amount of CO_2 removed from the atmosphere. Changes in water temperatures and changes in coastal salinities caused by changes in insolation, evaporation and rainfall, and land runoff patterns will result in the resetting of biogeographic boundaries. Some species will extend their ranges while others will contract – often with unpredictable consequences for regional community composition. Regional conservation strategies and practices may in turn therefore require incorporation of climate change scenarios, necessitating a clear understanding of changing ecological relationships.

Approaches to address threats to marine biodiversity: Marine conservation

The term 'marine conservation' has come to mean at least two rather different things. The dominant sense in which we shall use the term in this book is to mean *preservation of the components of marine biodiversity*, including their structures and processes, in a natural state. The key words here are 'preservation' and 'natural state'. Preservation of marine biodiversity entails the establishment and management of MPAs, and removing (or severely restricting) human influences on them. This will be the major theme of this book. Some would argue that natural states or pristine environments no longer exist, or that such conservation is no longer attainable in the face of human manipulations of the planet's resources. We shall leave this argument in abeyance, despite the high rates of species extirpations both on land and in the sea, and argue that we must make efforts to systematically conserve what we have.

A second sense of the term marine conservation is *the sustainable use of biological resources and ecosystems*. However, as we shall show, it has become evident that such conservation – by sustainable exploitation – still benefits greatly from the establishment of protected areas. Marine conservation has had a long history in many countries, much of it unrecorded and unsuccessful until recently. Not until the advent of scientifically controlled MPAs, which closed certain areas of the oceans to human activities, did the effects of marine conservation actually become apparent. The concept of zoning of the oceans (e.g. Agardy, 2010) – that the oceans are no longer to be conceived as a 'commons' or a 'free-for-all' but that human activities in the oceans must now be regulated – has now come of age. In part this has become feasible because of new technologies. Only in the last two decades has it become possible to know not only where everyone on the oceans actually is, but also – to a considerable extent – what they are doing.

Fisheries management by managing the behaviours of both suppliers (fishers while at sea) and consumers (in terms of product choices) have now become effective options. Even in the coastal zone, where effects and consequences of human actions are individually visible, management has been largely ineffective until education and public awareness have forced changes.

Marine protected areas (MPAs)

Marine conservation can be regarded as a multifaceted discipline that seeks to address both preservation of marine biodiversity and the regulation of use of exploited resources. The emphasis in this text is on the analysis of the components of marine biodiversity, and on marine protected areas and their role in preservation of marine biodiversity. Marine protected areas come by several names in the literature, but we shall refer to all marine protected areas by the generic term of MPAs.

It has been argued, several times, that if we could only restrict the spatial extent and intensity of fishing activities (e.g. bottom trawling), plus control the flow of pollutants to the oceans, we should not need to adopt any further form of marine conservation. In theory this may perhaps be a rational argument, but until a global consensus on such management of the oceans might be reached, the single most effective means of simultaneously preserving biodiversity and enhancing fisheries appears to be to locally establish protected areas – MPAs – where human activities are regulated.

Establishment of MPAs is not the only thing we need to do to accomplish sustainable management of the oceans. However, it has been repeatedly shown that MPAs are effective not only in protecting the various habitats of the marine environment – that is, they have a dominant role in marine preservation – but that they can also contribute significantly to the conservation of individual species – primarily of fish. That is, they have an important role in the sustainable exploitation of biological resources.

Sustainable exploitation of biological resources in the oceans – primarily through fisheries – is now generally considered to require BOTH the establishment of restricted fishing areas and the regulation of stocks through catch quotas. Marine conservation also entails protection of the coastal zone from effects of land-run-off, for example soil erosion and eutrophication. At least three things are therefore required for effective marine conservation: MPAs, pollution control and regulation of fisheries (both in terms of catch quotas and gear activities). Marine protected areas can be therefore thought of as a necessary but not a sufficient contribution to integrated marine conservation (e.g. Allison et al, 1998). MPAs are only one tool in a potential arsenal of approaches to marine conservation, but they are an essential tool. We can think of MPAs as a series of modern-day 'Noah's Arks' for at least the interim protection of selected areas.

How we select MPAs as a planning tool for conservation of the components of marine biodiversity, without being purely arbitrary, is a major theme of this book. MPAs are very effective in conserving certain types of habitats and certain types of biological communities, particularly if they have been chosen using a science-based representative framework. For example, coral reefs are particularly well suited to protected area status because they are physically defined areas harbouring a characteristic diversity of species (e.g. Thorn-Miller and Catena, 1991). Other benthic communities may also receive adequate protection from an MPA, but pelagic communities are less amenable to such methods. Similarly if MPAs are likely to be significantly influenced by impacts originating outside their boundaries (e.g. pollution from mainland runoff), then an individual MPA may have only limited benefits.

The effectiveness of the protection afforded by an MPA, or a set of MPAs, to marine animals and plants that occur within it is a critical concept to evaluate if conservation initiatives are to remain credible (see e.g. Leslie, 2005). Effectiveness of an MPA will depend on several considerations, including:

- The function of an MPA, e.g. Representation (Chapter 5) or to protect selected species, e.g. Distinctive areas (Chapter 7).
- The size of the area protected (see Chapter 14).
- The activities that are restricted and allowed within the MPA boundaries. This is the concept of zoning recently addressed by Agardy (2010).
- The MPA designation and whether it restricts polluting activities that occur outside the MPA but that threaten life within the MPA.
- Its ecological integrity, in terms of source–sink dynamics and recruitment to other MPAs within a network (see Chapters 16 and 17).

In protecting and conserving marine biodiversity it is important to recognize that biodiversity can be understood, conserved and managed at a range of spatial and temporal scales. Biodiversity occurs at the scale of large marine ecosystems, such as major oceanic ecosystems, and may be defined by large-scale oceanographic processes (i.e. currents and upwellings) and by trophodynamics, as well as coastal and oceanic physiography and topography. Biodiversity also occurs at other scales, whether considered as communities (see Chapter 6), habitats or specific sites. At these finer scales, patterns in biodiversity may be dominated by small-scale physical processes such as the type of substratum, cyclones, storm events, tidal range and changes in wave exposure, or by biological processes such as competition and predation. All these aspects are discussed more fully in subsequent chapters.

The importance of scientific knowledge of the oceans

In the face of human impacts on the oceans, the fundamental importance of scientific knowledge should be apparent. The necessity for conservation of the marine environment, its structures and processes, has never been more pressing. The sad reality is that we still know very little – in systematic terms – about the marine realm, its global significance and the impacts of human activities upon it.

> *Despite their importance to us, humankind is destroying marine populations, species and ecosystems. Leading marine scientists have concluded that the entire marine realm, from estuaries and coastal waters to the open ocean and the deep sea, is at risk.* (Norse, 1993)

Fortunately, several recent initiatives, including the Census of Marine Life (CoML) are now seeking to improve our knowledge of biodiversity in the oceans, and thus provide the basis for understanding the causes and consequences of changes in the diversity of life in marine waters. Examples of some of the significant recent advances, summarized by NRC (1995) include:

- The number of species
  - It is estimated that less than 10 per cent of marine species have been discovered. Consequently, measures of species richness and diversity may reflect the level of sampling effort in an area rather than true biological diversity.
  - Previous understanding of the ecology and evolution of deep-sea communities has been radically altered by the discovery that the diversity of deep-sea communities is much higher than previously thought.
  - Many undescribed species exist in 'familiar' environments, for example 158 species of polychaete worms were found in coral reef sediments from Hawaii, of which 112 species may be new.
  - New species and species assemblages have been discovered in novel habitats such as hydrothermal vents, whale carcasses and hydrocarbon seepage.
- Intraspecific genetic diversity
  - Seagrasses thought to be clonal have been found to possess high genetic diversity which has critical significance to community stability and management.
  - Recovery of threatened or endangered species whose abundance has been reduced to dangerously low levels may be at risk due to pronounced genetic 'bottlenecks' and reduced genetic variability, for example major inbreeding of humpback whales could have occurred if the international efforts to stop harvesting had not occurred when it did.
- Multispecies complexes
  - Cryptic sibling species have recently been discovered in important commercial species, including the oyster *Crassostrea*, the shrimp *Penaeus* and the stone crab *Menippe* with obvious implications for conservation and management. Similarly, the recent discovery that the US and Brazilian populations of Spanish mackerel were in fact two separate species that mature at different ages and sizes had dramatic implications for fisheries management.
- Novel groups
  - Immediately upon the introduction of new molecular techniques, previously unknown major bacterial groups were discovered in the sea. This, combined with the discovery of the widespread existence and abundance of marine viruses has fundamentally altered concepts of marine microbial diversity and the central role of microbes in global biogeochemical cycles.

Need for a systematic and integrated approach to marine conservation: Species, spaces, systems

The various present approaches to marine conservation – for example based on conservation of individual species, or habitats, or fisheries or ecosystem based management – are not at odds or in competition with one another. What we should seek is to integrate all the various approaches and initiatives within an overall ecologically logical framework. This is the fundamental attempt of this book.

Our basic question is: What should we aim to conserve? Our answer is: as many of the recognizable components of marine biodiversity as possible in networks of MPAs. Our problem then becomes: How do we decide WHAT we should conserve or preserve and how much of it (see Roff, 2009). Obviously, in the face of growing human use and exploitation of the marine environment, we cannot preserve everything; indeed we have already lost much.

However, there are certain principles that we can follow in order to develop coherent plans for marine conservation at global, national, regional and local levels, based on ecological concepts. With this as a foundation, individual groups, organizations and governments will be able to judge the importance, value and contributions of their conservation efforts and initiatives within a planning framework that spans the spatial hierarchy from global to local scales.

We believe that the important considerations for planning include the following:

Analysis of the spatial distribution of the components of biodiversity
Analysis of global biogeography
Understanding of relationships between habitats and communities
Conservation of Representative areas
Conservation of Distinctive areas
Analysis of the appropriate size of proposed MPAs
Proposal of candidate MPAs based on ecological principles
Definition of Coherent Sets of protected areas that encompass the above
Establishment of networks of MPAs
Attention to the coastal zone
Regulation of fisheries
Regulation of pollution

This list essentially defines the agenda for our book. Our presumption is that it is imperative to conserve as much as possible of the natural biodiversity of the oceans. In order to do this we need to recognize the components of marine biodiversity and how systematically to approach the complex business of marine conservation.

Some recurrent themes of the book

Certain themes will recur throughout this book. The first of these is the fundamental ecological hierarchy from genes to ecosystems – in fact, from genes to the biosphere as a whole. This hierarchy is just as natural to conservation ecologists as classification and taxonomy are to the biological systematist. Trying to preserve as many of the components of biodiversity as possible is a fundamental goal of marine conservation, even if many of the components at the ecosystem level are still beyond the present scope of human interference. The listings in Table 1.2 are not meant to be exhaustive or exclusive; but it presents a useful checklist against which to identify the important or irrelevant components of biodiversity at any spatial or temporal scale. Such a listing can therefore be useful to show how the components of biodiversity can be 'captured' in conservation planning. It can also be used to show at what level of the hierarchy or spatial scale conservation initiatives can be undertaken, from the local to the international scale.

Within the ecological hierarchy we can identify the structures and processes at each level (see Table 1.2). Structures are immediately recognized and measured (the number of organisms, the temperature of the water etc.), but processes present more problems. We generally infer processes from sequential measurements made at time intervals, or more likely simply from changes between separate observations. However, it is important to recognize the distinction between structures and processes because several different processes could in fact result in the same observed structures. Our explanations for the derivation of important structures may therefore be in error (a general cause for disagreement in science!) and consequent management decisions may be misinformed and misled.

The concept of scale has become important for all environmental and ecological enterprises, with the realization that a structure or process that is important at one scale may have little or no significance at another. For example, the process of diffusion is vital at a scale of millimetres to virtually all organisms (including respiration in humans!), but at larger scales it is overwhelmed by other processes of water motions. The important biological process of predation may be important in shaping population numbers and their distributions at local scales, but is generally replaced in significance at larger biogeographic scales by abiotic processes, or biotic processes of adaptations of individual species to their environment. The concept of scale, and judgement as to where and when a process may be of significance, is therefore always important in conservation planning. As we shall see, time and space scales tend to co-vary in the oceans, but relationships are often confused by the heterogeneity, variability and disturbances within natural systems.

The data we need to define 'natural regions' and their biota is often limited. Biological data at the required scale is sparse and temporally variable. Biological data is also expensive to collect and interpret. Recourse must therefore be had to spatially define both the ordinary (representative) and unique (distinctive) biological communities from geophysical surrogates. Physiographic and oceanographic variables, collected by a variety of means including remote sensing, can in fact quite well define biologically natural regions and their boundaries. With the growing realization that it is possible to draw lines on the oceans, this is a growing and vital area of research for marine conservation.

An overarching theme of this book will be the selection and establishment of MPAs based on sound ecological principles, and how these MPAs can be assembled into mutually supporting sets of protected areas. Ultimately this will culminate in showing how the goal – of networks of MPAs, promised by so many of the world's nations – can be achieved from global to local scales.

Conclusions and management implications

The oceans are of fundamental significance to the biological functioning of our planet. Life on earth without the 'goods and services' of the oceans is unthinkable. Biodiversity of life in the oceans runs the spectrum from the genetic to the ecosystem level. This ecological hierarchy allows us to appreciate the contribution of each level of organization to the structures and processes of marine life and their habitats.

The oceans are under threat from human activities and continue to degrade, causing loss of species and habitats. Measures to address these threats (legislation, education and awareness, international conventions, management tools etc.) are varied in their success.

Specific parts of the *Convention on Biological Diversity* refer to endangered species, threatened habitats and ecosystem management, including:

- Conserve biodiversity by establishing protected areas (Article 8).
- Recover endangered species and degraded ecosystems (Article 8).
- Protect traditional indigenous knowledge (Article 8).
- Integrate sustainable use principles into decision making (Article 10).
- Apply economic and social incentives for conservation (Article 11).

Marine conservation can be approached in a variety of ways but is fundamentally concerned with the preservation of marine biodiversity and the sustainable use of marine resources. Achieving a balance between preservation of biodiversity and resource utilization is the major challenge for marine conservation.

This book is primarily concerned with understanding the structure and function of marine environments in order to properly conserve and manage the world's oceans. There are many additional facets to marine conservation, including: law and policy; economic incentives; consumer education and awareness; property rights and so on, which are foundational to marine conservation efforts but not discussed in this volume.

This book is not primarily about marine management – only the ecological basis upon which management could be founded. Nevertheless, in each of the following chapters, we include a short section on conclusions and management implications of the ecological and environmental principles described. These sections should indicate how management could be achieved and at what spatial level or with what techniques.

References

Agardy, T. (2010) *Ocean Zoning: Making Marine Management More Effective*, Earthscan, London

Allison, G. W., Lubchenco, J. and Carr, M. H. (1998) 'Marine reserves are necessary but not sufficient for marine conservation', Ecological Applications, vol 8, supplement, ppS79–S92

Beaumont, N. J., Austen, M. C., Atkins, J. P., Burdon, D., Degraer, S., Dentinho, T. P., Derous, S., Holm, P., Horton, T., van Ierland, E., Marboe, A. H., Starkeyi, D. J., Townsend, M. and Zarzycki, T. (2007) 'Identification, definition and quantification of goods and services provided by marine biodiversity: Implications for the ecosystem approach', *Marine Pollution Bulletin*, vol 54, pp253–265

Carson, R. (1962) *Silent Spring*, Houghton Mifflin, NY

Census of Marine Life, www.coml.org/, accessed 30 August 2010

Christensen, V., Aiken, K. A. and Villanueva, M. C. (2007) 'Threats to the ocean: On the role of ecosystem approaches to fisheries', *Social Science Information sur les Sciences Sociales*, vol 46, no 1, pp67–86

Dobzhansky, T. (1973) 'Nothing in biology makes sense except in the light of evolution', *American Biology Teacher*, vol 35, pp125–129

Duarte, C. M. (2000) 'Marine biodiversity and ecosystem services: an elusive link', *Journal of Experimental Marine Biology and Ecology*, vol 250, pp117–131

Frank, K. T., Petrie, B., Choi, J. S. and Leggett, W. C. (2005) 'Trophic cascades in a formerly cod-dominated ecosystem', *Science*, vol 308, no 5728, pp1621–1623

Franks, F. (ed) (1972) *Water: A Comprehensive Treatise*, Plenum Press, New York

Gislason, H., Sinclair, M., Sainsbury, K. and O'Boyle, R. (2000) 'Symposium overview: Incorporating ecosystem objectives within fisheries management', *ICES Journal of Marine Science*, vol 57, pp468–475

Gray, J. S. (1997) 'Marine biodiversity: Patterns, threats and conservation needs', *Biodiversity and Conservation*, vol 6, pp153–175

Guerry, A. D. (2005) 'Icarus and Daedalus: Conceptual and tactical lessons for marine ecosystem-based management', *Frontiers in Ecology and the Environment*, vol 3, pp202–211

Hardin, G. (1968) 'The tragedy of the commons', *Science*, vol 162, pp1243–1248

Hoyt, E. (2005) *Marine Protected Areas for Whales, Dolphins and Porpoises*, Earthscan, London

Hughes, T. P., Bellwood, D. R., Folke, C., Steneck, R. S. and Wilson J. (2005) 'New paradigms for supporting the resilience of marine ecosystems', *TRENDS in Ecology and Evolution*, vol 20, pp380–386

Irish, K. E. and Norse, E. A. (1996) 'Scant emphasis on marine biodiversity', *Conservation Biology*, vol 10, p680

Leslie, H. M. (2005) 'A synthesis of marine conservation planning approaches', *Conservation Biology*, vol 19, pp1701–1713

McLuhan, M. (1962) *War and Peace in the Global Village*, Bantam, New York

NRC (National Research Council) (1995) *Understanding Marine Biodiversity*, National Academy Press, Washington, DC

Newman, D. J. and Cragg, G. M. (2007) 'Natural products as sources of new drugs over the last 25 years', *Journal of Natural Products*, vol 70, no 3, pp461–477

Norse, E. A. (ed) (1993) *Global Marine Biodiversity: A Strategy for Building Conservation into Decision Making*, Island Press, Washington, DC

Noss, R. F. (1990) 'Indicators for monitoring biodiversity: A hierarchical approach', *Conservation Biology*, vol 4, pp355–364

Pauly, D., Christensen, V., Dalsgaard, J., Froese, R. and Torres F. (1998) 'Fishing down marine food webs', *Science*, vol 279, no 5352, pp860–863

Roberts, C. (2007) *The Unnatural History of the Sea*, Island Press, Washington, DC

Roff, J. C. (2009) 'Conservation of marine biodiversity: How much is enough?', *Aquatic Conservation: Marine and Freshwater Ecosystems*, vol 19, pp249–251

Sadava, D., Heller, H. C., Orians, G. H., Purves, W. K. and Hillis, D. M. (2006, 8th edition) *Life: The Science of Biology*, Sinauer Associates Inc. and W. H. Freeman and Company

Sepkoski, J. J. (1997) 'Biodiversity: Past, present, and future', *Journal of Paleontology*, vol 71, pp533–539

Signor, P. W. (1994) 'Biodiversity in geological time', *American Zoologist*, vol 34, pp23–32

Sverdrup, H. U., Johnson, M. W. and Fleming, R. H. (1942) *The Oceans: Their Physics, Chemistry and General Biology*, Prentice-Hall, NY

Thorne-Miller, B. and Catena, J. (1991) *The Living Ocean: Understanding and Protecting Marine Biodiversity*, Island Press, Washington, DC

UNEP (United National Environment Programme) (1992) Convention on Biological Diversity, available at www.cbd.int/doc/legal/cbd-en.pdf, accessed 20 December 2010

Wilson, E. O. (ed) (1988) *Biodiversity*, National Academic Press, Washington, DC

Zacharias, M. A. and Roff, J. C. (2000) 'A hierarchical ecological approach to conserving marine biodiversity', *Conservation Biology*, vol 14, pp1327–1334

2

The Marine Environment: Physico-chemical Characteristics

Structures and processes – enduring and recurrent factors

But more wonderful than the lore of old men and the lore of books is the secret lore of ocean.

H. P. Lovecraft (1890–1937)

Introduction

Marine conservation is a relatively new discipline, lagging behind terrestrial conservation (see e.g. Irish and Norse, 1996), but interest in it has been growing substantially in recent years. However, many practitioners who become engaged in marine conservation come to it from interests developed primarily in the terrestrial environment, perhaps unaware that the principles that apply to terrestrial habitats and conservation may or may not be transportable to the marine environment. In order to develop appropriate strategies and frameworks for marine conservation, we must acknowledge the inherent structures and processes of the marine environment. We should clearly note where they differ from those of terrestrial environments, and discern where terrestrial paradigms and approaches will not apply to marine systems.

For these reasons, this chapter presents a brief examination of the major physico-chemical characteristics of marine environments – which are the structure and process components of marine biodiversity at the habitat/ecosystem level of organization. The following Chapter 3 presents a brief review of some biological/ecological features of marine environments – which are the structure and process components of marine biodiversity at the species/population and community levels of organization. Consideration of the genetic level of organization is deferred until Chapter 10. An appreciation of how marine ecosystems are similar to and different from terrestrial ecosystems, and indeed from other aquatic ecosystems, is essential in order to 'set the scene' for the following chapters and concepts. These similarities and differences – for abiotic, and for biological/ecological characteristics – are summarized for marine and terrestrial ecosystems, for marine and other aquatic ecosystems and for arctic, sub-arctic, temperate sub-tropical and tropical ecosystems in Chapter 3.

A full description of the marine environment and its oceanography lies beyond the scope of

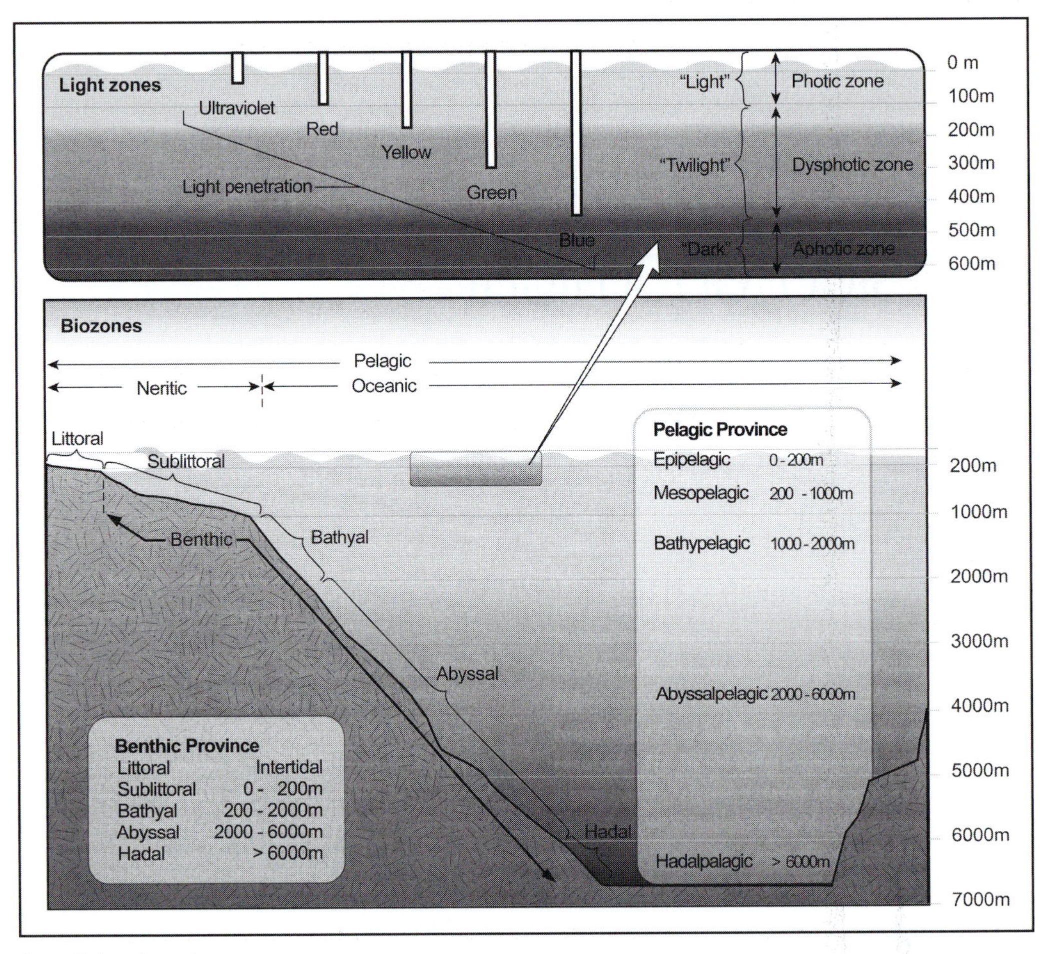

Source: Redrawn from various sources

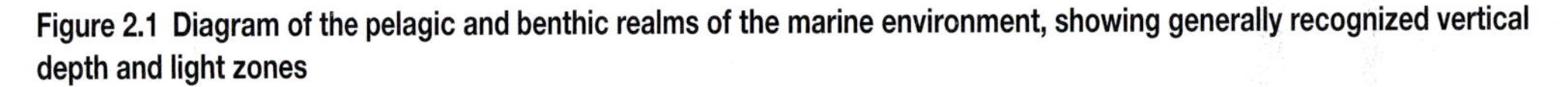

Figure 2.1 Diagram of the pelagic and benthic realms of the marine environment, showing generally recognized vertical depth and light zones

this text. The emphasis here will be to introduce concepts, and factors (variables – which can be measured directly; and parameters – composites of variables) both enduring and recurrent, that are involved in shaping the character of marine communities, relevant to the distribution of the components of marine biodiversity, and that will inform decisions for marine conservation planning. Enduring factors are those that persist at a given location over time (e.g. substrate type), and recurrent factors are those that periodically change in predictable ways (e.g. tides and currents). We shall consider these factors as belonging to two main types: structures and processes; and as belonging to two main categories: physiographic – pertaining to the ocean basin itself, and oceanographic – pertaining to the water column.

A fundamental division of the oceans, that affects all further considerations, is into two major realms: the *pelagic* and *benthic* realms (Figure 2.1). The pelagic realm is the water column itself and all the organisms that inhabit it. The benthic realm is the sea-floor with all the creatures that live within or upon it. The pelagic realm is a fully three-dimensional world while in comparison, to a first approximation, the benthic realm

Box 2.1 Selected recommended reading as background to the physics, chemistry and biology of the oceans

Barnes, R. S. K. and Hughes, R. N. (1988) *An Introduction to Marine Ecology*, Blackwell Scientific Publications, Oxford
Bertnes, M. D. (1999) *The Ecology of Atlantic Shorelines*, Sinauer Associates, Sunderland, MA
Bertnes, M. D., Gaines, S. D. and Hay, M. E. (2001) *Marine Community Ecology*, Sinauer Associates, Sunderland, MA
Knox, G. A. (2001) *The Ecology of Seashores*, CRC Press, London
Mann, K. H. and Lazier, J. R. N. (1996) *Dynamics of Marine Ecosystems: Biological-physical interactions in the oceans*, Blackwell Science, Cambridge, MA
Mann, K. H. (2000) *Ecology of Coastal Waters with Implications for Management*, Blackwell Scientific Publications, Oxford
Open University Course Team (1989) *Waves, Tides and Shallow-Water Processes*, Pergamon Press, Oxford
Open University Course Team (1989) *The Ocean Basins: Their Structure and Evolution*, Pergamon Press, Oxford
Open University Course Team (1989) *Seawater: Its Composition, Properties and Behaviour*, Pergamon Press, Oxford
Open University Course Team (1989) *Ocean Chemistry and Deep-Sea Sediments*, Pergamon Press, Oxford
Open University Course Team (1989) *Ocean Circulation*, Pergamon Press, Oxford
Ray, G. C. and McCormick-Ray, J. (2004) *Coastal Marine Conservation: Science and Policy*, Blackwell, Malden, MA
Sherman, K., Alexander, L. M. and Gold, B. D. (1992) *Large Marine Ecosystems: Patterns, Processes and Yields*, AAAS Press, Washington, DC
Sverdrup, H. U., Johnson, M. W. and Fleming, R. H. (1942) *The Oceans: Their Physics, Chemistry and General Biology*, Prentice-Hall, NY
Sverdrup, K. A., Duxbury, A. C. and Duxbury, A. B. (2003) *An Introduction to the World's Oceans*, McGraw-Hill, NY
Valiela, I. (1995) *Marine Ecological Processes*, Springer, NY

can be considered as nearly two dimensional. The two realms are intimately and continuously connected by a variety of physico-chemical and biological processes, but if we analyse these two realms separately (as it is often convenient and simpler to do), then we must consider the oceans as effectively comprising no less than five dimensions! Oceanographic factors apply primarily to the pelagic realm, with physiographic structures only becoming of significance where submarine topography affects oceanographic processes. Physiographic factors primarily apply to the benthic realm, but oceanographic structures and processes are also significant. Such classifications are very important for conservation planning, even though our human bias is naturally towards the familiar shoreline.

For a fuller consideration of the fundamentals of oceanography and marine biology/ecology, the reader is referred to – among others – the texts listed in Box 2.1.

Major features of the oceans - physiographic structures

Physiographic characteristics are those features broadly recognized as 'marine landform' – in essence relating to the topography and substrates of the sea-floor. The physiography of the shoreline and seabed determines the broad character of benthic biological communities, though there are overlying geological, biogeographical and oceanographic contexts, which are responsible

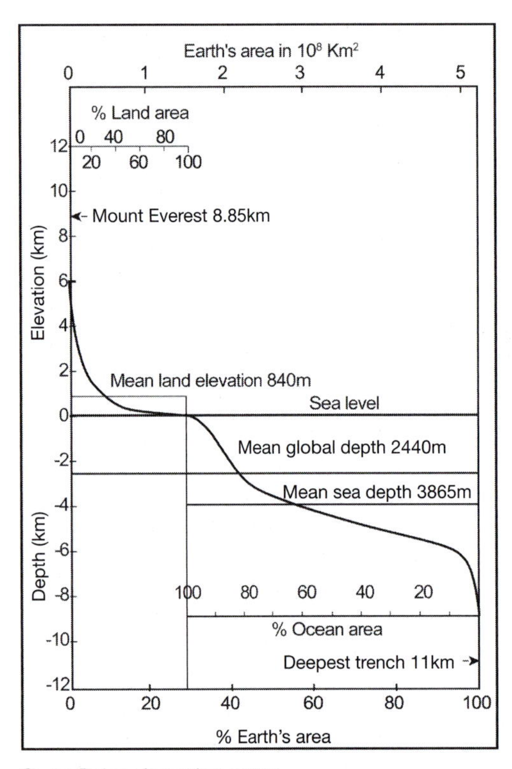

Source: Redrawn from various sources

Figure 2.2 Hypsographic curve of the terrestrial and marine environments, showing the distribution of elevations and depths on earth

for large-scale differences in biological communities. The physiography of coastlines and the sea-bottom is one of the easiest components to map, and with satellite, airborne and in situ sensing technologies, considerable accuracy is available or possible.

Area, depth and volume

Collectively, the oceans of the world cover 70.6 per cent of the surface of the globe. The average depth of the oceans is approximately 3.8km and maximum depths exceed 10km. The volume of all the oceans and seas combined is an immense 1370 x 10^6km^3. A hypsographic curve of lands and seas (Figure 2.2) clearly shows the preponderance of living volume in the oceans over that on land. Life on land (29.3 per cent of the area of the globe) extends from only a few metres below ground – in caves, plant roots and in geological fractures – to a few metres above ground – at the tops of trees (discounting birds, which are still land referenced). In contrast, life in the oceans extends from surface waters to over 10,000m in depth. Therefore well over 99 per cent of the habitable volume for the biota of our planet is in the oceans.

Horizontal divisions and bathymetry

In the horizontal dimension, the marine environment is generally regarded as divisible into several major provinces (Figure 2.1). Coastal waters comprising the *coastal zone* (variously defined – see Chapter 11 – but here regarded as less than 30m in depth) extend seaward from the high water level and include the *littoral zone*. Near to shore are various kinds of *inlets* including estuaries, bays and coves and associated wetlands. *Estuaries* are the meeting place of freshwaters with the ocean, where the salinity is measurably diluted by freshwater runoff (see below). *Bays* and *coves* are shoreline concavities of the ocean where salinity may not be diluted (unless they are bays *within* estuaries). The coastal regions run into the *neritic province* of *sub-littoral waters*, which is the region of the ocean that lies above the *continental shelf* out to depths of 200m. Although the edge of the continental shelf corresponds to the *exclusive economic zones* (EEZ) of nations in some places, in most places around the world there is no relationship between the two. Some implications of this will become clear in Chapter 12 on Deep Seas and High Seas. The neritic province then merges into the *oceanic province* at the *shelf edge* or the *shelf break*. The oceanic province comprises those vast areas of the oceans that physically lie beyond the edge of the continental shelves and whose waters exceed depths greater than 200m.

Depth, light and pressure

The next major division of the oceans is made in terms of depth, where a variety of terms is used to describe the habitats and their biological communities. The conventional descriptive divisions of the oceans with respect to depth, which have long been recognized, are shown for both the pelagic and benthic realms in Figure 2.1. Depth is an important factor in both the pelagic and benthic realms. In combination with temperature, salinity, light and pressure (with which it co-varies) it defines the distributions of major community types (e.g. Glenmarec, 1973).

Somewhat arbitrarily, the oceans are vertically subdivided into the *epipelagic* (down to 200m), the *mesopelagic* (200 to 1000m), and the *bathypelagic* zones (1000 to 2000m) and *abyssal/hadal* zones (>2000m). Similar terms are applied to the benthic realm (see Figure 2.1).

Light intensity diminishes exponentially with depth in the oceans. In the vertical dimension, and very fundamentally as far as the photosynthetic organisms are concerned, the oceans can therefore be subdivided as follows: the *euphotic* (= photic, or well lighted) zone is the region in which sufficient light penetrates to allow net photosynthesis and plant growth to occur; below this is the *dysphotic* (or poorly lighted) zone where light is still present, but its intensity is too low to support plant growth; below this again, the great majority of the oceans' depths lie within the *aphotic* zone, where no light penetrates (Figure 2.1).

Light provides the energy for photosynthesis and primary production in most marine ecosystems. The penetration of light within the water column is attenuated with both depth and turbidity, and both parameters are important determinants of the vertical distribution of pelagic and benthic vegetation. The euphotic, dysphotic and aphotic zones are real, functional zones which limit the development and types of biological communities. The division between the photic and non-photic zones is more significant than the further subdivisions (dysphotic and aphotic) within it. Beyond the limits of the euphotic zone lie the dysphotic and aphotic regions, defining communities which cannot photosynthesize. At these depths, energy for consumers is imported from other areas, predominantly by vertical settling of detritus from upper layers of the sea. Consequently the whole trophic structure of communities below the euphotic zone is different from those within it, and is dependent upon detrital carbon. With increasing depth, the amount of available food declines exponentially as a function of surface productivity (e.g. Suess, 1980).

The compensation point occurs at the bottom of the euphotic zone, a depth below which the rate of respiration exceeds the rate of photosynthesis. The actual depth of the euphotic zone increases with water depth itself, from the coast towards the edge of the shelf and into oceanic waters, and it also varies at different times of the year. For example, in estuaries, the euphotic zone may be less than 2m in depth, in average coastal waters it approximates 30–50m, while in oceanic waters it may exceed 200m. Similarly in the Arctic Ocean, the euphotic zone may exceed 100m during the spring, and suddenly decrease to only a few metres during the summer phytoplankton bloom. The biomass and productivity of phytoplankton can be estimated from ocean colour and water clarity (e.g. Bukata et al, 1995). Light penetration and turbidity are also important determinants of submergent vegetation.

Depth is also a surrogate variable for *pressure*. The increase in pressure with depth has a significant impact on organisms. With every ten metres of depth, the water pressure increases by approximately one atmosphere (with the greatest change from 0 to 1 atmosphere occurring in the top ten metres). Additional physical, chemical and biological changes lead to a decrease of dissolved oxygen and increase of dissolved carbon dioxide (see below). Organisms which live in the deeper regions of the oceans are adapted to these physical conditions of high pressure, low temperatures and dilute resource concentrations, and rarely move into the epipelagic region.

Temperature also decreases with depth, from ambient surface values to a nearly constant 0–4°C in deepest oceanic waters. Conversely, salinity typically increases with depth. Concentrations of particulate organic carbon (the detrital flux from the euphotic zone) also decrease exponentially with depth, while oxygen concentrations decline and carbon dioxide concentrations increase with depth. Depth is therefore an index of a variety of concurrently changing physical and chemical conditions, which collectively influence the nature of biological communities.

Basin morphometry and topography

The general topography or morphometry of a region (e.g. an estuary, inlet, basin) can exert a significant effect on the character of a coastal region. For example, at a large scale, an entire basin may have a natural period of oscillation that reinforces the local tidal frequency. In such a case, resonance occurs, and very high tides and rapid tidal currents are observed. A prime example of this in Canadian waters is the Bay of Fundy, Nova Scotia. In such conditions, the high tidal range and fast currents may determine that the local substrate consists of coarse particles or even bare rock for considerable distances. At the opposite extreme, where a local basin experiences low tides, sedimentary areas usually predominate. However, because the tidal amplitude and nature of the substrate can be independently obtained, and because substrate type is also a function of geology and wave action, these factors may be assessed directly, rather than interpreted from morphometric characters.

At finer scales, topography can have profound effects on marine fauna and flora. A particular marine phenomenon, which is unlikely to be captured in any regional study of the distribution of marine geophysical features, is the existence of underwater marine caves. Localized 'pockets' of organisms either not found elsewhere, or only sparsely existing in other locations, may thrive in marine caves. These are the kinds of habitats beloved of SCUBA divers, and essentially inaccessible by other means. Sampling from the surface generally does not reveal their existence. These constitute 'distinctive' faunas, exhibiting the phenomena of 'interiority' (Morton et al, 1991) and spatial heterogeneity (Bergeron and Bourget, 1986) at the smallest scales.

Relief and slope

Relief applies to the vertical change in height in relation to horizontal distance, and provides an indication of slope. The shoreline slope, in combination with local tidal amplitude, determines the extent of the intertidal zone. Slope and exposure also influence the substrate type in intertidal regions. Steep slopes and high exposure lead to bare rock, while intertidal mud-flats occur at the opposite end of the slope and energy spectrum.

The relief (also variously described as slope, rate of change of slope, rugosity or benthic complexity) at the shoreline and within coastal and marine waters is highly variable. While slope intervals are sometimes mapped, slope is more often inferred from vertical changes in height in relation to horizontal distance (i.e. steeper slope where bathymetric contours are closer together). Depth and hence slope is generally mapped in more detail in areas of navigable waters. However, it is important to remember that a calculated slope depends on the frequency of data points.

Areas of high relief tend to have irregular bottom morphologies and high elevation ranges; low relief areas have uniform slopes with small elevation gradients. High relief areas provide habitat for numerous species assemblages and may indicate high species richness, diversity and biomass (Lamb and Edgell, 1986). Relief may also be an indirect indication of mixing. Sediment stability is partly dependent upon slope, while the angle of repose of marine sediments depends on particle size and activity of water motion. Stable marine slopes for sediment accumulation are much lower than for terrestrial slopes of similar grain sizes.

Relief and slope characteristics are generally considered as secondary diagnostics, compared to direct knowledge of substrate type, current speed, exposure and so on. These factors may, however, become useful as predictors of local substrate type under some circumstances, where direct data on substrate type is not available (e.g. regions of the Arctic), but in most cases where substrate type and current velocities are known, these factors may add little extra information concerning biological community types. An exception to this may be at the edge of the continental shelf, where the change in substrate slope itself, in concert with current activity, may enhance local benthic production by processes not yet well understood.

Substrate type and particle size

Substrate particle size is a dominant influence on marine communities. It is frequently classified

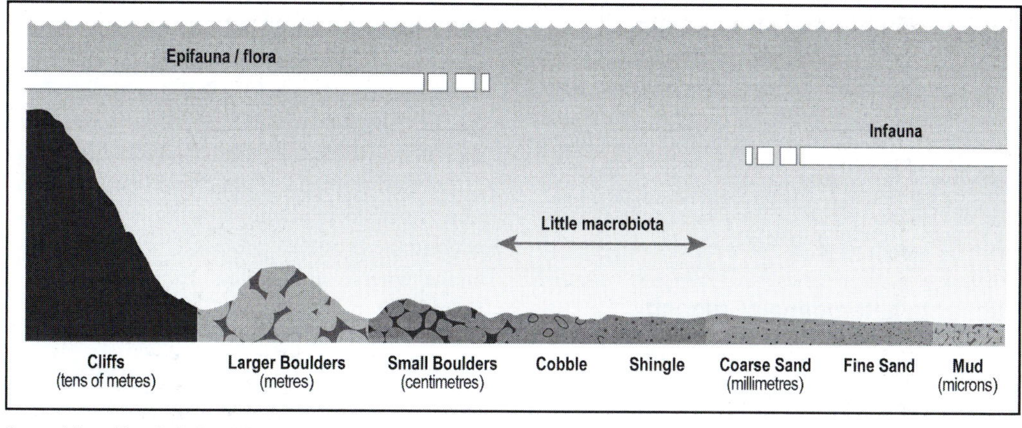

Source: Adapted from Raffaelli and Hawkins (1996)

Figure 2.3 Diagram showing how major benthic marine community types change along a gradient of substrate particle size, from epiflora and epifauna in areas of high exposure and solid substrates, to infauna of finer particle-sized substrates

according to a 'Wentworth' scale or mean particle-size scale, which is often reduced and simplified to various descriptive categories, such as rock and boulders, gravel, coarse sand, fine sand or mud/silt. Information on sediment composition is often available directly in shallow waters, or can be derived as a surrogate from some combination of depth, slope, bottom morphology, geological characteristics and current speeds.

There is considerable interplay between geological characteristics, current speed, depth and slope in determining the local bottom substrate type and consequent community type. In high energy (erosional) areas, of higher slope and current speed, exposed substrate or coarse sediments (stones, gravel) will predominate. In lower energy (depositional) areas having lower slope and current speed, finer muds and silts will be found. Grain size therefore depends primarily on depth and current speed which is predominantly tidally generated (e.g. Pingree, 1978; Eisma, 1988). As a first approximation, we can divide benthic community types into suspension feeders, associated with coarser sediments and higher current velocities of erosional areas, and deposit feeders, associated with finer sediments and lower current speeds in depositional areas (Wildish, 1977), although there is some overlap between these types.

The general trend with depth for substrates is from high variability of substrate type in coastal waters, especially in intertidal and immediate subtidal regions (from rock to muds), to relative uniformity of silts and muds in deeper waters off the continental shelves. Sediments of mud, sand or gravel may be transported considerable distances along a coast or within an estuary, prior to settling on the bottom. Sediment transport rates are dependent upon the direction and current velocity as well as grain size. Sediment stability is dependent upon slope and other variables, while the angle of repose of marine sediments depends on particle size and degree of water motion. The substrate type at the extremes of the erosional/depositional regimes tends to be relatively constant although there may be very significant changes in sediment accumulation or erosion between years as the result of storm activity in coastal waters. Thus sand and gravel environments typically experience the greatest changes in sediment quantity and mobility between years. For this reason, the species diversity of plants and animals tends to be highest on rocky shores and again on muddy shores, with the more mobile sandy/gravel substrates typically showing much lower species diversity (Figure 2.3). Within the benthic euphotic zone, productivity of plants is high in the macrophytic algae on rocky shores, low in sand and gravel, and high again in microphytic algae of mud-flats and salt marsh macrophytic angiosperms (e.g. Levinton, 1982).

In the deep regions of the sea, away from the influence of surface currents, wave motion and tidal effects, the velocities of currents are low and sedimentation of fine materials is the norm. Here, virtually all the seabed consists of some form of sediment. Bare rock is found mainly as basalt at oceanic ridges where we find the truly extraordinary vent communities (Ballard, 1977).

Substrate heterogeneity, rugosity

Substrate (or more broadly – environmental) heterogeneity is a highly complex subject, whose understanding is inherently dependent upon the scale of the process or phenomenon under examination. For example, at very small scales (millimetres to centimetres, corresponding to the habitat of the individual plant or animal), cracks and crevices in rock may be critical to the recruitment and survival of individuals. Substrate irregularities at only slightly larger scales (centimetres to tens of metres) may be irrelevant to such individuals and their species populations. However, it is precisely at these scales that the rugosity and variety of substrates generate new habitats and combinations of habitats that can lead to high species diversity. This phenomenon will be considered further in Chapters 6 and 11.

At larger scales again (tens to thousands of metres), caves (concavities) and seamounts (convexities) may be critical to the survival of certain species that rely on such features as refugia from predators, or as topographic 'lenses' or 'accumulators' of resources.

At broader spatial scales again (hundreds of metres to tens of kilometres), the environmental heterogeneity provided by local or regional structures such as mud-flats, upwelling areas and gyres that accumulate pelagic resources may be critical to marine mammals and birds that make long-distance migrations.

All these effects are extremely important to understand and document during the process of describing the ecological nature within regions, and in the process of selection of marine protected areas (MPAs). It becomes especially important to distinguish between areas that may be unique or *distinctive* in some way rather than *representative* because they occur at a range of spatial scales (see Chapters 5 and 7).

Geology and rock type

In the intertidal and sub-tidal zones, rock type may be sedimentary, metamorphic or igneous. This has an effect on particle size of the substrate of shorelines – because of relative erodabilities – and, in turn, on the type of associated benthic organisms. Sedimentary bedrock is most likely to be affected by erosional processes, while the crystalline igneous and metamorphic materials are more likely to be resistant. There may be significant differences in the types of intertidal communities that develop on rocky shores, related to the degree and orientation (relative to the shore) of rock faulting. This can give rise to what Morton (1991) has called 'interiority'. Thus a more faulted and locally convoluted rocky shore can give rise to a richer more diverse community than a smoother one. Bergeron and Bourget (1986) have studied such relationships between physical heterogeneity and diversity, but only at smaller scales than considered here.

Major features - physiographic processes

In general, oceanographic processes overshadow physiographic ones in terms of impact on biota, because many of their time frames are of the same order of magnitude as those of biological processes, whereas geological and physiographic processes occur over much longer time frames.

Crustal activity

Over the course of geological time, tectonic movements and other crustal movements have dramatically affected the size and shape of the world's oceans. In general the oceans around a landmass are much older than the landmass itself and the lakes contained within it. For example, ecosystems and their habitats within north temperate landmass have essentially been modified or created during the last glacial interval (10,000 to > 1,000,000 years ago) whereas the oceans have been forming and adjusting over thousands of millions of years. Not only have the oceanic areas and their associated ecosystems developed over a longer period, but there are significant age differences between the three major oceans. The greater geological age of the Pacific Ocean is believed to be the major reason for its higher biological diversity in marine communities compared to the Atlantic (see e.g. Levinton, 1982, and Chapter 8).

Modern day geological activities do naturally periodically have an impact on marine communities, but not in any regular or predictable way that could suitably be used to distinguish community types. Exceptions to this include the geochemical processes at deep sea vent areas that support the extraordinary deep sea vent communities. Other tectonic activities that cause sediment transport and blanketing of benthic communities come under the category of periodic disturbances (see Chapter 19).

Major features – oceanographic structures

Oceanographic factors are highly dynamic, and water characteristics are of major importance to the determination and development of representative and distinctive communities in both pelagic and benthic realms (e.g. Tremblay and Roff, 1983; Emerson et al, 1986; McLusky and McIntyre, 1988). Oceanographic factors are generally variable throughout the year at any location.

Temperature

Temperatures are relatively stable within the oceans, and there is far less seasonal fluctuation than in terrestrial systems. Seasonal variation in temperature is generally low at both high and low latitudes, but seasonal variation may be high at mid-latitudes – e.g. from 0 to +20°C. The temperature in the oceans varies from nearly -2°C in polar oceans to +32°C in tropical waters. The lower temperature is set by the freezing point of seawater. The freezing point of seawater is depressed below that of freshwater (by definition 0°C) because of its salt content, and it is variable and proportional to the salinity. The temperature of maximum density lies below the freezing point for all seawaters of salinity >24.7psu (i.e. all open ocean waters); it is not +4°C as for freshwaters. This means that seawater of a given salinity will continue to sink below warmer waters of the same salinity as it cools.

The oceans are strongly vertically thermally stratified by a 'permanent' thermocline. To a first approximation, ocean waters can be divided into an upper wind-mixed layer – from the surface to a depth of 600 to 1000m – where temperatures range from +8°C to +32°C. Temperatures of the lower layers – from 600m to the bottom – are generally well below those at the surface, averaging between 0 and 4°C. The temperature of the bulk of marine waters therefore lies between 0 and +4°C. Water temperatures in the deeper layers are also more constant than in surface waters, and may be significantly influenced by ocean currents. Bottom topography also has an influence on water temperature, with pockets of warmer (but more saline) waters often being trapped for periods of time in large basins. Deeper colder waters can also be transported towards the surface in upwellings.

Temperature is a major determinant of marine communities, because of its positive relation to growth and complex relations to components of biodiversity and nutrient supply (see below). Because large areas of the oceans experience approximately the same seasonal average and range of temperatures, temperature itself will tend to be a discriminant among community types, primarily at the largest scales. However, temperature may also be a discriminant at finer scales, for example within shallow sheltered coastal waters and estuaries, where temperatures may locally exceed those in surrounding waters (see below under *temperature gradients and anomalies*). In temperate coastal and neritic waters, the water column may become seasonally strongly stratified.

Ice cover and scour

Ice cover may be permanent, seasonal or absent; its extent and development is a major influence on community types over broad geographical areas. The presence of permanent ice greatly restricts marine productivity. Seasonal ice has a major impact in seasonally eliminating or enhancing production by a variety of mechanisms. The physical effects of ice scour and glaciers have the greatest impact on shallow water marine communities, especially in intertidal regions, primarily by reducing community diversity or restricting some organisms to crevices. Except for inshore areas – or in polar oceans where effects of rafted ice can reach to considerable depths – ice does not have an impact on the sub-tidal benthos.

Temperature gradients and anomalies

A common feature of marine areas exploited by migrant species is high productivity or accumulation of resources. This subject will be explored more fully in Chapter 7. One physical feature which many of these areas appear to have in common is some sort of temperature anomaly or sharp temperature gradient. Temperature anomalies are well documented as being associated with several types of region. Upwelling regions, carrying nutrient-rich waters to the surface, also carry a signature of surface temperature lower than surrounding waters. While the geographic location of upwelling events is predictable, the timing may not be, and may depend on unpredictable meteorological events. Upwellings may be seasonally or annually variable in development; the water column is typically vertically mixed, and stratification is weak or absent.

During the summer months, temperate intertidal mud-flats generate temperatures higher than surrounding areas and overlying waters, and become areas of high diatom production. Productive frontal systems between stratified warmer waters and non-stratified colder waters also show strong horizontal temperature gradients (see below).

Salinity

The salt content of virtually all parts of the oceans lies between 32 and 39psu (practical salinity units or parts per thousand – also written as °/oo, or ppt) and the chemical composition of the major ions of seawater remains virtually constant. This is because the rate of input and export of salts to and from the oceans is low compared to the rate of global mixing. The salinity of surface waters around the globe is strongly correlated to the difference between latitudinal variations in precipitation and evaporation, and to the rate of mixing with sub-surface waters. Salinity variations with depth also contribute to vertical stratification and stability of the water column (see below).

Variations in salt content over the range of 32 to 39psu have little discernible effect on marine organisms, the majority of which are stenohaline (adapted to constant salinity) and intolerant of salinity changes. Although in the past, considerable attention has been devoted to examination of relationships between distributions of organisms and salinity, in fact other factors such as temperature and the movements of water masses predominate in controlling the distributions of marine organisms.

Only in estuaries, estuarine bays or sheltered mountainous coastal regions experiencing high rainfall – where salinity varies from close to zero to over 33psu – does salinity have major effects on the distribution of aquatic species, because of its significance in osmotic and ionic regulation. The majority of aquatic species are adapted either to a life in freshwaters at low but relatively constant salt content, or to the much higher, nearly constant salinities and virtually constant composition of the oceans. Relatively few species (euryhaline) are well adapted to life in estuaries at intermediate and fluctuating salt content. In estuaries, salinity exhibits complex behaviour, both physically and in its impact on the distribution and abundance of organisms (see below).

Constancy of seawater composition and nutrients

Seawater is a complex solution containing virtually every element of the periodic table. This is a testament to the high dissolving power of water. Despite variations in total salt content (the salinity), the composition of seawater (that is the ratio of the major elements to one another) remains virtually constant. This is because the concentrations of most elements are high relative to their rate of use by the biota. The exception to the rule is for elements (or ionic compounds) called nutrients.

Nutrients in marine environments are those inorganic substances that are required for, and can control the rate of, aquatic plant growth, primarily: PO_4, NO_3, NH_4, Fe and Si. Only nutrients within, or being supplied to, the euphotic zone are of significance to our understanding of marine processes for conservation purposes. Nutrient concentrations below the euphotic zone are much higher than within it, but they cannot be effectively used by plants because of light limitation. Nutrients below the euphotic zone are therefore irrelevant to our considerations of biological functioning and conservation unless and until they reach the surface waters in upwellings. Organic nutrients, vitamins, antibiotics etc. and trace elements are more complex in behaviour; they may play a role in seasonal succession of plant species, but are not considered further here.

Nutrients in aquatic environments are derived from several sources: via the atmosphere, from land drainage, by recirculation of deep waters to the surface and from internal regeneration. Although all these processes are of some significance to marine conservation, a full examination of them is well beyond the scope of this book. However, a brief perspective is required, because various parts of the marine environment obtain their nutrients by different processes and from different predominant sources.

In the open water pelagic realm, nutrient inputs are predominantly a function of the mixing regime and/or internal regeneration by the food web. Atmospheric inputs are assumed to be uniform over macroclimatic regions. Terrestrial inputs to the oceans are low compared to internal recycling, except in estuaries and the nearshore coastal zone. Internal regeneration and recirculation of nutrients within the ocean thus far outweigh the significance of external inputs – except in the immediate coastal zone.

A major (internal) source of nutrients to coastal areas is upwelling or entrainment of subsurface water. While the geographic location of these events is largely predictable, the timing may depend on unpredictable meteorological events. Such water mixing regimes may be evident from local low-temperature anomalies.

Within water masses, the nutrient regime is a function of whether the water column is stratified or mixed, as described by the stratification parameter (see below).

Oxygen and other dissolved gases

Within the neritic and pelagic realms of the oceans and larger lakes, dissolved gases are normally at or close to saturation in surface waters. This is because physical processes, that cause exchange of gases between water and atmosphere, typically overshadow the biological processes of photosynthesis and respiration in magnitude. The wide diel variations of gas content, typical of smaller bodies of water, are generally not observed. Even below a thermocline, oxygen rarely becomes naturally seasonally depleted within the water column, except in stratified inlets and estuaries where there are major inputs of organic matter (Nixon, 1993). In oceanic tropical and subtropical waters, an oxygen minimum layer develops at a depth of between 600 and 1000m in both the Atlantic and Pacific oceans (more highly developed in the latter).

There is currently considerable concern for several regions worldwide because of low oxygen concentrations due to human impacts, caused by the inputs of dissolved nutrients and suspended solids from land runoff (Diaz and Rosenberg, 2008). Perhaps the largest of these areas that has been growing in extent for several years lies in the Gulf of Mexico. Here, large areas to the south and west of the Mississippi River have become anoxic following algal blooms caused by cultural eutrophication off coastal waters. However, it is important to realize that such anoxic regions can also be caused naturally. For example, the anoxic regions beneath the Benguela current (off south-west Africa) are caused by the aperiodic upwellings in that region (Chapman and Shannon, 1987). Here the high planktonic production is not efficiently used by higher trophic levels (because of its unpredictable periodicity), and much of the photosynthetically fixed energy sinks to the benthos and decays, causing oxygen depletion.

Within benthic sediments themselves, oxygen levels can be very low, approaching zero or even becoming anoxic and allowing reducing conditions. Thus oxygen concentration in sediments, along with sediment grain size, may be an important determinant of community type.

Water masses and density

The combination of temperature and salinity determines the density of seawater; higher temperature reduces density and higher salinity increases it. Outside of estuarine environments, density itself is of little direct biological consequence in the oceans because it varies only within narrow limits. Density is, however, of major global significance for physical oceanographic purposes, as a determinant of ocean thermohaline circulation (see below). There is no functional relationship between temperature and salinity, for example, temperature does NOT change the salinity of a water body as is sometimes stated. This is because salinity is a mass per unit mass measure not a mass per unit volume measure.

The combinations of temperatures and salinities – each as a retained separate variable – are, however, of vital importance to conservation. Each water mass of the oceans can be recognized, and its distribution mapped on the basis of its individual salinity and temperature characteristics. The combinations of temperature and salinity values act as signatures of individual water masses, and point to their geographic origins. Density (as derived from these two variables) is *not* used for this purpose because widely different combinations of temperature and salinity can yield exactly the same density value (see Figure 2.4).

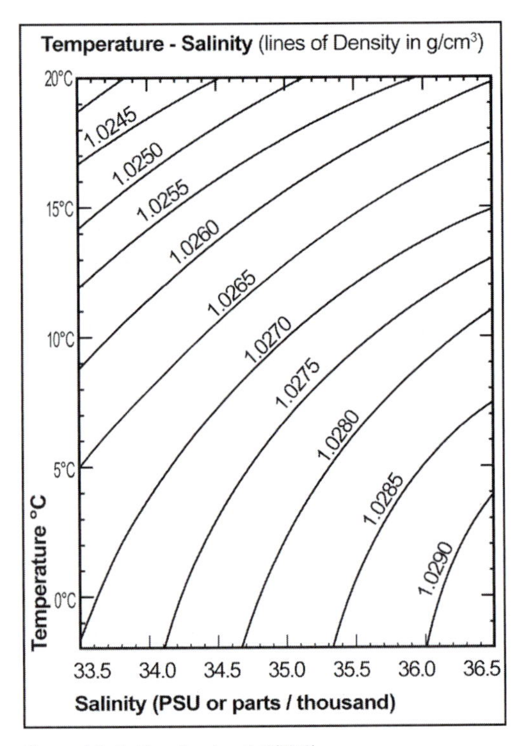

Source: Adapted from Sverdrup et al (1942)

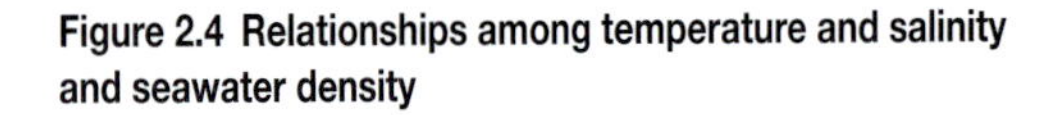

Figure 2.4 Relationships among temperature and salinity and seawater density

Certain 'usual' or 'habitual' combinations of temperature and salinity (defining the water masses) persistently occur, and the distribution of such associations can be traced over wide areas of the oceans, for thousands of kilometres. Water masses can in some respects be considered as analogues of the major climatic regions of terrestrial environments, and they define the extent and influence of major ocean currents. Both temperature and salinity need to be identified in order to define the origins, movements and extent of the distribution of water masses. A water mass will seasonally change its temperature (salinity is far less seasonally variable) because of atmospheric interactions. Temperature by itself, therefore, is not ideal as a measure or descriptor of water masses or origin of water masses.

At the broadest geographic scales, water masses are well correlated with major differences between biological community types in both the pelagic and benthic realms. However, it is important to understand the basic reasons why this is so. Water masses are indeed correlated with the geographical limits of species distributions, but largely because of the combined effects of geographic origins of individual water masses and their temperature effects. Major operative variables in marine biogeographic distributions include the combined result of water masses – indicating origins and transport of marine organisms, and temperature – indicating local physiological tolerance. Thus movements of water masses can be highly informative with respect to both the present day distribution of community types and their potential regional changes due to transport of propagules and colonization by expatriate and invasive species – for example under future global and climate changes.

Geographical position, latitude

Any geographical position in the oceans, in terms of latitude and longitude, is a site of interactions of structures and processes. The geographical context and latitude of an area are affected by prevailing currents, winds and connections to biota elsewhere, which are responsible for whole ocean scale differences among community types. The latitude of an area is an indication of seasonal temperature and the amount of light available for primary production. However, at any given latitude in the marine environment, the actual temperature of the water mass may be significantly at variance with that predicted from the flux of local solar radiation. The Labrador current and Gulf Stream are good examples of water masses transported across latitudes, and exhibiting, respectively, regionally lower and higher than expected temperatures. Definition of broad scale community features in terms of water masses is therefore more fundamental than latitude. At a finer scale, 'stratification' of water masses is also a better predictor of community type and timing of events than latitude (see below). Therefore the location of an area, whether in the Atlantic, Pacific or Indian oceans, affects the prevailing currents, winds and connections with biota elsewhere. The major consideration of longitude is in terms of ocean circulation patterns and westward intensified currents (see below).

Major features - oceanographic processes

Water motions

Natural bodies of water show movements of many types at all spatial and temporal scales (see Figure 2.5). Water motion is an essential feature of the oceans, and is essential for all life; a proper perspective on water movements is essential to an understanding of aquatic ecosystem functioning. Ultimately three 'forces' cause all water motion on the earth: the sun's radiation (causing heating and evaporation/precipitation); the earth's rotation on its own axis (a modifying action rather than a true force); gravitational effects of the sun and moon. A minor fourth factor is geological, which causes tsunamis. Proximately, water motion may be generated by winds, tides or density differences. The list of observable and distinguishable phenomena is extensive, and only the major resulting phenomena (rather than the causative mechanisms) associated with differences in community types will be considered here.

It is vital to understand the scale at which processes occur, in order to appreciate their ecological and biological significance, and their interactions (see Figure 2.6). Water movements vary in scale from simple molecular diffusion to the grand scale of circulation of entire ocean basins. At the smallest scales (μm to cms) molecular diffusion is a basic necessity for all life forms from bacteria to mammals. Even in complex multicellular mammals the final phase of gas exchange in the lungs is accomplished by simple diffusion across a liquid barrier. Nevertheless, fundamental as such processes are, they do not generally feature in conservation concerns.

The net effects of water motions are several. They sustain life by replenishing and stimulating the production of resources for organisms of every trophic level. They provide the passive transport 'highways' of the oceans for the dispersal of organisms or their larvae in the broad-scale or more local patterns of ocean connectivity (see Chapter 17). The junctions between the gyral systems and major ocean currents contribute to the barriers that separate species and higher taxa, and that define the major biogeographic provinces and regions of the oceans (see Chapters 5 and 12).

Surface ocean circulation

Water motion in the surface waters of the oceans (down to a depth of about 600m) is driven primarily by winds and tides. Below about 600m, water motion and horizontal currents are much slower and driven primarily by density differences due to temperature and salinity imbalances. The frequency and strength of water movements have profound effects on biological communities.

At the global scale, the basic patterns of surface ocean circulation are the same in the Pacific and Atlantic oceans, consisting of sets of currents which together comprise a sub-polar gyre and a

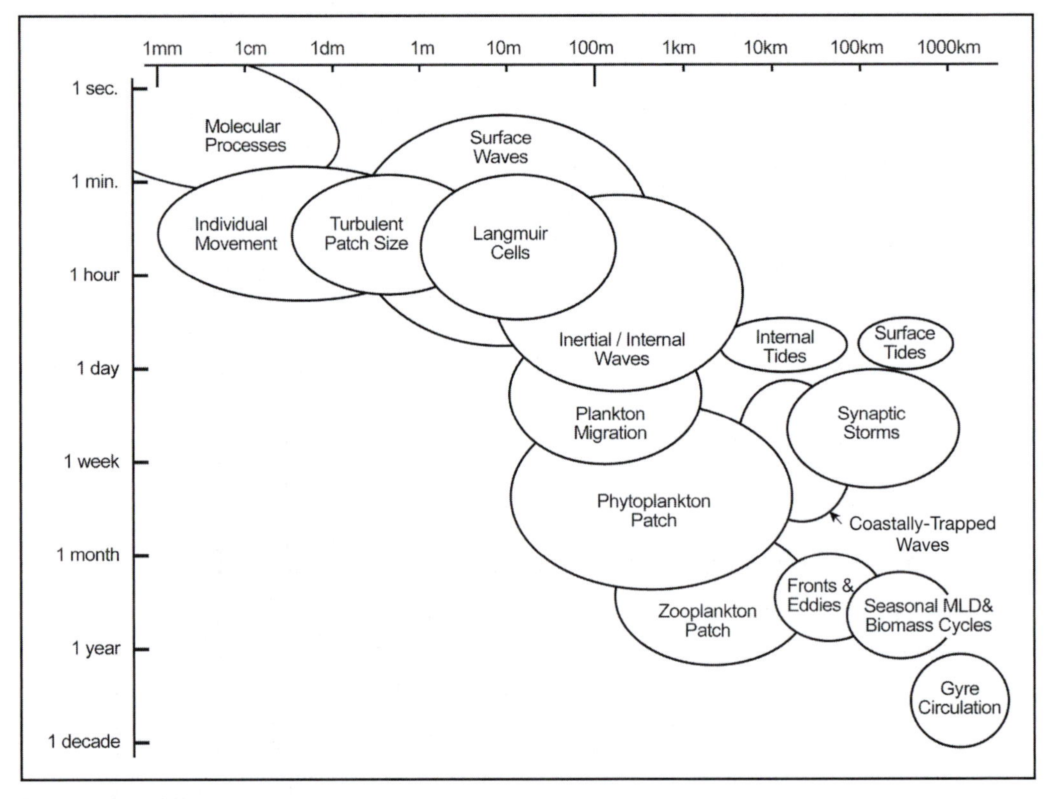

Source: After Dickey (1991)

Figure 2.5 Covariance of time and space scales of physical and biological processes

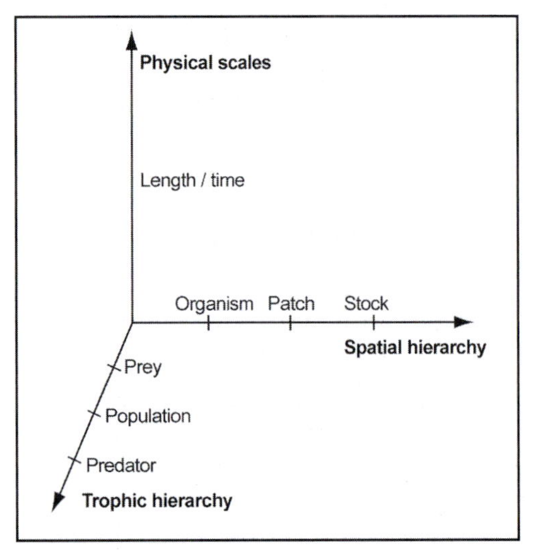

Source: Steele (1988)

Figure 2.6 Three sets of ecological variables plotted on a three-dimensional grid

subtropical gyre (Figure 2.7). The counterparts of each of these currents in the North Atlantic and North Pacific, and in the South Atlantic and South Pacific is very clear. For example, the counterparts of the sub-polar Labrador current and the subtropic Gulf Stream in the North Atlantic Ocean are the Kurile and Kuroshio currents in the North Pacific Ocean. In all oceans, circulation is eccentric and a westward intensified current develops. It is the effect of the earth's rotation (described in terms of the Coriolis 'force' or parameter) that makes the various ocean gyres eccentric with westward, intensified currents.

Each gyre revolves around a central but eccentric hub, simply referred to as a 'central gyre'. The only one named specifically is the Sargasso Sea of the North Atlantic. Each subtropical gyre encloses a relatively warm, salty, clear blue sea, poor in nutrients and of low productivity. Production is dependent primarily on internally re-

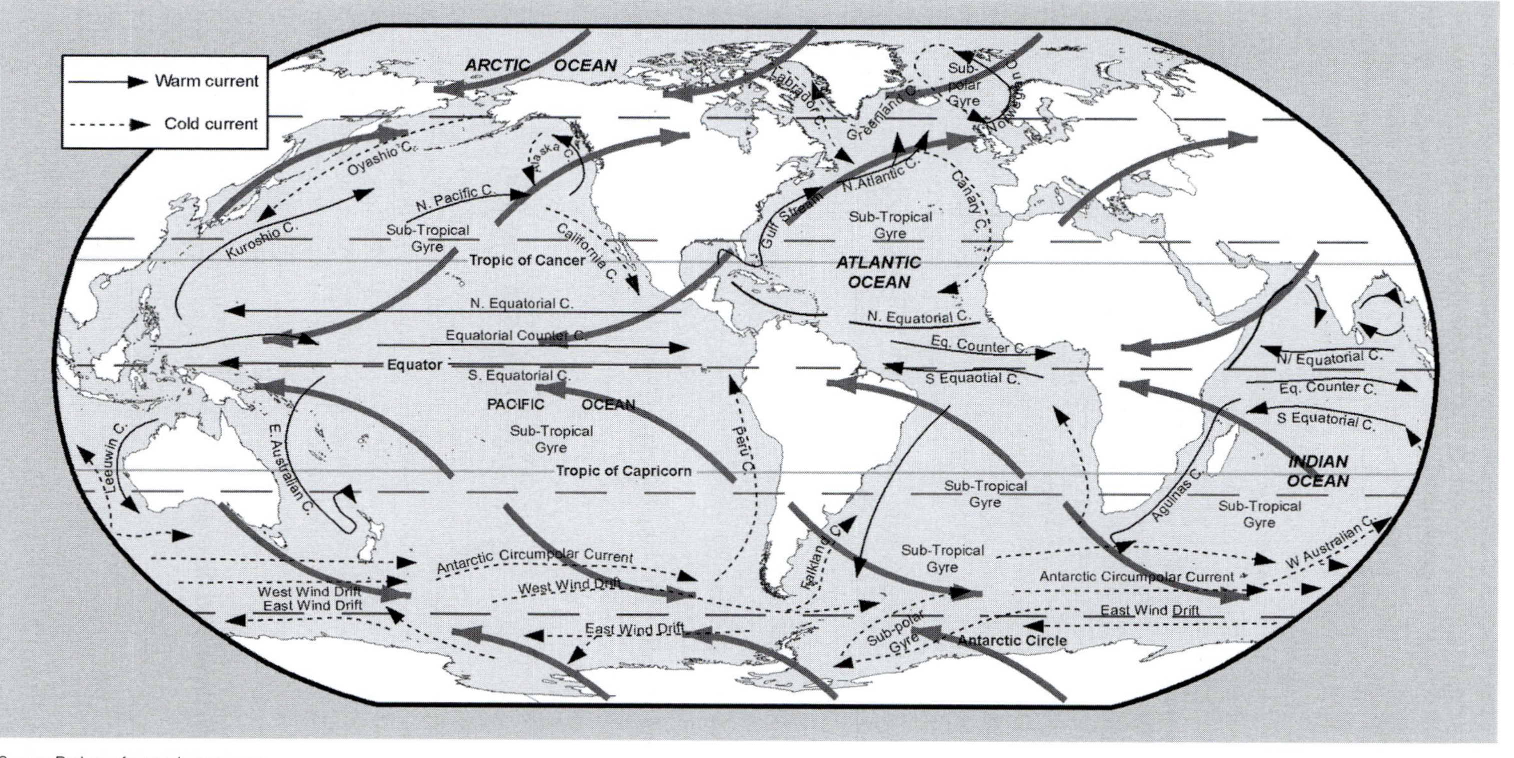

Source: Redrawn from various sources

Figure 2.7 The surface circulation pattern of the oceans, and the major surface ocean currents.

cycled NH_4. These are essentially the deep, desert areas of the ocean. Although the best known example is the Sargasso Sea, the eccentric hub of the North Atlantic circulation pattern, other analogous central gyre areas are found in the north and south Pacific oceans.

In the tropical areas of the oceans, both a north equatorial current and a south equatorial current can be recognized flowing from east to west. In the Atlantic and Pacific oceans, an equatorial countercurrent is found between these two, running from west to east. The Pacific equatorial countercurrent is 8000 nautical miles (nm) (= 12,000km) long and about 250nm (= 400km) wide. It is one of the impressive, steady currents of the world. The Atlantic equatorial countercurrent is more variable; it becomes stronger in summer, extending its influence westward towards South America.

The most pronounced and persistent ocean current is the Antarctic circumpolar current (West Wind Drift), which subsumes the sub-polar gyres of the southern oceans and completely circles the globe running from west to east.

The most variable surface ocean currents are found in the Indian Ocean. Here, even the major ocean currents show great variability in velocity, direction and latitudinal and longitudinal position. For example, the Somali current flows south in winter, but north in summer at high velocity. The north equatorial current reverses direction under monsoon influence, and flowing east in summer is now called the monsoon current. The equatorial countercurrent is only observed in winter, and disappears in summer when the monsoon current occurs. The major force which causes these changes is the variable winds of the monsoon season, clearly showing the dependence of surface ocean circulation on the atmospheric winds.

Oceanic convergences and divergences

Where the various oceanic gyral systems meet, corresponding to regions of reversal of major atmospheric wind systems, water masses either converge or diverge. That is, at the surface, the circulation pattern induces a series of convergences and divergences where water either descends below the surface, or is brought up from below the surface. The locations where oceanic water masses meet and descend (convergences) or rise to the surface and disperse (divergences) are present throughout the oceans, but are more pronounced in the southern oceans, and are of major importance to the overall rate of production. These are typically regions of much higher than average production, at trophic levels from phytoplankton to fish. However, these regions are much more clearly defined in the tropics, subtropics and southern oceans than at northern latitudes, and convergences and divergences proper are typically restricted to oceanic waters.

It is important to realize the differences in the character of oceanic convergences and divergences. Convergences are where surface waters meet and biota may be accumulated – thus biomass is increased. They may be areas of rich life which has physically accumulated – that is, it has been advected to the region, but not produced there. Rich fish feeding areas may nevertheless correspond to convergence zones. A divergence is not usually so well defined, though it carries a signature of reduced surface temperature. At divergences, nutrient-rich water is brought to the surface stimulating new production. Perhaps the best known example is the Antarctic divergence where extremely high levels of primary planktonic production are known.

Coastal upwelling

Production is also high in regions of near-coastal upwelling and divergence, where production is dependant primarily on inputs of NO_3 from deeper waters. If the coast lies to the left of the wind direction, then surface water is transported to the right in the northern hemisphere, as a result of deflection due to the Coriolis parameter. This necessitates a deep replacement current moving onshore and upwards – that is, upwelling. This upwelling water brings with it much higher concentrations of nutrients which have been biologically regenerated in deep water. The Coriolis 'force' is thus directly linked to areas of high production. Examples include the Grand Banks of Newfoundland where the Gulf Stream has moved offshore.

In the southern hemisphere, the situation is completely reversed. The best-known example here is the northward-flowing Peru–Chile current which leads into the Humbolt current. It progresses along the west coast of South America and leads to upwelling of cold deep nutrient-rich water, which in turn stimulates the planktonic production. Coastal upwellings may be relatively constant, periodically interrupted (e.g. as in the El Nino Southern Oscillation System off Peru and Chile), or more episodic as in the case of the Benguela current off southwestern Africa.

Deep ocean circulation

Deep ocean currents generally move at a slower pace than surface waters, and are driven by small differences in density caused by both temperature and salinity variations (thermohaline circulation). Formation of the deep and bottom waters of the oceans only occurs in two ways. Saline water is transported to high latitudes and is subsequently cooled without significant dilution and sinks. Cold surface water freezes and, as it does, it forms ice (above the eutectic point where solvent alone freezes and eliminates the salt solutes), so that the salinity of the remaining water is increased and consequently sinks. The first process is of greatest importance in the North Atlantic, between Iceland and Norway, and the second around Antarctica primarily in the Weddell Sea region – the only two regions of the world where deep and bottom waters are formed in any quantities.

Water sinking in the North Atlantic makes its way south towards the Antarctic. Here it meets water sinking from the Weddell Sea where it is recooled and moves off to the Indian Ocean, the North Pacific, and then the South Pacific. As deduced from carbon-14 dating techniques, this entire journey takes some 1500 years! Despite the long time that these waters have been out of contact with the atmosphere, they still retain significant levels of oxygen, which indicates the sparse nature of life in the depths of the oceans, and its low metabolic rates at low temperatures. There is still much concerning the movements of deep and bottom waters that is poorly known, leading to many uncertainties in terms of conservation strategies for these waters (see Chapter 12).

Tidal amplitude and currents

Tides are caused by the gravitational effects of the moon and the sun. In the absence of land the world's oceans would experience semi-diurnal tides of a periodicity of 12.25 and 24.50 hours, and a spring-neap cycle of variable amplitude depending on conjunction of moon and sun. Regionally, both tidal periodicity and tidal amplitude depend on the physical dimensions of the ocean basin. Some locations in the world experience only one tide per day (diurnal) and others up to four per day (quadridiurnal). Which occurs depends on the location of the amphidromic points around which tides circulate, and which of the sun's and moon's gravitational influences most nearly reinforce the natural period of oscillation of the regional ocean basin.

Tidal amplitude is a consequence of the natural period of oscillation of a regional ocean basin, sea or bay. The Bay of Fundy and its extension, the Minas Basin, experience the highest tides in the world, because the natural period of oscillation of the Bay of Fundy is approximately 13 hours. This corresponds closely to the moon's M2 gravitational period (12.25 hours), creating a natural resonance with the waters in the Bay of Fundy and spring tide amplitudes up to 16m. In contrast the natural periods of oscillation of the basins in the Mediterranean Sea are about 9.30 hours. Here, there is interference between gravitational period and natural period of oscillation, and tides of only a few centimetres result.

Currents in the oceans can derive from several processes, including general ocean circulation, density-driven origins of water masses, tides and winds. In coastal waters, currents due to tides are generally the most predictable and best documented. Tidal amplitude together with bottom slope determine the vertical and horizontal extent of the intertidal zone. Tidal currents and bottom slope determine, in part, the substrate characteristics, which in turn determine the extent and type of benthic communities. A major distinction between substrate types and associated benthic communities can be recognized as follows: erosional areas, where current speeds are high and material is removed from the substrate resulting in hard bottoms (rock, boulders, gravel etc.); and depositional areas, where water movement is sufficiently slow that particles sediment out, resulting in soft bottoms (mud, silt etc.).

Stratification and mixing regime

All temperate marine and fresh waters typically mix vertically in spring and autumn, and they may or may not become stratified in the summer. The seasonal cycle and timing of stratification has major implications for the productivity regime of a region. Two forces cause near-surface water mixing: tides and winds. The effects of wind mixing are difficult to formulate, but they are more uniformly spatially distributed than those of tides. In the oceans, when the water column depth exceeds ~ 50m, winds cannot prevent the water column from stratifying during the annual heating cycle; however, tidal action can. In freshwaters, tides are insignificant, and temperate lakes over about 50m in depth always stratify. In an initial simplification for marine waters therefore, it is appropriate to consider only mixing due to tides.

Arctic waters typically mix in early summer, and then become stratified mainly as a result of vertical salinity differences. In sub-Arctic waters, stratification can be due to a combination of salinity and temperature effects. In temperate and subtropical waters, stratification of the water column is primarily due to temperature; however, it may also result from a change of salinity. In coastal waters this may be associated with seasonal freshwater runoff from land, and it can also be related to the cycle of primary production. Seasonally, ice-free areas of the Arctic Ocean experience such effects (Roff and Legendre, 1986) as do coastal waters with high runoff (e.g. Hallfors et al, 1981; Kullenberg, 1981). In addition, some areas of the Pacific coast may respond to runoff events, although the timing and locations may or may not be annually predictable. Magnitude and locations of effects may be predictable from a combination of terrestrial landforms (mountains), rainfall distribution and drainage basin outflow to the sea.

In temperate waters, mixing and stabilization regimes are largely determined by two parameters – the stratification parameter and the water depth. The stratification parameter measures the tendency of the water column to stabilize under the influence of a surface heat flux, in the presence of mixing due to tidal stirring. It is defined as the ratio of potential energy from the sun – which tends to stratify the water column, to the rate of tidal energy dissipation – which maintains well mixed conditions within a water column. The length of the mixing period, and the timing and intensity of stratification during the seasonal heating cycle, can be effectively described by the stratification parameter 'S' where $S = \log_{10}[H/C_D.IUI^3]$, where H= water depth, U = tidal current velocity, C_D = friction/drag coefficient (Pingree, 1978).

The speeds of tidal currents in a region are a function of the natural period of oscillation of the basin, and the location of the amphidromic point. Tidal current speeds (at least the M2 or semidiurnal component) are generally well known, or can be modelled in a region. Simpson and Hunter (1974) have argued that for a constant surface heat flux, Q_o, the critical value of the stratification parameter, S = 1.5, will determine the position of summer boundary fronts separating well-mixed waters from well-stratified waters. Wind generated waves are never strong enough to prevent stratification at values of 'S' > 1.5 (Pingree, 1978). In addition to measurements at sea, infra-red satellite images of sea surface temperatures provide confirmatory evidence of the importance of the tidal stratification parameter as an index of mixing on the continental shelf. Thus regions characterized by high values of the stratification parameter may be associated with well stratified conditions in the summer months, whereas low values represent areas where the water column remains mixed throughout the year. The stratification scale spans, in effect, the range of conditions from where weather dominates mixing and stabilization, to conditions dominated by tidal stirring.

Values of 'S' are geographically almost fixed and are annually reproducible, leading to a defined spatial pattern of development of seasonal thermoclines. For open waters, although the rate of change of density with depth is a direct measure of intensity of water column stratification, it must be measured *in situ*. Frontal systems (where S = 1.5) are identifiable from their temperature gradients in satellite images (see temperature anomalies and gradients above). Frontal systems, that separate stratified and non-stratified waters, mark the boundary between areas where the tides are sufficiently strong to mix the water column, and regions where they are not, where seasonal stratification occurs. Mixed water means high nutrients but poor light conditions; stratified water means good light conditions but limited nutrient. Frontal regions typically exhibit the highest and seasonally most persistent production, because they represent areas of 'compromise' between sufficient light and sufficient nutrient supply for pelagic plants. Though they may vary from year to year and change in intensity, frontal systems can nevertheless be considered as recurrent oceanographic features. These physical characteristics define areas potentially important to migrant species.

'S' appears to be the only feature that horizontally delineates pelagic habitats at a scale below that of water masses, and which is reproducibly mapable from remotely sensed or modelled data. Stratification, depth and light penetration are the only features available to vertically delineate pelagic habitats. 'S' may also define population boundaries, recruitment cells, retention and gyral systems, and migration limits for at least some pelagic species and larvae of benthic species (Iles and Sinclair, 1982; Bradford and Iles, 1992). However, the delimitation of populations by physical circulation can potentially occur by several mechanisms. This is an active field of research that should be of interest to marine conservationists because of the significance of such areas for migrant and/or resident species, and implications for the concept of connectivity (see Chapter 17).

Local gyres and eddies

Various interactions between coastal topography, bathymetry and ocean currents can establish local gyres and eddies that act to retain and aggregate biological particles and are thus important considerations in coastal connectivity (see Chapter 17). These gyres and eddies also locally increase resources for higher trophic levels including fish and marine mammals (e.g. Archambault et al, 1998). Phenomena causing local organism retentions include Langmuir convergence cells, tidal fronts and headland gyres. Migratory animals and seabirds especially take advantage of such locally elevated resources (see Chapter 7).

Exposure: Atmosphere and waves

Water movements caused by tides, ocean currents and wind waves are also associated with the degree of exposure of littoral communities, and this is of major importance to the development of the various community types in both intertidal and sub-tidal regions. High energy areas tend to wash sediments out of the littoral zone into areas with lower energy regimes, either in deeper water or to more protected areas such as bays or gullies. Even well below the littoral zone, substrates are still affected by high energy currents which can determine the particle size distribution of sediments.

Exposure may be considered as occurring in two ways: through exposure to wave action; and through exposure to desiccation. Neither type of exposure is a discriminating variable for pelagic communities. Although wind waves may have some influence on pelagic organisms, they are unpredictable and random in space and time. In the benthos, exposure is only significant for the intertidal and immediate sub-tidal regions. The intensity of wave action that can develop is related to fetch, predominantly a function of the distance over which waves are propagated and the angle and 'development' of the shoreline (see e.g. Chapter 11). This can be determined from maps and meteorological records.

A different form of exposure for littoral communities is recurrent exposure to desiccation by the atmosphere during a tidal cycle. Exposure is strongly correlated with tidal amplitude, substrate slope and substrate particle size – along a spectrum from bedrock to mud (Levinton, 1982). Both types of exposure exert considerable influence on the type of littoral plant and animal communities that develop in a region.

Tsunamis, storm surges, hurricanes and water spouts

These massively destructive atmospheric events are now more-or-less predictable in terms of location of impact and scale of effect, at least on the time scale of days. Although considerable local destruction to coastal marine communities may result from these phenomena, over a period of years (most intertidal communities) to decades (coral reefs, mangroves), the local communities may generally become re-established in approximately their original composition. Such phenomena, although with profound local effects in shaping marine communities, should be thought of as resetting the ecological 'clock of succession'. They should not be considered as belonging in the same category of 'determinants' of community type as the remaining oceanographic factors, except in so far as they are 'recurrent' within time scales, and may determine whether certain community structures exist.

The major features of estuaries

Coastal zones, especially estuaries, are more complex morphologically, topographically, oceanographically and biologically than waters further offshore. They are also far more subject to immediate human impacts and are given special consideration in Chapter 11. Estuaries, along with their associated wetlands, are among the most ecologically complex regions of the coastal zone. Estuaries are the meeting place of fresh waters from land and salt waters from the sea. There are many definitions of an estuary, but perhaps the simplest is: 'An estuary is an inlet of the sea whose salinity is measurably diluted by freshwater' (Officer, 1976). They display characteristics intermediate between marine and freshwaters, but also have many unique properties and processes.

There are two main driving forces in estuarine circulation. Freshwater discharge drives surface waters towards the ocean and entrains subsurface salt water with it. This motion induces a return flow of salt water towards the head of the estuary. Superimposed on the freshwater flow is a tidally driven flow that circulates water both into and out of the estuary. The freshwater discharge leads to an altered distribution of mass which in turn induces thermohaline circulation. In most estuaries, despite their complex thermal and saline nature, the effect of temperature is small compared to salinity variations, and to a first approximation it is often ignored.

Nutrients are added to estuaries in both dissolved and particulate form from both marine and freshwaters; consequently estuaries are highly biologically productive. In addition, most estuaries display various mechanisms that aid in the retention and/or sedimentation of particles and regeneration of nutrients. In their biological characteristics, estuaries are strongly influenced by salinity and circulation mechanisms. Variations in salinity can result in low species diversity, because of the stress imposed on organisms due to variations in the external osmotic environment. Estuarine species are typically euryhaline and eurythermal, tolerating wide variations in temperature and salinity.

Estuaries are therefore typically regions characterized by strong circulation, high nutrient inputs, high biological productivity, but relatively low species diversity – at least in the central reaches, where salinity is most variable. Salinity, temperature and circulation patterns within estuaries produce complex sets of habitat types that require separate analyses of their oceanographic and physiographic features and processes (see Chapter 11).

Probably because of their high productivity, estuaries tend to be important nursery areas for commercially exploited populations of fish and invertebrate species. Estuaries have historically long been important areas of human settlement and commerce, typically developing ports, fishing communities and major towns and cities. Human impacts have therefore often reached their peak in estuaries, typically by adding to the nutrient inputs and reducing biological diversity; these areas are therefore often severely degraded in their biological characteristics and communities.

Classification of estuaries

Estuaries vary considerably in their characteristics and circulation patterns. Estuarine types

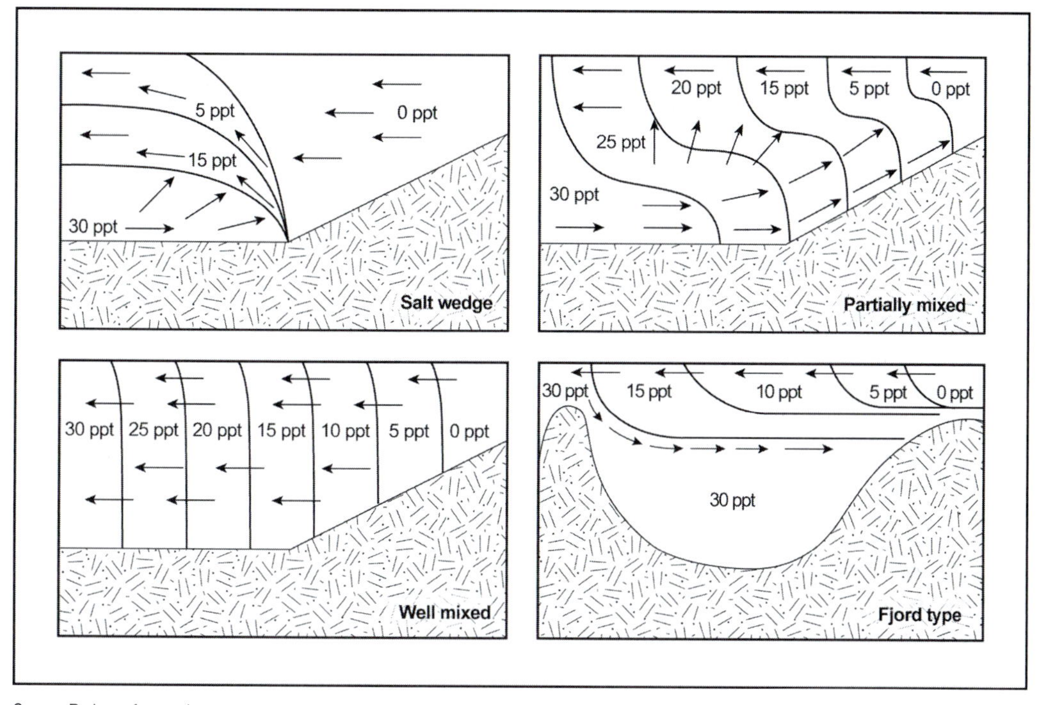

Source: Redrawn from various sources

Figure 2.8 Major types of estuaries and their salinity distributions

result from a combination of effects that include geomorphology, physical topography, tidal range and freshwater discharge. The fundamental physical character of an estuary can in fact be described by the interplay of freshwater discharge and tidal activity (e.g. Hansen and Rattray, 1966). No generally accepted classification of estuaries exists, but the following types are generally recognized (see also Figure 2.8):

- Well-mixed = little or no vertical gradient of salinity, only a horizontal gradient.
- Partially mixed = a change of a few ppt of salinity in a vertical gradient.
- Salt-wedge = considerable vertical change of salinity, with a wedge of fresher water overlying more dense water.
- Fjord type = usually strongly stratified with a seaward sill restricting circulation.
- Reverse = no freshwater discharge, with evaporation in shallow water, leading to a reversal of normal estuarine circulation.
- Plumes = everted estuaries extending far out to sea, observed from large rivers and under ice in Arctic regions.

Other classification schemes, based on freshwater discharge and the volume of tidal water flow – calculated from tidal amplitude and cross-section area of the estuary – yield similar categories of estuaries.

Two further parameters which can be used to classify estuaries and determine their type are both related to water retention time (i.e. inversely related to dispersion time). They are:

1. The **estuarine Richardson number**, which describes stratification vs. circulation as a ratio.
2. The **longitudinal dispersion coefficient**, which is a measure of dispersion or flushing rate (see e.g. Roff et al, 1980).

River plumes that discharge into the ocean forming an 'everted' estuary can develop seasonally in temperate areas subject to considerable land runoff, and under the ice in Arctic regions. Such plumes can be important mechanisms for dispersal of organisms, enrichment and nutrient additions to coastal waters, and in the development of coastal circulation patterns.

Conclusions and management implications

Separate recognition of the marine environment into pelagic and benthic realms is essential for conservation planning purposes. A prime reason for this segregation is that the two realms must be mapped using different combinations of factors. Without such a division, we would be obliged to classify aquatic environments in terms of ecosystems. It is essentially impossible to do this without being arbitrary as to what in fact constitutes an 'ecosystem' within the marine environment.

A knowledge of ocean structures and processes, from the global to the local scale, is essential for understanding the relationships between the physical environment and biological communities. Although the oceans may appear to be featureless compared to the terrestrial environment, the combination of physiographic and oceanographic factors can be used to map and define the array of marine habitats. This mapping of the marine environment is crucial to the development of comprehensive plans for conservation.

The physiographic and oceanographic characteristics of the oceans alone can be used at a broad (regional) scale to classify habitat types. These should in turn act as surrogates for biological community or assemblage types to some level of the environmental/ecological hierarchy (see Chapter 1). In Chapter 5 we shall explore this idea more fully in the context of representative areas.

Although the concepts presented in this book can be applied to all marine environments, we believe that the coastal zone, including estuaries, should be considered separately from other fully marine waters in conservation efforts.

References

Archambault, P., Roff, J. C. and Bourget, E. (1998) 'Nearshore abundance of zooplankton in relation to coastal topographic heterogeneity, and the mechanisms involved', *Journal of Plankton Research*, vol 20, pp671–690

Ballard, R. D. (1977) 'Notes on a major oceanographic find', *Oceanus*, vol 20, pp35–44

Barnes, R. S. K. and Hughes, R. N. (1988) *An Introduction to Marine Ecology*, Blackwell Scientific Publications, Oxford

Bergeron, P. and Bourget, E. (1986) 'Shore topography and spatial partitioning of crevice refuges by sessile epibenthos in an ice disturbed environment', *Marine Ecology Progress Series*, vol 28, pp129–145

Bertnes, M. D. (1999) *The Ecology of Atlantic Shorelines*, Sinauer Associates, Sunderland, MA

Bertnes, M. D., Gaines, S. D. and Hay, M. E. (2001) *Marine Community Ecology*, Sinauer Associates, Sunderland, MA

Bradford, R. G. and Iles, T. D. (1992) 'Retention of herring *Clupea harengus* larvae inside Minas Basin, inner Bay of Fundy', *Canadian Journal of Zoology*, vol 71, pp56–63

Bukata, R. P., Jerome, J. H., Kondratyev, K. Y. and Pzdnyakov, D.V. (1995) *Optical Properties and Remote Sensing of Inland and Coastal Waters*, CRC Press, Boca Raton, FL

Chapman P. and Shannon L.V. (1987) 'Seasonality in the oxygen minimum layers at the extremities of the Benguela system', in A. I. L. Payne, J. A. Gulland and K. H. Brink (eds) 'The Benguela and Comparable Ecosystems', *South African Journal of Marine Science*, vol 5, pp85–94

Diaz, R. J. and Rosenberg, R. (2008) 'Spreading dead zones and consequences for marine ecosystems', *Science*, vol 321, no 5891, pp926–929

Dickey, T. (1991) 'The emergence of concurrent high resolution physical and bio-optical measurements in the upper ocean and their applications', *Reviews of Geophysics*, vol 29, pp383–413

Eisma, D. (1988) 'An Introduction to the Geology of Continental Shelves', in H. Postma and J. J. Zijlstra (eds) *Ecosystems of the World: Continental Shelves*, vol 27, Elsevier, New York

Emerson, C. W., Roff, J. C. and Wildish, D. J. (1986) 'Pelagic-benthic coupling at the mouth of the Bay of Fundy', *Ophelia*, vol 26, pp165–180

Glenmarec, M. (1973) 'The benthic communities of the European North Atlantic continental shelf', *Oceanography and Marine Biology: An Annual Review*, vol 11, pp263–289

Hallfors, G., Niemi, A., Ackefors, H., Lassig, J. and Leppakoski, E. (1981) 'Biological Oceanography', in A. Voipio (ed) *The Baltic Sea*, Elsevier Oceanography Series, vol 30, Amsterdam, The Netherlands

Hansen, D.V. and Rattray, M. (1966) 'New dimensions in estuary classification', *Limnology and Oceanography*, vol 11, pp319–326

Iles, T. D. and Sinclair, M. (1982) 'Atlantic herring: Stock discreteness and abundance', *Science*, vol 215, no 4533, pp627–633

Irish, K. E. and Norse, E. A. (1996) 'Scant emphasis on marine biodiversity', *Conservation Biology*, vol 10, pp680

Knox, G. A. (2001) *The Ecology of Seashores*, CRC Press, London

Kullenberg, G. (1981) 'Physical Oceanography', in A. Voipio (ed) *The Baltic Sea*, Elsevier Oceanography Series, vol 30, Amsterdam, The Netherlands

Lamb, A. and Edgell, P. (1986) *Coastal Fishes of the Pacific Northwest*, Harbour Publishing, Madeira Park, British Columbia

Levinton, J. S. (1982) *Marine Ecology*, Prentice-Hall, Englewood Cliffs, NJ

Mann, K. H. (2000) *Ecology of Coastal Waters with Implications for Management*, Blackwell Scientific Publications, Oxford

Mann, K. H. and Lazier, J. R. N. (1996) *Dynamics of Marine Ecosystems: Biological-physical interactions in the oceans*, Blackwell Science, Cambridge, MA

McLuskey, D. S. and McIntyre, A. D. (1988) 'Characteristics of the benthic fauna', in H. Postma and J.J. Zijlstra (eds) *Ecosystems of the World: Continental Shelves*, vol 27, Elsevier, New York

Morton, J., Roff, J. C. and Burton, B. M. (1991) *Shorelife between Fundy Tides*, Canadian Scholars Press, Toronto

Nixon, S. W. (1993) 'Nutrients and coastal waters', *Oceanus*, vol 36, pp38–47

Officer, C. B. (1976) *Physical Oceanography of Estuaries, and Associated Coastal Waters*, John Wiley, New York

Open University Course Team (1989) *Waves, Tides and Shallow-Water Processes*, Pergamon Press, Oxford

Open University Course Team (1989) *The Ocean Basins: Their Structure and Evolution*, Pergamon Press, Oxford

Open University Course Team (1989) *Seawater: Its Composition, Properties and Behaviour*, Pergamon Press, Oxford

Open University Course Team (1989) *Ocean Chemistry and Deep-Sea Sediments*, Pergamon Press, Oxford

Open University Course Team (1989) *Ocean Circulation*, Pergamon Press, Oxford

Pingree, R. D. (1978) 'Mixing and Stabilization of Phytoplankton Distributions on the Northwest European Continental Shelf', in J. H. Steele (ed) *Spatial Patterns in Plankton Communities*, Plenum Press, New York

Raffaelli, D. and Hawkins, S. J. (1996) *Intertidal Ecology*, Chapman and Hall, London

Ray, G. C. and McCormick-Ray, J. (2004) *Coastal Marine Conservation: Science and Policy*, Blackwell, Malden, MA

Roff, J. C. and Legendre, L. (1986) 'Physico-chemical and Biological Oceanography of Hudson Bay', in I. P. Martini (ed) *Canadian Inland Seas*, Elsevier, New York

Sherman, K., Alexander, L. M. and Gold, B. D. (1992) *Large Marine Ecosystems: Patterns, Processes and Yields*, AAAS Press, Washington, DC

Steele, J. H. (1988) 'Scale Selection for Biodynamic Theories', in B. J. Rothschild (ed) *Towards a Theory on Biological–Physical Interactions in the World Ocean*, Kluwer, Amsterdam

Simpson J. H. and Hunter J. R. (1974) 'Fronts in the Irish Sea', *Nature*, vol 1250, pp404–406

Suess, E. (1980) 'Particulate organic carbon flux in the oceans: Surface productivity and oxygen utilization', *Nature*, vol 288, pp260–263

Sverdrup, H. U., Johnson, M. W. and Fleming, R. H. (1942) *The Oceans: Their Physics, Chemistry and General Biology*, Prentice-Hall, New York

Sverdrup, K. A., Duxbury, A. C. and Duxbury, A. B. (2003) *An Introduction to the World's Oceans*, McGraw-Hill, New York

Tremblay, M. J. and Roff, J. C. (1983) 'Community gradients in the Scotian Shelf zooplankton', *Canadian Journal of Fisheries and Aquatic Sciences*, vol 40, pp598–611

Valiela, I. (1995) *Marine Ecological Processes*, Springer, New York

Wildish, D. J. (1977) 'Factors controlling marine and estuarine sublittoral macrofauna', *Helgoland Wissenschaft Meeresuntersuchungen*, vol 30, pp445–454

3

The Marine Environment: Ecology and Biology

Pelagic and benthic realms and coastal fringing communities

I do not know what I may appear to the world; but to myself I seem to have been only like a boy playing on the seashore, and diverting myself in now and then finding a smoother pebble or a prettier shell than ordinary, whilst the great ocean of truth lay all undiscovered before me.

Isaac Newton (1642–1727)

Introduction

This chapter presents a brief review of some biological and ecological features of marine environments – which are the structure and process components of marine biodiversity at the species/population and community levels of organization. Consideration of the genetic level of organization is deferred until Chapter 10. Less is said here about processes than structures at these levels, because what we know about them comes mainly from experimentation or is inferred from structures.

In order to stress the uniqueness of the marine environment, comparisons are made between marine and terrestrial ecosystems, and among the various kinds of aquatic ecosystems in general. Differences between marine ecosystems across latitudes from tropical to polar are also stressed, and the contrasts between pelagic and benthic realms are introduced. Emphasis is on general taxonomy and ecological classification of biota in the major marine realms and communities. The theme of ecological classification of communities and their relationships to habitat characteristics will continue in subsequent chapters as we build the frameworks for conservation planning.

The major taxa of the marine environment

The flora and fauna of the oceans range in taxonomy and size from marine viruses and microbes to marine mammals, which are generally recognized as organized into a series of more-or-less definable communities. For a summary of the major animal phyla and plant divisions in the oceans, freshwaters and on land (which are constantly subject to new discoveries and revisions), see Table 3.1. It is a daunting task even to compile an inventory of species identities (even if we had a good definition of a 'species'!), let alone to devise plans for their conservation. This is one main reason why conservation emphasis has now tended to move away from individual species

Table 3.1 The major higher taxonomic groups of marine, freshwater and terrestrial flora and fauna

Predominantly unicellular forms

Kingdom or phylum	Marine	Freshwater	Terrestrial
Archaea	Y	Y	Y
Two main phyla of 'bacteria'			
Eubacteria	Y	Y	Y
Several phyla of true bacteria			
Fungi	Y	Y	Y
Four to six major phyla			
'Slime moulds'	Y	Y	
A polyphyletic group			
Protista	Y	Y	
Unicellular (and some multicellular) eukaryotes 30–40 separate phyla			

The major divisions of algae*

Division	Marine	Freshwater	Terrestrial
Bacilliarophyta	Y	Y	
Charophyta	Y	Y	
Chloroarachniophyta	Y	Y	
Chlorophyta	Y	Y	
Chrysophyta	Y	Y	
Cryptomonads	Y	Y	
Cryptophyta	Y	Y	
Cyanophyta	Y	Y	
Dinophyta	Y	Y	
Euglenophyta	Y	Y	
Eustigmatophyta	Y	Y	
Glaucophyta	Y	Y	
Haptophyta	Y	Y	
Phaeophyta	Y	Y	
Prasinophyta	Y	Y	
Rhodophyta	Y	Y	
Rhaphidophyta	Y	Y	
Xanthophyta	Y	Y	

The major divisions of 'higher' plants**

Division	Marine	Freshwater	Terrestrial
Anthocerotophyta		Y	Y
Bryophyta		Y	Y
Marchantiophyta			Y
Lycopodiophyta			Y
Pteridophyta			Y
Pteridospermatophyta			Y
Coniferophyta			Y
Cycadophyta			Y
Ginkgophyta			Y
Gnetophyta			Y
Anthophyta (or Magnoliophyta)			Y

The major phyla of multicellular animals

Phylum	Marine	Freshwater	Terrestrial
Acanthocephala	Y	Y	
Acoelomorpha	Y		
Annelida	Y	Y	Y
Arthropoda	Y	Y	Y
Brachiopoda	Y	Y	
Bryozoa	Y	Y	
Chaetognatha	Y		
Chordata	Y	Y	Y
Cnidaria	Y	Y	
Ctenophora	Y		
Cycliophora	Y		
Echinodermata	Y		
Echiura	Y		
Entoprocta	Y		
Gastrotricha	Y	Y	
Gnathostomulida	Y		
Hemichordata	Y		
Kinorhyncha	Y		
Loricifera	Y		
Micrognathozoa		Y	
Mollusca	Y	Y	Y
Nematoda	Y	Y	Y
Nematomorpha		Y	
Nemertea	Y	Y	Y
Onychophora			Y
Orthonectida	Y		
Phoronida	Y		
Placozoa	Y		
Platyhelminthes	Y	Y	
Porifera	Y	Y	
Priapulida	Y		
Rhombozoa	Y		
Rotifera	Y	Y	
Sipuncula	Y		
Tardigrada	Y		Y
Xenoturbellida	Y		

Note: The terms 'phylum' and 'division' (for animals and plants respectively) are equivalent. The taxonomy of all these groups is in a state of flux

*The Chlorophyta, Phaeophyta and Rhodophyta are the major groups of macrophytic algae. Most of the other groups in this part of the table are either unicellular forms or undifferentiated multicellular forms

** These are the familiar, almost exclusively terrestrial, macrophytic plants, the mosses, liverworts, ferns, conifers and flowering plants towards a focus on habitats, which we can think of as essentially acting as 'umbrella' communities (see Chapter 9).

Some general comments on the taxonomy of organisms are warranted. First, the lower groups of bacteria, fungi and protists are widely distributed among all environments. The lower plants – the algae, both microscopic single cell forms and multicellular macrophytic forms, are found in both marine and freshwaters, but not on land. The higher multicellular plants, with few exceptions, are limited to the terrestrial environment. The great majority of phyla of multicellular animals are found only in the marine environment, with fewer phyla in freshwaters and fewer still on land.

Our perception of species diversity is heavily biased towards the larger species (of both plants and animals) and towards the terrestrial flora and fauna, but it is very clearly in the oceans that the greatest diversity of life forms (in terms of phyla and divisions) is found. Nevertheless, it is generally accepted that species diversity overall is significantly higher on land, where species counts are dominated by insects and the vascular plants (May, 1994), than in the oceans (see e.g. Gray, 1997). However, the continued application of genetic techniques may change this picture, as the diversity of unicellular forms, including bacteria, becomes better documented.

The groups of marine animals most familiar to most people would include fish and marine mammals, both groups of the phylum Chordata. But an examination of Table 3.1 shows that the Chordata constitutes only one part of the diversity of marine phyla, even though the species diversity of fish is high. Some 20,000 species of fish are known and more are described every week.

Similarities and differences between marine and terrestrial environments

For marine conservation to be successful, the similarities and differences between marine and terrestrial ecosystems need to be understood. Marine systems are significantly different from terrestrial ones, and knowledge gained from terrestrial ecosystems will not transfer directly to marine contexts. There are major differences between marine and terrestrial ecosystems, habitats and their biotas. Any attempts to protect marine biodiversity simply by applying terrestrial conservation methods will encounter severe biases. Unfortunately, marine ecologists and terrestrial ecologists have largely worked in isolation from one another (see Steele, 1995; Stergiou and Browman, 2005), so that differences and commonalities in theories, paradigms, concepts and even observations have gone unexplored.

Marine ecosystems are complex and inherently more interconnected and physically influenced across a range of spatial-temporal scales than terrestrial ecosystems. Physical influences tend to propagate over broader areas and persist for longer periods of time in the oceans. Unlike terrestrial ecosystems, their fundamental attributes, their biological characteristics and indeed their species are not directly visible to us. A summary of similarities and differences between terrestrial and marine ecosystems is included here, adapted from Day and Roff (2000).

Similarities between terrestrial and marine ecosystems

At a very broad conceptual level, marine and terrestrial systems do have some similarities:

- Both are composed of interacting physical and biological components with energy from the sun driving most ecosystems.
- Both are complex patchworks of differing environments and habitats that are occupied by different communities and species.
- Both marine and terrestrial species show gradients in diversity with latitude – with species diversity generally increasing with decreasing latitude (Thorne-Miller and Catena, 1991).
- In both types of ecosystems, the primary zones of biological activity tends to be concentrated nearer the surface (i.e. the sea–air or land–air interface).

Differences between terrestrial and marine ecosystems

There are many more differences than similarities between marine and terrestrial ecosystems. These differences can be found in the dimensions of space and time scales of physical processes, mobile versus sessile lifestyles, size, growth rate and factors relating to trophic position (Steele et al, 1993; Steele, 1995). Differences are also related to the fundamental physical properties of water itself (e.g. Sverdrup et al, 1942).

Difference in size

The oceans are far larger in area than all the land masses combined, covering 70.6 per cent of the entire surface of the globe. The Pacific Ocean alone could easily contain all the continents; our planet should be called 'water' not 'earth'. Even more marked is the difference in vertical extent and volumes between marine and terrestrial habitats. Life on land generally extends from only a few metres underground to the tops of trees – a vertical extent of perhaps no more than 30–40 metres. Although birds, bats, insects and bacteria may periodically rise above these heights, above the trees is a medium of temporary dispersal only, not a habitat, because species must return to the terrestrial environment for resources, reproduction and shelter. In contrast, the average depth of the oceans is 3800 metres, all of it containing living creatures. The sea's habitable volume is therefore hundreds of times greater than that of the land.

Difference in physical properties

Terrestrial and marine ecosystems differ markedly in their physical properties. For example, the water overlying the seabed is 60 times more viscous than air and has greater surface tension. Water is also about 850 times denser than air; this provides buoyancy and allows organisms to survive without the need for powerful supporting structures. Buoyancy allows organisms to exist in the sea that are morphologically and anatomically very different from those on land. Seawater's buoyancy and viscosity keep food particles suspended and result in an environment – the pelagic realm – without analogue in the terrestrial environment. Specialized sub-components of this realm, just beneath or at the sea–air interface, are the plueston and neuston; here, in freshwaters, are found water striding insects, and in marine waters the Portuguese man-of-war and the larvae of many bottom- and water-column dwellers.

Differences in temperatures

The strong seasonal and inter-annual fluctuations in the terrestrial climate contrast with the much more moderate fluctuations in the marine environment. Seawater has a much greater heat capacity than air; therefore temperatures change more slowly than they do on land. The higher viscosity of seawater also causes it to circulate more slowly than air.

Light and vertical gradients

Compared to the terrestrial environment, a far greater proportion of the ocean is light-limited. Due to changes in availability of light and nutrients only the upper 50–200 metres of the oceans – the photic zone – will support the growth of primary producers. Light is necessary for primary production through photosynthesis, both within the water column and in the benthos, except for a few unique communities such as hydrothermal vents which are dependent on their own chemosynthetic producers. Therefore essentially no organic matter is produced within the vast unlit depths of the ocean (the dysphotic and aphotic zones), and in these dark areas, the entire biological economy of the seas is dependent on the flux of detrital organic matter from the productive surface layer.

Different realms

The oceans contain two distinct realms, each with its own array of communities: the pelagic realm – the water column from the surface to the bottom of the sea; and the benthic realm – the seabed and a layer of water immediately above it. The pelagic realm supports a diversity of organisms and life forms – collectively, the plankton and nekton – that have no dimensional or community counterparts on land. Pelagic organisms inhabit a fully three-dimensional realm; most of its species inhabit it permanently (among the planktonic organisms – the holoplankton) or some temporarily as larvae (e.g. the meroplankton). Within it conditions vary dramatically with depth, changing the array of organisms present and community structure. Pelagic species may also occupy different trophic levels and different stages of development. The 'fringing communities' of the benthic realm – intertidal and sub-tidal communities within the euphotic zone – which contain various photosynthetic plants, could be considered the functional equivalent of terrestrial communities, where species either live on the ground or are dependent on it for habitat and resources. However, a significant difference between the benthic realm and terrestrial environments is that most of the benthic realm lies below the photic zone and is therefore devoid of its own primary producers. It must therefore rely on detrital resources settling from the photic zone and hence has no terrestrial community counterpart. To a first approximation, terrestrial and benthic ecosystems can be considered as two-dimensional. Certainly at larger spatial scales and for conservation purposes, they are typically considered in this way. Marine ecosystems therefore can be considered as comprising five dimensions, two of the benthos and three of the pelagic realm!

The benthic realm is sometimes considered to be an extension of the terrestrial environment beneath the sea. In fact, in evolutionary terms and from a global view, the reverse perspective is more appropriate; the terrestrial environment is really an extension of the benthic realm – above water. Within the pelagic realm marine organisms, from viruses to whales, permanently inhabit the fluid environment itself. In terms of fluid mechanics and circulation processes, the terrestrial atmosphere and the oceans have much in common – indeed meteorologists and physical oceanographers study parallel arrays of phenomena. Biologically, however, there is no terrestrial equivalent to the pelagic realm. While several species of terrestrial organisms (for example insects and birds) may temporarily inhabit the atmosphere or use it to disseminate their propagules, the atmosphere has no permanent inhabitants or primary producers to provide a sustainable food resource.

Mobile and fluid nature of water

The fluid nature of the marine environment means that most marine species are widely dispersed, and individuals can be far ranging. In addition to enhancing cross-fertilization and dispersal of larvae and other propagules, water motion also enhances the migration and aggregation of marine species (especially those in pelagic systems). Water also dissolves and circulates nutrients (Denman and Powell, 1984; Thorne-Miller and Catena, 1991). Even marine species that can be considered static as mature benthic forms (e.g. many molluscs and seaweeds) usually have highly mobile larval or dispersive reproductive phases within the plankton of the pelagic realm, yet their populations may be controlled by mobile predators. This means that it may be extremely difficult for them to be managed as spatially defined populations. Similarly the sediments of the substrate themselves are often mobile to a degree which makes land erosion cycles seem slow and mild (Ballantine, 1991).

Circulation differences

Although both terrestrial and aquatic environments exhibit patterns of circulation (i.e. circulation of the water in the oceans and of the atmosphere above the land and water), the two are not strictly comparable. In the oceans, the medium – water – contains the organisms themselves; aquatic organisms live in the medium, flow with it and are subject to its physics and chemistry. In contrast, the presence of organisms in the atmosphere is strictly temporary. Thus in the oceans, recruitment failure in one region may be reversed by passive recruitment from another area. On land, a similar process may require active migration as well as passive dispersal (National Research Council, 1994). Even for benthic communities, movements of the overlying waters are essential for the transport of food resources; this is generally not the function of the atmosphere for terrestrial communities.

Differences in primary production

Perhaps the most obvious biological difference between terrestrial and marine ecosystems may be in the types and source of primary production (Steele, 1991b). In terrestrial ecosystems, primary producers (mainly vascular plants such as trees and grass) constitute the great majority of biomass, and individuals are often large. In contrast, the dominant primary producers in marine ecosystems – the phytoplankton – are generally microscopic and can reproduce rapidly. Consequently they have much higher turnover rates compared to the forests or prairies of terrestrial ecosystems. In terrestrial ecosystems, the biomass of primary producers tends to be highly conserved (e.g. in woody plants). In contrast, in marine ecosystems the biomass of primary producers is rapidly processed, either by consumers or reducers. As a consequence, marine sediments are typically much lower in organic carbon than terrestrial soils, and become progressively lower still with increasing depth. Even marine macrophytic plants (e.g. algae and seagrass angiosperms), though often perennial, have short generation times compared to terrestrial plants. No marine plants have the longevity of terrestrial gymnosperms and angiosperms. At lower latitudes the partially seawater-adapted mangroves (emergents) are recent re-invaders of marine waters.

Taxonomic differences

In the oceans, there is a high diversity of divisions among the 'algae', comprising many 'phyla' of microscopic unicellular or multicellular forms (see Table 3.1). This in fact represents a much greater diversity at the level of division (phylum), and therefore genetic diversity, than in the terrestrial environment. This is often not recognized because of our bias and perception of the overwhelming diversity of flowering plants on land – which nevertheless all belong to a single division – the Angiosperms (or Anthophyta). Bryophytes, Pteridophytes and Gymnosperms are all absent in marine waters, and Angiosperms are significantly under-represented, mainly by only a few species of seagrasses and mangroves.

Of the animal phyla – all are represented in the oceans in one way or another, while on land several of the major phyla are completely absent having failed to adapt to its more demanding climate (see Table 3.1). Although the greatest species-diversity undoubtedly lies within the terrestrial insects, the greatest phyletic (and therefore genetic) diversity is clearly among the flora and fauna of the oceans.

Temporal and spatial scales

Other important differences between terrestrial and marine ecosystems result from – or are associated with – the temporal and spatial scales of ecological responses to changes in the physical environment (see e.g. Steele, 1974, 1985, 1991a, 1991b; Longhurst, 1981; Cole et al, 1989; Parsons, 1991), and are related to the fundamental differences in the communities of primary producers, and to the properties of water. The high storage of biomass in terrestrial organisms and organic detritus tends to de-couple biological and physical processes to some degree. In the oceans, the space and time scales of physical and biological processes are nearly coincident (at least in the pelagic realm), such that biological communities can respond rapidly to physical processes. Spatial variation and distributions of primary producers on land are largely related to topography and soil, and they change slowly compared to marine primary producers at the same scale. On land, largely as a function of the longer generation times of larger organisms, population and community cycles may depend more on biological than on immediate physical processes (Steele et al, 1993). However, at all spatial scales, physical changes in the atmosphere are faster than in the ocean. Cyclonic systems in the atmosphere have typical spatial scales of about 1000km and last for about a week. Equivalent eddies in the ocean have diameters of about 100km and can last for months or years (Steele, 1995). Thus we have the apparent paradox of 'slowly responding' terrestrial communities buffeted by a faster moving atmosphere, while in the oceans we have 'quickly responding' communities insulated by a slowly reacting and insulating ocean!

Bedrock type plays a minimal role in marine systems

Unlike in terrestrial ecosystems, bedrock type and geological composition apparently have little influence on the biota of the marine benthos. This is at least partially a function of the global constancy of composition of seawater. Within the intertidal and sub-tidal euphotic zone, macrophytic algae are attached to rocks of all types. Within these same zones where soft substrates exist, it is tidal velocity and wave motion that determine substrate grain size and combine to become the main determinants of community composition.

Relatively short response times to environmental perturbations

As a result of the mobile and fluid nature and inter-connectedness of marine waters, any environmental perturbation (e.g. oil spill, toxic substances, toxic algal blooms) can be rapidly dispersed throughout the marine environment. Depending on the properties of any pollutant, this can readily occur in two or three dimensions.

Boundary differences

Terrestrial environments have more pronounced physical boundaries between ecosystems. Especially at finer scales, it may be difficult to identify distinct boundaries in marine systems because they are so dynamic, and because pelagic and benthic realms require separate consideration. This does not mean that there are no distinct marine ecosystems, but generally that their boundaries are more 'fuzzy' or transitional, and the concept is less useful in marine than in freshwater or terrestrial environments. Nevertheless, it is still appropriate to examine the broad marine ecosystem characteristics and to define representative and distinctive habitat types in terms of their enduring geophysical features or recurrent processes.

Longitudinal diversity gradients

In addition to the latitudinal gradients observed in both terrestrial and marine communities, there is also a longitudinal diversity gradient in the marine environment, with species diversity decreasing from west to east in both the Atlantic and Pacific oceans. As well, faunas in the Pacific Ocean (e.g. coral reefs) are on the whole more diverse than in the Atlantic Ocean (Thorne-Miller and Catena, 1991). The greater diversity in the western part of the basins can be related to the eccentric circulation pattern of the two oceans, whereby the west receives waters from low latitudes (with higher species diversity) at high velocity (e.g. the Gulf Stream and Kuroshio), while the eastern part of the basin receives lower velocity water from higher latitude (with lower species diversity). The greater overall species diversity of the Pacific is generally explained in terms of its greater geological age.

Diversity at higher taxonomic levels

At higher taxonomic levels the diversity of marine fauna is much greater than the diversity of terrestrial fauna (Ray and McCormick-Ray, 1992); all phyla of animals are represented in the oceans. Some taxa, for example the fishes, are extraordinarily diverse; others are less diverse although in 'lower' phyla undoubtedly many species remain to be described. Most marine communities are also highly patchy and variable in species composition.

Length of lifetime and body size relations

In the open ocean there is a regular increase in lifespan and in body size with increasing trophic level (Sheldon et al, 1972), whereas in terrestrial systems these patterns are not so clear (Steele, 1991b).

Human perspectives

Marine systems are hidden from sight

A very basic difference between marine and terrestrial ecosystems is that most of the marine environment is hidden from human sight resulting in major differences in our knowledge and understanding of these ecosystems. Our 'out of sight, out of mind' mentality and the vastness of our oceans have both contributed to the misguided belief that the oceans can harmlessly absorb whatever we drop into them. Similarly because the sea is remote and is an alien environment, demonstrating the need for marine protection has been difficult. The collapse of commercial fisheries in many regions of the world has begun to raise public awareness of the need for marine conservation.

Difficulties of marine research

Research and monitoring are far more difficult in and under the water than they are on land. The exponential increase in the cost of obtaining data from increasing water depths explains why, in relative terms, so little is known about the sea, especially the deep oceans. The difficulty and expense of maintaining even moderately stable working conditions at sea are so great that most marine research and management is virtually 'hit and run', after which the researchers and managers come home to land. Furthermore, far fewer people are employed to investigate or manage marine ecosystems, yet the areas to be investigated and administered are far larger than terrestrial ones.

Known extinction rates

While current estimates of species loss in terrestrial ecosystems are frightening, because of the lack of study little is known about, or even predicted for, species extinctions in marine ecosystems. Recent evidence shows that many marine species worldwide are in dire straits, especially those of commercial value (e.g. whales, sea turtles, dugongs and several fish species).

Ownership

Until very recently, it was commonly accepted that no one owned a piece of the sea or any of its resources. Certainly in the more developed countries, it has long been felt, and generally still is, that marine resources were there for the taking or using by anyone who had the skill and initiative to do so (Ballantine, 1991). Today, governments are taking on specific responsibilities for some marine areas, but unfortunately most marine environments remain subject to the 'tragedy of the commons'. Responsibility for international waters is hampered by the mobility of water masses and connectedness of oceanic and territorial seas.

Resource extraction

During terrestrial resource extraction, the physical landscape is often dramatically and visibly altered. It has recently been realized that the harvesting of marine resources can also dramatically alter the seafloor on a large scale. Much of the floor of the North Sea, for example, resembles a ploughed field, due to trawling activities. Marine conservation requires regulation of all types of resource extraction, including those that directly disrupt the physical seascape (e.g. oil drilling and mining), and activities such as commercial harvesting (e.g. bottom dredging, bottom trawling and dragging) and aquaculture which perturb the seafloor at smaller scales, but over broad areas.

Delineating park boundaries

On land it is far easier to mark park boundaries using signs or natural features. At sea, the absence of obvious geographical features makes it more difficult for marine users to determine if they are inside a protected area and when to curtail their harvesting activities. However, the increasingly common use of GPS and chart plotting systems hopefully makes this a diminishing concern.

Linkages between terrestrial and marine systems and the atmosphere

Biogeochemical linkages

The sea is biogeochemically 'downstream' from the land, thus virtually any substance 'loosed' into the ecosphere is carried seaward. This has obvious implications for marine pollution because many inputs into marine systems have terrestrial origins.

The natural hydrological cycle flows from the land to the sea, but water is returned via the atmosphere to precipitate on land. Erosion of the earth's crust is therefore a one-way process from land to receiving waters. The assumption is generally made that, over geological time, inputs to lakes and oceans are balanced by 'output' in the form of sedimentation. The ocean sediments, in turn, move laterally with the spreading seafloor, and eventually are subducted at tectonic plate margins to form new crustal rocks.

Because of these natural cycles and processes, it is often assumed that lakes and oceans are in fact dependent on such inputs from land, especially the input of plant nutrients, to drive biological production. This is only partially true even for lakes, and is not true for the oceans as a whole. Inputs to lakes and oceans come from drainage basin runoff, the atmosphere, internal recycling and animal migrations.

With the exception of smaller lakes receiving large populations of anadromous fish (e.g. salmonids), animal migration is biogeochemically unimportant. Inputs from the drainage basin are generally the major source of nutrients only for smaller lakes, or for larger lakes subject to considerable direct human influence (e.g. Lake Erie). However, for the larger Great Lakes, less affected by human populations (e.g. Lake Superior), the major external source of nutrients is from the atmosphere. Estuaries span the range from those greatly affected by their drainage basin and its nutrient inputs (e.g. San Francisco Bay), to those whose nutrients derive predominantly from the ocean itself (e.g. Chesterfield Inlet) by reverse entrainment.

In marine waters as a whole, it is calculated that less than 2 per cent of annual nutrient requirements for biological production derives from land drainage and the atmosphere. The actual proportion may be even lower (Harrison, 1980). Most nutrients come from internal recycling; for nitrogen (most often limiting to production in marine systems) in the forms of NH_4^+ and NO_3^-. In coastal waters, the seasonal cycle of production is modulated by the changing supply of these two major forms of nitrogen between mixed and stratified waters. In oceanic waters, over 90 per cent of nutrient requirements are derived from the *in situ* recycling of NH_4^+.

Thus, the general assumption that coastal waters are driven by land runoff is not correct, even though many estuaries are severely affected by inputs of nutrient derived from human activities. Unfortunately, in estuaries we rarely know the relative significance of natural nutrient supply from the drainage basin, versus nutrients derived from the ocean, versus nutrients from human influences.

Other biogeochemical linkages between the terrestrial environment and marine waters do, of course, exist (e.g. exchange of hydrocarbons including pollutants, loss and absorption of gases, and importantly metals including iron), but these are generally transported to the oceans predominantly by the atmosphere rather than in land runoff (e.g. Schlesinger, 1997).

Biological community linkages

The major marine communities of the oceans comprise the pelagic and benthic realms. Outside of the intertidal zone, the communities of neither of these realms show any strong linkages to the terrestrial environment. They are essentially independent entities, driven by their own cycles of production, consumption, decomposition and regeneration, which are seasonally related to the planetary seasonal oceanographic cycles of heating, cooling and circulation.

Within the intertidal zone itself, linkages to the atmosphere are as strong as to the land, because organisms must tolerate periodic exposure to desiccation. However, both in freshwaters and in marine ecosystems there are species that inhabit the transition zones between land and water. These range from lichens to salt-tolerant angiosperms in the plant world, and from insect larvae and molluscs to migrant birds in the animal world. Fringing communities in general (which include the intertidal zone) are zones of high productivity, though variable in species diversity, often supporting highly specialized or migrant populations of animals.

The coastal zone, or neritic ecosystem, lies between the terrestrial and oceanic systems. It has a terrestrial boundary defined in various ways by the tide, and a seaward boundary defined by the edge of the continental shelf (Ray, 1991, and see Chapter 11). It is influenced by the proximity of the terrestrial system, but differs substantially from oceanic systems. Nearshore ecosystems tend to be productive and complex, largely because terrestrial, oceanic and atmospheric processes all interact (Leigh et al, 1987; Ray and Hayden, 1992).

Similarities and differences between marine, estuarine and freshwater ecosystems

Natural bodies of water can be categorized as belonging to the following groups of environments:

- Freshwater lotic (streams and rivers).
- Freshwater lentic (lakes).
- Estuaries (the junctions of fresh and marine waters).
- Marine (oceans and seas).

Because freshwaters comprise such a small proportion of the total bodies of surface waters on the planet (< 0.5 per cent), the term 'marine' is sometimes used more or less synonymously with 'aquatic'. While aquatic environments have many features in common, each of them has its own unique set of properties, and its own suite of relationships between environmental (geophysical) characteristics and biological communities. Although some relationships may traverse the boundaries of environment type, it would be a mistake to uncritically extrapolate relationships from one to another. Each of these groups of aquatic environments requires its own classification of habitats and community types and its own approaches for conservation.

As a general principle, we should remember that smaller bodies of water are geologically 'younger', and that their floras and faunas are likely to be more individualistic or idiosyncratic. This follows from island biogeography theory (MacArthur and Wilson, 1963) because of random colonization and extinction effects in isolated geographical regions.

Marine and freshwater communities all owe something of their character to the fundamental properties of water and geophysical properties. However, overall it is probably a fair generalization to say that freshwater communities are more heavily influenced by biological interactions (including these random colonization and extinction processes) than are marine communities. This may be a consequence of the greater biological simplicity of food webs in freshwaters and resultant greater impacts of 'top-down' predator effects (e.g. McQueen et al, 1989). In addition, the aquatic communities of lakes and estuaries have been more heavily affected by human activities than have those in the oceans (effects on fisheries, coral reefs etc. notwithstanding).

A summary of similarities and differences among marine, estuarine and freshwater ecosystems, adapted from Day and Roff (2000), is included here.

Similarities among marine, estuarine and freshwater ecosystems

All aquatic environments share the same basic physical properties of water. High specific and latent heats lead to thermal stability, and confer protection against temperature extremes. The high density of water creates a buoyant medium in which organisms can remain suspended with little or no expenditure of energy. Resources are available both in solution (because of the high dissolving power of water) and in suspension.

Water transparency

Water is highly transparent, allowing light penetration for photosynthetic plants. The process of light attenuation is basically the same in all aquatic environments, but the rate of extinction of light depends on water clarity. This is very variable between the three environments. All the aquatic environments show comparable divisions into euphotic (much light), dysphotic (little light) and aphotic (no light) zones, though their extents are different.

Water motion

Water motion can be characterized as diffusive (non-directional, local-scale mixing) or advective (broad-scale, directional mixing). Diffusive motions are essentially similar for all aquatic environments. At the molecular level, the basic processes of diffusion of gases and solutes are comparable in scale and effects among aquatic environments. Eddy diffusive processes (of bulk property transfer), for example convective streaming and Langmuir circulation, are also basically comparable among aquatic environments.

Stratification and mixing

The seasonal cycles of stratification and vertical mixing of the water column are also fundamentally comparable in all bodies of water. Similar cycles of stratification and mixing tend to be followed at comparable latitudes and for similar depths of water column. Temperate lakes are typically dimictic, meaning that they mix vertically in early spring and again in autumn, becoming stratified (to produce a thermocline of rapid vertical temperature change) throughout the summer, and ice covered during winter. Comparable summer stratification due to the annual heating cycle also occurs at temperate latitudes in the oceans, but there are additional complications due to tides, salt effects and greater depth.

Internal waves can develop in all stratified waters at the thermocline. In lakes these are called internal seiches; in the oceans they are called internal waves. The phenomenon is the same in each case, but internal waves in the oceans are of several sorts and various periodicities, and are generally more complex. In all aquatic environments atmospheric circulation creates wind waves. These are most pronounced in the oceans due to greater fetch (proportional to distance to shoreline). All shorelines are subject to exposure from effects of wind waves, although again effects are most pronounced around the oceans' shorelines.

Differences among marine, estuarine and freshwater ecosystems

Geological age

Many fundamental differences between aquatic environments can be related to their geological age. Lakes are typically very young, 10^3–10^5 years; the Great Lakes of North America are only 10^4 years old. Lakes appear and disappear quite rapidly in geological terms. In contrast, the oceans are some 10^8–10^9 years old, and have been in existence in some form since the planet first differentiated into land, atmosphere and water. Estuaries are intermediate in age, 10^4+ years, but being connected to the oceans, their floras and faunas are geologically older than those of lakes. Lakes have many different types of geological origins; estuaries originated in fewer ways.

Size, depth, volume

There are also fundamental differences in size, depth, surface areas and volumes. Maximum ocean depth exceeds 10,000m, and pressure increases by one atmosphere for every 10m. Because of the vastly greater depth, pressure in the oceans also greatly exceeds that of other aquatic environments, leading to specialized faunas.

Largely as a function of their size, different bodies of water also exhibit a wide range of water replacement times (volume exchange, turnover times, or residence times). Estuaries, because of their tidal amplitudes, generally replace their volumes in days, weeks or months. Lakes generally have water replacement times of a few years to decades. The oceans have vastly longer residence times of water. Here, the deep and bottom waters of the ocean can take millennia to circumnavigate the globe. Exchange time between the atmosphere and the oceans varies from days – for surface waters – to millennia for bottom ocean waters.

Light attenuation

Although light is attenuated according to the same physical principles in all aquatic environments, they differ greatly in the extent of light penetration. In the clearest oceanic water, photosynthesis can still occur at 200m depth. The euphotic zones in lakes and estuaries are typically much less, often only 20m or less.

Water motion

Advective processes differ greatly among aquatic environments; they are of fundamental significance to the biological communities of the oceans. Planet-wide surface ocean circulation (down to about 600m) is largely driven by atmospheric circulation, and modified by the earth's rotation on its own axis (the Coriolis effect). Except in a few very large lakes, there is no lake equivalent to such circulation, and circulation in estuaries is driven primarily by the tides. Circulation of deep and bottom waters of the oceans is driven by thermohaline mechanisms (a combination of temperature and salt concentration) and geostrophic effects. Again there is no equivalent in lakes, but estuarine circulation is driven by equivalent phenomena involving salt density differences and Coriolis effects. The oceans exhibit surface convergence and divergence zones, of considerable significance to production; in lakes there is no equivalent circulation, and mechanisms of circulation in estuaries are complex. Large lakes may show basin-wide circulation, but this is generally driven by atmospheric forces or by inflowing rivers. Larger lakes may also exhibit internal seiches, resulting in local shoreline upwelling zones and cold surface water anomalies. Such motion in the Great Lakes is often unpredictable in timing, though it may be predictable spatially. Large-scale circulation processes are therefore essentially different between marine and freshwater systems. Streams and rivers are unique because of the one-way flow of water into lakes or oceans, inducing special behaviours in their biota.

Stratification and tides

Although most natural bodies of water can become stratified, the additional effects of tides must be considered in the oceans. Tidal effects in freshwaters, even in the Great Lakes, are negligible. In temperate regions, lakes greater in depth than about 20m always become seasonally stratified. In the oceans, high tidal activity may totally prevent stratification in some areas, but lower tidal ranges in other regions permits stratification. Tropical and subtropical areas of the oceans remain permanently stratified. The spatial extent and seasonal timing of stratification in lakes and in the oceans is highly predictable, but for different reasons.

Chemical composition

There are major differences in the chemical composition of fresh and marine waters. The concentration of total dissolved solids (= salts) in lakes is generally low and variable, from < 5 to > 400 parts per million (ppm). In contrast, in the oceans, total salt concentration is highly invariant, from 33 to 39 parts per thousand (ppt or ‰ or psu), with a mean of close to 35ppt. Estuaries span the range between lakes and oceans.

The composition of lake waters is also highly variable, and basically different from that of the oceans. The ions Ca^{2+}, Mg^{2+}, HCO_3^- tend to dominate in freshwaters, though relative composition varies greatly, while in the oceans, relative composition is extremely uniform and stable, and Na^+ and Cl^- dominate. Estuaries are intermediate in chemical composition and variability. In all aquatic environments the major nutrient salts for primary producers are the same – PO_4, NO_3, NH_4 and SiO_4, but their relative importance and effects in limiting production vary among environments, predictably but differently in marine and freshwaters.

These variations in total salt content and composition of freshwaters arise because lakes are responsive on an individual basis to geological and morphoedaphic influences of the surrounding watershed. In contrast, the oceans globally integrate all these effects. Because the circulation and mixing times of the oceans are fast (although they may take centuries to eons!) compared to rates of inputs of substances, their composition is almost homogeneous throughout. In contrast to the oceans where internal recycling of nutrients overwhelmingly predominates over the contributions from external sources, nutrients enter larger lakes primarily via the atmosphere and land drainage. Which predominates is a function of the ratio of lake surface area to drainage basin area. For example, in the Great Lakes of North America, there is a gradient from Lake Superior to Lake Erie of greater atmospheric to greater drainage basin input of nutrients.

Variations in total salt content and composition present osmotic and ionic challenges to aquatic organisms. The greatest osmotic differences between internal and external medium are for freshwater organism, but the greatest variability, even on a daily basis, occurs in estuaries with the ebb and flow of the tides and corresponding changes in total salt content.

Post-glacial history

The most basic characteristics of lakes in northern North America are determined by their post-glacial history and their morphoedaphic index (though this has been disputed – see Ryder, 1982). According to the combination of these two criteria, the Great Lakes comprise a distinct group, separate from other North American lakes.

Sediment types in lakes vary greatly, from bare rocks (predominantly in young lakes) to organic muds. Estuarine and river substrates can be highly variable, mainly as a function of current speeds. The oceans have well established gradients of sediment types that can be related to both physical circulation and biological production rates (which in turn follow physical effects).

Community types

All aquatic environments have the same fundamental community types: pelagic and benthic, including fringing communities. The exceptions to this are streams and rivers, which do not develop a true pelagic realm.

Pelagic and benthic communities are very differently developed in freshwaters and in the oceans. The oceans experience a greater or lesser tidal range along the shores. Lakes lack such tides but estuaries may experience very high tidal effects. The whole rich intertidal zone is therefore absent in lakes. As a further consequence of the lack of tides, fringing communities in lakes do not show the great diversity of type found in the oceans. In freshwaters, fringing communities tend to consist of submergent macrophyte stands, or emergent wetlands or semi-terrestrial marshes. Estuaries tend to show salt marshes and mud-flats or mangrove assemblages. The oceans show a great diversity of fringing communities, including all those types found in lakes and estuaries, plus seagrasses, coral reefs and macrophytic algae.

Species diversity

Of all aquatic environments, the oceans show the greatest species diversity of animals. All major phyla are represented. Lakes exhibit a significant reduction in diversity at higher taxonomic levels, and many of the major marine phyla are simply not represented in freshwaters (see Table 3.1). In partial compensation at lower taxonomic levels, freshwater habitats do show a high diversity of insect larvae and oligochaete annelids – which are essentially absent in the oceans. Estuaries might be expected to show a high diversity of life because they are the junctions of fresh and marine waters. In fact, the intermediate and variable salinity poses the constant challenge of osmotic regulation. Estuaries therefore show significant changes in species and community composition along a salinity gradient, and are very variable in their species diversity. Species diversity tends to be lowest at ~ 5ppt salinity. Many marine species (stenohaline forms) cannot tolerate salinities below ~ 17ppt (e.g. most echinoderms). Other specialist euryhaline species thrive at the intermediate and variable salinities of estuaries, and their populations may become very dense and productive. At the interfaces between fresh and marine waters, estuaries are perhaps the most physically complex and the most biologically productive of all aquatic ecosystems.

Similarities and differences between Arctic, Antarctic, sub-Arctic, temperate, subtropical and tropical environments

Temperature is often regarded as the most important single factor governing the distribution and behaviour of organisms on earth (e.g. Gunther, 1957). There is considerable scientific literature on the effects of temperature on factors such as the distribution, physiology and behaviour of aquatic organisms. All marine organisms show some pattern of limitation of their distribution according to temperature. It will become clear that temperature is an important determinant of both representative communities (at broad spatial scales) and distinctive biological communities (at local scales). Very few species have geographical ranges covering the entire range of observed aquatic temperatures. Those that apparently do, so-called cosmopolitan species, will probably be found to consist of species complexes (see Chapter 10). Within the oceans, but not in freshwaters, the distributions of temperature and salinity are partially confounded, and are often used in combination to describe the origins and movements of water masses, and their associated fauna and floras.

All oceans exhibit the same fundamental array of ecosystem-level processes, such as mixing regimes, nutrient cycles, trophodynamics, sedimentation, water–atmosphere–land exchange and so on. All exhibit a similar array of community types, whether classified according to taxonomy (i.e. fish, marine mammals etc.) or according to oceanographic and physiographic factors (i.e. pelagic, benthic etc.). In addition, all the oceans show significant similarities in taxonomic composition at the generic or higher taxonomic levels.

The oceans as a whole have been forming and readjusting their shapes and connections over thousands of millions of years. Not only have the oceanic areas and their associated ecosystems developed over a longer period than the terrestrial environment, but there are also significant age differences between the three oceans. The much greater geological age of the Pacific Ocean is believed to be the major reason for its generally higher biological diversity in marine communities compared to the Atlantic (see e.g. Levinton, 1982). This is usually invoked as part of the 'stability-time hypothesis' (Sanders, 1968). However, differences between the oceans are largely at the generic and specific level rather than at higher taxonomic levels. This can be seen for example in the major fisheries on the Atlantic and Pacific coasts; the differences are essentially in species composition, and the relative significance of anadromous versus migratory or local fisheries. In terms of overall species diversity, the Pacific is highest, followed by the Atlantic and then the Arctic Ocean, at least for the great majority of higher taxa.

In physical terms, the Arctic Ocean has the lowest salinity, largely as a consequence of the huge amount of seasonal freshwater runoff which cannot easily escape due to the surrounding land masses. Stratification in the Arctic Ocean is therefore at least as much a consequence of salinity as it is of vertical temperature gradients. The greater ice cover of the Arctic also tends to uncouple it from (at least) seasonal interaction with the atmosphere, although, perhaps surprisingly, it has become clear that atmospheric transport is a major source of pollutants to the Arctic.

Largely because of its strong vertical stratification and ice cover, both resisting the vertical transport of nutrients, the Arctic Ocean has the lowest production among the oceans (see e.g. Dunbar, 1968). In strong contrast to this is the high seasonal production of regions of the Antarctic Ocean, especially around the divergence zones. Here, upwelling brings elevated nutrient concentrations to the surface waters, stimulating some of the highest phytoplankton production rates anywhere in the oceans, and leading to high production at higher trophic levels as well.

In the ecological literature it is often stated that communities in polar regions are primarily physically accommodated, whereas tropical ones are predominantly biologically accommodated. This is generally deduced from the greater fluctuations in seasonal cycles of biological production and abundance in the Arctic as compared to the temperate and tropical regions, which in turn is taken as evidence of lack of strong biological control between 'trophic levels'. However, it should be realized that temperature fluctuations themselves are no greater in polar waters than they are in tropical waters. Further comparisons between tropical and temperate marine ecosystems are summarized in Table 3.2.

Pelagic and benthic realms: Their habitats and communities

Hierarchies of habitats

A hierarchical classification of the habitats and communities of marine environments is an absolute necessity for effective marine conservation. Without such a classification it is not possible to derive a systematic plan for the conservation of marine biodiversity, and specifically to ensure that all the types of habitats and their communities and species are represented in protected areas. However, hierarchical classifications can be devised in several ways, and any one applied to the natural world will be subject to constant

Table 3.2 A summary of differences between tropical and temperate ecosystems

Factors	Tropical compared to temperate waters
PHYSICAL and CHEMICAL	
Temperature	Higher mean; lower annual range; thermal maximum closer to ambient
Light	Higher total annual; lower annual range in day length
Dissolved oxygen	Lower
Total dissolved CO_2	Lower
Dissolved nitrogen and phosphorus	Lower
Water clarity	Higher (except locally)
Rainfall and coastal runoff	More variable seasonally
Tides	Lower mean amplitude
Sediments	Calcareous sediments more characteristic of tropical waters
Seasons	Typically two monsoonal or trade wind seasons rather than four; distinguished on basis of wind, rainfall and currents rather than temperature
Incidence of storms, hurricanes, typhoons	Greater
COMMUNITY STRUCTURE	
Species diversity	Higher in major groups
Diversity in genera	Higher
Mean organism size	Smaller in major groups
Biomass	Lower in major groups
Population density of individual species	Lower in major groups
Population size	Smaller
Predators	Higher percentage in documented groups
Colonial life forms	Greater incidence
Abundance of herbivorous fishes	Greater
Eggs	Smaller with less yolk in documented groups
Larvae	Higher percentage planktonic in invertebrates and fish
Lipid content	Generally lower
External anatomy	More adaptations for predator defence
Colour polymorphism	More conspicuous
Genetic variability	Higher
BIOLOGICAL FUNCTIONS	
Metabolic rates	Higher in all groups
Primary production	Higher in coral reefs, mangroves, seagrasses. Lower in phytoplankton except in upwelling areas
Growth rates	Higher
Thermal tolerance	Narrower range
Frequency of reproduction	Greater
Reproduction potential	Greater
Breeding seasons	Longer and spread more broadly over the year
Asexual reproduction	Higher incidence in invertebrates
Larval development	Faster
Feeding habits	More specializations
Natural mortality rate	Higher in fishes
Niche breadth	Narrower
Lower oxygen limit	Closer to ambient levels
Toxic and venomous forms	Greater incidence
Symbioses and parasitism	Greater incidence
Life span	Shorter in documented groups
Degree of skeletal calcification	Greater
Endemism	Higher in invertebrates
Hermaphroditism	Greater incidence in fishes
$CaCO_3$ biological precipitation	Much greater
Rates of evolution	Higher

Source: Adapted from Hatcher et al (1989)

Table 3.3 Ecological taxonomic types or 'ecospecies'

VIRUSES		
ARCHEBACTERIA		
EUBACTERIA		
FUNGI		
PLANTS		
Ecological Type	Habitat	Life Cycle/ Reproduction
Microphytic algae planktonic	Water column	Asexual fission; some sexual
Microphytic algae benthic	Hard substrates encrusting	Asexual fission; some sexual
Microphytic algae benthic	Soft substrates adherent	Asexual fission; some sexual
Microphytic algae -symbiotic	E.g. corals	Symbionts with coral
Macrophytic algae benthic	Hard substrates attached	Asexual; sexual motile spores
Angiosperms mangroves	Estuaries, bays	Sexual; dispersing Propagules
Angiosperms seagrasses	Sandy bottom	Asexual; sexual seed dispersal
Angiosperms saltmarsh	Saltmarshes	Asexual; sexual seed dispersal
PROTISTS		
Protists planktonic	Water column	Asexual fission; some sexual
Protists benthic	Encrusting or burrowing	Asexual fission; some sexual
ANIMALS		
Marine mammals	Pelagic water column migratory	Sexual; live birth
Marine mammals	Shore-referenced	Sexual; live birth
Marine birds	Oceanic	Sexual; eggs nesting
Marine birds	Shore-referenced	Sexual; eggs nesting
Marine reptiles	Pelagic water column migratory and shore-referenced	Sexual; eggs nesting
Fish	Pelagic highly migratory	Sexual; free larvae
Fish	Anadromous	Sexual; benthic larvae
Fish	Catadromous	Sexual; free larvae
Fish	Pelagic regionally defined	Sexual; larvae
Fish	Demersal regionally defined	Sexual; larvae
Fish	Territorial	Sexual; protected larvae
Nekton	Water column regionally defined	Sexual; free larvae
Zooplankton	Water column regionally defined	Sexual; free larvae
Zoobenthos	Regionally migratory/ mobile	Sexual; free larvae
Zoobenthos corals	Epibenthic attached	Asexual fragmentation; sexual planula larva
Zoobenthos	Epibenthic attached	Sexual; free larvae or 'direct'
Zoobenthos	Burrowing infauna	Sexual; free larvae or 'direct'

revision, just as the systematic classifications of plants and animals undergo constant revision. For example, species can be classified according to their ecological type (Table 3.3, sometimes called 'ecospecies' or 'trophospecies'), and communities can be classified according to their ecological type – recognized in all marine ecology textbooks (Table 3.4). Movements of animals are highly variable. Marine conservation needs to take all these categories of species and communities into account.

It could be argued, rationally and defensibly, that any hierarchical classification of aquatic environments should recognize the split into pelagic and benthic realms even before they are separated into marine, freshwater and estuarine types, and certainly before separation by biogeographic and temperature regimes. The differences in body type and form between pelagic and benthic realms are extreme. In some respects, and in some groups of plants and animals, the relationship between freshwater benthic and terrestrial

Table 3.4 Summary of some major marine communities

Biological realm	Major community 'units'	Community type
Fringing communities	Estuaries	
	Non-biogenic communities	Rocky shores
		Tidal flats
		Sub-tidal soft bottoms
		Shorebirds
	Biogenic communities	Biogenic reefs
		Corals
		Kelp bed Macrophytic algae
		Mangroves
		Seagrasses
		Salt marshes
Pelagic	Pelagic communities	Phytoplankton
		Holozooplankton
		Meroplankton
		Fish
		Diadromous fish
		Seabirds / waterfowl
		Marine reptiles
		Marine mammals
Benthic	Benthic communities	Demersal fish
		Territorial fish
		Epi-benthic zoobenthos
		In fauna burrowing zoobenthos
		Cold seeps
Deep sea		Deep sea corals
		Sponge beds
		Soft corals (Gorgonians)
	Seamounts	
	Hydrothermal vents	
	Abyssal plains	
	Trenches	

Notes: This does not include:
Most geomorphological units
Focal species
Pelagic processes in distinctive areas
Biogenic communities are ones in which the organisms themselves form a major component of the substrate.

taxa is closer than between pelagic and benthic forms. For example, aquatic macrophytic plants and aquatic insects in freshwaters are closely related to (indeed, in the case of many insects, may metamorphose into) terrestrial taxa. However, marine and estuarine benthic plants and animals are typically not closely related to terrestrial taxa. The counter-argument is that in marine waters, many organisms move between the pelagic and benthic realms, indeed the pelagic realm harbours the larvae of the same species that are adults in the benthic realm.

Separation into pelagic and benthic represents the first step in the recognition of hierarchies of community types within the marine environment, and within each of these two major realms. Segregation into pelagic and benthic realms is perhaps not an immediately obvious division to most people. When constructing an ecological hierarchy of environment types, divisions into land versus water, then into the major environment types (fresh, estuarine, marine), then into approximate latitudinal ranges (arctic, subarctic, temperate, subtropical, etc.), are likely to appeal as more 'natural' and to be considered at a higher level than division into pelagic and benthic realms.

The type of community that develops in a region is the result of the interactions of a series of biological and geophysical factors (Table 3.5). These operate differently within the pelagic and benthic realms. Within the pelagic realm, associations between organisms and environment are largely to water masses of the upper water column, depth, inshore/offshore gradients and to salinity/temperature gradients. Within the benthos, associations are to bottom water masses, depth, substrate types, temperature and currents. A major determinant of community type for both pelagic and benthic communities is the extent of water movements; this involves the interactions of winds and tidal activity in determining the stratification and mixing regimes. For the benthos, it also involves the degree of exposure to wave action, and the vertical range of the tides in determining the character of intertidal communities. Therefore we need to classify these two realms separately, because we must rely on geophysical factors to differentiate among community types. These relationships will be explored more fully in subsequent chapters.

Pelagic and benthic realms

Marine systems are significantly different from terrestrial ones in having a three-dimensional structure that has no recognized analogue in

Table 3.5 Factors associated with or controlling the nature and distribution of marine biota

Biological	Oceanographic	Physiographic
predation	tsunamis, storm surges, hurricanes	tectonic, seismic, volcanic activity
resources (nutrients/food)	tornadoes	geographical position/latitude
resource selectivity	salinity	geological history of ocean basin
competition (density dependent/ independent)	water masses	bathymetry (water depth)
life-history patterns	temperature	relief/slope
mutualisms	ice cover/ ice scour	substrate/sediment particle size
opportunist/equilibrium species	temperature gradients/ anomalies	basin morphometry/topography
recruitment mechanisms	water motion (many types)	geology/rock type
migrant species	convergences/divergences	substrate heterogeneity
vagility/larval dispersal	upwellings	
buoyancy/ sinking	stratification/mixing regime	
desiccation resistance	nutrients	
osmotic tolerance	light penetration/turbidity	
spatial/dimensional use	depth/ pressure	
patchiness	tidal amplitude/currents	
seasonal cycles	exposure (to atmosphere/waves)	
biological succession	oxygen and other dissolved gases	
human activities		
productivity		
plant species (associations)		
animal species (associations)		

Note: NB – some factors may appear in more than one category, or under more than one title

terrestrial systems. The ecosystems of oceans and seas can be broadly divided into two realms, or sets of habitats and communities, which are dimensionally different:

1 Pelagic communities contain all those organisms that inhabit the water column itself and inhabit a three-dimensional world.
2 Benthic communities contain all those organisms that inhabit the substrates of marine basins, in an effectively two-dimensional world. (Note the term 'demersal', which means from a functional viewpoint 'bottom-referenced'.)

This is a basic division of habitat and community types, recognized in all marine biology and marine ecology texts (see Box 2.1), that recognizes the fundamental differences between the organisms inhabiting the water column and those that live in or on substrates.

From the perspective of ecological representation, differences in both physical oceanographic and biological oceanographic conditions are highly significant between the pelagic and benthic realms. There are species and indeed whole communities that occupy only the pelagic realm and have no reference to the underlying benthic realm for their entire life cycle. Conversely it is also known that many benthic organisms utilize the pelagic zone for dispersal stages and pelagic species may use some aspect of the benthos for reproduction.

Benthic communities are sometimes further subdivided (e.g. Hedgpeth, 1957) into:

- Fringing communities (comprising the primary producers and consumers of the euphotic zones).
- Benthic proper (those below the euphotic zone dependent on a detrital economy).

There are pros and cons to recognizing such a distinction. Here fringing and benthic communities are considered together, in order to avoid artificial separation of producers and their attendant consumers.

The pelagic realm - definition and description

The pelagic realm is a fully three-dimensional world. At scales from metres to hundreds of kilometres it may appear to be largely homogeneous, until more closely examined. A major feature of the pelagic realm is in fact its spatial heterogeneity, generally referred to as 'patchiness'. However, specialized sampling apparatus is often needed to reveal these quantitative inhomogeneities. Pelagic species demonstrate a great diversity of body size and form and taxonomic type, and a high diversity of both plant and animal phyla (Table 3.1). In fact, a single plankton collection can contain a higher diversity of either plant divisions or animal phyla (depending upon the sampling apparatus) than any other single collection made anywhere else on the planet. Nevertheless, at the species level diversity in the pelagic realm tends to be lower than in the benthos (Gray, 1997), perhaps as a function of greater environmental heterogeneity in the benthos.

For descriptive purposes the pelagic realm can be subdivided into the *planktonic* community of smaller organisms, and the *nektonic* assemblage of larger, more mobile species. The plankton comprises all those aquatic organisms that are 'suspended' freely in the water mass. They drift passively in the water currents, and their powers of locomotion are insufficient to enable them to move against the horizontal motion of the water. However, many planktonic species may make extensive vertical movements. One specialized component of the plankton is the neuston which comprises a diversity of organisms inhabiting the immediate water surface.

The nekton comprises all those actively swimming consumers that constitute the middle and upper trophic levels of marine ecosystems. Among these, the dominant organisms are many types of fish (e.g. herring, tuna), increasing in size up to and including marine reptiles and mammals. Some species of fish, reptiles and marine mammals make extensive seasonal migrations, while other fish species make regional movements. The smaller components of the nekton, which may also make seasonal migrations, comprise invertebrates such as euphausiids. The chief feature separating nekton from plankton is their relative mobility, and their ability to swim or migrate contrary to ocean currents. The distinction between plankton and nekton is by no means absolute. For example, while being nektonic as adults, the larvae of fish are members of the plankton.

Because of their ubiquity, plankton have generally not been considered as worthy of conservation attention. Nevertheless, an appreciation of their global significance is warranted. As in other communities, plankton comprise primary producers, primary and secondary consumers and decomposers. The primary producers, the *phytoplankton* (Colour plate 1b), are unicellular or chain-forming micro-algae ranging from < 1μm to > 1mm in diameter. Functionally, they are often divided into: picoplankton (< 2μm), nanoplankton (2–20μm) and netplankton (> 20μm). Taxonomically the phytoplankton comprise some 12–18+ divisions of plant life (Table 3.1), a far greater phyletic diversity than on land. These are the dominant photosynthetic organisms on the planet, yet they remain virtually unknown and unseen by the general public.

Within the oceans themselves the phytoplankton are responsible for > 95 per cent of the annual primary production. In temperate waters, the seasonal cycle of primary production within the plankton is typically highly pulsed. It typically peaks in the spring with an outburst of diatom growth, as the water column becomes *stratified*. Seasonal cycles are generally more extreme at high latitudes (e.g. in the Arctic) and in coastal waters, and are less modulated in the tropics and in offshore oceanic waters. Within the pelagic realm, the rate of primary production is controlled by the availablity of light and nutrients (typically: nitrogen, phosphorus, iron, silica).

Net carbon fixation can only occur within the euphotic zone, where the rate of photosynthesis exceeds the rate of respiration. The depth of this zone varies greatly even among natural aquatic environments. In those affected by human activities, the depth of the euphotic zone is typically decreased, due to elevated plant production (*eutrophication*). Some examples of the depths of euphotic zones would be: smaller lakes 5–10m; larger lakes 10–30m; estuaries 5–15m;

coastal and neritic waters 10–60m; oceanic waters up to 200m. The biomass and productivity of phytoplankton can be estimated *in situ* from the uptake rate of radioisotopes such as ^{14}C, or from remotely sensed ocean colour and water clarity (e.g. Bukata et al, 1995).

The presence of floating ice at high latitudes also affords new habitat for micro-algae. A major new community of primary producers – the epontic or ice-algae community – develops seasonally in the bottom few centimetres of ice. We can consider this as a sort of 'upside-down benthos' complete with plants, animals and even polar bears!

The zooplankton comprises a wide diversity of phyla, body forms and modes of life. *Holoplankton* are the predominant members which spend their entire life cycle within the water column. They are distinguished from the *meroplankton*, the larval forms of benthos, which recolonize the benthos after a period in the plankton. A category of *ichthyoplankton*, comprising the larvae of many fish species, is also often recognized. Individuals may be herbivores, omnivores or strict carnivores, or function ontogenically at different 'trophic levels'. Every phylum of the animal world is represented somewhere in the marine plankton; however, they typically do not exhibit a high diversity of species, compared to benthic or terrestrial communities. Zooplankton are the most abundant animals (metazoa) in the world, yet again they are largely unknown to the public. The most abundant zooplankton are the copepods (Colour plate 1c). Their nauplius larva is the most common and abundant type of animal body plan on our planet.

Zooplankton range in size from < 50μm (nauplii) to over 2m (lion's mane jellyfish *Cyanea* spp). Despite being at the mercy of horizontal water motions, many zooplankton can make extensive vertical migrations. The extent of these migrations generally increases with increasing body size and with depth. In neritic and oceanic waters this typically leads to larger species inhabiting deeper levels of the water column, and overlapping in their migrations with smaller species living closer to the surface.

The phytoplankton are grazed by organisms of two major food webs. The *classical food web* consists of the larger phytoplankton (nano- and netplankton), grazed by zooplankton which in turn are consumed by the nekton (euphausiids, fish, marine mammals etc.). This food web has been recognized for many years because the major commercial fisheries of the world are dependent upon it. The classical food web, fuelled largely by nutrients from seasonally mixed or upwelling waters, is most highly developed in coastal (neritic) regions; it becomes progressively diluted into oceanic waters. A second pelagic food web, the *microbial food web*, has more recently been recognized (e.g. Azam, 1998), though in evolutionary terms it is much more ancient. It is spatially less variable from neritic to oceanic waters, and ecologically it dominates in oceanic waters. It consists of the smaller primary producers (pico- and nanoplankton), bacteria, flagellates and ciliates. The ubiquitous nauplii may be an important interface between these two food webs (Turner and Roff, 1993).

The planktonic community is pre-eminently a region where organic production is directly available to grazing consumers. There is no equivalent to the resistant woody tissues of terrestrial plants. Therefore, a large proportion of the production of the phytoplankton is consumed immediately. Ungrazed organic material and fecal pellets sink towards the ocean floor. Depending on depth, seasonal timing of cycles and so on, this material may be efficiently recycled several times within the water column, before the residue finally becomes available to the zoobenthos.

Both pelagic and benthic communities are strongly related to depth in the water column, as a function of light and pressure (Figure 2.1). The epi-pelagic lies within the uppermost part of the water column, with characteristic communities of plankton and fish. Organisms which live in the mesopelagic and bathypelagic regions are adapted to these physical conditions and dilute resource concentrations, having evolved in response to the enduring factor of pressure. However, some fish and invertebrates in the bathypelagic zone (such as the sergestid prawns and myctophid fishes) migrate into the upper two layers at night. Where bathypelagic zones are occupied by specially adapted fish or invertebrates, the actual populations may be low except in areas where food is more abundant such as around sea mounts. Vertical changes in these biological communities with depth are real, but especially in the pelagic realm, where organisms form a series of overlapping distributions and vertical migration ranges, divisions should be regarded as typically gradual rather than distinct.

The benthic realm - definition and description

For practical purposes, the benthic realm can be considered as two-dimensional. While it can be argued that benthic organisms use their space three-dimensionally (e.g. plants extend well into the water column, animals burrow into the substrate etc.), their mode of life and physical adaptations lie in sharp contrast to organisms of the pelagic realm. Benthic plants and animals are bottom referenced; their distributions are also strongly related to water depth.

Fringing communities

In the shallowest euphotic regions of aquatic habitats, photosynthetic fixation of carbon is carried out by several types of fringing floras. Despite their diversity and the high biomasses that they can attain, their overall contribution to primary production of marine ecosystems is low compared to the phytoplankton, except in shallow coastal waters. Rates of primary production for all types of marine communities can be found in Mann (2000).

Several community types, many of them familiar to us because of superficial similarities to terrestrial forms, can be recognized:

- Dense stands of sub-tidal (and therefore submergent) macrophytic angiosperm vegetation (basically of terrestrial ancestry). In marine habitats these are represented by species such as the truly marine seagrass *Thalassia*.
- Salt marshes with species such as *Spartina* around the ocean margins are intertidal (rarely extending below mean tide level) and essentially semi-terrestrial in nature (Colour plate 2a).
- In the wetlands, marshes and coastal regions of lakes, both submergent and emergent macrophyte plant communities develop, whose species diversity is greater than those of similar marine communities. These communities are a mixture of truly aquatic and semi-terrestrial species.
- The seaweed communities are a dominant feature of rocky marine shores (Colour plate 2b). High biomasses of these macrophytic algae may develop both intertidally where they are temporarily exposed to the air during the phases of the tide, and sub-tidally within the euphotic zone. This community type is poorly represented in freshwaters, usually being replaced by macrophytic angiosperms.
- The intertidal and sub-tidal micro-algae are often forgotten in comparison with the more visually striking macrophytic algae of the shorelines. These microscopic algal communities (single cells, chains or mats) are often dominated by the benthic diatoms which form mats on the bottom sediments or rock or may form epiphytic growths on the macro-algae themselves.
- The hermatypic corals forming reef structures. Although the corals themselves are of course animals, their symbiotic micro-algae (zooxanthellae) are responsible for high rates of photosynthesis. Reef forming photosynthetic corals are not represented in temperate waters, though non-photosynthetic 'hard corals' are found in deeper waters and extend into the high Arctic.
- Mangrove swamps typically develop in hot wet regions of the ocean margins, where tidal amplitude is low (Colour plate 2c).

The rate of primary production in these fringing communities may be extremely high. It is usually much higher per unit area per unit time than that of the planktonic algae. However, the greater vertical extent and horizontal coverage of the phytoplankton more than compensates in terms of overall production. A major distinction between planktonic and benthic producers is that the macrophytic plants tend to conserve their biomass. The carbon fixed by them tends to enter detritus pathways (as is generally the case in terrestrial ecosystems) rather than to be directly grazed (Mann, 2000).

Associated with the plants of the fringing communities is a wide diversity of bottom-dwelling invertebrates. These are considered next under zoobenthos communities.

Zoobenthos communities

Fundamentally, unlike the pelagic realm, the benthos can be considered to a first approximation as a two-dimensional world. Animal life extends from the highest limit of the high tides to the abyssal depths of the oceans. Life is either attached to or roams the substrate, or is buried only a few centimetres within it (Colour plate 3).

In the shallow waters of the euphotic zone, benthic and nektonic grazers can graze directly on a diet of macrophytic or microphytic algae. Within the intertidal zone, the upper limits to distributions are set by tolerance to atmospheric desiccation, while lower extents are often set by less physically tolerant predators. There are major differences between the communities that develop on rocky shores versus those consisting of silts and muds.

In the immediate sub-tidal, plants are still present within the euphotic zone. However, in deeper waters the living plants are absent below the photic zone, and major changes occur in the animal communities. Below the euphotic zone, the substrate has generally changed to a relatively uniform sand/silt/mud in composition. The benthos in these regions can only be supported by the rain of detrital material which is derived from the other photosynthetic aquatic communities. The benthos in the dysphotic and aphotic zones is essentially a region of detritus-based food webs, and of heterotrophic bacterial activity. It is also a dominant site of nutrient regeneration. In the ocean as a whole, detritus-based food chains are extremely significant; in fact they are the dominant food webs of the oceanic province.

Specialized members of the fish community also exist at various depths within the benthos. For example, flatfishes, skates, dogfish and many other species are essentially bottom dependent, or at least bottom referenced, and may be relatively static, or highly localized and territorial. Other bottom-dwelling species are periodically mobile or migratory, for example crabs and lobsters, while others are either permanently attached to a substrate or burrow within it.

Influences on benthic community types

The general trend in benthic communities is that both substrate type and community type become less variable with increasing depth. In the depths of the oceans the zoobenthos tends to converge to only a few community types, and many species may be cosmopolitan. The tendency is for the bottom itself to be composed largely of silty mud, for current speeds to be low and for the community to become dominated by deposit feeders.

Substrate particle size is a dominant influence on benthic and demersal communities. The type of marine benthic community and the types of organisms that can live within or on the substrate are substantially determined by particle size along a spectrum from solid rock to mud (e.g. Barnes and Hughes, 1988). These distinctions between the communities of hard versus soft bottoms have been recognized for many years; indeed, whole ecological volumes are devoted solely to one type of community or to the other – for example, Stephenson and Stephenson (1971) for rocky shores, Eltringham (1971) for soft bottom communities. Hard-bottom communities, both intertidal and in the immediate sub-tidal, are generally dominated by macrophytic algae and a variety of epibenthic molluscs and crustacea, with few species managing to successfully burrow into the substrate itself. Soft-bottom communities, in sharp contrast, are dominated by burrowing species of worms and molluscs, and plant life may be restricted to microphytic algae on the sediment itself. Although there are various gradients between these extremes of community type, and mixtures may be found in regions where the substrate itself is mixed in character, nevertheless, these major types of biotope are readily recognizable worldwide (Table 3.4).

An important distinction between soft-bottom community types is strongly related to the sedimentary regime and water motion. Communities characterized by suspension feeders, which take particles directly from the water column, dominate in areas of relatively coarse sediments and high current velocities (which bring food to the zone). In contrast, communities characterized by deposit feeders dominate in regions of fine sediments and low current velocities, where fine particulate matter settles directly on to the bottom. Thus, as a first approximation, communities can be divided into 'suspension feeders' (associated with coarser sediments and higher current velocities) and 'deposit feeders' (associated with finer sediments and lower current speeds) (Wildish, 1977), although there is some overlap between these community types. Tidal current speeds appear to be the predominant agent determining community type (Warwick and Uncles, 1980).

The type of benthic community that develops in a region is also strongly related to degree of exposure. Exposure may be considered as occurring in two ways:

1 Through exposure to wave action.
2 Through exposure to desiccation.

In the benthos, exposure is only significant for the intertidal and immediate sub-tidal regions. High exposure coastlines have different intertidal and nearshore biota than protected shorelines, as a result of the mechanical wave action on the shore and the shallow seabed. The intensity of wave action that can develop is related to fetch, predominantly a function of the distance over which waves are propagated and the angle and 'development' of the shoreline. This can be determined from maps and meteorological records. Note also that differing factors become more critical depending on location; for example, within intertidal areas:

- Physical determinants drive the upper intertidal levels.
- Biological determinants drive the lower intertidal levels.

A different form of exposure for littoral communities is recurrent exposure to desiccation by the atmosphere during a tidal cycle. Exposure is strongly correlated with tidal amplitude, substrate slope and substrate particle size – along a spectrum from bedrock to mud. Both types of exposure exert considerable influence on the type of benthic plant and animal communities that develop in a marine region.

On hard substrates, there is generally little correlation between rock type and community composition. The reasons for this are as follows: The major marine angiosperms are rooted in soft sediments. Here they are buffered by the composition of seawater itself, which overshadows any residual effects of substrate particle type. Macrophytic algae possess holdfasts that simply anchor them to the hard substrate, irrespective of geological type, and their resources are derived from the water itself.

There are, however, more subtle differences in the communities which may develop on various types of rock substrate. These are perhaps best categorized as differences in community 'species richness' rather than as differences in community type (i.e. the characteristic biotope remains the same, but species composition varies), although this view requires quantitative support. For example, on softer substrates (e.g. limestone) boring invertebrates may be much more abundantly developed, whereas barnacles may be less common. Soft substrates that retain moisture at low tide may also harbour species not tolerant of the same degree of exposure in regions of harder substrates (Stephenson and Stephenson, 1972). There may also be significant differences in the type of intertidal communities that develop on rocky shores, related to the degree and orientation (relative to the shore) of rock faulting, and the cracks and crevices that develop. This can give rise to what Morton (1991) has called 'interiority'. Thus a more faulted and locally convoluted rocky shore can give rise to a richer more diverse community than a smoother one (Bergeron and Bourget, 1986).

Within the sediments themselves, oxygen levels can become very low, approaching zero or even becoming anoxic and allowing reducing conditions. In marine waters, the lowest oxygen levels are typically found in intertidal mudflats, especially in estuarine regions. Here, organic matter from productive overlying waters becomes trapped in the mud/silt interstices, and bacterial decomposition reduces oxygen levels because of poor exchange rates between mud and overlying waters. Areas experiencing such low oxygen levels within the sediments usually develop high populations of physiologically tolerant burrowing invertebrates.

Pelagic–benthic interactions

Pelagic and benthic communities are often spoken of as being 'coupled'. However, it is perhaps better to refer to them as 'interdependent', because the degree of coupling varies in space and time. Typical pelagic and benthic communities do not share adult species members, except temporarily; however, there are significant two-way interactions between the communities. Below the euphotic zone, benthic communities are entirely dependent on the rain of detrital material from the pelagic realm. The only exception to this is found in the extraordinary hydrothermal vent communities around the oceanic ridges, which are essentially independent of resources from the overlying water column, and are self-reliant on a chemosynthetic sulphur economy.

The two major realms can be considered to interact in perhaps three ways:

1 By exchange of energy.
2 By recruitment of larvae.
3 By temporary exchange of members.

The rate of sedimentation of primary energy (ungrazed algal cells) and recycled organic matter (faecal pellets and other detritus) to the benthos decreases with depth (e.g. Suess, 1980). Pelagic–benthic interactions therefore 'weaken' with depth because the flux of organic matter diminishes and the benthos must become food limited. In shallow waters, however, the benthos can graze the phytoplankton of unstratified water columns directly.

The flux of organic matter to the benthos is also highly seasonal, being greatest during and immediately following the spring diatom increase, and lowest during the summer stratified period. Losses from the plankton to the benthos are probably proportionately highest in the Arctic and lowest in the tropics (Roff and Legendre, 1986). This is because the greater the degree of modulation of the seasonal cycle, the less efficiently energy is used within the water column, and a greater proportion sinks to the benthos. At lower latitudes at higher temperatures in the benthos a greater proportion of this energy must be lost to bacterial decomposition. The net effect of these ecological differences and seasonal and latitudinal changes in processes, is that the planktonic realm tends to be proportionately more productive than the benthos at lower latitudes and in systems whose production is not highly modulated, while the reverse is true (i.e. the benthos tends to be proportionately more productive) at high latitudes or in strongly pulsed ecosystems (e.g. Longhurst, 1995).

This is by no means the end of the complexities of pelagic–benthic interactions. Lateral transport of energy can also dominate in a region where tidal currents are strong. For example, the planktonic and benthic systems in the outer Bay of Fundy have been described as 'uncoupled', because production of the zoobenthos is heavily dependent on tidal transport of detritus (Emerson et al, 1986).

The planktonic world is also heavily, but temporarily, colonized by the larvae of benthic forms. Lobsters, crabs, molluscs and such are all temporarily members of the meroplankton as larval stages. In this form they have direct access to the richer pelagic food resources, and may use this realm as a means of dispersal. The proportion of benthic species having meroplanktonic larvae tends to decrease with depth (because of greater distance to and from recruitment), and towards higher latitudes (because of shorter and more pulsed seasonal production cycles).

Lastly, the pelagic and benthic realms may temporarily exchange members. Several species may inhabit the water column – for example at night – but become bottom-referenced during the day, or may predominantly inhabit the water column, but use benthic resources, or vice versa. Mysids and euphausiids and several fish species do this. Such invertebrate species are often referred to as *epibenthic*, while such fish may be referred to as *demersal*.

Conclusions and management implications

Despite our still very incomplete understanding and knowledge of the flora and fauna of the oceans, we can nevertheless proceed to examine patterns in their distributions in terms of biogeography and species diversity at global, regional and local scales. These are necessary prerequisites to conservation planning. These subjects – biogeography, ecological geography and the distribution of species diversity – are considered in subsequent chapters.

The major implications of the differences between the pelagic and benthic realms are clear. There is no counterpart to the pelagic realm in the terrestrial world, and for conservation purposes – especially for mapping and analysis – they must be considered separately. Eventual decisions regarding selection of MPAs will of necessity involve evaluation of both the pelagic and benthic realms together.

Management of marine areas or marine species is a complex undertaking and may require a knowledge of both terrestrial and marine environments. For example, some marine species also spend time in terrestrial environments; furthermore, many species spend large parts of their life cycle outside protected areas. Consider for example the green turtle: within the Great Barrier Reef region of Australia, green turtles lay their eggs on the mainland or islands (i.e. outside the marine ecosystem). Once they hatch (and if they survive), they then move into nearshore marine areas feeding on seaweed and seagrasses. They then migrate thousands of kilometres across the open sea to other countries where they are frequently hunted and caught; those females that survive return to the same stretch of beach back in Australia every 2–8 years to nest. This means that effective conservation of this species alone needs to consider state jurisdictions (both terrestrial and marine) and federal jurisdictions, as well as a number of international jurisdictions (both in the open sea and within other countries). Even an MPA as large as the Great Barrier Reef Marine Park is still not large enough to encompass the full life cycle of the green turtle.

References

Azam, F. (1998) 'Microbial control of oceanic carbon flux: The plot thickens', *Science*, vol 280, no 5364, pp694–696

Ballantine, W. J. (1991) 'Marine reserves for New Zealand', *Leigh Laboratory Bulletin*, no 25, University of Auckland, New Zealand

Barnes, R. S. K. and Hughes, R. N. (1988) *An Introduction to Marine Ecology*, Blackwell Scientific Publications, Oxford

Bergeron, P. and Bourget, E. (1986) 'Shore topography and spatial partitioning of crevice refuges by sessile epibenthos in an ice disturbed environment', *Marine Ecology Progress Series*, vol 28, pp129–145

Bukata, R. P., Jerome, J. H., Kondratyev, K. Y. and Pozdnyakov, D.V. (1995) *Optical Properties and Remote Sensing of Inland and Coastal Waters*, CRC Press, Boca Raton, FL

Cole, J., Lovett G. and Findlay S. (1989) *Comparative Analyses of Ecosystems: Patterns Mechanisms and Theories*, Springer-Verlag, New York

Day, J. C. and Roff, J. C. (2000) *Planning for Representative Marine Protected Areas: A Framework for Canada's Oceans*, World Wildlife Fund Canada, Toronto

Denman, K. L. and Powell, T. M. (1984) 'Effects of physical processes on planktonic ecosystems in the coastal ocean', *Oceanography and Marine Biology Annual Review*, vol 22, pp125–168

Dunbar, M. J. (1968) *Ecological Development in Polar Regions: A Study in Evolution*, Prentice-Hall, Upper Saddle River, NJ

Eltringham, S. K. (1971) *Life in Mud and Sand*, English Universities Press, London

Emerson, C. W., Roff, J. C. and Wildish, D. J. (1986) 'Pelagic benthic coupling at the mouth of the Bay of Fundy, Atlantic Canada', *Ophelia*, vol 26, pp165–180

Gray, J. S. (1997) 'Marine biodiversity: Patterns, threats and conservation needs', *Biodiversity and Conservation*, vol 6, pp153–175

Gunther, G. (1957) 'Temperature', *Memorandum of the Geological Society of America*, no 67, vol 1, pp159–184

Harrison, W. G. (1980) 'Nutrient Regeneration and Primary Production in the Sea', in P. G. Falkowski (ed) *Primary Productivity in the Sea*, Plenum Press, New York

Hatcher, B. G., Johanned, R. E. and Robertson, A. I. (1989) 'Review of research relevant to the conservation of shallow trophic marine ecosystems', *Oceanography and Marine Biology Annual Review*, vol 27, pp337–414

Hedgpeth, J. W. (1957) 'Marine biogeography', in J. W. Hedgpeth (ed) *Treatise on Marine Ecology and Paleoecology. I. Ecology.*, Geological Society of America, no 67, New York

Leigh, E. G. Jr., Paine, R. T., Quinn, J. F. and Suchanek, T. H. (1987) 'Wave energy and intertidal productivity', *Proceedings of the National Academy of Sciences*, vol 84, pp1314–1318

Levinton, J. S. (1982) *Marine Ecology*, Prentice-Hall, Upper Saddle River, NJ

Longhurst, A. R. (1981) *Analysis of Marine Ecosystems*, Academic Press, New York

Longhurst, A. R. (1995) 'Seasonal cycles of pelagic production and consumption', *Progress in Oceanography*, vol 22, pp47–123

MacArthur, R. H. and Wilson, E. O. (1963) 'The theory of island biogeography', *Monographs in Population Biology*, Princeton University Press, Princeton, NJ

Mann, K. H. (2000) *Ecology of Coastal Waters*, Blackwell Science, Oxford

May, R. M. (1994) 'Biological diversity: Differences between land and sea', *Philosophical Transactions of the Royal Society of London B*, vol 343, pp105–111

McQueen, D. J., Johannes, M. R. S., Post, J. R., Stewart, T. J. and Lean, D. R. S. (1989) 'Bottom–up and top–down impacts on freshwater pelagic community structure', *Ecological Monographs*, vol 59, pp289–309

Morton, J., Roff, J. C. and Burton, B. M. (1991) *Shorelife between Fundy Tides,* Canadian Scholars Press, Toronto

National Research Council (1994) *Priorities for Coastal Ecosystem Science*, National Academy Press, Washington, DC

Parsons, T. R. (1991) 'Trophic Relationships in Marine Pelagic Ecosystems', in J. Mauchline and T. Nemoto (eds) *Marine Biology: Its Accomplishment and Future Prospect*, Elsevier, Amsterdam, The Netherlands

Ray, G. C. (1991) 'Coastal-zone biodiversity patterns', *BioScience*, vol 41, pp490–498

Ray, G. C. and Hayden, B. P. (1992) 'Coastal Zone Ecotones', in A. Hansen and F. di Castilla (eds) *Landscape Boundaries*, Springer Verlag, New York

Ray, G. C. and McCormick-Ray, M. G. (1992) *Marine and Estuarine Protected Areas: A Strategy for a National Representative System within Australian Coastal and Marine Environments*, Australian National Parks and Wildlife Service, Canberra, Australia

Roff, J. C. and Legendre, L. (1986) 'Physico-chemical and Biological Oceanography of Hudson Bay', in I. P. Martini (ed) *Canadian Inland Arctic Seas*, Elsevier Oceanography Series, vol 44, pp265–291

Ryder, R. A. (1982) 'The morphoedaphic index: Use, abuse, and fundamental concepts', *Transactions of the American Fisheries Society*, vol 111, pp154–164

Sanders, H. L. (1968) 'Marine benthic diversity: A comparative study', *American Naturalist*, vol 102, pp243–282

Schlesinger, W. H. (1997) *Biogeochemistry*, Academic Press, San Diego, CA

Sheldon, R. W., Prakash, A. and Sutcliffe, W. (1972) 'The size distribution of particles in the ocean', *Limnology and Oceanography*, vol 17, pp327–340

Steele, J. H. (1991a) 'Marine functional diversity', *BioScience*, vol 41, pp470–474

Steele, J. H. (1991b) 'Ecological Explanations in Marine and Terrestrial Systems', in T. Machline and T. Nemoto (eds) *Marine Biology: Its Accomplishment and Future Prospect*, Elsevier, Amsterdam, The Netherlands

Steele, J. H. (1974) *The Structure of Marine Ecosystems*, Harvard University Press, Cambridge, MA.

Steele, J. H. (1985) 'A comparison of terrestrial and marine ecological systems', *Nature*, vol 313, pp355–358

Steele, J. H. (1995) 'Can Ecological Concepts Span the Land and Ocean Domains?', in T. M. Powell and J. H. Steele (eds) *Ecological Time Series*, Chapman & Hall, New York

Steele, J. H., Carpenter, S. R., Cohen, J. E., Dayton, P. K. and Ricklefs, R. E. (1993) 'Comparing Terrestrial and Marine Ecological Systems', in S. A. Levin, T. M. Powell and J. H. Steele (eds) *Patch Dynamics*, Springer-Verlag, New York

Stephenson, T. A. and Stephenson, A. (1972) *Life between Tidemarks on Rocky Shores*', W. H. Freeman and Co., San Francisco, CA

Stergiou, K. I. and Browman, H. I. (2005) 'Bridging the gap between aquatic and terrestrial ecology', *Marine Ecology Progress Series*, vol 304, pp271–272

Suess, E. (1980) 'Particulate organic carbon flux in the oceans: Surface productivity and oxygen utilization', *Nature*, vol 288, pp260–263

Sverdrup, H. U., Johnson, M. W. and Fleming, R. H. (1942) *The Oceans: Their Physics, Chemistry and General Biology*, Prentice-Hall, Upper Saddle River, NJ

Thorne-Miller, B. and Catena, J. (1991) *The Living Ocean: Understanding and Protecting Marine Biodiversity*, Island Press, Washington, DC

Turner, J. T. and Roff, J. C. (1993) 'Trophic levels and trophospecies in marine plankton: Lessons from the microbial food web', *Marine Microbial Food Webs*, vol 7, no 2, pp225–248

Warwick, R. M. and Uncles, R. J. (1980) 'Distribution of benthic macrofauna associations in the Bristol Channel in relation to tidal stress', *Marine Ecology Progress Series*, vol 3, pp97–103

Wildish, D. J. (1977) 'Factors controlling marine and estuarine sublittoral macrofauna', *Helgoland Marine Research*, vol 30, pp445–454

4

Approaches to Marine Conservation

Traditional strategies and ecological frameworks

There is a coherent plan to the universe, though I do not know what it is a plan for.

Fred Hoyle (1915–2001)

'Species' and 'spaces'

Given our comprehensive definition of marine biodiversity and its hierarchical components, there are obviously many ways, ecologically, in which we could approach and plan for its conservation. Marine conservation initiatives have been increasing in recent years, from local to international, and many new ways of managing and conserving marine environments have been suggested. Many of those responsible for these efforts, however, may be unaware of the options available to them and may view the various approaches to marine conservation simply in terms of the individual disciplines that have emerged to address various threats to marine environments.

The well-known approaches to marine conservation and management include fisheries management, coastal zone management, ecosystem approaches to management and marine protected areas (MPAs). These approaches can be broadly divided into 'species' approaches, which focus on a single species and its requirements for management and conservation, and 'spaces' approaches, which focus on the management or conservation of ecological structures and processes within a defined geographic area. Over the past 20 years marine conservation has experienced a shift from an emphasis on the conservation of species to the conservation of spaces (National Research Council, 1995). Early marine conservation efforts targeted specific species (primarily harvested species such as fisheries and marine mammals) through quotas or prohibitions on their harvest or through limiting pollution or preserving habitat in order to ensure their survival. The failure of fisheries management and associated pollution legislation and conventions to protect marine resources has driven efforts to protect spaces in the form of marine reserves or MPAs. The current focus on the establishment of MPAs has, however, often distracted scientists and managers from examining other approaches to conservation of marine environments.

Considerable literature exists on how conservation objectives can be met using various approaches, but there have been recent arguments that neither the species nor spaces approach has been applied successfully towards the conservation of biodiversity in the marine environment (National Research Council, 1995; Allison et al, 1998; Simberloff, 1998). Each of the approaches is often applied in an ad hoc manner – not based on application of any ecological or environmental principles. Fisheries management decisions are often driven by political agendas and predetermined outcomes. Coastal zone management is an often-used buzzword for jurisdictions to claim that they are managing coastal environments in a more 'sustainable' manner. Establishment of protected areas has been driven 'more by opportunity than design, scenery rather than science' (Hackman, 1995).

There are, however, signs of improvement and of more systematic approaches to marine conservation. At the global and international levels, biogeographical and geophysical concepts have become embodied in studies of: pelagic biomes of the world (Longhurst, 1998); the 'marine ecoregions of the world' (MEoW, Spalding et al, 2007); and large marine ecosystems (Sherman et al, 1980). All these initiatives are promising for the assessment and management of global marine resources and biodiversity. At the national level, however, there have been few attempts to approach systematically the issues of marine conservation and environmental health of marine and coastal waters (Laffoley et al, 2000).

Overall, however, marine conservation is a largely undisciplined field that lags a generation behind terrestrial conservation both in terms of understanding the structures and functions of marine systems, and in understanding the options available to implement their conservation (National Research Council, 1995).

In order to move the discipline of marine conservation forward, it is vital to define the relationships among the various possible approaches to it. One reason that the field is presently in some disarray is precisely because different groups have adopted different conservation strategies (conservation of individual species, establishment of MPAs based on aesthetically pleasing coastal scenery and so on). We need to define the relationships among the different approaches to marine conservation, and ask how they reinforce and complement one another and show redundancies in efforts.

An immediate goal of marine conservation should be to 'codify' the discipline in ecological, environmental and socio-economic terms. By this we mean: to bring order to the components, contributing disciplines and activities, and to show how fundamental ecological knowledge and its application can lead to effective conservation. This requires frameworks for the conservation and management of marine and coastal habitats and their communities. Such frameworks need to be developed for global, international, regional and local levels. Some specific objectives would include the following:

- To define the scales and sources of environmental influences on marine communities, so as to reduce, eliminate or contain undesired effects.
- To determine the relationships between the structures and functions of marine entities for all levels of the ecological hierarchy from genes to ecosystems.
- To ensure protection of marine biodiversity in coherent sets and networks of protected areas through the management and conservation of representative and distinctive areas (see Chapters 5 and 7 respectively), selected according to defined ecological criteria.
- To define the concept of 'ecological integrity' of marine communities in both theoretical and practical management terms (e.g. Müller et al, 2000), so as to allow evaluation of human uses and permitted activities in and around designated MPAs, and maintain 'connectivity' among them.

In this chapter we review the existing approaches to marine conservation and briefly examine the concepts and application of fisheries management, coastal zone management, ecosystem approaches to management and MPAs. Also discussed are approaches that distinguish between representative and distinctive habitats and communities which are gaining recognition as methods to manage and conserve biodiversity in marine environments. We then examine the hierarchical ecological framework (see Chapter 1) of the various biotic and abiotic components required to conserve biodiversity, and address the objectives stated above. The framework can be used to identify and suggest how conservation strategies could be implemented in various types of marine environments depending on the objectives of conservation and data availability. Lastly, traditional methods of marine conservation are then discussed in terms of this framework and how they can be applied in a manner that reflects the structure and function of marine environments. We also discuss the concepts of representation and distinctiveness in terms of their relationship to the traditional approaches to marine conservation and how these concepts relate to the ecological framework discussed above. Each of the levels of the hierarchical framework is then considered in Chapters 5 onwards.

The 'traditional' strategies for marine conservation

The purpose of the following sections is to provide a brief overview of the 'traditional' approaches to marine conservation. The more common approaches, including fisheries management (a species approach), coastal zone management (a species and spaces approach), ecosystem approaches to management (a species and spaces approach) and MPAs (a spaces approach) will be described, as well as new ways to examine marine environments in terms of representative and distinctive habitats and their communities. In each section the major objectives of each approach will be discussed with the primary methods and techniques applied to address these objectives, and the success or failure of these approaches in addressing marine conservation issues. Readers interested in more detailed treatments of these subjects should refer to the reference section at the end of this chapter, and to following chapters.

Fisheries management overview

With the exception of intertidal and nearshore invertebrate species harvested by traditional cultures, there was little perceived need for fisheries management prior to the 1900s. Marine fisheries (including marine mammals) were thought to be inexhaustible and therefore there was little impetus to allocate resources towards their management. Most fisheries (including nearshore fisheries) were 'commons' resources, which were open to anyone with the equipment to exploit them. With the advent of internal combustion engines and increasing human populations, fisheries that were easily accessible from land were quickly depleted and nations realized that they needed to be managed and regulated. A suite of national and international legislation and conventions was developed to enforce fisheries regulations, which included International Convention for the Regulation of Whaling (1946) and the Law of the Sea (1958). A number of other conventions, such as MARPOL (1973), the London Dumping Convention (1972) and the Convention for the Prevention of Marine Pollution by Dumping of Wastes and Other Matter (1993) were designed to protect marine environments – and therefore fisheries – from land- and marine-based pollutants. The purpose of this section is not to chronicle the plight of specific fisheries or dwell on the failure of fisheries management, but to present some of the principles of fisheries management and how fisheries management has evolved as a discipline since its inception in the early 20th century.

While humans have utilized almost all marine taxa for food, fuel and medicine, focus has traditionally been on fishing efforts for finfish (e.g. class/order Agnatha, Chondrichthyes, Osteichthyes), crustaceans (e.g. class/order Amphipoda, Decapoda, Euphausiacea, Mysidacea), molluscs (e.g. class/order Bivalvia, Cephalopoda, Gastropoda) and marine mammals (e.g. class/order Odontoceti, Mysticeti, Pinnipedia, Sirenia). Fisheries for marine algae are important in many regions, and many species from other marine taxa (e.g. Echinodermata) are also intensely harvested.

Fisheries can be categorized in a number of ways and employ a variety of techniques (Box 4.1). However, four types of fisheries are generally recognized: 'food', 'industrial', 'shellfish' and 'recreational'.

Food fisheries are generally operated by and for local peoples and – while declining over the past 30 years – they continue to be important in both developed and developing nations. Food fisheries generally harvest nearshore finfish and shellfish, with some remaining subsistence whaling by indigenous peoples in Arctic areas. Industrial fisheries have been the fastest growing in recent decades, due to developments in boat and gear technology along with technological improvements

Box 4.1 Types of fisheries based on gear types

Trolling. Use of one or more baited fishing lines often pulled slowly through the water column behind a boat. Trolling targets pelagic species including Albacore tuna.
Drift gillnets. A gillnet is a panel of netting suspended vertically in the water by floats, with weights along the bottom. Fish are entangled in the net. Drift gillnet gear is anchored to a vessel, and drifts along with the current. It is usually used to target swordfish and common thresher shark.
Harpoon. A larger spear, which may contain explosives, that is thrown or launched at individual animals. The harpoon fishery mainly targets swordfish and marine mammals.
Pelagic longline. Pelagic longline gear consists of a main horizontal line that has shorter lines with baited hooks attached to it. The gear is used at various depths and at different times of day, depending on the species being targeted.
Cable longline. This gear consists of heavy wire for the mainline instead of monofilament. It has been used to target mako and blue sharks.
Coastal purse seine. A purse seine is an encircling net that is closed by means of a purse line threaded through rings on the bottom of the net. This gear is effective in catching schooled tunas. 'Coastal' purse seiners are smaller vessels that fish close to the shore. They mainly harvest coastal pelagic species (sardines, anchovies, mackerel), but they also fish for bluefin and other tunas when they are present.
Large purse seine. Large purse seine gear is used in major high seas fisheries.
Recreational fisheries. The recreational fisheries for resident and migratory species consist of private vessels and charter vessels using hook-and-line gear.
Bottom trawl. A fishing net is dragged across the seafloor (benthic trawling) or immediately above the seafloor (demersal trawling). Bottom trawling targets shrimp, squid and various types of groundfish.
Midwater trawl. A fishing net is suspended in the pelagic zone targeting tunas and small, schooling fish.
Dynamite. Explosives are used to concuss vertebrate fish where they are subsequently harvested. Most frequently applied in tropical coral reef environments.
Poison. Poison, most often sodium cyanide or bleach, are used to kill or stun fish. Most frequently applied in tropical coral reef environments.

(e.g. satellite images, global positioning systems), which increases the likelihood of finding fish. Industrial fisheries are the dominant harvesters on the high seas and are increasingly harvesting pelagic invertebrates (e.g. squid, krill). Shellfish fisheries target invertebrates that are primarily on or within the benthos. Finally, recreational fisheries operate throughout the world and there are limited statistics on the total amount of such landings.

To date, fisheries management has essentially been limited to assessment of populations or stocks of individual species. The basic goal of fishery management is to estimate the amount of fish that can be harvested (total allowable catch – TAC) so as to impose catch limits (Box 4.2), while maintaining viable fish populations. These estimates may be modified by political, economic and social considerations. Overly conservative management can result in under-utilized fisheries production due to under-harvesting, while too liberal or no management may result in over-harvesting and severely reduced populations.

Fisheries management is generally composed of four components: the fish(es) that are harvested; the non-harvested species; the environment or habitat where the fish live; and the human use and interactions with the fish and their habitat (Lackey and Nielsen, 1980). Each of these aspects of fisheries management has its own set of theories, concepts and methods to address these issues. While the ideal ecological unit of fisheries management would be a population, for practical reasons fisheries are managed on the basis of stocks. Sometimes more than one species is included in a stock because they are harvested together as though they were one species. In other

Box 4.2 Catch limits

Total allowable catch (TAC) is a management measure that limits the total output from a fishery by setting the maximum weight or number of fish that can be harvested over a period of time. TAC-based management requires that landings be monitored and that fishing operations stop when the TAC for the fishery is met. A TAC is based on stock assessments (discussed below) and other indicators of biological productivity, usually derived from both fishery-dependent (catch) and fishery-independent (biological survey) data. Data collected from fishermen, processors or dockside sampling can be combined with at-sea observations and independent fishery survey cruises to provide information about the total biomass, age distribution and number of fish harvested. Typically, the TAC is determined on an annual basis, but then partitioned across seasons. To the extent that a TAC is well estimated and enforced, it can control total fishing mortality on a stock (e.g. Pacific halibut).

Trip limits and bag limits are measures that pace landings by limiting the amount of harvest of a species in a given trip. Trip limits are applied in commercial fisheries when there is interest in spacing out the landings over time or a desire to specify maximum landings sizes, and they are usually accompanied by a limit on the frequency of landings.

Individual fishing quotas (IFQs) are a fishery management tool used in the Alaska halibut and sablefish, wreckfish and surf clam/ocean quahog fisheries in the United States, and other fisheries throughout the world, that allocates a certain portion of the TAC to individual vessels, fishermen or other eligible recipients, based on initial qualifying criteria.

Individual vessel quotas (IVQs) are used in a number of fisheries worldwide, including some Canadian and Norwegian fisheries. IVQs are similar to IFQs, excepted that they divide the TAC among vessels registered in a fishery, rather than among individuals.

cases, different species may be managed together for convenience. A more detailed description of fisheries management is provided in Chapter 13.

Since the early 20th century, fisheries have generally been managed using the concept of maximum sustained yield (MSY). MSY concepts were iteratively developed over time. Early efforts by Baranov (1918) used growth rates of fish combined with their natural mortality rate to identify the peak biomass of a cohort. Russell (1931) incorporated recruitment and mortality from fishing and natural causes. Graham (1939) developed models that suggested that low rates of harvesting yielded smaller catches of larger fish while high rates of harvesting yielded similar catches of smaller fish. Therefore, harvesting less than maximum was an inefficient use of fish and harvesting more than the maximum was an inefficient use of effort. MSY was adopted as the primary fisheries management tool from the 1940s until the 1970s. In addition to MSY, many fisheries managers believed the stocks could not become extinct as the cost of harvesting the few remaining fish would be such that fisheries would shift to other, more readily available species. Economic theory, however, did not reflect reality as fisheries – for whatever reasons – continued to pursue stocks to ecological and economic extinction (Taylor, 1951).

With the failure of MSY as a fisheries management tool, the concept of optimum sustained yield (OSY) was developed. OSY is essentially MSY modified to reflect relevant economic, ecological or social consideration and is more of a management construct than an empirically derived number. The vagueness of OSY, however, has led to many difficulties and debates into how the concept should be applied.

Fisheries managers have a number of tools and techniques to manage a stock. These techniques can be broadly separated into input controls, which are an indirect form of control because they do not limit the catch, and output controls, which directly limit the catch.

Input controls are designed to limit either the number of people fishing or the efficiency of fishing and are adopted when a fishery is first managed. Input controls include restrictions on gear types, number and size of vessels, the area fished, the time fished or the numbers of fishers. They can be applied to both commercial and sport fisheries, and may be applied to an entire fishery or to segments of it. Popular input controls include licences and licence endorsements, which may be used to certify fishermen or vessels, or used as a management measure to limit the number and types of vessels or fishermen that can participate in the fishery. The licensing system is designed to limit fishing capacity and effort, but their effect on either is indirect. However, if licences do not stipulate a maximum vessel size or other limits on fishing power or capacity, the capacity of the fleet can drift upwards as small vessels are replaced with larger ones. The problem arises because size is only one dimension of fishing power. Also, attempts to control size can lead to adaptations that are inefficient or unseaworthy.

Output controls directly limit catch and therefore a significant component of fishing mortality (which also includes mortality from by-catch, ghost fishing and habitat degradation due to fishing). Output controls can be used to set catch limits for an entire fleet or fishery, such as a TAC – see Box 4.2 for this and other types of limits. They can also be used to set catch limits for specific vessels (trip limits, IVQs), owners or operators (IFQs), so that the sum of the catch limits for individuals or vessels equals the TAC for the entire fishery. Output controls rely on the ability to monitor total catch. This can be achieved by either (1) measuring total landed catch with reliable landings records, port-sampling data and some estimates of discarded or unreported catch, or (2) measuring the actual total catch with at-sea observer coverage or verifiable logbook data.

One of the most important aspects of fisheries management is the ability to describe the condition or status of a stock, also known as stock assessment. Stock assessments provide information on the size, age structure and health of a stock and recommendations on the management of the stock. There are two components of stock assessment. The first is to study the biology and life history of the species and the second is to understand the effects and impacts of fishing on the stock. For many stocks throughout the globe there is little information available on the life history of the species, therefore fisheries management decisions are often made based on information obtained from the landing (catch) of fish from the stock. This is a classic dilemma in fisheries mangement; fisheries are often initiated without any knowledge of the life history of the targeted species or estimates of pre-exploitation population abundance and age structure. Without any baseline information on the natural state and variability of the stock prior to fishing, fisheries management decisions must either be based on patterns in the landings of fish or must attempt to build a 'picture' of the stock prior to exploitation from which to base management decisions.

In a perfect world with perfect information, full and accurate stock assessments would be made using the following information:

- The number of fishers and types of gear used in the fishery (e.g. longline, trawl, seine, etc.).
- Annual catches by gear type.
- Annual effort expended by gear type.
- Age structure of the fish caught by gear type.
- Ratio of males to females in the catch.
- How the fish are marketed (preferred size, etc.).
- Value of fish to the different groups of fishermen.
- Timing and location of best catches.

The biological information would include:

- The age structure of the stock.
- The age at first spawning.
- Fecundity (average number of eggs each age fish can produce).
- Ratio of males to females in the stock.
- Natural mortality (the rate at which fish die from natural causes).
- Fishing mortality (the rate at which fish die from being harvested).
- Growth rate of the fish.
- Spawning behaviour (time and place).
- Habitats of recently hatched fish (larvae), of juveniles and of adults.

- Migratory habits.
- Food habits for all ages of fish in the stock.

When the above information is collected by examining the landings of fishermen, it is called fishery-dependent data. When the information is collected by biologists through their own sampling programme, it is called fishery-independent data. Both methods contribute valuable information to the stock assessment.

In the real world, however, very little of the above information is available, particularly for newer fisheries that target poorly understood species (often high seas species). Depending on how much information is available on the stock, there are a number of ways to manage a fishery, which include estimating populations of stocks based on the ease of catching fish, protecting fish until they can spawn, ensuring a proper age structure in the stock and through the development of models of population survivabilty. The following paragraphs briefly discuss some of these methods.

The simplest stock assessments are made by calculating catch per unit effort (CPUE) using a combination of the history of landings (catches) for the stock and the level of effort expended in the harvest of the stock. CPUE, therefore, is essentially an indicator of stock abundance. At the start of new fisheries, CPUE tends to be high as catches are high while the effort required to catch the fish is low. As more fishers participate in the fishery, or technological improvements result in the ability to harvest more fish, CPUE generally either stabilizes in a well managed fishery, or continues to decline in a poorly managed fishery. CPUE is a relatively simple and intuitive way to manage a fishery. However, CPUE has some serious limitations and therefore is no longer solely used as a management tool. Problems with CPUE include insufficient information on landings (catch), insufficient information on fishing effort, and technological advances which make comparisons with past practices difficult. The primary problem is that by the time reliable estimates of CPUE have been generated, the fishery is often in decline.

A somewhat more reliable way to assess a stock is to determine at what age(s) fish spawn and then structure the fishery in such a way that fish are not harvested before they can recruit. The objective is to protect fish until they are old enough to spawn. The harvest of fish before they can recruit and replace themselves is termed 'recruitment overfishing' and has had serious consequences on many fisheries where the life history of the stock was unknown. Once spawning age(s) are determined, the fishery can be managed through specifications on gear type (e.g. mesh size in nets) or size limits. Protection against recruitment fishing does not protect a stock from overharvesting, as more fish can still be removed than can recruit, and recent evidence has shown that larger fish are disproportionately more fecund than smaller fish in relation to body sizes.

Another method to estimate stocks is to develop a mortality and spawning potential ratio (SPR). An SPR is the number of eggs that could be produced by an average recruit over its lifetime when the stock is fished, divided by the number of eggs that could be produced by an average recruit over its lifetime when the stock is unfished. In other words, SPR compares the spawning ability of a stock in the fished condition to the stock's spawning ability in the unfished condition. SPR can also be calculated using the biomass (weight) of the entire adult stock, the biomass of mature females in the stock, or the biomass of the eggs they produce. These measures are called spawning stock biomass (SSB), and when they are put on a per-recruit basis they are called spawning stock biomass per recruit (SSBR). SPRs are based on a knowledge of the age structure of a stock collected from either estimating ages from the lengths and weights of a fish or from the examination of the ear bones from fish, termed otoliths, which – like trees – develop annual growth rings that can be counted.

In summary, fisheries management is a complex field with considerable uncertainty. Probably the most difficult aspect of fisheries management is uncertainty in stock assessment, which leads to uncertainty in setting harvesting rates. While certain stocks are relatively easy to census and assess, other species have life histories that are either cryptic or highly mobile. Fish populations are affected by a number of abiotic variables, including changes in the oceanographic or physiographic aspects of their habitat, as well as biotic variables, including food resources, competition, predation, pathogens and parasitism. Fisheries management is also hindered by changes in technology, which affect CPUE and therefore make CPUE difficult to interpret. Traditional fisheries management makes no allowances for variability in these biotic or abiotic aspects of the marine environment. More recent attempts to manage fisheries in terms of multi-species fisheries and ecosystem-based management will be considered in Chapter 13.

Coastal zone management overview

Coastal zone management (CZM) emerged in response to the failure of multiple individual management and conservation strategies (including fisheries management and MPAs) to achieve their objectives. It was realized that an integrated approach was required to restore ecological processes in the coastal zone (see Box 4.3), to prevent further degradation of coastal resources, and address the varied and often interconnected issues that directly or indirectly affect the coastal zone. The need for new ways of managing coastal resources was underscored by estimates that by 2020 up to 75 per cent of the world's population may be living within 60km of the coastal zone. Coastal management also became necessary as nations began to exert jurisdiction over their coastal resources. This trend began with the Truman Proclamation in 1945, which declared sovereign rights over the US continental shelf. A number of countries followed suit and in 1958 the first United Nations Convention on the Law of the Sea (UNCLOS) produced treaties on the territorial sea and contiguous zone, the continental shelf, the high seas, and high seas fisheries.

While a recognition that an integrated approach to the management of coastal areas became apparent in debates surrounding the application of the Law of the Sea in the 1960s, coastal management as a discipline was formally launched by the introduction of the US Coastal Zone Management Act in 1972. The Act defined the objectives, principles, concepts and guidelines for coastal zone management in the US. The second UNCLOS meeting in 1982 created a formal legal structure for regulating the use of the sea and mandated that CZM be used as a tool for managing coastal areas. Under the 1982 UNCLOS agreement, the basic obligation of every nation is to 'protect and preserve the marine environment'. The agreement follows the basic principles of international environmental law and states that activities within a state's jurisdiction should be conducted in a manner that does not cause damage to other countries. UNCLOS also extends these principles to protect areas beyond national jurisdictions (i.e. the high seas and deep seabed).

UNCLOS clarified several important jurisdictional issues, which resulted in additional impetus for nations to implement CZM. UNCLOS extended the sovereignty of the territorial seas for coastal states to 12 nautical miles (nmi). Utilization of this area remains subject to the law and regulations of the coastal state, with few intrusions by international law except the right of innocent passage. UNCLOS also bestowed sovereignty and jurisdiction over a 200nmi distance for coastal nations (the exclusive economic zone, or EEZ). In return, the state must perform a number of obligations, including the conservation and management of living resources in the zone, which includes determining the allowable catch of the living resources in its EEZ and ensuring that the maintenance of the living resource within the EEZ is not endangered by over-exploitation. The UNCLOS agreement therefore supercedes the former doctrine of freedom of fishing, which now applies only to the high seas. Lastly, UNCLOS granted coastal states sovereignty over the natural resources of the continental shelf and gave coastal states exclusive rights to the sedentary species of the seabed.

With their newly affirmed sovereignty under UNCLOS, many nations developed CZM programmes in the 1980s. Also contributing to the rise in CZM programmes was the 1987 publication of *Our Common Future* by the World Commission on Environment and Development, which outlined a number of 'sustainable development principles', which included:

- Integration of conservation and development.
- Maintenance of ecological integrity.

Box 4.3 Definitions of the coastal zone

The coastal zone is defined as

> *...the strip of land and adjacent lake or ocean (water and submerged land) in which the land ecology and land use affect lake and ocean space ecology and vice versa. Functionally, it is a broad interface between land and water where production, consumption and exchange processes occur at high rates of intensity. Ecologically, it is an area of dynamic biochemical activity but with limited capacity for supporting various forms of human use. Geographically, the outermost boundary is defined as the extent to which land-based activities have measurable influence on the chemistry of the water or on the ecology or biota. The innermost boundary is one kilometer from the shoreline except at places where recognizable indicators for marine influences exist, like mangroves, nipa swamp, beach vegetation, sand dunes, salt beds, marshlands, bayous, recent marine deposits, beach and sand deposits, and deltaic deposits in which case the one-kilometer distance shall be reckoned from the edges of such features.* (National Environmental Protection Council, 1980)

> *Contains both land and ocean components; has land and sea boundaries that are determined by the degree of influence of the land on the ocean and the ocean on the land; and is not of uniform width, depth or height.* (Kay and Alder, 1992)

> *All areas to the landward side of the coastal waters in which there are physical features, ecological or natural processes or human activities that affect or potentially affect, the coast or coastal resources.* (Queensland Coastal Protection and Management Act)

- Economic efficiency.
- Satisfaction of basic human needs.
- Opportunities to fill other non-material human needs.
- Progress towards equity (present/future, cultural/economic) and social justice.
- Respect and support for cultural diversity.
- Social self-determination.

The final international agreement which set the stage for CZM was the United Nations Conference on Environment and Development (UNCED), also known as the Earth Summit or Rio Declaration. UNCED contained Agenda 21, which was a global commitment by coastal nations to apply integrated management and sustainable development principles to coastal and marine environments under their jurisdiction. Since UNCED, The Global Conference on Sustainable Development of Small Island Developing States (1994) and the International Coral Reef Initiative (1994) have also reinforced the need for coastal management (Cicin-Sain et al, 1995).

Broadly stated, CZM is an attempt to integrate a number of often disparate ecological, political and socio-economic methods to move away from a sectoral management model (e.g. managing fisheries and foreshore development separately) and towards an integrated model which reflects the complex and integrated ecological nature of coastal environments, while also reducing duplication and overlap in traditional management methods.

CZM is more akin to a philosophy on how coastal environments can be managed rather than a specific set of instructions that are required to be applied in all instances (see Box 4.4). The application of CZM depends on the type of problems to be addressed (e.g. pollution, overharvesting, etc.), the characteristics of the environment under consideration (e.g. tropical, temperate, etc.) and the social and political will and resources available to address these problems. However, there are some common goals of all CZM initiatives, which are as follows:

Box 4.4 Definitions of coastal zone management and approaches

Coastal resource management (CRM) is a participatory process of planning, implementing and monitoring sustainable uses of coastal resources through collective action and sound decision-making.
Integrated coastal management (ICM) comprises those activities that achieve sustainable use and management of economically and ecologically valuable resources in coastal areas, and that consider interaction among and within resource systems as well as interaction between humans and their environment (White and Lopez, 1991). ICM encompasses 'CRM' being a broader set of activities that emphasize integration within government, non-government and environmental realms.
Collaborative management or **co-management** is based on the participation of all individuals and groups that have a stake in the management of the resource. Important elements include (White et al, 1994):

- All stakeholders have a say in the management of a resource on which they depend.
- The sharing of the management responsibility varies according to conditions of authority between local community organizations and government. However, in virtually all cases, a level of government continues to assume responsibility for overall policy and coordination functions.
- Social, cultural and economic objectives are an integral part of the management framework. Particular attention is paid to the needs of those who depend on the resource and to equity and participation.

Community-based coastal resource management (CBCRM) implies that individuals, groups and organizations have a major role, responsibility and share in the resource management and decision-making process. Community-based management is consistent with the tenets of collaborative management since government is always part of the management process.
Coastal zone management (CZM) comprises those activities that achieve sustainable use and management of valuable resources and land uses in coastal areas as defined through CRM or ICM but with an emphasis on a specified coastal geographical area or zone.
Management of natural resources is the set of rules, labour, finance, and technologies that determines the location, extent and conditions of human utilization of these resources; management, consequently, determines the rate of resource depletion and renewal (Renard et al, 1991).
Resource stakeholders include all those who define and apply some rule, labour, finance or technology and assume part of the management responsibility. The users of a resource, together with the concerned owners or agencies, are also its managers and stakeholders.
Source: After Department of Environment and Natural Resources et al (2001)

- Achieve sustainable development of coastal and marine areas.
- Reduce vulnerability of coastal areas and their inhabitants to natural hazards.
- Maintain essential ecological processes, life support systems and biological diversity in coastal and marine areas.

Regardless of the definition of CZM, there are a number of widely accepted principles of CZM which must be adhered to in order for it to be effective. CZM must:

- Be holistic, integrated and multi-sectoral in approach.
- Be consistent with, and integrated into, development plans.
- Be consistent with the national environmental and fisheries policies.
- Build on, and integrate into, existing institutionalized programmes.
- Be participatory.
- Build on local/community capacity for sustained implementation.
- Build self-reliant financing mechanisms for sustained implementation.

- Address quality of life issues of local communities as well as conservation issues.

Ecosystem approaches to management overview

The notion that the constituent parts of an ocean area can be simultaneously managed in a coordinated fashion is over a century old (Baird, 1873; Link, 2005). However, it wasn't until the concept of the 'ecosystem' was championed by Tansley (1935) that scientists and decision-makers began to determine whether and how ecosystems should be 'managed', how to define an ecosystem and what aspects of traditional management paradigms and policies needed to be changed to make them 'ecosystem-based' (Leopold, 1949; Larkin, 1996).

The recognition that a global and holistic approach to marine management that simultaneously considers multiple ecological and socio-economic objectives in the management of either a geographic area (e.g. protected area) or ecosystem (however defined) is the central premise in the development of a number of what are generally termed ecosystem approaches to management. There are over a dozen terms in use that fundamentally denote an approach to natural resource management that focuses on sustaining ecosystems to meet both ecological and human needs for the future. For the purpose of this book we will use the term ecosystem approaches to management (EAM), as it is applied widely throughout both the terrestrial and marine environments and is broader in scope than certain other, similar terms in use (e.g. ecosystem-based management) in the marine environment.

The term EAM, however, has no single, agreed-upon definition that is shared across jurisdictions; as such, confusion arises over what exactly EAM is, and what it entails for the marine environment. A myriad of definitions of the concept exist, including: ecosystem-based management; integrated management; integrated oceans management; coastal zone management; integrated coastal zone management; and sustainable development.

Our understanding and use of the term marine EAM broadly seeks to incorporate socio-economic, cultural and ecological inputs into management, conservation and decision-making processes. In this regard, for the purposes of this book we will utilize a 2005 definition of EAM developed by an American environmental non-government organization (COMPASS, 2005) which is perhaps the most complete EAM definition to date:

> *An integrated approach to management that considers the entire ecosystem, including humans with the goal to maintain an ecosystem in a healthy, productive, and resilient condition so that it can provide the services we want and need.*

Many EAM principles have existed in international soft law for decades; however, it was not until the 1995 Food and Agriculture Organization (FAO) Code of Conduct for Responsible Fisheries and the UN Fish Stocks Agreement that soft law or voluntary agreements began to outline the principles and operational procedures for what would become marine EAM. Specifically, the UN Fish Stocks Agreement states that there is a 'conscious need to avoid adverse impacts on the marine environment, preserve biodiversity, and maintain the integrity of marine ecosystems' (FAO, 2005).

Marine EAM concepts have a growing presence in the scientific, peer-reviewed literature and a significant presence in the grey literature of the applied conservation and management disciplines. Marine EAM academic- and application-related papers and reports are found in disciplines as broad as biology, economics, social science and indigenous cultures. Some papers are descriptive, some are model-based, but few follow the traditional empirical development and testing of hypotheses. Ecosystem approaches are also a global phenomenon, with over 100 countries either participating in EAM efforts or hosting EAM programmes within their borders. Simple searches of 'marine ecosystem management' on most internet search engines demonstrates the following: there is a significant cannon of marine EAM literature or similarly termed subjects; certain jurisdictions, most notably Australia, Canada and the United States, have EAM principles enshrined in certain marine statutes; marine EAM has a long, if somewhat obscured, history in international law; EAM has been used as a way to rebrand status quo, or slightly improved, fisheries management practices and is thus used incorrectly; there is no single, authoritative guide to EAM; and finally, marine EAM is being 'applied' throughout the world.

Ecosystem approaches will be explored further in Chapter 13 in terms of their definition, purpose and applications to conserving marine biodiversity while maintaining goods and services to humanity. In particular, the 'ecosystem approach to fisheries' (EAF) will be discussed as fisheries are the most challenging aspect of implementing 'ecosystem approaches' in marine environments.

Marine protected areas overview

The need to protect marine areas is a comparatively recent realization compared to the use of protected areas for terrestrial conservation and management. The general benefits of MPAs are outlined in Table 4.1.

Some countries had established MPAs prior to the first World Conference on National Parks in 1962, but this conference was probably the first time the need for protection of coastal and marine areas was internationally recognized. However, the need for a systematic and representative approach to establishing protected areas in marine environments was only clearly articulated at the International Conference on Marine Parks and Protected Areas, convened by the International Union for the Conservation of Nature (IUCN) in Tokyo in 1975 (Kenchington, 1996). MPAs are now internationally recognized as being an essential and fundamental component of marine conservation, and are referred to within many international agreements, including: UNCLOS, the Convention on Biological Diversity and the accompanying Jakarta Mandate; the Global Program of Action, MARPOL 73/78; and the more recent IMO Guidelines, and the World Heritage Convention.

Despite the fact that there are currently over 1800 MPAs worldwide in over 80 countries, there continues to be some debate surrounding the definition of an MPA and ergo the purpose(s) of their establishment. A widely quoted definition of an MPA used commonly throughout the world is that of the IUCN, which defines an MPA as:

> *Any area of intertidal or subtidal terrain, together with its overlying water and associated flora, fauna, historical and cultural features which has been reserved by law or other effective means to protect part or all of the enclosed environment.* (Kelleher and Kenchington, 1992)

Some users have found certain difficulties in applying this definition; for example, Nijkamp and Peet (1994) suggest that the definition refers primarily to terrain rather than to marine waters, which seems to emphasize the value of seabeds rather than the value of the overlying water or associated flora and fauna. They also find that the reference to fauna and flora is too restrictive as it might exclude marine features such as ocean vents and upwelling areas. Lastly, they note that an area that is reserved by law is not necessarily protected by law. Nijkamp and Peet (1994) therefore suggest a modified definition of an MPA to be:

> *Any area of sea or ocean - where appropriate in combination with contiguous intertidal areas - together with associated natural and cultural features in the water column within, or on top of the seabed, for which measures have been taken for the purpose of protecting part or all of the enclosed environment.*

Using either definition, there is a plethora of areas throughout the world which could be termed MPAs. Ballantine (1991) cites some 40 names which are in common use around the globe for areas that are currently set aside for the protection of parts of the sea (see Table 4.2). This multitude of 'labels', definitions and terminologies had the potential to confuse debate through misunderstanding and uncertainty. MPAs also range from small, highly protected 'no-take' reserves that sustain species and maintain natural resources, through to very large multiple-use areas in which the use and removal of resources is permitted, but controlled to ensure that conservation goals are achieved. There is a need therefore to ensure that the intention underlying the use of specific terminology is clarified.

Table 4.1 The benefits that can be reasonably expected with an appropriate system of 'no-take' marine reserves

1. Protection of ecosystem structure, function and integrity	
Protect physical habitat structure from:	Maintain high quality feeding areas for fish and wildlife
– fishing gear impacts	Restore population size and age structure
– other anthropogenic and incidental impacts	Leave less room for irresponsible development
Protect biodiversity at all levels	Promote ecosystem management
Restore community composition (species presence and abundance)	Encourage holistic approach to management
Protect genetics from direct and indirect fisheries selection	Allows the distinction of natural from anthropogenic changes
Protect ecological processes:	
– keystone species	
– cascading effects	
– threshold effects	
– second order effects	
– food web and trophic structure	
– system resilience to stress	
2. Increased knowledge and understanding of marine systems	
Provide long-term monitoring sites	Provide focus for study
Provide continuity of knowledge in undisturbed sites	Reduce risks to long-term experiments
Provide opportunity to restore or maintain natural behaviours	Provide experimental sites needing natural areas
Provide synergism of knowledge and cumulative understanding	Provide undisturbed natural sites for certain experiments
Provide natural reference areas for assessing anthropogenic impacts (including fisheries)	
3. Improved non-consumptive opportunities	
Enhance and diversify:	Provide wilderness opportunities
– economic opportunities	Enhance aesthetic experiences
– social activities	Enhance spiritual connection
Improve peace of mind	Promote ecotourism
Enhance non-consumptive recreation	Improve appreciation of conservation
Enhance educational opportunities	Stabilize economy
Increase sustainable employment opportunities	Create public awareness about environment
4. Potential fishery benefits of 'no-take' marine reserves	
Protection of endangered species	Protection and recovery of habitat
Increased species abundance	Increased size and age of species
Increased reproductive output	Enhanced recruitment
Maintenance or increase in genetic diversity	Enhanced fishery yields
Increased species diversity	Increased habitat complexity and quality
Increased community stability	Provision of baseline areas for research

Note: For some benefits, the degree of benefit will be variable for individual species and dependent on both their life histories and reserve design
Source: Adapted from a listing produced by the Center for Marine Conservation in 1995 (Sobel, 1996)

For the purposes of this book, we assume the following very basic and generic definition of an MPA:

A marine protected area is any marine area set aside under legislation to protect marine values.

There exists a wide variety of types of MPAs designed to accomplish various purposes (see Table 4.3). As outlined below and in subsequent chapters, MPAs may be created to satisfy a wide range of values and the goal or purpose of an MPA influences the 'type' of MPA and consequently its design and selection and ultimately its name.

Table 4.2 Various names in common use around the globe for areas that are currently set aside for the protection of parts of the sea, and the 'values' they may protect

NAMES		
Marine Park	Marine Protected Area	Maritime Park
Marine Reserve	National Seashore	Marine Sanctuary
Marine Nature Reserve	Marine Wildlife Reserve	Marine Life Refuge
Marine Habitat Reserve	Marine Wilderness Area	Marine Conservation Area

VALUES	
Conservation values	Recreational values
Commercial values	Scenery/aesthetics
Enhancement of special species	Unique features
Scientific importance	Cultural values
Historic features	Traditional uses

Source: 'Names' after Ballantine (1991)

These different 'types' of MPAs are not necessarily autonomous; for example, within the Great Barrier Reef Marine Park (which itself is regarded as a huge 'multiple-use' MPA), various zones equate to both the 'no-take' and WWF minimum standard [i.e. 'no-take' is basically the same as the Marine National Park 'B' Zone, whilst the WWF standard equates to the Marine National Park 'A' Zone and other more restrictive zones].

Representative and distinctive habitats overview

We can conceive of marine environments as fundamentally comprising a set of habitat types each containing a particular type of community. Habitats and their communities can be viewed as either representative or distinctive. The term 'community' is used here in the neutral sense of a species assemblage at any scale (see Box 1.4), approximating the term 'biome' or 'étage' at higher levels of the hierarchy. Here we use the term 'representative' to mean that a habitat type is typical of its surroundings, at some scale (Roff and Taylor, 2000; and see Chapter 5). This term is deliberately vague as to scale because of the heterogeneity of marine environments at all scales. Correspondingly, but in contrast, a 'distinctive' habitat is one that is atypical of its surroundings at some given scale; the significance of such distinctive habitats will become clear in Chapter 7. Note that in planning a system of MPAs, we clearly desire to capture not only the major representative types of communities in our protected spaces, but also to capture the distinctive (= locally unique) communities as well.

Conservation efforts have traditionally been directed to the rare, endangered, large animals and their associated habitats. This is especially true in the marine environment where 'focal species' (any species on which our attention is focused for some reason) such as 'charismatic megafauna' (whales, seals, migratory birds etc.) capture public attention and may be used as 'flagships' to enlist public support (Zacharias and Roff, 2001). Often these flagship species seasonally occupy distinctive habitats, whose communities are substantially different from those in surrounding representative habitats. However, a conservation strategy that addresses only individual species or only distinctive habitats leaves the majority of species and their habitats at risk, and does not address the need to protect the representative or 'ordinary' parts of the biosphere.

Accordingly, emphasis in conservation is now directed both to 'species' and distinctive habitats, and to 'spaces' and representative habitats. In terrestrial environments this change in emphasis has emerged as the 'enduring features' approach, and in the marine environment as the 'geophysical' approach based on enduring features and recurrent processes (see Roff and Taylor, 2000, for review). This change in strategy has probably occurred for two main reasons: first, no species can survive without its habitat; second, recognition that effort should be expended not only to protect the rare and endangered, but the ordinary, representative (and great majority of) species as well.

In the terrestrial environment, approaches based on species conservation and landscape ecology are now merging into practical analytical frameworks, but this is not yet the case in the marine environment. Here, no one has yet defined the relationship between the two strategies. Are the two approaches mutually compatible,

Table 4.3 Some examples of types and purposes of MPAs

In general, a **marine protected area (MPA)** is any marine area set aside under legislation to protect marine values.

A **representative MPA** is a special kind of MPA that has been designed to maximize representation based on an ecological or biogeographical framework.

A **'no-take' MPA (or harvest refugia)** is a special type of MPA (or a zone within a multiple use MPA) where:

- any removal of marine species and modification or extraction of marine resources is prohibited (by such means as fishing, trawling, dredging, mining, drilling, etc.);
- other human disturbance is restricted.

A **'multiple-use' MPA** is a particular type of MPA in which the use and removal of resources may be permitted, but where such use is controlled to ensure that long-term conservation goals are not compromised. Multiple-use MPAs generally have a spectrum of 'zones' within them, with some zones allowing greater use and removal of resources than other zones (e.g. no-take zones are commonly designated as one of the zones).

A **'biosphere reserve'** is a particular type of protected area, generally large, that has totally protected core area(s),surrounded by partially protected buffer areas, and an outer zone/ transition area. Both the buffer and the outer zone may allow some ecologically sustainable resource use (e.g. low key and traditional uses) and should also have facilities for research and monitoring, but activities within these areas should therefore not compromise the integrity of the core area. This concept was developed for terrestrial reserves but is considered by some to also have application for marine environments and hence MPAs (Price and Humphrey, 1993).

[1]Various authors (Batisse, 1990; Kenchington and Agardy, 1990; Brunckhorst et al, 1997) suggest that, if the UNESCO Biosphere Reserve Programme were to be redesigned and properly planned and implemented, it would provide a very useful tool for the wider aspects of marine conservation. See also Agardy (2010) on Ocean Zoning.

or are they mutually exclusive? One pragmatic approach to marine conservation would be to capture some defined proportion of each representative habitat type, plus all known distinctive habitats. This would require investigation of the roles of focal species (with which the public primarily identifies), investigation of processes in distinctive habitats, and the distribution of marine biodiversity.

An analysis of representative marine habitats is vital for many reasons (see Roff et al, 2003). If marine environments are to be systematically protected from the adverse effects of human activities, then identification of the types of marine habitats and the communities they contain, and delineation of their boundaries in a consistent classification is required. Human impacts on defined communities can then be assessed, candidate marine protected areas can be designated and the 'health' of these communities can be monitored.

Within any region there will also be distinctive habitats containing special features that (by definition) are not revealed by any analysis of representative habitats. We need to recognize and make an inventory of such habitats, define their boundaries and examine their geophysical and biological properties (both structures – observable entities, and processes – rates of functions) that make them distinct. Distinctive habitats often appear to be distinguished because special oceanographic processes are occurring within them on a regional or local scale, whereas representative habitats are not notable in this way.

A general ecological and environmental framework for conservation of marine biodiversity

A primary limitation of both the species and spaces approaches discussed in the previous section is an inadequate understanding of the structures and processes of marine environments. Compared to terrestrial systems, marine environments exhibit considerable variability and connectivity, where threats such as habitat loss, climate change, pollution and introduced species operate on what Ricklefs (1987) terms

Box 4.5 Objectives for the development of a hierarchical ecological framework for the conservation of marine biodiversity

Objectives	Description
Standardize terminology	Identify common terminology, clarify ambiguous language and attempt to equate terrestrial terminology with marine equivalents.
Clarify objectives	Identify the components (e.g. habitat versus community) of a marine environment requiring protection. Develop marine equivalents to the population, community, ecosystem and landscape levels of organization.
Relate scales	Relate spatial and temporal scales of community and ecosystem organization to those of conservation efforts.
Identify knowledge and data gaps	Determine information and knowledge required to properly implement conservation strategies.
Organize existing research	Identify whether research has been based on biotic or abiotic approaches, and what scale(s) the research has been applied at.
Identify conservation methodologies	Identify possible techniques or approaches to conserve marine environments.
Direct the collection of new data	Direct the collection of new information in a manner that facilitates conservation efforts.

'...processes beyond the normal scale of consideration' that cannot be mitigated using traditional marine conservation measures. Many conservation measures, including MPAs and fisheries management, have been implemented in marine environments without careful consideration of their overall objectives or temporal scales and variability over time and space (see Box 4.5).

Although there is a growing realization that efforts to conserve marine biodiversity are often inadequate, few studies have examined what measures are required to properly address the major conservation issues. The inability to progress beyond species or spaces approaches to marine conservation can in part be attributed to a lack of understanding of the mechanisms structuring marine biodiversity. An example of the fundamental knowledge gaps is demonstrated in the debate surrounding the degree to which biological or physical processes structure various types of marine communities (May, 1992; National Research Council, 1995). The conservation implications of this debate are clear: How can an environment be conserved when the components that support it are undefined?

This difficulty is not unique to marine environments; it has in part been addressed in terrestrial conservation through the development of ecological models of biodiversity. The purpose of these models is to provide an understanding of the various components required to conserve biodiversity, and to reconcile objectives and methods between those interested in the conservation of species and those advocating the conservation of spaces. These models are also used to outline the structure and function of various habitats and communities and the scales at which they operate. One of the better known models – developed in the Pacific Northwest United States – is the framework (introduced in Chapter 1) that conceptualizes biodiversity into compositional, structural and functional (process) attributes at the genetic, population, community/ecosystem and landscape levels (Franklin et al, 1981; Norse et al, 1986; OTA, 1987; Noss, 1990). Compositional components include the genetic composition of a population, the composition of a community or ecosystem and the spatial and temporal distribution of these communities throughout a landscape. Structural attributes are composed of biotic and abiotic features that contribute to biodiversity by providing various habitats and patchiness at different levels of organization. Process attributes include those required to sustain biodiversity, which include climatic, geologic, hydrologic, ecological and evolutionary processes (Huston, 1994; Noss, 1990; and see Table 1.2).

While designed for terrestrial environments, the framework by Noss (1990) and others can be applied to marine environments (Franklin et al, 1981; Norse et al, 1986; OTA, 1987). The separation of structural and process attributes and the division of biological organization into four hierarchical levels is consistent with the function of marine environments (Mann and Lazier, 1996; Nybakken, 1997). Where this work needs to be adapted for application to conserving marine biodiversity is in the translation of communities, ecosystems and landscape types into meaningful marine equivalents.

In the marine environment, the genetic and species/population levels of organization can be utilized in the same manner as in terrestrial environments, but communities and ecosystems have different connotations. In marine environments, communities are generally perceived as biological entities and ecosystems as physically and chemically defined systems (May, 1992). Given the importance of abiotic (ecosystem) components for marine biodiversity, and the fact that the term *ecosystem* is already used to denote abiotic processes, our framework largely separates the community (biotic) from the ecosystem (abiotic) levels.

The framework we have adapted for marine environments (Table 1.2) is fundamentally similar to the terrestrial framework of Noss (1990) with the following exceptions: subdivision of the community and ecosystem levels; changes to the structural and process attributes to reflect the biotic and abiotic nature of these two respective levels; and elimination of the landscape level. This proposed ecological classification should be considered in parallel with the spatial classification of Butler et al (2001) and Beaman (2005) as modified in Table 5.1, the significance of which will become clear in later chapters.

The landscape level has no generally acknowledged counterpart in the marine environment (and is subsumed into our ecosystem level), but the analogous term 'seascape' is gaining currency in use. By seascapes is meant: either a set of more than one kind of habitat type, or multiple units of the same kind of habitat types. Because seascapes comprise sets of habitats (whether homogeneoous or heterogeneous in composition) their structures and processes can be considered either at the community or ecosystem levels.

According to conventional definition, ecosystems themselves are non-hierarchical and should be delineated by clear geographic (but not necessarily biogeographic) boundaries (Tansley, 1935). In the marine environment, ecosystems themselves are poorly (if at all) defined, but ecosystem-level structures and processes are nevertheless important to explicitly recognize. In contrast, the habitat level is hierarchical (see Table 5.1), spanning from the community level to the ecosystem level; it is therefore easiest to speak of 'habitat types'. Relationships between habitat types and community types will be considered further in Chapter 6. Since terminology is not yet fully standardized in marine conservation, it is best to use terms and specify meanings carefully.

Although it may appear so at first sight, the marine ecological hierarchy is not also simply a spatial one. Spatial and temporal scales do not dictate the levels of organization to the same extent as in the terrestrial framework. The ecological hierarchy is in fact confusingly confounded across time and space scales in both pelagic and benthic realms (see Figures 2.5 and 2.6). From the genetic to the ecosystem levels we can be dealing with any spatial scale from global to local. For example, genetic studies are revising our concepts of global biogeography, large mammal species exploit entire oceans, and ecosystem-level structures and processes can determine local species diversity. The biological processes identified under the community approach, however, generally operate at smaller spatial and temporal scales than do ecosystem processes. There are some exceptions – including succession (temporal) and migration (spatial).

Separating attributes at the population, community and ecosystem levels in the marine environment is important because there are conservation implications at each level of the hierarchy. Attributes at the population and ecosystem levels tend to be easier to observe (e.g. migration or water movement) than community attributes (e.g. competition). Of all the levels, ecosystem attributes – such as depth – tend to be the easiest to observe. In addition to the relative ease of monitoring population and ecosystem attributes, abiotic attributes (e.g. water movements and temperature) tend to be easier to observe and predict than biotic attributes (e.g. disease).

Ecosystem processes such as productivity, however, involve both biotic and abiotic components, and therefore have different implications for conservation than strictly abiotic attributes. Water motion, for example, is an ecosystem process driven by forces that (with the possible exception of global climate change) are generally immutable from human activities. Other ecosystem processes such as biogeochemical cycles, events and productivity may be more sensitive to human activities than many community processes such as predation and competition.

A representative sample of marine research has been evaluated against the framework to identify the compositional level at which these studies were undertaken (Table 4.4). Those research efforts that utilized a combination of community and ecosystem levels have been identified as a fourth level of organization. All four approaches have been implemented during the past 50 years at various scales and using different terminology, suggesting that there are no standard conservation approaches for different types of environments.

Levels of the framework for marine conservation

The hierarchical framework of Zacharias and Roff (2000) can be used to show how different conservation approaches can be applied to the marine environment under the genetic, population, community and ecosystem levels of spatial, temporal, taxonomic and functional organization. Research and planning at each of these levels can be, and has been, applied to enact conservation measures. However, a much greater degree of integration is required. An overall approach to the conservation of the components of marine biodiversity can be appreciated from a comparison of Tables 4.5 and 4.6. This is simply a preliminary illustration of how the various approaches to marine conservation can define their individual roles and integrate efforts and responsibilities. Briefly here, and in more detail in following chapters, we provide examples of how this framework can be applied.

The global level

At the global level, studies of biogeography of the earth's flora and fauna have a long history and many contributors, and have attempted to define and delineate the 'natural regions' of the world and their biota. In the oceans, major advances were made in the early part of the 20th century with expeditions such as those of the *Meteor* and *Discovery* (see e.g. Hedgpeth, 1957; Sverdrup et al, 1942 for reviews). Older studies in the oceans (see e.g. Pierrot-Bults et al, 1986) have generally concentrated on individual taxonomic groups of organisms, often with varying results. It is not our intention to review all this material here. More recently three systems of global biogeographic classifications have been proposed, each with its strengths and weaknesses.

Table 4.4 A representative sample of marine conservation literature assessed as to whether it falls under population, community or ecosystem approaches as we defined them

Approach	Studies	Environment /species studied	Key terminology	Scale
Population (biotic)	Paine (1969)	rocky intertidal shorelines (*Pisaster sp.*)	keystone species	metres
	Estes and Palmisano (1974)	rocky subtidal pelagic (*Enhydrus lutris*)	umbrella species	metres
Community (biotic)	Augier (1982)	benthic communities in the Mediterranean	biocoenoses	continental
	Thorsen (1957)	global inventory of benthic communities	isoparallel communities	oceanic
	Peres and Picard (1964)	Mediterranean	facies	100s km
	Ekman (1953)	global distribution of fauna	faunistic regions	oceanic
	Glemarec (1973)	European north Atlantic	etages	continental
Community/ecosystem (biotic and abiotic)	Connor (1997)	intertidal environments	biotopes	100s km
	Dauvin et al (1994)	French coastlines	biocoenoses	100s km
	Pielou (1979)	zoogeographic communities	biotic provinces	oceanic
	Menge (1992)	rocky intertidal shorelines	bottom-up influences	metres
	Cowardin et al (1979) and Dethier (1992)	intertidal and shallow subtidal environments	habitat types	metres
	Metaxas and Scheibling (1996)	rocky shore tide pools		metres
	Briggs (1974)	global faunal assemblages	realms	oceanic
Ecosystem (abiotic)				
	Hayden et al (1984)	hierarchical abiotic classification	provinces	oceanic
	Dolan et al (1972)	coastal classification		continental
	Hesse et al (1951)	delineation of water masses	domains	oceanic
	Sherman et al (1980)	global coastal	large marine ecosystems	oceanic
	Caddy and Bakun (1994)	regional studies of nutrient enrichment	marine catchment basins	oceanic

Longhurst's (1998) *Ecological Geography of the Sea* is an analysis of marine biomes that have the same or similar fundamental ecosystem level processes. It is based on interpretation of seasonal changes of temperature and colour (a surrogate for chlorophyll-*a* concentration) from remote sensing scanners. The analysis applies only to the regional oceanography of the epipelagic realm and does not consider the benthos at all. Although the analysis is valuable from the viewpoint of marine trophodynamics, in view of the near absence of taxonomic coverage, it has limited direct applicability to marine conservation. However, it has relevance in terms of comparisons of management strategies for MPAs within comparable biomes.

We may argue whether the concept of ecosystems is a valuable one for the marine environment, but the concept of large marine ecosystems (LMEs) by Sherman et al (1980) has clearly taken hold from a fisheries management perspective. Interest in LMEs is predominantly from a fisheries perspective (including fish species composition, catches and fisheries conservation),

Table 4.5 A reprise of Table 1.2. An expansion of the marine biodiversity framework from Zacharias and Roff (2000), showing attributes of structures (statics) and processes (function or dynamics) arranged under the population, community and ecosystem levels of organization

Genetic		Species/Population		Community		Ecosystem	
Structures	Processes	Structures	Processes	Structures	Processes	Structures	Processes
1. Genetic structure	1. Mutation	1. Population structure	1. Migrations	1. Community structure	1. Succession	1. Water masses	1. Ocean currents
2. Genotypes	2. Genotype differentiation	2. Population abundance	2. Dispersion	2. Species diversity	2. Predation	2. Temperature	2. Tidal currents
3. Fitness	3. Genetic drift	3. Distribution	3. Retention	3. Species richness	3. Competition	3. Salinity	3. Physical disturbances
4. Genetic diversity	4. Gene flow	4. Focal species	4. Migration/ counter drift	4. Species evenness	4. Parasitism	4. Water properties	4. Gyres
5. Stock discrimination	5. Natural selection	5. Keystone species	5. Growth/ production	5. Species abundance	5. Mutualism	5. Boundaries	5. Retention mechanisms
	6. Inbreeding	6A. Indicator species – condition	6. Reproduction	6A. Representative communities	6. Disease	6. Depth/pressure	6. Pelagic–benthic coupling
	7. Non-random mating/ sexual selection	6B. Indicator species – composition	7. Recruitment	6B. Distinctive communities	7. Production	7. Light intensity	7. Entrainment
	8. Directional selection	7. Umbrella species		7. Biome types	8. Decomposition	8. Stratification	8. Biogeochemical cycles (inc. nutrient dynamics/ energy flow)
	9. Stabilizing selection	8. Charismatic species		8. Biocoenoses		9. Bottom topography	9. Seasonal cycles (physical and biological)
	10. Disruptive selection	9. Vulnerable species		9. Species–area relationships		10. Substrate type	10. Productivity
	11. Micro-evolution	10. Economic species		10. Transition areas		11. Geophysical anomalies (inc. frontal systems)	11. Hydrosphere–atmosphere equilibria
	12. Genetic erosion	11. Phenotypes		11. Functional groups		12. Wave exposure	12. Hydrosphere–lithosphere equilibria
	13. Speciation	12. Population fragmentation		12. Heterogeneity		13. Patchiness	13. Eddy diffusion/ turbulence/internal waves
	14. Macro-evolution	13. Meta-populations		13. Endemism		14. Nutrients	14. Mixing/stabilization
				14. Alternate stable states		15. Dissolved gasses	15. Upwelling/ convergences
				15. Symbioses		16. Anoxic regions	16. Divergences
				16. Biomass			17. Ecological integrity
							18. Erosion/ sedimentation
							19. Desiccation

Table 4.6 Showing how the structure and process elements of the ecological hierarchy are primarily captured, or should be considered, in various conservation approaches

Ecological level →	Genetic level		Species/Population level		Community level		Ecosystem level	
Conservation approach ↓	Structures	Processes	Structures	Processes	Structures	Processes	Structures	Processes
Distinctive habitats	1 2 3 4	All inferred from genetic structures	4 5 6A 6B 7 8 9 10 11	1 3 4 5 6 7	1 2 3 5 6B Distinctive communities (by inference from anomalies) 12 13 16	Mostly assumed – not measured at planning scales	5 11 13 16	1 2 3 4 5 6 7 8 9 10 15 16 18
Representative habitats	1 2 3 4 5	All inferred from genetic structures	1 2 3 5 6A 6B 9 10	5 6 7	1 2 3 4 5 6A Representative communities (by inference from habitat–community relationships) 7 8 9 10 11 13 14 (assumed) 15 (assumed) 16	Mostly assumed – not measured at planning scales 7 8	1 2 3 4 5 6 7 8 9 10 12 13 14 15 17	10 11 12 13 14 18 19
Fisheries management	1 2 3 4 5 For fish community only	All inferred from genetic structures	1 2 3 6B 10 For fish community only	1 2 3 4 5 6 7 For fish community only	6A 6B 9 16 For fish community only	Mostly assumed – not measured at planning scales. For fish community only	NOT APPLICABLE	NOT APPLICABLE
Coastal zone management	NOT APPLICABLE	NOT APPLICABLE	NOT APPLICABLE	NOT APPLICABLE	NOT APPLICABLE	NOT APPLICABLE	2 3 4 5 6 9 10 12 14	2 3 4 5 7 8 9 10 18

Notes: 1 Biodiversity features may be captured in more than one way

2 Numbers in the columns in this table refer to numbered biodiversity components in the corresponding column of structures and processes in Table 4.5

but the various publications on LMEs also contain a wealth of regional oceanographic information. However, LMEs are too large and too loosely defined biogeographically to aid in overall conservation of marine biodiversity, nor is this their primary purpose.

In contrast to the above two global classifications – neither of which is hierarchical – stands the recent 'marine ecoregions of the world' (MEoW) of Spalding et al (2007), which probably represents the best we have at present as a foundation for planning for conservation of marine biodiversity, although it does not yet extend to deep or high seas (see Chapter 12). The MEoW system uses a combination of geophysical and biological data (probably unavoidable at the global level because of the uneven availability of data) to define 'natural regions' of biota at the ecoprovince and ecoregion levels. It can in fact be used as the starting point for national and regional conservation planning, where most initiatives are undertaken. Its value is that it is inclusive and hierarchical, though it will undoubtedly undergo revisions in coming years as more data becomes available.

The ecosystem level

Within ecoprovinces or ecoregions defined at the global spatial level, the highest level in the ecological framework is represented by ecosystem structures and processes. The advantages of this approach are that ecosystem structures and processes are relatively easy to observe and monitor, often indicate the presence of large areas of productivity or diversity (e.g. upwellings or anomalies), and can often can be correlated with biological communities. This approach has been advocated by Hayden et al (1984) to classify coastal environments and Caddy and Bakun (1994) among others to classify marine catchment basins. Chapters 5 and 7 present detailed accounts on the incorporation of ecosystem information into marine conservation strategies for representative and distinctive habitats.

The community level

Although conservation at the community level may require more knowledge of structure and process than at the population level, community-level approaches are considered more robust because the conservation of an environment does not rest on a few key species (Simberloff, 1998). Community-level approaches to conservation have been applied in all marine environments, but predominantly to benthic environments, where communities are either sessile or slow moving and are easier to census than in pelagic environments (Thorsen, 1957; Augier, 1982).

Several studies have combined ecosystem structure and process with community and/or population approaches to build a biophysical framework of a community and its abiotic environment. Ecosystem structure and process are integral parts of this type of analysis to determine which biotic or abiotic variables or combinations of variables can be used as foundations for planning to conserve biodiversity. Included in this community/ecosystem approach are studies that relate community composition to what is often termed habitat. For the purposes of this framework, habitat is the combination of ecosystem structures and processes listed in Table 4.5 that supports a recognizable community. This combined approach has been used extensively in the intertidal, where community and ecosystem data have been used to describe 'biotopes' or 'habitat types' (Menge, 1992; Connor, 1997). Chapter 6 presents a review of the relationships between community types and habitat types for marine conservation strategies.

The species/population level

Patterns in the distributions of species have been the historical foundation of biodiversity science; the species level is considered in more detail in Chapters 8 and 9. Population-level techniques have been used extensively in marine environments primarily under the guise of single-species fisheries management or conservation of marine mammals. Considerable effort has been expended to understand population processes (migration, dispersion, retention, growth/production, reproduction and recruitment), especially for commerically and ecologically important species as well as species that are endangered or threatened. Structural applications of the population level have recently been concerned with focal species – indicators, keystones, umbrellas, flagships – and their potential applications to marine conservation (e.g. Paine, 1966; Estes and Palmisano, 1974). Their potential contributions to marine conservation will be considered in Chapter 9.

The genetic level

The final level of the framework addresses the genetic level, where processes run all the way from genetic drift to evolution. A detailed discussion of population and conservation genetics lies outside the scope of this book but it is probably fair to say that the discipline of genetics has not yet contributed to anything like its full potential for marine conservation. Its implications impinge upon and inform all the other levels of the ecological hierarchy, and in spatial terms run from the global to the local level.

Genetic applications include an array of problems from reassessments of global biogeography to assessment of local patterns of the connectivity (or conversely, the isolation) of populations. This knowledge is then used for a variety of purposes, which include the siting and design of MPAs and their boundaries, the allocation of fisheries resources such that certain genetically distinct populations are maintained (Utter and Ryman, 1993), and the identification of 'sources' of recruits such that these areas can be managed and/or protected (Cowen et al, 2000).

We typically do not study genetic processes as they occur, but rather infer them (at various time scales) from genetic structures. Important genetic considerations at the structural level include genetic structure, genotypes, fitness, genetic diversity and stock discrimination. Chapter 10 presents an account of the incorporation of genetic information into marine conservation strategies.

Conclusions and management implications

Each of the traditional approaches to marine conservation has its virtues. There is little point in debating the relative merits of each one separately, because each is necessary – but not by itself sufficient – to achieve the overall objective of conservation of marine biodiversity. For example, Longhurst's (1998) *Ecological Geography of the Sea* is valuable because it recognizes natural boundaries of the pelagic realm. The biomes so defined will then have similar ecosystem-level structures and processes and can be treated in analogous ways; however, their species complements are (by definition) different. Biomes clearly have relevance in terms of comparisons of management strategies for MPAs, but are not appropriate as a foundation for conservation of biodiversity. The concept of LMEs (Sherman et al, 1980) is also clearly consistent with growing interest in 'ecosystem-based' management, although the size of the units means an international perspective in most cases. But it is the MEoW system (Spalding et al, 2007) that provides an explicit global biogeographic basis for conservation of marine biodiversity as a whole within each of the defined ecoregions.

Fisheries conservation efforts alone, directed at individual species management, leaves the entire issues of marine environmental quality and biodiversity conservation unaddressed. Fishing practices themselves may be major contributors to environmental degradation (see Chapter 13). Coastal zone management is traditionally undertaken in a local context for purposes of engineering works, erosion control, pollution abatement and so forth. Although such efforts may be locally integrated with fisheries regulation, they are generally not coordinated with regional efforts for conservation of biodiversity. Recognition and mapping of representative and distinctive areas can readily become the foundation for integrated marine planning, involving both protection and regulation of fisheries and coastal zone management.

What is clearly required is a synthesis and integration of *all* these approaches, showing how they can be combined to reinforce and complement one another. In this respect, a further application of the ecological framework is immediately apparent. Tables 4.5 and 4.6 show how the structure and process components of biodiversity are 'captured' – or should be considered, or are NOT considered – at all levels of the ecological hierarchy, in the various conservation approaches discussed in this chapter. These tables therefore provide a checklist (here in very preliminary form) against which the various approaches and initiatives for marine conservation can be judged in terms of their contributions to, or responsibilities for, the conservation of the components of biodiversity. This framework can therefore be used to assess the various conservation options for marine environments as well as to judge how a conservation programme might be structured given the type of marine environment under study. The framework can also be used to provide options on what components of marine biodiversity can be observed, measured or applied to the modelling, inventory or monitoring of marine environments. It is also a basis for real 'ecosystem-based' management.

The following chapters discuss the contributions of various approaches and the application of the framework to the ecosystem, community, population and genetic levels in more detail. They also consider the fundamental principles necessary to develop integrated frameworks for the conservation of as many components of marine biodiversity as possible, within comprehensive regional and national networks of MPAs, namely:

- Definition of biogeographic regions (global, regional, provincial and ecoregional)
- Definition of representative areas (as habitat types or seascapes)
- Definition of distinctive areas and hotspots of species diversity
- Separate consideration of coastal zones
- Separate consideration of deep seas and high seas
- Integration of fisheries and biodiversity conservation
- Criteria for establishing size of MPAs
- Integration of representative and distinctive areas into coherent sets of candidate MPAs
- Defining the connectivity among candidate MPAs
- Establishment of monitoring and environmental assessment programmes
- Assessment, evaluation and priorities
- Definition of remaining problems.

References

Agardy, T. (2010) *Ocean Zoning:* Making Marine Management More Effective, Earthscan, London

Allison, G. W., Lubchenco, J. and Carr, M. H. (1998) 'Marine reserves are necessary but not sufficient for marine conservation', *Ecological Applications*, vol 8, no 1, pp79–92

Augier, H. (1982) 'Inventory and classification of marine benthic biocoenoses of the Mediterranean', *Nature and Environment Series*, vol 25, Council of Europe, Strasbourg

Baird, S. F. (1873) 'Report on the condition of the sea fisheries of the south coast of New England in 1871 and 1872', *Report of the United States Fish Commission*, vol 1, GPO, Washington, DC

Ballantine, W. J. (1991) 'Marine reserves for New Zealand', *Leigh Laboratory Bulletin*, no 25, University of Auckland, New Zealand

Baranov, F. I. (1918) 'On the question of the biological basis of fisheries', *Nauch Issled Ikhtiol Inst Izu*, vol 1, no 1, pp81–128

Batisse, M. (1990) 'Development and implementation of the biosphere reserve concept and its applicability to coastal regions', *Environmental Conservation*, vol 17, pp111–116

Beaman, R. J. (2005) *A GIS Study of Australia's Marine Benthic Habitats*, University of Tasmania, Australia

Briggs, J. C. (1974) *Marine Zoogeography*, McGraw-Hill Books, New York

Brunckhorst, D. J., Bridgewater, P. and Parker, P. (1997) 'The UNESCO Biosphere Reserve Program Comes of Age: Learning by doing, landscape models for a sustainable conservation and resource', in P. Hale and D. Lamb (eds) *Conservation Outside Reserves*, University of Queensland Press, Brisbane

Butler, A., Harris, P. T., Lyne, V., Heap, A., Passlow, V. and Smith, R. (2001) 'An interim, draft of bioregionalisation for the continental slope and deeper waters of the south-east marine region of Australia', Report to the National Oceans Office, CSIRO Marine Research, Geoscience Australia, Hobart, Australia

Caddy, J. F. and Bakun, A. (1994) 'A tentative classification of coastal marine ecosystems based on dominant processes in nutrient supply', *Ocean and Coastal Management*, vol 23, pp201–211

Cicin-Sain, B., Knecht, R. W. and Fisk, G. W. (1995) 'Growth in capacity for integrated coastal management since UNCED: An international

perspective', *Ocean and Coastal Management*, vol 29, nos 1–3, pp93–123
Communication Partnership for Science and the Sea (COMPASS) (2005) 'EBM consensus statement', www.compassonline.org/sites/all/files/document_files/EBM_Consensus_Statement_v12.pdf, accessed December 23
Connor, D. W. (1997) 'Marine biotope classification for Britain and Ireland', *Joint Nature Conservation Committee Report Series*, Peterborough
Cowardin, L. M., Carter, V., Golet, F. C. and LaRoe, E. T. (1979) 'Classification of wetlands and deepwater habitats of the United States', FWS/OBS-79/31, Fish and Wildlife Service, Washington, DC
Cowen, R. K., Lwiza, K. M. M., Sponaugle, S., Paris, C. B. and Olsen, D. B. (2000) 'Connectivity of marine populations: Open or closed?', *Science*, vol 287, no 5454, pp857–859
Dauvin, J. C., Bellan, G., Bellan-Santini, D., Castric, A., Comolet-Tirman, J., Francour, P., Gentil, F., Girard, A., Gofas, S., Mahe, C., Noel, P. and de Reviers, B. (1994) 'Typologie des ZNIEFF-Mer: Liste des parameters et des biocoenoses des cotes francaises metroplitaines', *Collection Patrimoines Naturels*, vol 12, Secretariat Faune-Flore Museum National d'Histoire Naturelle, Paris
Department of Environment and Natural Resources, Bureau of Fisheries and Aquatic Resources of the Department of Agriculture, and Department of the Interior and Local Government (2001) *Philippine Coastal Management Guidebook No. 1: Coastal Management Orientation and Overview*, Department of Environment and Natural Resources, Cebu City, Philippines
Dethier, M. N. (1992) 'Classifying marine and estuarine natural communities: An alternative to the Cowardin system', *Natural Areas Journal*, vol 12, no 2, pp90–99
Dolan, R., Hayden, B. P., Hornberger, G., Zieman, J. and Vincent, M. (1972) 'Classification of the coastal environments of the world, Part I: The Americas', *Technical Report 1*, Office of Naval Research, University of Virginia, Charlottesville
Ekman, S. (1953) *Zoogeography of the Sea*, Sidgwick & Jackson, London
Estes, J. A., and Palmisano, J. F. (1974) 'Sea otters: Their role in structuring nearshore communities', *Science*, vol 185, pp1058–1060
FAO (2005) *Progress in the Implementation of the Code of Conduct for Responsible Fisheries and Related Plans of Action*, UN Food and Agriculture Organisation, Rome
Franklin, J. F., Cromack, K., Denison, W., McKee, A., Maser, C., Sedell, J., Swanson, F. and Juday, G. (1981) 'Ecological characteristics of old-growth Douglas fir forests', *General Technical Report PNW-118*, US Forest Service, Portland, OR
Glemarec, M. (1973) 'The benthic communities of the European north Atlantic continental shelf', *Oceanography and Marine Biology Annual Review*, vol 11, pp263–289
Graham, M. M. (1939) 'The sigmoid curve and the over-fishing problem', *Rapp. P.-v. Réun. Cons. perm. int. Explor. Mer*, vol 110, no 2, pp15–20
Hackman, A. (1995) 'Preface', in K. Kavanagh and T. Iacobelli (eds) *A Protected Areas Gap Analysis Methodology: Planning for the Conservation of Biodiversity*, World Wildlife Fund Canada, Toronto, Canada
Hayden, B. P., Ray, G. C. and Dolan, R. (1984) 'Classification of coastal and marine environments', *Environmental Conservation*, vol 11, no 3, pp199–207
Hedgpeth, J. W. (1957) 'Marine biogeography', in J. W. Hedgpeth (ed) *Treatise of Marine Ecology and Paleoecology. Vol. 1: Ecology*, Geological Society of America, Washington, DC
Hesse, R., Allee, W. C. and Schmidt, K. P. (1951) *Ecological Animal Geography*, John Wiley & Sons, New York
Huston, M. A. (1994) *Biological Diversity: The Coexistence of Species on Changing Landscapes*, Cambridge University Press, New York
Kay, R. and Alder, J. (1999) *Coastal Planning and Management*, EF&N Spoon, London
Kenchington, R. A. and Agardy, M. T. (1990) 'Applying the biosphere reserve concept in marine conservation', *Environmental Conservation*, vol 17, no 1, pp39–44
Kenchington, R. A. (1996) 'A global representative system of marine protected areas', in R. Thackway (ed) *Developing Australia's Representative System of Marine Protected Areas,* Department of the Environment, Sport and Territories, Canberra, Australia
Kelleher, G. and Kenchington, R. A. (1992) *Guidelines for Establishing Marine Protected Areas*, IUCN, Gland, Switzerland
Laffoley D., Connor, D. W., Tasker, M. L. and Bines T. (2000) 'Nationally important seascapes, habitats and species: A recommended approach to their identification, conservation and protection', *English Nature Research Report*, no 392
Lackey, R. T. and Nielsen, L. A. (1980) *Fisheries Management*, John Wiley and Sons, New York

Larkin, P. A. (1996) 'Concepts and issues in marine ecosystem management', *Reviews in Fish Biology and Fisheries*, vol 6, pp139–164

Leopold, A. (1949) *A Sand County Almanac and Sketches Here and There*, Oxford University Press, Oxford

Link, J. S. (2005) 'Translating ecosystem indicators into decision criteria', *Journal du Conseil*, vol 62, no 3, pp569–576

Longhurst, A. (1998) *Ecological Geography of the Sea*, Academic Press, San Diego, CA

Lourie, S. A. and Vincent, A. C. J. (2004) 'Using biogeography to help set priorities in marine conservation', *Conservation Biology*, vol 18, pp1004–1020

Mann, K. H., and Lazier, J. R. N. (1996) *Dynamics of Marine Ecosystems: Biological–Physical Interactions in the Oceans*, Blackwell Science, London

May, R. M. (1992) 'Biodiversity: Bottoms up for the oceans', *Nature*, vol 357, pp278–279

Menge, B. A. (1992) 'Community regulation: Under what conditions are bottom up factors important on rocky shores', *Ecology*, vol 73, pp755–765

Metaxas, A. and Scheibling, R. E. (1996) 'Top down and bottom up regulations of phytoplankton assemblages in tidepools', *Marine Ecology Progress Series*, vol 145, pp161–177

Müller, F., Hoffmann-Kroll, R. and Wiggering, H. (2000) 'Indicating ecosystem integrity: Theoretical concepts and environmental requirements', *Ecological Modelling*, vol 130, pp13–23

National Research Council (1995) *Understanding Marine Biodiversity*, National Academy Press, Washington, DC

Nijkamp, H. and Peet, G. (1994) *Marine Protected Areas in Europe*, Commission of European Communities, Amsterdam

Norse, E. A., Rosenbaum, K. L., Wilcove, D. S., Wilcox, B. A., Romme, W. H., Johnston, D. W. and Stout, M. L. (1986) *Conserving Biological Diversity in Our National Forests*, The Wilderness Society, Washington, DC

Noss, R. (1990) 'Indicators for monitoring biodiversity: A hierarchical approach', *Conservation Biology*, vol 4, pp355–364

Nybakken, J. (1997) *Marine Biology: An Ecological Approach*, Addison Wesley Longman, New York

Olson, D. M. and Dinerstein, E. (2002) 'The global 200: Priority ecoregions for conservation', *Annals of the Missouri Botanical Garden*, vol 89, pp199–224

One Ocean (2000) *Legal and Jurisdictional Guidebook for Coastal Resource Management in the Philippines*, www.oneocean.org/download/20000215/annex_a.pdf, accessed December 23 2010

OTA (Office of Technology Assessment) (1987) *Technologies to Maintain Biodiversity*, US Government Printing Office, Washington DC

Paine, R. T. (1966) 'Food web complexity and species diversity', *American Naturalist*, vol 100, pp65–75

Paine, R. T. (1969) 'A note on trophic complexity and community stability', *American Naturalist*, vol 103, pp91–93

Peres, J. M. and Picard, J. (1964) 'Nouveau manel de bionomie de la mer', *Mediterranee. Recl. Trav. Stn mar. Endoume, Bull*, vol 31, no 47, pp1–147

Pielou, E. C. (1979) *Biogeography*, Wiley-Interscience, New York

Pierrot-Bults, A. C., van der Spoel, S., Zahuranec, B. J. and Johnson, R. K. (1986) 'Pelagic biogeography', proceedings of international conference, 'UNESCO Technical Papers in Marine Science', Paris

Price, A. R. G. and Humphrey, S. L. (1993) *Application of the Biosphere Reserve Concept to Coastal Marine Areas*, IUCN, Gland, Switzerland

Renard, Y., Walters, B. B. and Smith, A. H. (1991) 'Community-based approaches to conservation and resource management in the Caribbean', International Congress for the Conservation of Caribbean Biodiversity, Santo Domingo, Dominican Republic, January 14–17

Ricklefs, R. E. (1987) 'Community diversity: Relative roles of local and regional processes', *Science*, vol 235, pp167–171

Roff, J. C. and Taylor, M. E. (2000) 'National frameworks for marine conservation: A hierarchical geophysical approach', *Aquatic Conservation: Marine and Freshwater Ecosystems*, vol 10, pp209–223

Roff, J.C. and Evans, S. (2002) 'Frameworks for marine conservation; Non-hierarchical approaches and distinctive habitats', *Aquatic Conservation: Marine and Freshwater Ecosystems*, vol 12, pp635–648

Roff, J. C., Taylor, M. E. and Laughren, J. (2003) 'Geophysical approaches to the classification, delineation and monitoring of marine habitats and their communities', *Aquatic Conservation, Marine and Freshwater Ecosystems*, vol 13, pp77–90

Russell, E. S. (1931) 'Some theoretical considerations on the "overfishing" problem', *Journal de Conseil International pour l'Explaration de la Mer*, vol 6, pp3–20

Sherman, K. L., Alexander, M. and Gold, B. D. (1980) *Large Marine Ecosystems: Patterns, Processes, and Yields*', American Association for the Advancement of Science, Washington DC

Sherman, K., Sissenwine, M., Christensen, V., Duda, A., Hempel, G., Ibe, C., Levin, S., Lluch-Belda, D., Matishov, V., McGlade, G., O'Toole, M., Seitzinger, S., Serra, R., Skjoldal, H. R., Tang, Q., Thulin, J., Vanderweerd, V. and Zwanenburg, K. (2005) 'A global movement toward an ecosystem approach to management of marine resources', *Marine Ecology Progress Series*, vol 300, pp275–279

Simberloff, D. (1998) 'Flagships, umbrellas, and keystones: Is single-species management passe in the landscape era?', *Biological Conservation*, vol 83, pp247–257

Sobel, J. (1996) 'Marine reserves: Necessary tools for biodiversity conservation?', *Global Biodiversity*, vol 6, no 1, Canadian Museum of Nature, Ottawa, Canada

Spalding, M. D., Fox, H. E., Allen, G. R., Davidson, N., Ferdana, Z. A., Finlayson, M., Halpern, B. S., Jorge, M. A., Lombana, A. and Lourie, S. A. (2007) 'Marine ecoregions of the world: A bioregionalization of coastal and shelf areas', *Bioscience*, vol 57, no 7, pp573–584

Sverdrup, H. U., Johnson, M. W. and Fleming, R. H. (1942) *The Oceans: Their Physics, Chemistry and General Biology*, Prentice-Hall, Upper Saddle River, NJ

Tansley, A. G. (1935) 'The use and abuse of vegetational terms and concepts', *Ecology*, vol 16, pp284–307

Taylor, H. F. (1951) *Survey of Marine Fisheries of North Carolina*, North Carolina University Press, Chapel Hill, NC

Thorsen, G. (1957) 'Bottom communities (sublittoral or shallow shelf)', *Memorandum of the Geographical Society of America*, vol 67, pp461–534

Urban, D. L., O'Neill, R.V. and Shugart, H. H. (1987) 'Landscape ecology', *BioScience*, vol 37, pp119–127

Utter, F. and Ryman, N. (1993) 'Genetic markers and mixed stock fisheries', *Fisheries*, vol 18, pp11–21

White, A. T. and Lopez, N. (1991) 'Coastal resources management planning and implementation for the Fishery Sector Program of the Philippines', *Proceedings of the 7th Symposium on Coastal and Ocean Management*, pp762–775

White, A. T. , Hale, L. Z., Renard, Y. and Cortesi, L. (1994) *Collaborative and Community Based Management of Coral Reefs*, Kumarian Press, Hastford, CT

Zacharias, M. A. and Roff, J. C. (2000) 'A hierarchical ecological approach to conserving marine biodiversity', *Conservation Biology*, vol 13, no 5, pp1327–1334

Zacharias, M. A. and Roff, J. C. (2001) 'Use of focal species in marine conservation and management: A review and critique', *Aquatic Conservation: Marine and Freshwater Ecosystems*, vol 11, pp59–76

5

Representative Areas: Global to Ecoregional

Marine conservation at the ecosystem/habitat level

No conservation without representation.

John Roff, 2010

Introduction: Hierarchical classification approach to conservation

The previous chapters outlined the major geophysical and biological features of marine environments and various approaches to marine conservation. Whatever approach is adopted, a prime requirement is to map and spatially define the natural biogeographical patterns of marine distributions. Although the highest species diversity per unit area is found in 'hotspots' (see Chapter 8), it is nevertheless the far more spatially extensive – global – set of 'ordinary' or representative habitats that house the greatest number of species on the planet. Adequate protection of these representative areas requires analysis of their distributions. The concept of representation can in fact be applied across the entire spatial hierarchy, from the global to the micro-community level (Table 5.1), following the kind of classification suggested by Butler et al (2001).

Mapping the marine environment – 'drawing lines on the oceans' – may at first encounter seem a self-defeating exercise in a fluid and motile medium. Yet environmental mapping is essential for many reasons (summarized in Table 5.2 and Roff et al, 2003) and especially for marine conservation. Very few marine species are unrestricted in their distributions, only some marine mammals, reptiles and fish. The vast majority of species are limited in their distributions by some combination of environmental parameters/variables such as temperature, salinity, depth and, in the case of benthos, substrate type (Table 3.5). This means that in order to preserve as many species as possible, the approach should be to preserve as many recognizably different habitat types as possible, from the global to the local level. A hierarchical classification of representative habitat types is clearly essential, but many approaches are possible (see Table 5.3). Here, emphasis is on physiographic and oceanographic features.

In this chapter we review how the oceans' environments, habitats and biotas can be mapped, from the global to the local scale (Table 5.1). Conservation initiatives may be undertaken at several levels of the biological hierarchy and should consider both structural and functional attributes of biodiversity and the environment (see Chapter 3). In this chapter, we shall focus on the identification of representative habitats, especially as they can be defined by oceanographic and physiographic characteristics, or what Zacharias and Roff (2000) have termed the 'ecosystem' (abiotic) level.

Table 5.1 A spatial hierarchical classification scheme for marine environments

Level	Unit name	Scale	Description
1	Realm	Ocean realm (1000s + of km)	Very large regions of coastal, benthic or pelagic ocean across which biotas are internally coherent at higher taxonomic levels, as a result of a shared and unique evolutionary history. Realms have high levels of endemism, including unique taxa at generic and family levels. Distinguishing factors include water temperature and broad-scale isolation. Note that 'realm' is used to describe BOTH the largest spatial marine areas, and to distinguish the pelagic and benthic environments.
2	Province	Province (1000s of km)	Large areas defined by the presence of distinct biotas that have at least some cohesion over evolutionary time frames. Provinces will hold some level of endemism, principally at the level of species. Although historical isolation will play a role, many of these distinct biotas have arisen as a result of distinctive abiotic features that circumscribe their boundaries. These may include: geomorphological features, such as continental blocks, basins and abyssal plains, isolated island and shelf systems, and semi-enclosed seas; hydrographic features, such as currents, upwellings and ice dynamics; or geochemical influences, such as broadest-scale elements of nutrient supply and salinity.
3	Ecoregion	Regional (100s to 1000s of km)	Areas of relatively homogeneous species composition, clearly distinct from adjacent systems. The species composition is likely to be determined by the predominance of a distinct suite of oceanographic or topographic features. The dominant biogeographic forcing agents defining the ecoregions vary from location to location but may include isolation, upwelling, nutrient inputs, freshwater influx, temperature regimes, ice regimes, exposure, sediments, currents, and bathymetric or coastal complexity.
4	Region	Regional (100s to 1000s of km)	Broad-scale gross geomorphology nested within provinces, e.g. continental shelf, slope, abyssal plain and offshore continental blocks.
5	Geomorphic units	Regional (10s to 100s of km)	Areas with similar seabed geomorphology and usually with distinct biotas, e.g. seamounts, canyons, rocky banks, inlets, submarine canyons and sand wave fields.
6	Primary habitats (seascapes)	Local (kms to 10s of km)	Nested within geomorphic units are soft, hard or mixed substrate-based units, together with their associated substrate-based units and their associated biological communities.
7	Secondary habitats (seascapes)	Site (10s of m to km)	Generalized types of biological and physical substrate within the soft, hard or mixed substrate, e.g. limestone, granite, shelly sand and muddy sand.
8	Biotope	Site (m to 10s of m)	The combination of a specific assemblage (community or biocoenosis) of species associated with a defined habitat type.
9	Biological facies	Site (cms to m)	Biological indicator or suite of species used as a surrogate for a community, e.g. species of seagrass, group of hard corals or sponges.
10	Microcommunities (Inquilines, Symbionts, Epibionts)	Site (mm to cms)	Assemblages of species that depend on member species of the biological facies, e.g. holdfast communities in giant kelp.

Note: This table is a rather uneasy mix of geographic, geophysical, ecological and biological components, but is nevertheless a very useful checklist to guide marine conservation efforts

Source: Modified from Butler et al (2001) as seen in Beaman (2005), and incorporating definitions of Spalding et al (2007)

Table 5.2 Potential applications of habitat classification schemes

Definition of habitat–community type associations
Assessment of habitat suitability for defined purposes, e.g. fisheries enhancement/aquaculture
Assessment of conflicts between actual or intended resource uses
Examining patterns of biodiversity distribution
Judging potential impact of invading species
Evaluation of candidate representative Marine Protected Areas (MPAs)
Assessment of the potential role of focal species (e.g. umbrella and flagship species) in marine conservation
Guide to environmental monitoring programmes
Guide to habitat management and management practices
Guide to selection of unaffected reference areas for environmental monitoring
Framework for assessment and evaluation of ecosystem level processes
Framework for assessment of global warming effects
Foundation for ecosystem-based management
Assessment of the rarity or prevalence of each habitat type
Assessment of the number and sizes of each representative habitat type
Assessment of the habitat heterogeneity of subregions

Source: From Roff et al (2003)

Chapter 11 applies these principles to the coastal zone, and Chapter 12 will expand on the concepts and principles for classifications of global biogeography, especially as applied to the open oceans (high seas) and deep seas.

Habitat classification has now become widely adopted, and the advantages of a hierarchical approach to landscape ecology have been considered by Urban et al (1987) among others. Hierarchical classification systems owe their power and attraction, among other things to the fact that missing components can be identified and the classification can be modified as required. For conservation purposes, the objective is to establish a system within which all natural communities and habitats can be recognized. The 'units' at the lower levels of such a hierarchy should correspond to the fundamental biotopes, communities or facies themselves. The hierarchy itself should discriminate first between the most broadly different spatial and ecological units while, at the lower levels of the hierarchy, habitat and community types are progressively more closely related. In this way our hierarchy is an analogue of the taxonomic 'Natural System of Classification'.

From global to ecoregions

A desire to describe and understand the geography of life on earth extends back to antiquity. For obvious reasons of access, such studies have been much slower to develop in the oceans than on land, and our understanding is still far from complete. In the marine environment, global mapping and classification systems have been limited in spatial resolution, qualitative in nature and often inconsistent in spatial coverage or methodological approach. Until recently the most comprehensive approach, based on stratified sampling theory and biogeography, has been that adopted by the International Union for the Conservation of Nature (IUCN) as a basis for the creation of a global representative system of marine protected areas (Kelleher and Kenchington, 1992).

The key concept underlying the term 'representative' is the intent to protect a full range of biodiversity components worldwide – genes, species and higher taxa, along with the communities, evolutionary patterns and ecological processes that sustain this diversity (see Spalding et al, 2007). Biogeographic classifications therefore provide a crucial foundation for the assessment of representation (Olson and Dinerstein, 2002; Lourie and Vincent, 2004). Because previous attempts to use biogeographic regionalization for planning of global marine conservation have been qualitative, and because there has been widespread concern about the lack of an adequate global classification, Spalding et al (2007) undertook the task of developing a new comprehensive system of bioregionalization for marine coastal and shelf areas. Note that this system does not extend to deep or high seas – which will be considered in Chapter 12. Note also that it does not include distinctive areas which will be examined in Chapter 7.

A variety of criteria have been used for defining global biogeographic regions, including: degree of endemism (e.g. Briggs, 1995); temperature or productivity boundaries (e.g. Longhurst, 1998); bathymetry, hydrography, oceanographic features and processes, and the major fisheries

Table 5.3 A summary of some possible approaches to biogeography and mapping for the marine environment, its habitats and communities (a classification of classifications)

APPROACH	BASIS	SUB-BASIS	FACTORS
TAXONOMIC	Genetic differences		ESUs
('Conventional' biogeography)	Species – distributions and ranges		Taxa themselves
	Genera – distributions and ranges		Taxa themselves
	Families – ditto		Taxa themselves
	Migrant/flagship species – distributions		Feeding, breeding areas
	Community distributions and ranges		Biocoenoses, biotopes
	Charismatic communities		Vents, sponges
PHYSIOGNOMIC	Geophysical	Oceanographic properties	Temperature, salinity, water masses, nutrient regime, O_2 minimum layer, lysocline
		Physiographic	Depth and depth categories, substrate type, sediments
	Geomorphology	Topographic features	Ridges, seamounts, abyssal plains, continental slope etc.
ECOLOGICAL GEOGRAPHY	Combined biological and physical factors	Biomes	Ocean basin, ocean gyres, water masses, sea colour (chlorophyll) productivity regimes, latitude, longitude, temperature regimes, community types
		Ecosystems	Oceanographic features, gyres, boundary currents, convergence zones, divergences, ocean currents
	Geological history and palaeontology	Evolution of ecological boundaries	Plate tectonics, ocean ridges
SOCIO-ECONOMICS	Ecosystem-based management	Fisheries economics	Historical fishing areas, catch quotas, productivity regime
		LOMAs	
		Fishing areas	
	Resource exploitation	Non-renewable resources	

ESU – evolutionary significant unit (see Chapter 10); LOMA – large ocean management area

of the word (e.g. Sherman et al, 2005; see also Chapter 4). However, a new synthesis has clearly been needed for conservation purposes. Ideally such a synthesis would recognize and define the 'natural barriers' (albeit rather leaky ones) in the oceans, that foster allopatric speciation, and lead to the development of endemicity and identifiably different biotas (floras and faunas) in different regions of the oceans. An ideal system would also be hierarchical and nested, and would allow for multiscale conservation planning and management from global to local levels.

In developing their classification, Spalding et al (2007) conducted extensive reviews and consultations, examined the underlying data and the process of identification and definition of biogeographic units, and required that their classification should: have a strong biogeographic basis; offer practical utility; and be characterized by parsimony. The classification was especially informed by composite studies that combined multiple taxa and oceanographic factors to define boundaries, as these were deemed more likely to capture robust or recurring patterns

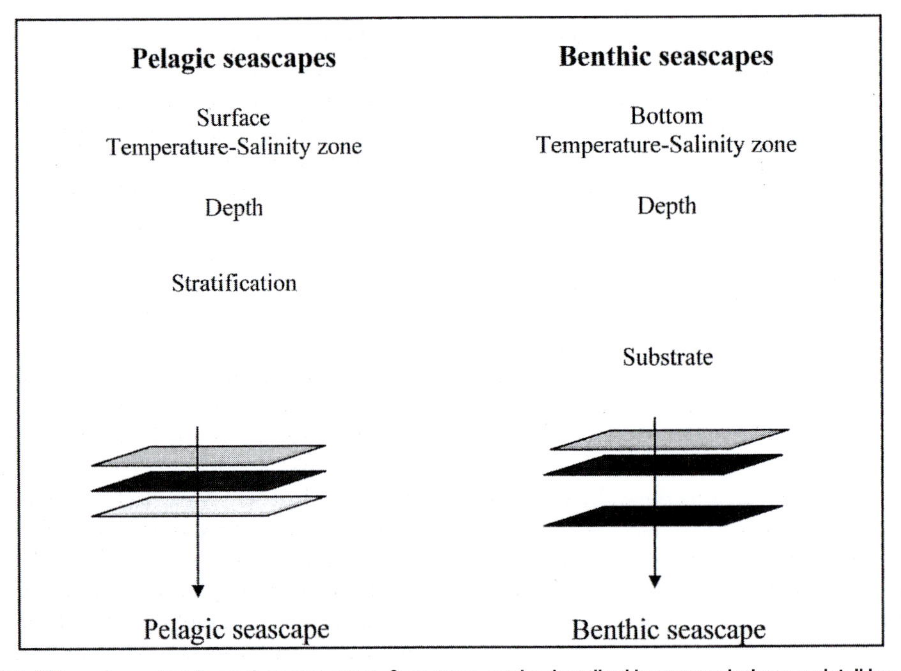

Figure 5.1 GIS overlay system to produce seascapes. Seascapes can be described in progressively more detail by adding more factors and variables. Mapping is done separately for defining seascapes within pelagic and benthic realms.

in overall biodiversity. The resulting three-stage classification is a nested system of 12 realms, 62 provinces and 232 ecoregions. The system provides considerably better spatial resolution than earlier ones and probably represents the best biogeographic classification we currently have as a global to ecoregional basis for marine conservation planning. Definitions and depictions of realms, provinces and ecoregions are summarized in Table 5.1 and Colour Plates 4a and b.

There will always be some degree of arbitrariness in defining the levels of any hierarchical biogeographic/ecological classification. This is as inevitable as the designation of taxonomic levels (from phyla to species) in the 'natural system of classification' for organisms. However, the pragmatic advantages of such a classification far outweigh the disadvantages of arbitrariness, and as more knowledge of global biodiversity accrues, existing biogeographic classifications will be constantly refined, as are our taxonomic and systematic ones.

Regional level and geomorphic units

Regions themselves can be defined primarily in terms of broad-scale gross geomorphology and water depth, from coastal systems to ocean depths. It has long been recognized that marine faunas, both pelagic and benthic, change significantly with depth and other factors, as described in Chapters 2 and 3.

Within ecoregions and regions, there are at least two recognized ways to proceed to define the lower levels of the hierarchy: by mapping geomorphic features and by mapping geophysical features using a system of overlays to produce seascapes (Figure 5.1). Both ways have been advocated and used and it is important to recognize their relationships. Table 5.1 indicates that seascapes lie within geomorphic units, but in fact they can be mapped separately. The two approaches may overlap but are not mutually exclusive; rather they can complement one another.

Mapping representative areas as geomorphic units produces maps that contain several types of habitats – that is, each of these units will display some level of β diversity (see Chapter 8) of primary habitats. There may be repeated examples of such geomorphic units within a region (e.g. sandbanks), or some types may be unique. Geomorphic units have become progressively better known with the advent of in situ monitoring devices such as echo-sounders and multibeam and sidescan sonars (see Box 5.1 for summary of techniques). Recognition and mapping of geomorphic features such as banks, basins, seamounts and canyons can be highly instructive, especially where they are known or suspected to be associated with ecosystem-level processes characteristic of distinctive areas (see Chapter 7). Such processes may include important physical water movements that accumulate biomass or stimulate production, such as gyral retention of planktonic organisms including larval fish on banks, local topographic upwellings in canyons and so on (see Chapters 7 and 8).

In mapping representative areas as seascapes by an overlay system, we are trying to define particular kinds of primary or secondary habitats – or at least regions within which a particular kind of habitat predominates. Inevitably at this level of definition, each seascape is still heterogeneous and will show β diversity of different secondary habitats and biotopes. There will generally be many repeat units of each type of seascape within a region.

The two approaches are therefore subtly but clearly different, but not competing. The geomorphological unit approach does not explicitly take into account the natural variation of biota as a function of temperature, salinity, depth, water masses and so on – which the seascape approach does. Whichever approach is used may depend on the availability of data (see below and Chapter 6). In fact the next lower step in the hierarchy is the same for both approaches, namely to use in situ data either from direct sampling or techniques such as multibeam surveys, to further refine the definition of habitat types and biotopes within a region. At the level of secondary habitats and biotopes therefore the approaches converge.

There are other potential problems with restricting ecological mapping to geomorphic features alone. For example, areas outside recognized features may be missed or not assigned in this broad-scale approach (see Colour plate 5), for example – where does a bank end and a basin begin? Missing some areas from a classification would mean that the biotas not included – which would differ in character and species composition from those within a geomorphic feature – are unaccounted for. In contrast, all regions of the oceans belong within some seascape type.

An additional complication between geomorphic features, seascapes and habitats is as follows. Because the term 'habitat' itself is used hierarchically (and often indiscriminantly), habitat classifications can mix units across spatial scales. For example, the Natura 2000 classification of marine 'habitat' types is in fact a composite of at least two and as many as five levels of the spatial hierarchical system (compare Tables 5.1 and 5.4). In order to ensure proper classification and adequate representation of all marine habitats, care must be taken to include all habitat types in a spatially defined hierarchical classification.

Within ecoregions – geophysical factors – seascapes

Each ecoregion will contain many different kinds of areas and habitats, both distinctive and representative. 'Distinctive' means that an area, habitat or community type is atypical of its surroundings, at some defined scale of reference, and 'representative' means that an area, habitat or community type is typical of its surroundings at some scale (see Roff and Taylor, 2000; Roff and Evans, 2002). Distinctive areas are isolated, and not contiguous; they are examined in more detail in Chapter 7. In contrast, representative areas are continuous and contiguous; all marine environments belong to some type of representative area.

A system of classification for representative marine habitats and their associated communities should be logical, easy to use and stable (or naturally adaptable) through time. We can conceive of three broad types of classification: (i) using biological features and species only; (ii) using physical features or processes only; (iii) a mixed system using both biological and physical features. The advantage of

Box 5.1 Applications of remote sensing techniques for mapping the marine environment and variables that can be identified / quantified through remotely sensed approaches

Remote sensing refers to the detection of electromagnetic energy from aircraft, satellites or boats. Typical parts of the electromagnetic spectrum detected by remote sensing techniques are the optical and microwave regions.

SHIPBORNE

Single beam – Use acoustic energy to collect measurements of seafloor depth
Multibeam – Active sensors that use acoustic energy to measure seafloor depth and character

AIRBORNE

LiDAR – An active sensory that transmits laser pulses to a target and records the time it takes for the pulse to return to the sensor receiver
Thermal infrared radiometer – Collects thermal imagery
Multispectral – Collect stereo black and white, colour infrared and true-colour imagery
Hyperspectral – Hyperspectral sensors are passive sensors that acquire simultaneous images in many relatively narrow continuous and/or non-continuous spectral bands through the ultraviolet, visible and infrared portions of the electromagnetic spectrum, e.g. CASI, AVIRIS, AISA and Probe-1

SATELLITE

Panchromatic (PAN) – Sensitive to the visible spectrum, normally shown in black and white
Multispectral (MS) – Collects data in two or more spectral bands of the electromagnetic spectrum. The bands may be located in the visible and infrared parts of the spectrum
Moderate resolution satellite (MR) – 1km to 30m, e.g. Landsat TM/ETM, SeaWifs, AVHRR, MODIS, CERES, AVISO, ENVISAT\MERIS
High resolution satellite (HR) – 60cm to 15m, e.g. SPOT, Quickbird, IKONOS, ASTER, LANDSAT Panchromatic
Radar altimetry – Measurement of ocean heights used to understand patterns of ocean circulation, e.g. TOPEX
Microwave sensors – Measurement of precipitation, oceanic water vapour, cloud water, near-surface wind speed, sea surface temperature, soil moisture, snow cover and sea ice parameters, e.g. AMSR-E (passive), RADARSAT (active)

(from Turner et al, 2003; NOAA Coastal Services Centre, 2009)

Variables	Remote sensing methods for measurement
Sediment composition	**Shipboard:** Multibeam Backscatter (Sutherland et al, 2007)
Depth	**Shipboard**: Multibeam (Smith and Sandwell, 1997) Single Beam Sonar (Smith and Sandwell, 1997) **Airborne**: LiDAR (intertidal) (Parrott et al, 2008)
Temperature	**Satellite:** AVHRR (Fiedler et al, 1984; Hendiarti et al, 2002)
Turbidity	**Satellite**: SeaWifs attenuation at k490 (not good nearshore) **Airborne**: CASI (Herut et al, 1999)
Chlorophyll/ Net primary productivity/ Particulate organic carbon	**Satellite**: SeaWIFS (Hendiarti et al, 2002; Biggs et al, 2008) AVHRR (Fiedler et al, 1984) Terra/MODIS (Mauri et al, 2007) Terra/ASTER (Sakuno et al, 2002; Nas et al, 2009) IKONOS (Ormeci et al, 2008) SPOT (Yang et al, 2000) MERIS (Fournier-Sicre et al, 2002; Lunetta et al, 2009) **Airborne**: AVIRIS (Lunetta et al, 2009) CASI (Herut et al, 1999)

Circulation and current	**Satellite:** TOPEX (Wunsch and Stammer, 1998) AVHRR (Strub and James, 1995) MODIS (Yuan et al, 2008) SeaWiFS (Singhruk, 2001) ENVISAT/MERIS (Zainuddin et al, 2006)
Climate/Rainfall	**Satellite:** Terra/CERES (Kato et al, 2006) AMSR-E (Kelly et al, 2003)
Nutrients	**Airborne:** Flourescent LiDAR
Pollutants	**Airborne:** Flourescent LiDAR (Babichenki et al, 2006)
Actual biology	**Airborne:** Aerial photography CASI (Mumby et al, 1997) Airborne multispectral system Thermal infrared radiometer (marine mammal surveys) **Satellite:** Landsat TM (Zainal et al, 1993; Mumby et al, 1997) Terra/ASTER (Capolsini et al, 2003) SPOT XS high res 2.5–10m (Cuq, 1993; Mumby et al, 1997) IKONOS high res 1–4m (Mumby and Edwards, 2002) Quickbird 60cm–2.8m, panchromatic and multispectral (monitoring walrus populations in the Bering and Chukchi seas)

Table 5.4 A partial list of 'habitat' types from the NATURA 2000 set, indicating the level they occupy according to the spatial classification of Butler et al (2001)

NATURA 'habitat' number	NATURA 'habitat' type	Level of spatial hierarchy of Butler et al (2001)
1110	Sandbanks	Secondary habitat (seascape)
1120	*Posidonia* beds	Biological facies
1130	Estuaries	Geomorphic unit
1140	Mud-flats and sand-flats not covered by seawater at low tide	Secondary habitat (seascape)
1150	Coastal lagoons	Primary habitat
1160	Large shallow inlets and bays	Geomorphic unit
1170	Reefs	Secondary habitat (seascape)
1180	Submarine structures made by leaking gases	Biological facies?
8330	Submerged or partially submerged caves	Primary habitat

a biological system is that there is no need to seek the relationships between organisms and physical parameters. However, most marine organisms are not visible either directly, or by remote sensing, and mapping marine communities by direct sampling is a daunting task.

The classification of representative areas from the level of realms to that of ecoregions is in fact a mixed one, using both biological and physical data from a sparsely sampled global scale. However, from the ecoregion level to the level of secondary habitats (levels 3 to 7) available biological data is often very sparse. It is generally only in shallow waters of the immediate coastal zone where biological communities can be directly appreciated. Thus distributions of communities such as corals, mangroves and seagrasses can be directly mapped, for example from aerial surveys or boats and video cameras. Beyond these obvious and easily sampled communities, the rest of the representative habitats and communities of the world, in deeper waters, are hidden from direct view, and the daunting task of sampling their biotas at appropriate scales has not been undertaken.

Because of the general paucity of data on biological marine resources, the use of geophysical features to describe marine habitats and to act as surrogates for marine communities has become widely applied. The habitat/ecosystem level of the biodiversity framework is thus the most appropriate one for development of hierarchical frameworks at large (e.g. regional and national) scales where data on the distributions of populations and communities is generally sparse. Indeed, in many cases it is the only practicable approach to marine conservation at these scales. There is now considerable research that suggests that populations and communities are strongly correlated with habitat types at a series of spatial scales (e.g. Connor, 1997), and this important issue will be examined in more detail in the following chapter.

At the lowest levels of the spatial hierarchy (8 to 9 in Table 5.1), direct sampling of biota again becomes possible at scales of centimetres to metres. However, such direct sampling can only be undertaken in restricted areas because of considerations of time, cost and taxonomic expertise. We should appreciate the interesting situation in which – from the global to the very local level – a spatial classification must rely upon variable combinations of biological and geophysical data. From the level of realms to ecoregions, a combination of geophysical and biological data is used; then from the ecoregion through habitat levels frequently only geophysical data is available – where biological data simply becomes too sparse (except in coastal areas where some communities such as corals and mangroves can be directly mapped). Finally the lowest levels of the spatial hierarchy revert to the use of biological data for local descriptions – or cross-calibration of geophysical data for verification of habitat/community associations (e.g. Maxwell et al, 1995).

Rationale and ecological principles for a geophysical system of classification

The biological communities of a region are the result of many interacting physical, chemical and biological parameters (Table 3.5); rarely do we know the relative significance of each parameter. However, in marine environments, the abiotic geophysical (physiographic and oceanographic) attributes of habitats will at least partially define local biological communities, particularly at mid-latitudes.

The term 'enduring features' is applied to terrestrial ecosystems to encompass those elements of the landscape that do not change (at least in terms of human life spans), and which are known to control and influence the diversity of biological systems. For the purposes of identifying terrestrial protected spaces, 'enduring features' are closely associated with terrestrial abiotic features (Kavanagh and Iacobelli, 1995). This includes such stable features as landforms, physiography and soils that have been demonstrated to play a significant role in the distribution and diversity of flora and fauna at small to medium scales. These features can be readily mapped in a two-dimensional format.

Marine systems are basically different from terrestrial ones, but they also consist of interacting physical and biological components that can be used as indicators to separate marine representative units (MRUs) or habitats. Marine systems are more dynamic and exhibit more complex spatial–temporal relationships than terrestrial systems, which makes the development of comparable techniques for marine systems more difficult. Unlike terrestrial features, the equivalent ecological attributes of marine ecosystems may change spatially and temporally in predictable or unpredictable ways, particularly in pelagic seascapes.

For marine systems the term 'enduring features' has been modified to 'marine enduring features and recurrent processes' (Roff and Taylor, 2000), in order to specify those non-biotic (geophysical) components of marine ecosystems that are observable and measurable, and which control the distribution and diversity of marine species and communities (though sometimes in a dynamic fashion), and which can serve as surrogates to identify the major MRUs and their types of communities. Recurrent processes, such as daily tides, seasonal stratification, development of frontal systems and ocean currents, are largely predictable, though variable. These *recurrent* oceanographic processes constitute additional components to the enduring features of terrestrial ecosystems. The boundaries of marine communities can be more dynamic than in terrestrial ecosystems (e.g. Tremblay and Roff, 1983), but it is possible to capture representative marine habitats and their communities in terms of the related enduring features or recurrent processes. If a system of classification is based on enduring and recurrent geophysical features that predict the biological communities, and if these factors can be mapped, then a complete system of classification can be undertaken for all regions.

The advantage of using geophysical features is that at large scales they control the distribution of organisms. While individual species are locally prone to population change, invasion or extinction, the community will persist (albeit as different biocoenoses – see Chapter 6), and can be represented by its enduring or recurrent physical correlates. The preference for using physical features alone also lies in several realities. First, geophysical data is always far more available over broad geographic regions than biological survey data; remote sensing technologies allow the ready acquisition of details on geography, geology, shoreline topography, surface plant biomass (e.g. from chlorophyll-*a*/ colour), temperature and other features (Box 5.1). Second, habitats are far more temporally stable and ecologically fundamental than the communities they support, as will be argued in Chapter 6. Third, even if the environment changes – for example in terms of distribution of temperature or coastal salinity due to land changes in rainfall and runoff – it is far easier to remap geophysical data than to resurvey an area for biota. Fourth, while no classification can account for large-scale or unpredictable disturbances, like hurricanes, communities generally re-establish themselves in due course and in species compositions based on the local enduring physiographic and recurrent oceanographic factors.

A set of guiding principles for the classification of marine habitats (also referred to as MRUs) is presented in Box 5.2.

Geophysical factors used to define habitats (marine representative units – MRUs)

The major oceanographic and physiographic factors, their influences on the marine biota and various roles in shaping marine communities have been considered in Chapter 2, and by Roff et al (2003). The geophysical factors that are ultimately selected as indicators of marine representative habitats and their communities will depend substantially on the level of detail and extent of the data available for each of them.

A problem which immediately confronts us is that the features controlling pelagic communities are different from those controlling benthic ones (effectively three- and two-dimensional realms respectively). This is one reason for segregating oceanographic and physiographic features. Oceanographic features are of significance to both pelagic and benthic communities, whereas physiographic features will be primarily of significance to benthic communities. Because the life cycles of many marine organisms involve both benthic and pelagic phases, there are cross-linkages and scalar differences that may be difficult to quantify and map. However, most initiatives for marine conservation, especially those directed at conservation of biodiversity, are primarily concerned with the benthic realm. Migratory species (predominantly within the pelagic realm) present multi-scalar problems that require separate consideration (see Chapter 7).

Given the task of mapping marine habitat types according to their geophysical properties, from a global perspective, most marine ecologists would probably identify a set of factors similar to (or at least including) those in Table 5.5 (see also Roff and Taylor, 2000). However, schemes to classify habitats at a local (tens to hundreds of kilometres) or regional (hundreds to thousands of kilometres) scale may identify different sets of actual factors as determinants, and use them in different sequences in a hierarchical classification (e.g. Dethier, 1992; Connor, 1997; Zacharias et al, 1998; Roff and Taylor, 2000). It is important to determine the reasons for these differences in local or regional applications of a global concept, and examine whether a common scheme for habitat classification could be developed among geographic regions. If it could, then such a generalized scheme would considerably facilitate conservation and environmental efforts at both regional and international levels. There are a number of important considerations for the development of a generalized habitat classification scheme.

Box 5.2 A set of guiding principles for the classification of marine habitats (or marine representative units – MRUs)

By **guiding principles**, we mean a series of statements designed to:

- guide the development and implementation of the marine representation methodology and framework;
- guide the requirements for 'ecological integrity' for marine endangered species and spaces and develop true networks of marine protected areas (see Chapters 16 and 17);
- enable a 'gap analysis' approach to identify those areas that should be declared as marine protected areas.

1. The system should have a global perspective, in which the higher levels of classification are defined by global processes.
2. Given the obvious and profound differences between pelagic and benthic communities, it will be necessary to consider these realms separately in conservation planning.
3. An approach based on 'marine enduring and recurrent features' (i.e. both structures and processes) should be used to classify and evaluate representative marine environments.
4. The factors chosen as the basis for a classification should be the ones most appropriate for each level of the hierarchy. Each variable is used only once in the classification. This implies that this variable is the most significant physico-chemical 'controlling' parameter at that scale.
5. Any classification should explicitly recognize and distinguish between representative and distinctive areas.
6. Recognition of biomes and ecosystems may be useful for management purposes, but such units do not form the basis of a comprehensive framework for the conservation of the components of biodiversity or MRUs.
7. These MRUs should directly or indirectly represent important components of marine biological diversity.
8. Mixed physical/biological classifications are not desirable at the ecoregion scale, but if unavoidable should be rationalized. Biological parameters themselves can be used at finer scales for local verification of physical correlates. It is to be expected that physical environmental parameters and biological community types will converge at the finest scales, as joint descriptors of representative spaces defined as recognizable biotopes.
9. The classification of marine representative areas should be hierarchical, so that description occurs on different spatial scales, allowing for the identification and ultimate protection of lower level classification units in the system.
10. The hierarchy should be constructed so that the upper levels discriminate first between the most broadly different community types; progressing to lower levels, community types should be increasingly more closely related. A system such as that of Butler et al (2001) (Table 5.1) is presently the preferred one.
11. The classification system should clearly delineate repeating community or habitat types based on scalable and non-scalable marine enduring features, incorporating a minimal set of key physiographic and oceanographic factors.
12. Data describing marine enduring features which are utilized for delineating MRUs should be available either directly or via suitable surrogates at comparable scales for all areas to be evaluated to ensure consistency of interpretation and comparison.
13. The system should have predictive power, describing the relationships between physical environments and biotic communities (see Chapter 6).
14. The system should be logical, easy to use and stable (or naturally adaptable) through time.
15. Selection of marine protected areas (MPAs) will require a further analysis of ecological information (for example of individual species and ecological processes (see Chapter 16).
16. Further analysis of oceanographic and genetic data is required to plan for networks of MPAs.
17. Subsequent planning steps are required to evaluate socio-economic implications and the evaluation of alternative MRUs as candidate sites for MPAs.

Note: See also Watson (1997)

Table 5.5 Some geophysical factors that can be used to classify habitat types

General list	Factors used in this classification	Predominant scale of effect of factors
Oceanographic	**Oceanographic**	
Ice cover	Ice cover	Global/Regional
Temperature	Temperature	Global/Regional/Local
Salinity		
Water masses (temperature and salinity signatures)		
Temperature anomalies		
Temperature gradients (e.g. fronts)		
Light penetration		
Water column stratification	Water column stratification	Regional
Nutrient concentrations		
Tidal amplitude		
Tidal currents		
Exposure (to wave action or desiccation)	Exposure (to wave action or desiccation)	Regional/Local
Current speed		
Oxygen concentration		
Physiographic	**Physiographic**	
Tectonic motion		
Latitude		
Depth (bathymetry)	Depth (bathymetry)	Regional
Relief/bottom slope	Relief/bottom slope	Regional
Rate of change of slope/ heterogeneity		
Substrate particle size	Substrate particle size	Regional/Local
Rock type		

Notes: By 'regional' is meant sub-national to national jurisdiction or scales of 100s to 1000s of km; by 'local' is meant sub-national jurisdiction or scales of 10s to 100s of km.
For further interpretation see Roff and Taylor (2000)

First, the potential (global) set of factors that can be used to discriminate among habitat types must be determined by what can be mapped from available geophysical data and what can be readily obtained by remote or in situ sensing. Fortunately our ability to map geophysical factors is improving as sensors become more sophisticated (see e.g. Box 5.1).

Second, there may be some redundancy between factors or they may need to be computed in different ways. Thus some combinations of factors may be used as surrogates for others; for example slope and current speed may be used as a surrogate for substrate type. Also, water column stratification may be predicted and modelled from the Simpson-Hunter stratification parameter (h/U^3, where h = water depth and U = tidal velocity; see Pingree, 1978), but only where the lunar tidal component dominates. In other areas, actual data on water column stratification may be needed (e.g. $\Delta\sigma t/\Delta z$; the change in density - $\Delta\sigma t$ – with change in depth Δz). However, the two parameters can be cross-calibrated to show the same thing.

Third, the actual set of factors chosen for a classification hierarchy within any region will depend upon the natural range of variation in each one. Some factors may not be applicable within a particular region because they show little variation. For example, outside of estuaries (which comprise a set of habitats separate from marine ones and should be separately classified, see Chapter 11) salinity is an important determinant of community type in the Straits of Georgia, British Columbia (Roff et al, unpublished data) and in the Baltic Sea (Hallfors et al, 1981), but not on the east coast of Canada (Day and Roff, 2000) or in the Mediterranean (Connor et al, 1995). This is because the first two of these regions are partially landlocked and subject to considerable land runoff and hence salinity variation. Thus they are estuarine in character. In the second two examples, salinity does not vary sufficiently to act as a major determinant of community type. Thus although a common set of factors can be envisaged for a hierarchy (at least in temperate areas throughout the world), in any particular region – where some factor may be relatively homogeneous (in other words it shows low variation) – it may not aid in discriminating among habitat types. Further, it follows that in tropical and subtropical regions, where there is little variation in several geophysical factors (e.g. temperature, salinity, stratification), we may need to place greater reliance on direct mapping of the biological communities themselves – as is indeed typically the practice in such regions.

Fourth, the sequence in which factors enter a hierarchy should ideally be determined by the same principle that applies in the 'Natural System of Biological Classification', that is: it should depend upon which has the greatest ability to discriminate among habitat types (and by implication their communities and array of taxa). This means that habitat types at the upper levels of a hierarchy should be more distinct from one another, while those within categories at lower levels of a hierarchy should be more similar. By implication and association, this also means that community types should become more similar to one another within the lower categories of a hierarchy, just as taxa do in a taxonomic classification system. In practice, this calls for some expert judgement and knowledge of the taxonomy, biogeography and physiology of the whole array of marine organisms. We shall not include a full defence of the sequence of a hierarchy here, which may be expected to evolve just as has the Natural System of Biological Classification. However, the sequence will also be based partly upon the scale at which each factor exerts its predominant effect. Table 5.5 indicates how a series of factors have been spatially ranked, and the sequence in which they may enter a hierarchy (see Allen and Starr, 1982; Roff and Taylor, 2000, Roff et al, 2003).

Following from this reasoning, it should be clear that – at the continental or international level – a common set of factors could be applied in a defined and defensible sequence to produce a common hierarchy of habitat types. However, it is to be expected that some factors will not apply in some regions – that is, they do not help to discriminate among habitat types. In such cases, these parts of the hierarchy are simply empty, and we pass on to the next factor and the next level.

Fifth, the number of levels in a hierarchical classification depends on data availability, its spatial resolution and spatial coverage and statistical considerations. Thus more levels than shown in Table 5.1 may be required at the habitat level to improve the degree of discrimination.

Examples of hierarchical conservation planning for representative areas at national and regional levels

Most planning for marine conservation will be undertaken at the national level, where boundaries may cross global ecoregional designations. The following two examples demonstrate regional to continental efforts to identify representative areas for conservation planning purposes.

The Marine Ecoregions of North America

The purpose of the Centre for Environmental Cooperation's Marine Ecoregions of North America is to classify the marine and estuarine exclusive economic zones of North America (Wilkinson et al, 2009). The outputs from the project will inform conservation and management efforts as well as develop a public understanding and awareness of the variety of North American marine ecosystems. Key design aspects of the Marine Ecoregions of North America include: the classification must be scalable (hierarchical) in order that smaller units nest within larger units for ease of management and environmental reporting; classification units must be based on biological, physiographic and oceanographic criteria such that they represent the 'real-world' distribution of ecological communities; and the system must be linked with existing maps and classifications.

The 'rules' for developing the classification were established by a tri-national committee who determined that the classification must:

- Make sense and be useful at a North American scale.
- Include three, nested, hierarchical levels bounded by 'hard lines', which are assumed to be approximate and do not consider political boundaries.
- Be based on the best available biological, oceanographic and physiographic scientific data and expert knowledge while building on past efforts and nomenclature.
- Support ecological interpretations through characterizing ecological regions.
- To the extent possible, reflect the three-dimensional nature of marine systems within a two-dimensional map.
- Adopt terminology based on places/locations that also reflect the level and major variables used to define that level.

Level I of the classification (Colour Plate 6) is primarily based on water mass characteristics (e.g. sea surface temperature, ice cover) and major physiographic features (e.g. enclosed seas, major currents, gyres, upwellings) to define 24 regions that are loosely considered analogous to 'marine ecosystems'. Level I extends from the coasts to the deep oceans, although biogeographic patterns and processes in the deeper regions are still poorly understood.

The main variables in Level II of the classification (Colour Plate 7) are primarily physiographic. Ocean features used in determining Level II units include the continental shelf, slope and abyssal plain, areas of seamounts, borderlands, trenches and ridges. These features are used as surrogates and predictors for the oceanographic features including current flows and upwelling. The broad purpose of Level II is to capture the break between neritic and oceanic areas as well as the diversity found within the benthic realm (Wilkinson et al, 2009).

Level III variables combine oceanographic, physiographic and biological factors to represent coastal and neritic environments, primarily from the coastline to the edge of the continental shelf. Specific variables included at this level include salinity-estuarine influenced areas, substrate type, community type, or community sub-type.

Geophysical maps of the Scotian Shelf and Gulf of Maine

The Scotian Shelf and Gulf of Maine includes the marine waters off the Atlantic coast and off Nova Scotia, Canada and the northern United States, extending out to the 200 nautical mile limit. It totals an area of approximately 277,388km^2 (80,886nmi^2 or 107,100mi^2). The region and its biogeographic boundaries are considered again in more detail in Chapter 16 in the CLF/WWF (2006) study on selecting sites for a network of MPAs. The northern limit of the study area lies west of Cape Breton Island (i.e. the Laurentian Channel) while the southern and western limits extend beyond Georges Bank; the Bay of Fundy is also within the study area.

The region is one of strong contrasts in temperature, due to the inshore influence of the cold Labrador Current and the warm offshore Gulf Stream. Outside of estuaries, salinity variations on the shelf are only important in the uppermost Bay of Fundy and inner Gulf of St. Lawrence (Petrie et al, 1996). The main pelagic temperature and salinity zones of the upper water masses are shown in Colour Plate 8a. Marine waters are continuous, therefore any limits set by temperatures must have some element of arbitrariness. However, in plotting temperatures on the shelf, it became apparent that there was marked congruence between the 5°C winter isotherm and the 18°C summer isotherm, marking an average boundary between temperate slope waters and the subtropical waters of the Gulf Stream, irrespective of season. The presence or absence of winter ice is significant for many organisms, and distinguishes cold (boreal) from temperate waters. The approximate location of the seasonal ice limit is adequately described by the winter 0°C surface isotherm, which runs into the northern part of Cape Breton. Depth classes in the region are shown in Colour Plate 8b.

Most of the Scotian Shelf is influenced by a mixed tidal regime of low amplitude leading to a stratified water column during the summer months (J. Loder, Bedford Institute of Oceanography, pers. comm.). From southwest Nova Scotia into the Gulf of Maine, the area is progressively dominated by the M2 tidal component, and resonance in the Bay of Fundy produces exceptionally high tides. These regions are therefore either unstratified or include frontal regions (Colour Plate 8c). The vertical stratification of a water column separates its communities both spatially and temporally, and stratified and non-stratified waters differ both in their annual productivity regimes and in their community structures (Pingree, 1978). Thus banks and coastal regions of high tidal amplitude will remain unstratified, and may retain or accumulate populations of important pelagic species such as larval fish (Jeffrey and Taggart, 1999). Transitional (frontal) areas can represent boundaries between populations or communities and may persist as regions of high production throughout the summer months (Pingree, 1978; Iles and Sinclair, 1982). The combination of temperature and salinity zones, depth classes and stratification regime results in the pelagic seascapes shown in Colour Plate 8d.

The Scotian Shelf and Gulf of Maine region becomes generally stratified during the summer months and is under the influence of the cold south-flowing Labrador waters, hence the bottom temperature remains cold over much of the shelf. Higher temperature and higher salinity water (Colour Plate 9a) may displace this colder but less saline water in the basins along the shelf.

The type of substrate (Colour Plate 9b), which results from a combination of factors including current velocity, depth, exposure and slope, has a major effect on the type of benthic community that develops in a region (e.g. Barnes and Hughes, 1988). The benthic seascapes that result from the combination of temperature and salinity zones, depth (Colour plate 9c) and substrate type are shown in Colour Plate 9d.

Conclusions and management implications

The process of mapping the marine environment, based largely on enduring and recurrent geophysical factors to depict representative areas, has now been undertaken by or is in process in many countries, including the UK and other EU countries, Australia, North America and elsewhere. In combination, and with international collaboration, such mapping can form the foundation for a truly comprehensive approach to global representative networks of MPAs.

Ideally, such maps – based initially on geophysical factors – would be calibrated by biological sampling at the appropriate scale and locations, in order to verify or adjust ecological boundaries. The maps so produced have multiple uses, perhaps foremost among them is that they form an indispensible foundation for marine conservation and ecosystem-based management. Perhaps most importantly for conservation planning, the numbers and types of representative areas can be appreciated, and the proportion of a region that each occupies (its rarity or prevalence) can be seen, along with a measure of environmental heterogeneity.

The factors chosen in this chapter are generally amenable to mapping using existing data. They define habitat types at each level, and should hold representative communities. The habitat types in the classification hierarchy should become progressively better defined from higher to lower levels, and at the lowest levels the identification of biotopes should provide ecological verification of such geophysical classifications (see Chapter 6). The scale at which a factor has its dominant effect is where it should be used in a hierarchical classification system. At the ecoregional scales considered here, a top-down approach, using physiographic and oceanographic features, is the only feasible approach, because of the difficulty of obtaining sufficient biological data for direct mapping of community types.

In terrestrial conservation biology, the concept of landscapes is well understood and recognizes that certain types of communities occur within them. We have proposed the term 'seascapes' (or for the benthos – marine landscapes) in an analogous manner. Seascapes are linked to their typical plant and animal communities, and would be the units of marine conservation. However, it remains to be determined at what level in a hierarchy the term should be used; it could be used for example as a multilevel analogue of the term habitat. It is not clear how to subdivide the pelagic realm beyond the fourth level, and this may be the most appropriate level to define 'seascapes' for conservation purposes. However, for the benthic environment there are still significant differences in habitat types and seascapes at finer scales. This is not nearly so tidy as in terrestrial systems and, because of scalar differences, it may be necessary to consider two maps for conservation purposes, for pelagic seascapes at a coarser scale and benthic seascapes at a more detailed scale.

Given such mapping, the relative merits of existing and proposed marine protected areas can be evaluated in terms of the representative habitats (or marine representative units) and ecosystem level processes they may capture, and their contributions to the protection of biodiversity, fisheries and migrant species. Thus it should be possible to develop a national conservation strategy that would define an ecologically rationalized set of MPAs, within and among which conservation goals can be achieved. Ideally, plans for marine conservation developed by individual nations – largely at the ecoregion level and below – would be reconciled and integrated with global plans of marine representation such as those developed by Spalding et al (2007).

We should clearly note, however, that the procedures we describe here comprise only one possible ecological approach to marine conservation, and only one step in the process of actual selection of MPAs. Nevertheless, habitat classification and mapping is an indispensable step. Several further steps in the planning processes for marine conservation must still be undertaken. We have not yet considered distinctive areas, nor how MPA sites would be selected, nor evaluated relationships among the members of any set of candidate MPAs, nor the concept of what constitutes a 'network' of MPAs. Our goal in the following chapters is to try to define the procedures, involving the fewest number of arbitrary decisions that would lead to the optimal selection of regional, national and international networks of MPAs.

References

Allen, T. F. H. and Starr, T. B. (1982) *Hierarchy: Perspectives for Ecological Complexity*, University of Chicago Press, Chicago, IL

Babichenki, S., Dudelzak, A., Lapimaa, J., Lisin, A., Poryvkina, L. and Vorobiev, A. (2006) 'Locating water pollution and shore discharges in coastal zone and inland waters with FLS LiDAR', *EARSeL EProceedings*, vol 5, pp32–42

Barnes, R. S. K. and Hughes, R. N. (1988) *An Introduction to Marine Ecology*, Blackwell Scientific Publications, Oxford

Beaman, R. (2005) *A GIS Study of Australia's Marine Benthic Habitats*, eprints.utas.edu.au/419/, accessed 23 December 2010

Biggs, D. C., Hu, C. and Miuller-Karger, F. E. (2008) 'Remotely sensed sea-surface chlorophyll and POC flux at deep Gulf of Mexico benthos sampling stations', *Deep Sea Research Part II: Topical Studies in Oceanography*, vol 55, nos 24–26, pp2555–2562

Briggs, J. C. (1995) *Global Biogeography*, Elsevier, Amsterdam

Butler, A., Harris, P. T., Lyne, V., Heap, A., Passlow, V. and Smith, R. (2001) 'An interim, draft bioregionalisation for the continental slope and deeper waters of the South-East Marine Region of Australia', Report to the National Oceans Office, CSIRO Marine Research, Geoscience Australia, Hobart, Australia

Capolsini, P., Andrefouet, S., Rion, C. and Payri, C. (2003) 'A comparison of Landsat ETM+, SPOT HRV, Ikonos, ASTER, and airborne MASTER data for coral reef habitat mapping in South Pacific Islands', *Canadian Journal of Remote Sensing*, vol 29, vol 2, pp187–200

Connor, D. W. (1997) *Marine Biotope Classification for Britain and Ireland*, Joint Nature Conservation Review, Peterborough

Connor, D. W., Hiscock, K., Foster-Smith, R. L. and Covey, R. (1995) 'A Classification System for Benthic Marine Biotopes', in A. Eleftheriou, A. D. Ansell and C. J. Smith (eds) *Biology and Ecology of Shallow Coastal Waters*, Olsen and Olsen, Fredensborg, Denmark

Cuq, F. (1993) 'Remote sensing of sea and surface features in the area of Golfe d'Arguin, Mauritania', *Hydrobiologia*, vol 258, pp33–40

Day, J. and Roff, J. C. (2000) *Planning for Representative Marine Protected Areas: A Framework for Canada's Oceans*, World Wildlife Fund Canada, Toronto

Dethier, M. N. (1992) 'Classifying marine and estuarine natural communities: An alternative to the Cowardin system', *Natural Areas Journal*, vol 12, pp90–99

Fiedler, P. C., Smith, G. B. and Laurs, R. M. (1984) 'Fisheries applications of satellite data in the eastern north Pacific', *Marine Fisheries Review*, vol 46, pp1–12

Fournier-Sicre, V. and Belanger, S. (2002) 'Intercomparison of SeaWiFS and MERIS marine products on case 1 waters', http://envisat.esa.int/workshops/validation_12_02/proceedings/meris/30_fournier.pdf, accessed 8 November 2009

Hallfors, G., Niemi, A., Ackefors, H., Lassig, J. and Leppakoski, E. (1981) 'Biological Oceanography', in A. Voipio (ed) *The Baltic Sea*, Elsevier Oceanography Series 30, Elsevier, Amsterdam

Hendiarti, N., Siegel, H. and Ohde, T. (2002) 'Investigation of different coastal processes in Indonesian waters using SeaWiFS data', *Deep Sea Research Part II: Topical Studies in Oceanography*, vol 51, nos 1–3, pp85–97

Herut, B., Tibor, G., Yacobi, Y. Z. and Kress, N. (1999) 'Synoptic measurements of chlorophyll-a and suspended particulate matter in a transitional zone from polluted to clean seawater utilizing airborne remote sensing and ground measurements, Haifa Bay (SE Mediterranean)', *Marine Pollution Bulletin*, vol 38, no 9, pp762–772

Iles, T. D. and Sinclair, M. (1982) 'Atlantic herring: Stock discreteness and abundance', *Science*, vol 215, pp627–633

Jeffrey, J. S. and Taggart, C. T. (1999) 'Growth variation and water mass associations of larval silver hake (*Merluccius bilinearis*) on the Scotian Shelf', *Canadian Journal of Fisheries and Aquatic Sciences*, vol 57, pp1728–1738

Kato, S., Loeb, N., Minnis, P., Francis, J. A., Charlock, T. P. and Rutan, D. A. (2006) 'Seasonal and interannual variations of top-of-atmosphere irradiance and cloud cover over polar regions derived from the CERES data set', *Geophysical Research Letters*, vol 33, L19804

Kavanagh, K. and Iacobelli, T. (1995) *Protected Areas Gap Analysis Methodology*, World Wildlife Fund Canada, Toronto

Kelleher, G. and Kenchington, R. (1992) *Guidelines for Establishing Marine Protected Areas. A Marine Conservation and Development Report*, IUCN, Gland, Switzerland

Kelly, R. E., Chang, A. T., Tsang, L. and Foster, J. L. (2003) 'A prototype AMSR-E global snow areas and snow depth algorithm', *IEEE Transactions on Geoscience and Remote Sensing*, vol 41, no 2, pp230–242

Longhurst, A. (1998) *Ecological Geography of the Sea*, Academic Press, San Diego, CA

Lourie, S. A. and Vincent, A. C. J. (2004) 'Using biogeography to help set priorities in marine conservation', *Conservation Biology*, vol 18, pp1004–1020

Lunetta, R., Knight, J., Paerl, H., Streichner, J., Peierls, B. and Gallo, T. (2009) 'Measurement of water colour using AVIRIS imagery to assess the potential for an operational monitoring capability in the Pamlico Sound Estuary, USA', *International Journal of Remote Sensing*, vol 30, no 13, pp3291–3314

Mauri, E., Poulain, P. M. and Juznic-Zonta, Z. (2007) 'MODIS chlorophyll variability in the northern Adriatic Sea and relationship with forcing parameters', *Journal of Geophysical Research*, vol 112

Maxwell, J. R., Edwards, C. J., Jensen, M. E., Paustian, S. J., Parrott, H. and Hill, D. M. (1995) 'A hierarchical framework of aquatic ecological units in North America (nearctic zone)', US Department of Agriculture, Forest Service, North Central Forest Experiment Station General Technical Report NC-176, St. Paul, MN

Mumby, P. J. and Edwards, A. J. (2002) 'Mapping marine environments with IKONOS imagery: Enhanced spatial resolution can deliver greater thematic accuracy', *Remote Sensing of the Environment*, vol 82, nos 2–3, pp248–257

Mumby, P. J., Green, E. P., Edwards, A. J. and Clark, C. D. (1997) 'Measurement of seagrass standing crop using satellite and airborne digital remote sensing', *Marine Ecology Progress Series*, vol 159, pp51–60

Nas, B., Karabork, H., Ekercin, S. and Berktay, A. (2009) 'Mapping chlorophyll-a through in-situ measurements and terra ASTER satellite data', *Environmental Monitoring and Assessment*, vol 157, nos 1–4, pp375–382

NOAA Coastal Services Centre (2009) 'Remote sensing for coastal management', www.csc.noaa.gov/crs/rs_apps/, accessed 8 November 2009

Olson, D. M. and Dinerstein, E. (2002) 'The global 200: Priority ecoregions for conservation', *Annals of the Missouri Botanical Garden*, vol 89, pp199–224

Ormeci, C., Sertel, E. and Sarikaya, O. (2008) 'Determination of chlorophyll-a amount in Golden Horn, Istanbul, Turkey using IKONOS and in situ data', *Environmental Monitoring and Assessment*, vol 155, nos 1–4, pp83–90

Parrott, D. R., Todd, B. J., Shaw, J., Hughes Clarke, J. E., Griffin, J. and MacGowan, B. (2008) 'Integration of multibeam bathymetry and LiDAR surveys of the Bay of Fundy, Canada', *Proceedings of the Canadian Hydrographic Conference* and *National Surveys Conference*, Victoria, British Columbia, Canada

Petrie, B., Drinkwater, K., Gregory, D., Pettipas, R. and Sandstrom, A. (1996) 'Temperature and salinity atlas for the Scotian Shelf and Gulf of Maine', *Canadian Technical Report of Hydrography and Ocean Sciences*, vol 171

Pingree, R. D. (1978) 'Mixing and Stabilisation of Phytoplankton Distributions on the Northwest European Shelf', in J. H. Steele (ed) *Spatial Patterns in Plankton Communities*, Plenum Press, New York

Roff, J.C. and Evans, S. (2002) 'Frameworks for marine conservation; Non-hierarchical approaches and distinctive habitats', *Aquatic Conservation: Marine and Freshwater Ecosystems*, vol 12, pp635–648

Roff, J. C. and Taylor, M. (2000) 'A geophysical classification system for marine conservation', *Journal of Aquatic Conservation: Marine and Freshwater Ecosystems*, vol 10, pp209–223

Roff, J. C., Taylor, M. E. and Laughren, J. (2003) 'Geophysical approaches to the classification, delineation and monitoring of marine habitats and their communities', *Aquatic Conservation, Marine and Freshwater Ecosystems*, vol 13, pp77–90

Sakuno, Y., Matsunga, T., Kozu, T. and Takayasu, K. (2002) 'Preliminary study of the monitoring for turbid coastal waters using a new satellite sensor, "ASTER"', Twelfth International Offshore and Polar Engineering Conference, Kyushu, Japan

Sherman, K., Sissenwine, M., Christensen, V., Duda, A., Hempel, G., Ibe, C., Levin, S., Lluch-Belda, D., Matishov, V., McGlade, G., O'Toole, M., Seitzinger, S., Serra, R., Skjoldal, H. R., Tang, Q., Thulin, J., Vanderweerd, V. and Zwanenburg, K. (2005) 'A global movement toward an ecosystem approach to management of marine resources', *Marine Ecology Progress Series*, vol 300, pp275–279

Singhruk, P. (2001) 'Circulation features in the Gulf of Thailand inferred from SeaWIFS data', www.aars-acrs.org/acrs/proceeding/ACRS2001/Papers/OCW-05.pdf, accessed 8 November 2009

Smith, W. and Sandwell, D. (1997) 'Global sea floor topography from satellite altimetry and ship depth soundings', *Science*, vol 26, no 5334, pp1956–1962

Spalding, M. D., Fox, H. E., Allen, G. R., Davidson, N., Ferdana, Z. A., Finlayson, M., Halpern, B. S., Jorge, M. A., Lombana, A. and Lourie, S. A. (2007) 'Marine ecoregions of the world: A bioregionalization of coastal and shelf areas', *Bioscience*, vol 57, no 7, pp573–584

Strub, P. T. and James, C. (1995) 'The large-scale summer circulation of the California current', *Geophysical Research Letters*, vol 22, no 3, pp207–210

Sutherland, T. F., Galloway, J., Loschiavo, R., Levings, C. D. and Hare, R. (2007) 'Calibration techniques and sampling resolution requirements for groundtruthing multibeam acoustic backscatter (EM3000) and QTC VIEW™ classification technology', *Estuarine, Coastal and Shelf Science*, vol 75, no 4, pp447–458

Tremblay, M. J. and Roff, J. C. (1983) 'Community gradients in the Scotian Shelf zooplankton', *Canadian Journal of Fisheries and Aquatic Sciences*, vol 40, pp598–611

Turner, W., Spector, S., Gardiner, N., Fladeland, M., Sterling, E. and Steininger, M. (2003) 'Remote sensing for biodiversity science and conservation', *Trends in Ecology and Evolution*, vol 18, no 6, pp306–314

Urban, D. L., O'Neill, R.V. and Shugart, H. H. (1987) 'Landscape ecology', *BioScience*, vol 37, pp119–127

Watson, J. (1997) 'A review of ecosystem classification: Delineating the Strait of Georgia', Department of Fisheries and Oceans, North Vancouver, British Columbia, Canada

Wilkinson, T., Wiken, E., Bezaury-Creel, J., Hourigan, T., Agardi, T., Herrmann, H., Janishevski, L., Madden, C., Morgan, L. and Padilla, M. (2009) *Marine Ecoregions of North America*, Commission for Environmental Cooperation, Montreal, Canada

Wunsch, C. and Stammer, D. (1998) 'Satellite altimetry, the marine geoid, and the oceanic general circulation', *Annual Review of Earth and Planetary Sciences*, vol 26, pp219–253

Yang, M. D., Sykes, R. M. and Merry, C. J. (2000) 'Estimation of algal biological parameters using water quality modeling and SPOT satellite data', *Ecological Modelling*, vol 125, no 1, pp1–13

Yuan, D., Zhu, J., Li, C. and Hu, D. (2008) 'Cross-shelf circulation in the Yellow and East China seas indicated by MODIS satellite observations', *Journal of Marine Systems*, vol 70, nos 1–2, pp134–149

Zacharias, M. A., Howes, D. E., Harper, J. R. and Wainwright, P. (1998) 'The British Columbia marine ecosystem classification: Rationale, development, and verification', *Coastal Management*, vol 26, pp105–124

Zacharias, M. A. and Roff, J. C. (2000) 'A hierarchical ecological approach to conserving marine biodiversity', *Conservation Biology*, vol 13, no 5, pp1327–1334

Zainal, A. J. M., Dalby, D. H. and Robinson, I. S. (1993) 'Monitoring marine ecological changes on the East Coast of Bahrain with Landsat TM', *Photogrammetric Engineering & Remote Sensing*, vol 59, pp415–421

Zainuddin, M., Kiyofuji, H., Saitoh, K. and Saitoh, S. (2006) 'Using multi-sensor satellite remote sensing and catch data to detect ocean hot spots for albacore (Thussus alulunga) in the northwestern north Pacific', *Deep Sea Research Part II: Topical Studies in Oceanography*, vol 53, nos 3–4, pp419–431

6

Habitats and Communities: Ecoregional to Local

Reality, variability and scales of relationships

We started off trying to set up a small anarchist community, but people wouldn't obey the rules.

Alan Bennett (1934–)

Introduction

A major goal of marine conservation is the preservation of all elements of marine biodiversity, including both representative areas and distinctive areas (see Chapter 7). However, almost everywhere in the oceans, we lack sufficient direct information to make informed decisions on such conservation goals. Consequently a series of methods, using surrogate measures of some aspects of biodiversity or characteristics (factors) of the marine environment, have been developed to map marine habitats and their presumed communities, based on broad-scale geophysical data, or regional geomorphic features, or in more local areas based on the use of multibeam and sidescan sonar techniques. Nevertheless, there is still limited experience of how to use such data in marine conservation planning, how to 'calibrate' it at the appropriate spatial scale, and how to develop a comprehensive strategy for marine conservation employing surrogate relationships.

As we have previously argued, mapping of habitats and community types is an indispensible prelude to almost any kind of planning for the marine environment – indeed it is the basis of marine ecosystem-based management. However, for comprehensive planning for marine conservation we need to go from the global to the local scale of mapping and to descriptions of communities and habitats. This can present several problems.

At a global scale, relationships between communities and their habitats becomes global biogeography – considered as provinces and ecoregions in the previous chapter. In this chapter we continue to follow the spatial hierarchical classification (Table 5.1) to lower levels and scales, to pursue the search for habitat–biological relationships. This takes us from the ecoregional to the microcommunity level as we consider the problems encountered.

Time and space scales and the types of ecological species interact strongly (see Figures 2.5 and 2.6, and Tables 3.3 and 3.4). By comparing Tables 3.3 and 3.4 with Table 5.1, it will also be immediately apparent that species types are referenced to different levels of the spatial hierarchical classification. Large migrant whales and fish pay little attention to any kind of marine ecological boundary, and are driven in their migrations presumably by considerations of seasonal reproduction and resources. These organisms are considered more fully in Chapters 7 and 9. Planktonic organisms are essentially defined by large scale ecoregions (e.g. Longhurst, 1998), and other species types are defined by readily recognizable community types – for example, coral and salt marshes. Attention in this chapter will focus largely on demersal fish and benthic invertebrates from levels 3 and lower of Table 5.1.

In the previous chapter we have shown how geophysical factors can be used to describe and map habitat types using remotely sensed and in situ data. We are generally forced to rely on ecosystem-level structures to guide us to differentiate among habitat types and their probable community types because of the general paucity of data on the distributions of organisms themselves at the needed scale of observation.

The reliability of geophysical factors to describe habitats, which in turn act as surrogates for marine communities, hinges on how well physical and biological systems are correlated. Most importantly it depends on:

- The precise meanings we give to terminology applied to communities.
- How far down the spatial ecological hierarchy (as summarized in Table 5.1) habitat-community relationships are reliable.
- The scale at which observations are made and sampling is conducted.
- The heterogeneity of the environment.
- The natural variability of marine communities (biotopes) in time and space.
- Aliasing of physical and biological observations.
- Whether observations are quantitative or qualitative in nature.

Fundamental ecological relationships between habitat characteristics and biological communities are often taken for granted in the marine ecological literature, and conventional descriptions of marine habitats and their communities do not sufficiently acknowledge the complications and uncertainties. However, these issues become important for the purposes of planning for the conservation of marine biodiversity. In this chapter we need to examine more closely the relationships between habitats and communities, to ensure that we are proceeding on firm ecological grounds. The community level is important because this is the level of the ecological hierarchy where significant conservation efforts have been expended. It is also the level that describes a major component of marine biodiversity, namely species diversity itself – which is considered in Chapter 8.

Although there is significant current interest in biological 'hotspots' (see Chapter 8) – which regionally or locally exhibit high species diversity – we should recognize that collectively, the ordinary or representative areas of the oceans account for far higher species diversity than the hotspots themselves. The areas most difficult to account for in mapping marine communities are the regions away from the continental shelves that are not amenable to direct observation. We shall deal with these high seas/deep seas areas in Chapter 12.

What is a marine community?

Between the ecosystem and the species levels of ecological organization lies the community. The term 'community' can be used in a multitude of ways: taxonomically – for example as a community of bird or marine mammal or fish species; trophically – for example the plant community or the community of predators on some species or group of species; spatially – for example as a localized set of species; ecologically – for example a benthic community; hierarchically – for example a pelagic → planktonic → zooplankton → crustacean → copepod community. These categories may also be used in combinations; thus we may speak of the 'planktonic community of the oceans as a whole', or the *Fucus vesiculosus* community of rocky shores, or a local community of birds or marine mammals. Humpty Dumpty (who said: 'Words mean whatever I want them to mean') would have been proud of the term; it can mean almost anything we want it to mean.

There has been much discussion – mainly in the terrestrial ecological literature – as to whether communities are in fact 'real' (see e.g. Krebs, 1972; Austin, 1985 for a review). The question of what, precisely, a marine benthic community is dates back at least to Petersen (1913) followed by MacGinitie (1939). Since those times, in localized studies in coastal and continental shelf waters, benthic species composition has been conventionally described in terms of community types, defined as preferred species assemblages, associated with specific habitat characteristics – especially substrate type – and characterized by 'indicator species'. The identity of organism communities as 'ecological units' was examined by Thorsen (1957) who reviewed the then competing concepts of a community as statistical units determined only by requirements for the same environmental factors (e.g. Petersen, 1913), or as assemblages with biological interactions (e.g. MacGinitie, 1939).

The maximum requirement for a community is that its component species should be interacting according to some ecological processes (competition, predation etc.) though such interactions can generally only be demonstrated under experimental conditions (see e.g. Lalli, 1990). The minimum requirement for a community is that it is simply a set of several species within an area – that is, it is equivalent to a species assemblage – or a statistically 'recurrent group' – defined by its species (structures) only.

Some ecologists claim that collections of almost the same species recur in the same kind of habitat in space and time. A third group would claim that communities exhibit homeostasis, that is, a tendency to a steady state following disturbance. Other ecologists regard communities as more flexible in composition. This is the essential difference between the 'niche-assembled' and 'dispersal-assembled' concepts of community composition (Hubbell, 2001).

The key ecological question is: Is a community anything more than an abstraction made by ecologists from continuously varying distributions of biota? In other words, the central issue was (and still is) whether marine benthic communities are simply a statistical entity with species distributed largely independently each in relation to its tolerance of environmental factors, or whether significant biological interactions shape such distributions, leading to preferred associations and species indicators of community composition. Some of these issues are summarized in Box 6.1.

As is often the case with scientific arguments, all groups of advocates may be correct under some circumstances (see Roff, 2004); it is simply that often the 'circumstances' are not well defined. Here we shall largely avoid this argument, though later in this chapter suggesting a resolution to it, and use the term community in the neutral sense of a species assemblage (not necessarily constant in composition) that typifies a kind of (nearly) definable habitat (see Pimm, 1991).

Thus there are several problems with the term community as it is often used. We never know all the species in any community and little or nothing about most of their interactions and dependencies; mostly we make assumptions or extrapolations of processes and interactions.

As we typically recognize them, communities are not necessarily continuous or contiguous, and do not even occupy all space as habitats do. Despite these (and other) problems, the term community is useful because people instinctively recognize community types, especially biogenic communities (also called foundational or engineering species, Colour plates 10a and b). In fact habitats and communities may be recognized interchangeably – for example, coral reefs and mangrove forests.

In contrast to a community, a habitat seems much simpler to define. We define it simply as a type of marine environment, characterized by its geophysical structures. However, in Chapter 8 it will become apparent that even this definition is not straightforward. Finally, it is important to note that, in contrast to ecosystems, both communities and habitats are hierarchical. Thus we can define communities within communities and habitats within habitats. For example: the pelagic community contains planktonic communities which in turn contain phytoplankton communities of different cell sizes. Also a rocky shore habitat may be exposed or sheltered and have a greater or lesser slope, and so on. Because biotopes are the associations of communities and their habitats, this means that biotopes in turn are also hierarchical, as indicated in Table 5.1.

The conventional view of marine communities

The recognition of different marine communities as entities containing distinct assemblages of

Box 6.1 Some (interim) conclusions on community structure, important for marine conservation planning

1 Communities may or may not appear to be 'real', for several reasons:
   - Different communities of organisms will be limited in their distributions by different factors at different scales. Because the individual component species will have different geographic ranges and dispersal (or migration) abilities, this automatically implies that any given community type will vary in species composition across its range.
   - Community types overlap in their distributions, producing ecozones and ecotones. Consequently the most useful description of community distributions is some measure of community predominance at any location. Dominance can be described in terms of biomass, numerical abundance, percentage cover and so on.
   - Community 'structure' is determined either by niche-based assembly or by dispersal assembly (or some variable combination of the two). The second better explains the variability of species composition of natural communities (see Hubbell, 2001).
   - A community may be of relatively constant composition where the range of substrate types is restricted and the number of species able to colonize is also restricted (e.g. temperate intertidal zones). A community may have very variable composition where the range of substrate types is restricted but the number of species able to colonize is large (e.g. deep sea soft-bottom substrates).
   - Community composition can change due to several factors including biological and physical disturbances. After a disturbance, a community may be reset in more-or-less the original composition, or otherwise. It may re-attain the original state after a period of succession.
   - Community composition is variable over space, time and seasons.
   - The scale of observation may extend over several communities which merge into one another.

2 Biogenic communities (Table 3.4) are the ones most readily recognized and generally most consistent in species composition.

3 Indicator species (composition indicators) may be unreliable depending upon:
   - the method of sampling, and whether data is quantitative or qualitative (e.g. presence-absence data);
   - the movement or dispersal ability of individual species;
   - the geographic range of a species and whether it is restricted to a particular biotope or a wider realized habitat (Figure 6.1);
   - the scale of observation which may merge biotopes.

species has a long history. Möbius may have been the first to recognize discrete assemblages of marine organisms as communities (or biocoenoses) in his 1877 study of oyster reefs in the North Sea. This was followed by a general recognition of many community types according to the assemblage of species represented (see e.g. Hedgpeth, 1957). However, a framework to define representative areas based solely on knowledge and mapping of the biological communities present in the oceans would require extensive studies on biological distributions and would be prohibitively expensive to produce. The biological data necessary for such analyses simply does not exist at appropriate scales.

However, the paradigm of a community level of biological organization has become firmly entrenched in the ecological literature. Conventional descriptions of marine communities and their associated habitat types are to be found in all marine ecology text books (e.g. Barnes and Hughes, 1988; Kingsford and Battershill, 2000; Bertnes et al, 2001; see also Chapter 3). The impression that these and other similar texts convey is that marine communities are of defined and distinct types, clear and unmixed with defined species compositions that are strongly related to geophysically specified habitat types – that is, they are niche-assembled (Hubbell, 2001); reality can be more muddled.

In tropical and subtropical seas, many types of marine communities are clearly recognizable in coastal waters. Here, directly visible biogenic communities may dominate a particular region. These are perhaps the easiest to define and map in shallow waters – usually by direct observation. Particularly in tropical and subtropical waters, where geophysical variables such as temperature and salinity are relatively constant over wide areas, mapping of community types by direct observation, diving surveys or deployment of instruments in situ in shallow waters may be the only option to establish representative areas (e.g. Roff et al, 2003). Such biogenic communities include mangroves, coral reefs, seagrasses, saltmarshes and macro-algal beds. Distributions of coral reefs of different kinds can also be mapped for example from degree of exposure, depth or light penetration as determining variables. In fact, all these biogenic communities are relatively easy to define in terms of their actual or potential distributions.

In contrast, in temperate areas and deeper waters, where direct appreciation of community types is not possible, our geophysical recognition and classification of habitats, starting at larger scales, should ultimately coincide with recognition of community types (at some level of the spatial hierarchical scale – Table 5.1) at regional to local scales.

Community structures and processes

Biological and physical influences on communities

Both physical and biological effects shape natural communities (Table 3.5). There is little point in debating the relative importance of each set of factors, because any community is at some point determined by a composite of both sets. In general, at higher levels of the spatial hierarchy (Table 5.1), where physical effects predominate, it is assumed that they will largely control the type and composition of the community present. At the lower levels of the hierarchy, and in more physically benign environments, biological factors are generally assumed to become dominant in shaping communities.

Despite the fact that biological factors are obviously important in determining local species compositions, we still have a poor understanding of how these factors operate at larger ecological scales. The essential problems are: How do these factors operate to structure marine communities, and how can we measure and map them in space and time? Although the influence of several biological factors at the community level can be investigated experimentally, and although marine ecologists have good ideas as to how these processes act locally to structure marine communities, they may provide little aid for making conservation decisions.

Variability of composition in marine communities

In its simplest form, the analysis of biological communities involves consideration of the ***statics*** of marine systems, for example species presence/absence, relative species abundances, and distribution patterns. Communities themselves are of course not static in composition. Any community type will vary in its species composition depending upon the relative and combined disturbances and influences of several physical and biological factors (Box 6.1).

All marine biological communities are subject to disturbances on various time and space scales (Mann and Lazier, 1996; Sousa, 2001). These disturbances may be: physically induced by storms, tides, upwelling events, stratification and other seasonal cycles, atmosphere–ocean regime shifts (e.g. Steele, 1988) and so on; biologically induced by competition, predation, keystone effects, species invasions and so on; and induced by human actions from fishing, dumping, pollution effects and so on. The effects of any disturbance regime are a function of magnitude, type, periodicity and organism size. Disturbance may be a significant contribution to total mortality in many species assemblages (Woodin, 1978), and sessile communities are more severely affected than motile ones. Modelling exercises (e.g. Caswell and Cohen, 1991) make clear that the diversity of metapopulations reflects the interactions among organisms with respect to the scales of disturbance and dispersal.

The intermediate disturbance hypothesis (Connell, 1978) has significant explanatory power for the diversity of marine communities (see e.g. Zacharias and Roff, 2001), with both low and high frequencies and severities of disturbance leading to lower species diversity. Accordingly, the disturbance regime (whether physical, biological or human) and the scale of the effects experienced by a particular kind of habitat have important consequences for the determination of the numbers, locations and sizes of marine protection areas (MPAs) within a region.

Physical effects

Physical effects due to storms and hurricanes can severely affect areas of the intertidal and immediate sub-tidal for tens or hundreds of kilometres in extent. Physical disturbances may occur at sufficiently frequent intervals, and recolonization events may be sufficiently slow, that intertidal and sub-tidal communities may not reach a climax assemblage between disturbance events. This appears to be especially true of structurally complex communities, with slow-growing member species, such as the coral reef communities of the Great Barrier Reef (Tanner et al, 1994). Other physical coastal oceanographic processes, such as upwelling events, can also dramatically affect the structure of regional communities (e.g. Menge et al, 2003).

Biological effects

Biological disturbance effects are predominantly due to predators and generally affect smaller areas than physical disturbances (e.g. Thrush, 1999); however, they may be far more frequent. Recolonization events may occur quickly, within days to months. The extent of effects range from small-scale effects of invertebrate predators in the intertidal zone, to metres to tens of metres in extent in sub-tidal regions due to vertebrate predators. Here marine mammals such as sea-otters (Kvitek et al, 1988), walrus (Oliver et al, 1985) and grey whales (Oliver and Slattery, 1985) disturb the substrate during their feeding excursions and excavations.

The presence or absence of a keystone species (see Chapter 9) may have especially dramatic effects upon the relative abundances of species within a community. For example the sub-tidal community of temperate rocky shores can vary greatly in relative species composition from dominance by macrophytic algae (Laminarians) to dominance by sea urchins, to almost bare rock, depending on local predator–prey and disease dynamics (Scheibling et al, 1999, Colour plates 11a, b and c). However, such apparently different communities are in fact phases of evolution of community structure (in other words succession) within a specific habitat type. This is a biotope shift along highly variable axes of relative species composition. Such variations in relative abundances of component species should not distract us from the recognition of biotopes and habitats.

Species may often be limited in their distributions by some combination of predation and resources, or by competition with other taxonomically related or unrelated species, and it is to be expected that biological factors limiting a species distribution will fluctuate in space and time. There is still considerable debate, for example, as to whether competition or predation is the major biological factor in determining the sets of species found together – that is, the community composition. We have no overall theory that we could apply to marine communities as to how these two major factors may interact or how each might become dominant or secondary in space and time. We cannot therefore hope to find any general relationships in these factors on which to base a conservation strategy.

Competition itself is clearly an important force in shaping community structure (see e.g. Valiela, 1995). In general, however, what we witness is not the effect of competition in action, but rather the end result of its action – that is, the 'ghost of competition past' (Sale, 2004). It is therefore not clear how we might effectively allow for the effects of competition in our conservation strategies.

Human effects

Human effects in coastal marine environments have at least two important competition-related effects. First, by the introduction of exotic species which can often have a competitive advantage (sometimes due to lack of their own competitors or predators) over the endemic species. Second, by physical changes in the environment humans make it preferentially more conducive to one species giving it a competitive edge. Thus competition can be a resultant effect that could be indirectly controlled but not directly managed.

It is now evident that predator effects may also be important in at least two different ways: either as natural succession effects, or as a result of human actions (largely in fisheries). Human effects on natural marine communities are becoming severe and can extensively alter the local species composition of marine communities. For example, a review by Auster and Langton (1999) reports that heavily fished areas of the North Sea and Middle Atlantic Bight may be disturbed by trawling activity from once to 50 times per year. There is also some indirect evidence that effects of removing top marine predators are having substantial effects on the structure of marine communities (Frank et al, 2005).

Other aspects of fishing activities are considered further in Chapter 13. Such human effects will not be fully considered here because our objective is to recognize and distinguish among the fundamental marine community types as far as possible, irrespective of human actions.

Community–habitat relationships: General considerations

Marine community types (as opposed to defined communities of specific species) are recognizable across widely separated geographic areas. Thus we can recognize sub-tidal communities dominated by macrophytic algae in the Atlantic, Pacific and Arctic oceans. In each geographic region, the species present and species assemblages are different, but the same types of communities with different species playing equivalent ecological roles – the biomes – are recognizable, irrespective of the specific geographic location.

Carl Petersen (1860–1928) was among the first to appreciate that marine benthic communities, as recurrent assemblages of species, can be related to the character of the environment. These relationships gave rise to the concept of the *biotope*, which links community species composition and habitat types. The significance of this linkage between the type of community and environmental characteristics is the foundation of marine spatial planning, and environmental classifications and frameworks. Marine conservation efforts seek to recognize distinguishable marine habitats, their environmental characteristics and their contained communities (in other words marine biotopes). At the finest scale of community analysis, these should ultimately represent the recurrent groups of marine species (groups of repeatedly co-occurring species), closely related to a particular type of physical environment – a habitat.

Most of our perceptions of marine communities have been shaped by observations and experimental manipulations in the intertidal and immediate sub-tidal zones. Here the community types are relatively well defined, and following disturbances they generally reset themselves in much the same species composition as before – at least in terms of the dominant species – because the pool of species able to live here is relatively limited. Physical factors (e.g. desiccation) and biological factors (e.g. predation and competition) interact to determine community composition and structure.

In sub-tidal regions, predominantly on soft bottoms, a series of community types and indicator species have been described – largely from (semi-) quantitative samples taken with a variety of grabs or dredges. There is a generally recognized correspondence between substrate type and community type, with strong separation between suspension feeders in regions of coarser substrates and stronger currents, and deposit feeders (which rework the sediments making them unsuitable for suspension feeders) in regions of finer substrates and weaker currents (Wildish 1977).

However, a confusing array of relationships has been described in the literature between habitat variables and community types; relationships in one region or study may not be consistent in other regions, depths and so forth, or at other scales. Undoubtedly some of the apparent discrepancies can be accounted for by the degree to which individual physical variables actually vary in any given study. Also natural marine communities are fragmented rather than continuous, and community types do not tend to follow a convenient and continuously decreasing spatial scale in the hierarchy. For example, along a coastline we find bits of rocky intertidal communities interspersed with the communities of muddy bays. Each of these habitats may occur at a very local scale, often only hundreds of metres or less. For older surveys, precise locations of biological and companion environmental data were often not available, leading to possible 'aliasing' of physical and biological data. The degree to which this may be a problem also depends on the heterogeneity of the region itself.

Because of this apparent variability we have at present no general paradigm of relationships between communities and habitat types for conservation purposes, only broad descriptions and correlations. Indeed a measure of our collective confusion is the fact that the descriptors of the units of the marine spatial hierarchy (Table 5.1) wander between geophysical and biological factors and their combinations. Although we can describe community distributions in relation to habitat variables, we generally have little idea what the limiting factors really are, or even whether physical or biological. Ideally we would have predictive relationships among geophysical variables and biological communities.

The arguments summarized by Zajac (2008), Gray (1994) and Snelgrove (1999) among others seem to imply that benthic species are very 'fussy' in their 'choice' of substrate features, including patch size and characteristics and neighbouring species; that is – community membership is niche-assembled. However, this leads us to a never-ending search to describe the fine details of species' habitat requirements – in a way analogous to the unattainable description of niche as an 'abstractly inhabited hypervolume' (Hutchinson, 1957). The distinction between 'niche' and 'habitat' is discussed by Whittaker et al (1973).

The notion of communities as niche-assembled is essentially untestable (Hubbell, 2001) because we cannot know how many niches are available within any habitat. Given also that (by definition) each niche is occupied by a single species, all we can say about a particular habitat is how many niches are presently occupied (because this equals the number of species!). The description of a niche is therefore simply the description of a species (see Krebs, 1972). The probability is that many benthic species are quite catholic or opportunistic with respect to their habitat requirements, while others are strict and particular protestants; what we do not know at present is the relative proportions of the two types. The long-held ecological distinction between R and k selected species is perhaps a parallel to these ideas.

How many habitats are there in the sea?

The argument with habitats is nearly (but not) as difficult as that for niches. The advantage for habitats is that we can geophysically describe them, even though they are often not in practice adequately described, whereas we cannot describe niches. Habitats may be over-described – that is, a recognized biocoenosis occupies more than one habitat, or under-described – that is, more than one recognized biocoenosis occupies a habitat. In fact both these instances occur, depending on the level of taxonomy, and at what scale and in what detail habitats are described.

The distribution of habitat types is unknown in most marine areas (Halpern et al, 2008), and most MPAs have been established without a proper knowledge of the distribution patterns of habitat diversity. Indeed, the seminal question 'How many habitats are there in the sea?' is only just being asked (Fraschetti et al, 2008). Various approaches to this important question have been suggested (see e.g. Orpin and Kostylev, 2006; Post, 2008; Verfaillie et al, 2009). Diaz et al (2004) suggest that the disparity between information in physical and biological factors is hindering the applicability and acceptability of benthic habitat mapping efforts. In addition to the lack of basic information on the biological and environmental tolerances of benthic species, the problem of data mismatch between physical and biological methods does not yet permit the resolution of the elusive components that make a physical substrate a definable habitat. The choice of geophysical surrogates to indicate community types and habitats is therefore still something of an art, because relationships seem to vary among regions and across scales.

By analogy to the concept of a fundamental versus a realized niche, a similar idea may be applied to a species habitat, except that it may have three components (Figure 6.1). These are: the predominant habitat (where a species attains its greatest numerical or biomass abundance); the realized habitat (where a species still exists but in reduced numbers); and the fundamental or potential habitat which a species could inhabit but for some restriction (generally biological such as predation). The significance of this suggestion will become evident later in this chapter. A biotope then becomes simply a region of the environment where the predominant habitats of several species (collectively a biocoenosis) spatially coincide. Such 'communities' are predominantly dispersal assembled, and limited by the availability of habitat.

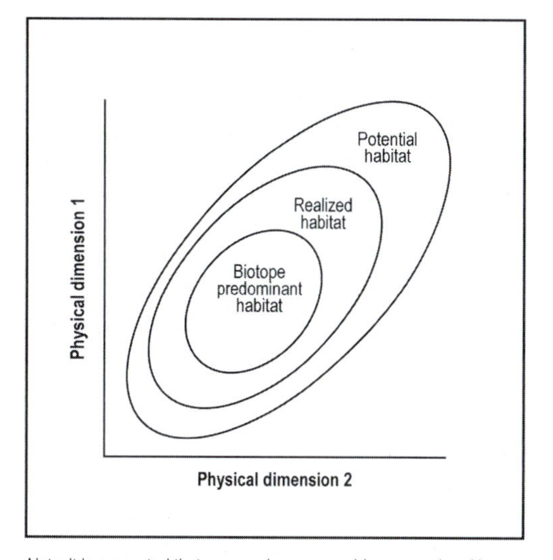

Note: It is suggested that any species or assemblage occupies either: a region where it numerically predominates, or a further region where it may occur in lower numbers – a realized habitat. A biotope or species assemblage, perhaps characterized by indicator species, occurs where a set of species each reaches its population maximum.

Figure 6.1 Visualization of the distribution of an individual species or a species assemblage in habitat space by analogy to the concept of a niche (see text).

Habitats become available after some disturbance creates colonizable space. Given that recolonization of the benthos can only occur after some disturbance – which must reset and presumably homogenize environmental conditions at some scale – a more parsimonious and testable null hypothesis suggests itself, based on Neutral Theory (see Hubbell, 2001). Following disturbance, the local area of the benthos – of some habitat type – is randomly recolonized by any combination of those species which can potentially inhabit that habitat, and which happen to be reproducing just at or just after that time (Munguia et al, 2010). This is essentially the larval lottery of Sale (2004), and is also consistent with our analyses of the Marine Recorder data (see below). The idea favours the concept of a community as dispersal-assembled rather than the dominant view of communities as niche-assembled. It is becoming clear that interactions between disturbance regimes and the dispersal abilities of organisms are vital determinants of the character of benthic communities (Lundquist et al, 2010).

In intertidal communities there may be little practical difference between niche-assembled and dispersal-assembled communities. However, in deeper waters as species diversity increases, and space becomes randomly available, the larval lottery yields variable and patchy community compositions (see Chapter 8).

Within the pelagic realm, physical conditions are relatively uniform over broad areas, and temperature, salinity and even depth may be relatively constant. Monitoring of water column conditions, for example by satellite data, may therefore be sufficient to map the pelagic realm over an entire region. However, such uniformity of the environment is clearly not the case for the benthos. Much of the following chapter therefore deals with examples of relationships between demersal fish and benthos and the geophysical features of their environments and habitats.

These discussions of niches, biocoenoses and habitats may seem overly arcane, but they have very serious practical consequences. Despite all these complications, we adhere to the pragmatic argument that habitats are more fundamental units for conservation planning than communities. Indeed this is the practical basis for ecosystem-based decision-making and conservation. Habitats have measurable and definable characteristics, even though at some scale or level of definition the biotas inhabiting them may be variable in community composition.

In this chapter we shall try to draw out some general principles of community–habitat relationships applicable to conservation, stressing the importance of scale of observation, the potential for aliasing of data and that different community types require different approaches and techniques. We are now dealing with levels of the spatial hierarchical classification from the ecoregion and below (Table 5.1).

Community–habitat relationships: Ecoregion and regions

At the ecoregion level, largely below the euphotic zone or the thermocline, the fish community is the one most likely to have available quantitative data in most areas of the world. This is also the community with which the public tends to identify and the one often of greatest commercial significance. Fortunately many studies around the world have described and mapped the species assemblages of demersal fishes, and sought to relate those assemblages to geophysical environmental factors and to establish biogeographical boundaries.

Typical of such studies is the paper by Mahon et al (1998), in which demersal fish trawl survey data for a 20-year period up to 1994 were analysed from a broad region of Atlantic Canada, over an area comprising ecoregion 39 of Spalding et al (2007) and parts of their ecoregions 37, 38 and 40. Visual classification of distribution maps for the 108 most abundant demersal species revealed nine species groups, based on both geography and depth distribution. Using multivariate statistical techniques, 18 assemblage groups – which explained 56.3 per cent of the variance in distribution of the species – were identified. These species assemblages were persistent in composition through time, but appeared to shift in location. However, because there was considerable overlap among the ranges of the species assemblages (in other words there were ecotones or ecozones where species assemblages graded into one another), we cannot judge where the biogeographical boundaries are or which assemblage dominates at any location. The predominant habitat (biotope) of any species assemblage was therefore not evident from this analysis.

A more revealing analysis of a similar data set to that of Mahon et al (1998) – but covering the years 1970 to 2001 in four time blocks, and comprising predominantly ecoregion 39 of Spalding et al (2007) – was carried out by Zwanenburg and Jaureguizar (unpublished study), in order to illustrate regional biogeographic boundaries, where predominance passes from one species assemblage to another. Zwanenburg and Jaureguizar proceeded as follows:

Species composition (expressed as mean weight per species per standard survey trawl haul) for each survey sampling station formed the basic observations for the analyses. For each sampling station, simultaneously collected information on environmental conditions (depth, bottom temperature and salinity) was used to determine which factors were most influential in determining species composition and distribution. The fish assemblage that predominated at each sampling station (numerically or as biomass), and the associated environmental variables were identified using de-trended canonical correspondence analysis (DCCA). To characterize differences in the species composition within assemblage areas among time blocks, Zwanenburg and Jaureguizar further defined *resident* species as those that were typical in all four time periods and *opportunistic* species as those that were not consistently common in an area and were likely periodic migrants.

The two axes of the DCCA representing linear combinations of depth, salinity and temperature together explained between 61.3 and 69.5 per cent of total variation in species composition in each time block. The first axis was most highly correlated with depth and salinity, while the second axis was most highly correlated with bottom temperature. Zwanenburg and Jaureguizar identified fish assemblages by identifying groups of stations in each time block that clustered along the environmental axes derived by the DCCA. Groups of stations were separated from one another by identifying discontinuities between the clusters of stations (see Jaureguizar et al, 2003). These in turn delineated geographically contiguous and adjacent groups of stations when mapped back on the geographic space of the Scotian Shelf and Bay of Fundy.

The geographic areas occupied by these groupings in each time block were compared, and the four most consistent groups were selected to provide the most reasonable and interpretable assemblage areas (Colour Plate 12a). These distributions of non-overlapping species assemblages now essentially specify the fish biotopes or predominant habitat of each assemblage (refer to the concept of Figure 6.1).

Changes in the abundances of resident species generally contributed less than 50 per cent (29–50 per cent) to overall variations across time blocks, while opportunistic species contributed the remainder. However, these assemblage areas were consistently distinguishable from each other by virtue of both their environmental conditions and the mix of fish species that inhabited them. Their boundaries showed a remarkable degree of temporal consistency, even though their exact location wandered over time. The assemblage areas were therefore quite clearly distinguishable from each other in all time blocks, and occupied relatively the same geography in each time period. They also show strong relation to the bottom water masses of the Scotian Shelf, mapped quite independently by Roff and Alidina (unpublished data – see Colour Plate 12b).

A comparison of the results of Zwanenburg and Jaureguizar with Mahon et al (1998) is essentially the difference between defining the biotopes of a region (as the predominant habitat of an assemblage or community) and defining the realized habitat of an assemblage – which it spatially occupies but where it does not predominate (Figure 6.1). The analysis of non-overlapping, predominant habitats therefore yields a very clear picture of the divisions of the marine environment and relationships between species assemblages and environmental factors.

Note that these areas or regions identified by geophysical factors and fish species assemblages do not relate to substrate characteristics – that is, to the lower levels of the spatial hierarchy as defined in Table 5.1 (which emphasizes the benthos). However, they can be considered at least as 'regions' according to the spatial hierarchy, or as pelagic (demersal referenced) seascapes or as primary habitats – for fish. These fish assemblages (biocoenoses) clearly also meet the definition of a biotope (a specific assemblage or community of species associated with a definable habitat). This should indicate to us that the concepts of 'habitat' and 'community' in fact span the spatial hierarchy of Table 5.1, and that they are a function of the size and mobility of the organisms with which we elect to define these terms.

Community-habitat relationships: Geomorphic units and habitats

In shallow coastal waters, the major habitats and their communities – especially the major ecosystem engineering organisms and their attendant communities – such as mangroves, coral reefs and seagrass beds can be directly mapped by aerial photographic surveys, bathymetry, drop camera surveys and so on. In many regions of the world such data has been accumulated and mapped.

Beyond these self-evident habitats and communities and the fish community, most planning for marine conservation in deeper coastal and shelf waters will recognize geomorphic units and/or seascapes, and is likely to emphasize the benthos and the component primary and secondary habitats. Ultimately, for any purpose of ecosystem-based management within a region, we need to map as precisely as possible the distributions of biotopes (species assemblages) themselves. This has historically been done by direct sampling of marine organisms (fish and benthos) by various techniques.

Mapping of the distributions of habitats or their communities at lower levels of the spatial hierarchy generally involves collection of data in situ and a direct examination of relationships between biota and environmental characteristics. The major problem is that such survey work is time and labour intensive. What is required is a faster way to predict benthic communities from geophysical data.

One technique now commonly in use is to conduct multibeam surveys, which operate by ensonifying a narrow strip of sea floor, detecting the resulting bottom echo, moving on to survey the next neighbouring strip and then integrating the data. These multibeam surveys are often interpreted in combination with measurements of backscatter intensity, seismic reflection, sidescan sonar sonograms and geological grab samples of seafloor materials, together with photographic images. The advent of multibeam and sidescan sonar technologies has revolutionized geological mapping (Courtney and Shaw, 2000). Multibeam data and derived images can reveal previously unrecognized seafloor morphological and sediment textural attributes, and can illustrate the mosaic nature and habitat complexity of a region (e.g. Auster et al, 1998).

Taken together these efficient survey techniques can be highly informative, can cover wide areas of the benthos and can provide a series of representations of the topography, substrate type and particle sizes of the seafloor. Analysis of the resultant data can lead to much improved description and understanding of habitat–community relationships in the benthos. Such surveys, employing combinations of physical and biological sampling techniques, have now become common around the world. A single example from Kostylev et al (2001) will illustrate the sampling process and interpretation of data. It is reported here in some detail because it represents a novel and informative combination of biological and geophysical techniques.

Georges Bank: A geomorphic unit

Kostylev et al (2001) undertook an interdisciplinary habitat mapping study based on analysis of macrobenthos (organisms larger than 1cm in linear dimensions) identified from seafloor photographs integrated with an interpretation of multibeam bathymetric data and associated geoscientific information. The biological objectives were to discriminate distinct assemblages of macrofauna benthic species, to understand and correlate the relationship between seafloor surficial sediments and biota, and to classify and map the defined benthic habitats. The study area was an entire geomorphic unit – Browns Bank – located at the southwestern end of the Scotian Shelf (see Figure 16.1B). The Scotian Shelf is a formerly glaciated continental shelf characterized by a series of large, shallow banks on the outer shelf, of which Browns Bank is one example. Oceanographic characteristics, circulation patterns and current speeds are known from previous studies in the area.

Survey methods

Multibeam bathymetric data were collected by the Canadian Hydrographic Service on Browns Bank in 1996 and 1997 using a ship equipped with a Simrad EM1000 multibeam bathymetric system. The swath of seafloor imaged on each survey line was five to six times the water depth. Line spacing was about three to four times water depth to provide ensonification overlap between adjacent lines. A differential global positioning system was used for navigation, providing positional accuracy of ±3m. Survey speeds averaged 14 knots resulting in an average data collection rate of about 5.0km^2 h^{-1} in water depths of 35–70m.

Multibeam bathymetric data were extracted from the datagrams and were gridded in 10m (horizontal) bins and shaded with artificial illumination using software developed by the Ocean Mapping Group at the Geological Survey of Canada (Atlantic). Relief maps, colour-coded to depth, were also developed. In addition to the bathymetric data, backscatter strengths ranging from 0 to 128 decibels (dB) were logged by the Simrad EM1000 system (see Urick, 1983; Mitchell and Somers, 1989).

To complement the multibeam survey, high-resolution geophysical profiles were also collected over Browns Bank in 1998, by deploying a deep tow seismic boomer, and a Simrad MS992 sidescan sonar (120 and 330kHz). The geophysical survey investigated different seafloor types and features identified from the multibeam bathymetric and backscatter data. Using an Institutt for Kontinentalsokkelundersøkelser (IKU) grab, 24 seafloor sediment samples were also collected at different seafloor types interpreted from geophysical profiles. The grab sampler penetrated the seafloor up to 0.5m and preserved the integrity of the layering within the surficial sediments. The sites chosen for collecting seafloor sediment samples were representative of broad areas of similar geomorphology and acoustic backscatter response.

With locations chosen from the multibeam bathymetric data and the geophysical profiles, 26 seafloor photographic sites were occupied in 1998. The 1998 set of seafloor photographs was augmented by 515 single and stereo photographs collected at 90 stations on Browns Bank by the Department of Fisheries and Oceans Canada (DFO) in 1984 and 1985. A total of 24 grab stations and 115 photographic stations were studied. The relative seafloor cover of boulders (> 256mm), gravel (2 to 256mm), sand (0.0625 to 2mm) and shell hash was estimated from each photograph, and the presence of fine-grained sediments was ranked on a scale from 0 (none) to 100 per cent (very abundant).

Identification of benthic fauna to the lowest possible taxonomic level, along with local geological description, was undertaken for the camera stations and the sediment samples. Subsamples of the grab samples were sieved through a 1mm mesh sieve and analysed for the presence or absence of species in order to augment the identification of fauna from seafloor photographs. Species were grouped into 22 major taxa, chosen on the basis of similarity of habitats and life history traits of benthic organisms.

Sediment–benthos relationships

The general trend of the distribution of benthic macrofauna on Browns Bank showed a predominance of suspension-feeders (e.g. the scallop *Placopecten*, Sabellidae) on the western, shallower part of the bank and an increase in abundance of deposit feeders with increasing depth towards the east. No macrofauna was observed on large sand bedforms. Structurally complex gravel habitats (in other words exhibiting a wide grain size variability) on the central and eastern parts of the bank are the most diverse and have the greatest abundance of sessile epifauna. Some sediments contain fine-grained fractions of silt and clay, which are trapped in crevices and depressions among boulders and other gravel, contributing to the overall heterogeneity of the substrates.

Dissimilarity analysis had shown that a combination of sediment type and depth yields the highest correlation with community structure suggesting that these two are the most important factors for the definition of habitats. A principal component analysis (PCA) showed that Factor 1 was related to sediment type, with high values corresponding to gravel and negative values corresponding to sand. Factor 2 was significantly correlated with depth, and alternatively to hydrodynamic intensity. Figure 6.2 shows how the various taxa and community structures were allocated between these two factors.

Based on the seafloor sediment map and statistical analysis of benthos, six habitats and corresponding associations of benthic animals were mapped. Each of the habitats was distinguished on the basis of substrate, habitat complexity, relative current strength and water depth. A composite conceptual map of the Browns Bank habitats (Colour Plate 13a) represents a first attempt at integration of geological, biological and dynamic conditions over the region, and depiction of habitat types and their biota.

The habitat types and biota recognized are summarized in Colour Plate 13a as follows:

1 Shallow water areas with sand substrate are characterized by a very low abundance and diversity of visible macrofauna, and are high-energy environments with mobile sediment where it is difficult for epifauna to establish and proliferate.
2 Low diversity and abundance of visible macrofauna characterize deep water areas with sand substrate surrounding Browns Bank. These relict sands are considered a more stable environment than modern sands, with

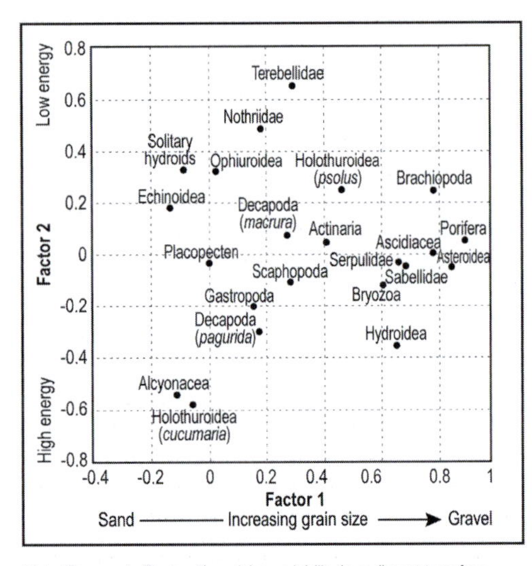

Note: The x-axis (Factor 1) explains variability in sediment type from sand (negative values) to gravel (positive values). The y-axis is related to depth and hydrodynamics

Source: Redrawn after Kostylev et al (2001)

Figure 6.2 A principal components analysis of the distribution of taxonomic groups of epifauna on Browns Bank, as identified from bottom photographs.

less sedimentation and well-developed macrofauna (Rhoads, 1976).

3 Soft coral and sea cucumber habitat occurs on gravel substrate on the western, shallow part of Browns Bank, a region dominated by strong currents. The presence of these large suspension feeders suggests that this habitat is rich in plankton and suspended organic matter.
4 Scallops are found in highest densities on gravel substrate on the western part of Browns Bank. These currents also provide scallops an abundant supply of phytoplankton, which is a primary source of their nutrition (Cranford and Grant, 1990). The scallop habitat is generally poor in other macrofauna species.
5 *Terebratulina* community habitat. Noble et al (1976) have described a distinct sub-tidal community from the Bay of Fundy, represented by this widespread and conspicuous brachiopod. On Browns Bank, this brachiopod-dominated community typically occurs on gravel substrate with boulder-sized particles in water depths of ~90m, mainly located in the central and eastern parts of the bank. Most of the animals in the community are suspension-feeders.
6 Deposit-feeder habitat. Several stations on Browns Bank show a distinct association of species, including high abundances of tube-dwelling deposit-feeding polychaetes. Photograph and grain size analysis indicate that surficial sediments in these areas consist of an accumulation of silt on gravel.

Summary and significance

In this study Kostylev et al (2001) defined benthic habitats on the basis of sediment characteristics, water depth and dominant benthic associations. This information was interpreted from multibeam, geophysical, geological and photographic data.

The use of bottom photographs instead of grab samples significantly enhances the speed of data analysis and allows the collection of more information with less expense. The major drawback of this approach is the lack of information about benthic infauna and associated detailed sediment stratigraphy; the presence of infauna can only be assessed on the basis of the occurrence of tubes, burrows and other bioturbation features on the seafloor.

Eighty taxa of macrobenthos were distinguished from the bottom photographs and are comparable to the data provided by Thouzeau et al (1991), who described a total of 106 species of epifaunal macrobenthos on Georges Bank identified from dredge samples. However, this is only 15 per cent of the total number of macrobenthic taxa identified by Wildish et al (1989, 1990) from grab samples. Disregarding infauna may lead to undercounting of bivalves and polychaete species; these are the most diverse group on Georges Bank (Theroux and Wigley, 1998). However, even with coarse taxonomic resolution Kostylev et al (2001) were able to distinguish associations of species and to outline general trends and relationships between biota and sediments.

Relationship between sediments and biota has been a recurrent descriptive theme for benthic marine ecologists, but sedimentary grain size alone may not be a primary determinant of species distributions (Snelgrove and Butman, 1994). Local currents appear to play a major role in defining both sediment grain size and community structure (e.g. Jumars, 1993; Wildish and Kristmanson, 1997 and Newell et al, 1998) suggest that community composition is also related to particle size via mobility and current speed at the sediment–water interface, as indicated in Figure 6.3 more than 70 years ago by Hjulstrom (1935). The search for relationships continues with further statistical analyses of the structures and processes in benthic habitats (Orpin and Kostylev, 2006).

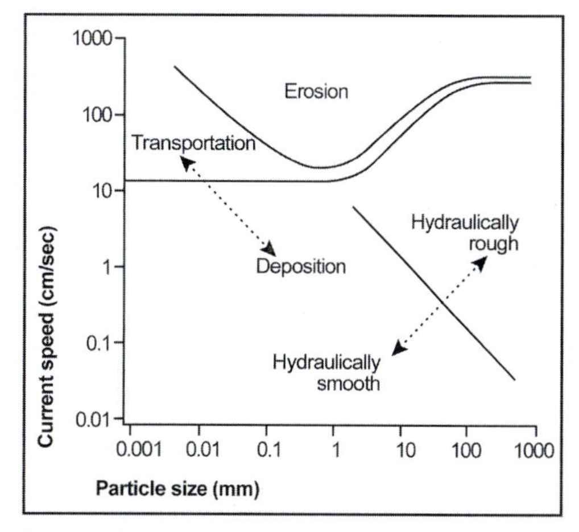

Source: Modified and simplified from Hjulsrom (1935)

Figure 6.3 Diagram of relationships between substrate particle size and current speed, showing zones of different sediment dynamics

Community–habitat relationships: Primary habitats – seascapes

The process of describing and mapping primary habitats or seascapes from geophysical data has been considered in Chapter 5. Seascapes, like geomorphological units, are comprised of primary and secondary habitats. However, planning for marine conservation may be sufficient even at this spatial level (see conclusions section).

Community–habitat relationships: Secondary habitats and biotopes

At these levels of the hierarchy there is an abundant literature, both historical and modern observational and experimental, that describes and documents the various kinds of marine communities. The concept of 'community types' has become entrenched in the biological literature even though both practically and theoretically we cannot define it adequately (see Pimm, 1991). Rather than trying to review all this conventional literature here, we take a particular tack prompted by examination of Marine Recorder data – probably the most comprehensive accumulation of records of benthic marine species in the world.

We (Roff, Bryan, Connor, unpublished) conducted a preliminary analysis of selected parts of the Marine Recorder data set, briefly reviewed here, to address some basic questions important for marine benthic ecology and conservation. Specifically we investigated the following:

- How well marine benthic community types can be related to habitat characteristics, especially substrate type.
- The evidence for the role of individual species as indicators of community type.
- The evidence for biological interactions among benthic species.
- The relationship between environmental complexity – as variability in substrate type – and species richness.
- How large a protected area would need to be to represent 'all' marine benthic species in a region.

Marine Recorder data

Data was extracted from Marine Recorder (v3.05), originally developed by the Marine Nature Conservation Review (MNCR) and published in 1997 (Connor et al,1997a, b; Connor et al, 2003). The database contains information on marine habitats around Great Britain and includes general information on broad habitat types and biotopes, and species lists of both epifauna and infauna as presence/absence data. Marine Recorder was developed through the compilation of empirical data sets, the review of other classifications and scientific literature, and in collaboration with a wide range of marine scientists and conservation managers. In photographic records of the Marine Recorder surveys, the separate biotopes are listed and described in some detail (see www.jncc.gov.uk).

Database regions selected included: west of Scotland; southwest Scotland; northwest Scotland; northern North Sea; north Scotland; Liverpool Bay; east Scotland; east English Channel; east England; and Clyde. Data were further filtered to include only samples ≥35psu, and specific substrate types that could be categorized as 'single' substrates (bedrock, boulder, cobble, gravel, pebble, sand and mud) containing ≥ 90 per cent of the targeted substrate, while the 'mixed' category contained a minimum of 20 per cent to a maximum of 40 per cent of at least three different substrates. Due to a lack of samples in some of the substrate categories, results were also binned into broader substrate categories: rock (combines bedrock and boulder categories), sand (combines cobble, gravel, pebble and sand categories), mud and mixed. Results were also binned based on depth categories (0–10m; 10–30m; 30–50m).

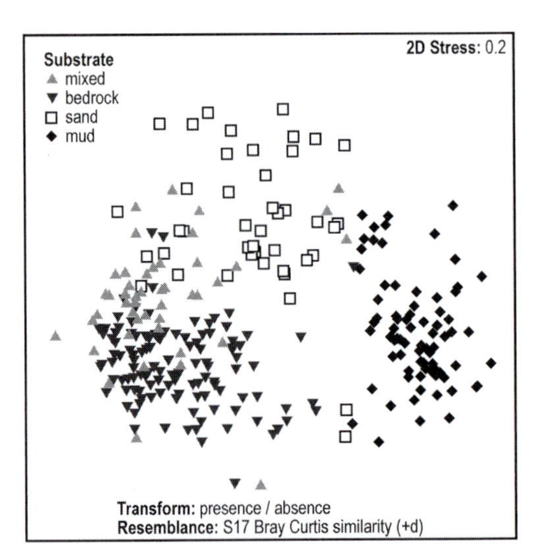

Figure 6.4 Results of multi-dimensional Scaling analysis of Marine Recorder benthic data from 10m depth including all substrate categories, showing differentiation into community types according to substrate size. See text for details.

Selected results

The various data sets examined contained as many as 450 species of epifauna and infauna species from the intertidal zone to about 30m depth.

Multidimensional scaling (MDS) plots (based on Bray-Curtis similarity matrices in Primer v6) showed a clear separation of collections into primary habitat-community types (or seascape types, Table 5.1), with samples from mixed substrates scattered between these groupings (Figure 6.4). The Marine Recorder substrate types cover the full array of benthic habitats around the British Isles and the North Sea. As has long been recognized in the ecological literature, such separation of community types is robust over broad geographic areas.

However, MDS plots of each of the individual substrate categories (primary habitats) showed no clear separations into further community types (in other words secondary habitats of species assemblages), only a broad scatter of data. This finding may be primarily a function of the nature of the data. We believe it is likely due to the facts that individual collections are of various types and made according to different protocols, and that many collections in fact contain heterogeneous substrates and therefore a mix of different species assemblages. Because of the mix of substrate types in all collections, and because the data record only presence or absence of species, we could not conduct further reliable analyses of relationships between species assemblages and substrate types, or between 'indicator species' and substrate types.

Nevertheless, indicator species can be examined with respect to each other. Indicator species groups for the North Sea were obtained from the literature, primarily Thorson (in Hedgpeth, 1957), and Zijlstra (1988). Samples containing pairs of indicator species were extracted from Marine Recorder data to determine the frequency at which they occurred in combination and separately. In all pair-wise combinations of species examined, individual members of any indicator group occurred more frequently without other members as with them. In other words, there was no indication of preferred group affinity among indicator species.

It is clear that the simple presence of an indicator species, by itself, means nothing in terms of the 'community' that may surround it. The Marine Recorder data consists of presence/absence records only, which means that it is not possible to tell the difference between a site where an individual species may be predominant and a site where it is rare but still present (see Figure 6.1). Presumably it is only when an indicator species surpasses some threshold of relative abundance that we may be able to infer the presence of other members of its assemblage or even some characteristics of the immediate habitat itself.

Our interpretation of all these results is that many marine benthic species probably have rather broad tolerances of environmental variables. However, in some areas an individual species may be heavily 'favoured' because of a subtle (but unknown) combination of environmental conditions. Under these conditions, an individual species may become so abundant that we would describe it as a 'community indicator species' although it is still a (minor) member of many other communities. Further, such a combination of favourable conditions may simultaneously apply to several species – which thereby comprise a biocoenosis. Nevertheless, individual members of the biocoenosis extend their distributions well beyond the limit of their 'typical' habitat. However, this does not mean that evidence for species interactions is entirely lacking.

Although we found no clear evidence of community structure within the categories of 'single' substrates, a very evident pattern appeared when we compared the species present in single substrate types with those in mixed substrates. Some species present in the single substrate collections were completely absent from the mixed substrates, and other species were found only in the mixed substrates and not in single substrates. These species demonstrated an absolute presence or absence, and we call these 'excluded' and 'edgist' species respectively. They are listed in Table 6.1. From such observations we cannot know the reasons for these differences, which could result from either physical or biological interactions. On balance, we believe that species exclusions are more likely mediated by biological interactions (competition or predation), while the presence of edgist species is more likely due to some unique combinations of physical characteristics of the environment.

The heterogeneity of the Marine Recorder data did permit asking a different question, namely: is there any relationship between the species diversity of a collection and its substrate heterogeneity – as a function of the proportions of different substrate types within a single collection? Provisionally the answer appears to be no. Species diversity was highly heteroscedastic when examined in relation to the proportions of different substrates within a single collection. However, see Chapter 8 for further consideration of species diversity and environmental heterogeneity in terms of topography.

Species accumulation curves were constructed for each of the substrates separately and for mixed substrates. None of these reached an asymptote, despite the inclusion of hundreds of species and collections. A Chao correction factor (Colwell and Coddington, 1994) was calculated to give the expected number of collections and the predicted number of species at an asymptote. Based on an average sample size of $10m^2$ (although this varies among the data contributed to the Marine Recorder), this calculation also yields estimates of the size of mosaic areas that should contain all benthic species within each type of primary habitat (Table 6.2). The estimated sizes are very modest, and range from 0.015 to about $0.6km^2$, but note that these size estimates are based on species-accumulation curves not on species–area relationships. See Ugland et al (2003) and Neigel (2003) for cautions in the use and interpretation of species-accumulation and species-area curves in conservation. The process of estimating the appropriate sizes for MPAs is considered further in Chapter 14.

In conclusion then: a community is found in a habitat whose environmental conditions are conducive to the differential development of populations of one or more species that numerically dominate the organisms (of the size fraction or taxonomic group quantitatively sampled) of that habitat; such organisms we call community indicators and collectively a biocoenosis.

Table 6.1 A partial listing of benthic species common to all three depth categories (0–10m, 10–20m, 20–30m) from the Marine Recorder data, which are found only in single substrate types (excluded) or found only in mixed substrate types (edgist) (see text for further explanation)

'Edgist' species Mixed substrates	'Excluded' species Single substrates
Aglaophenia pluma	*Alaria esculenta*
Amphipholis squamata	*Amphiura chiajei/filiformis*
Axinella dissimilis	*Angulus tenuis*
Axinellidae	*Audouinella spp*
Bispira volutacornis	*Corystes cassivelaunus*
Botrylloides leachi	*Diatoms – film*
Bugula plumose	*Edwardsia claparedii*
Cellepora pumicosa	*Hormathia coronata*
Ciocalypta penicillus	*Lesueurigobius friesii*
Diazona violacea	*Porphyra spp*
Dysidea fragilis	*Tellimya ferruginosa*
Epizoanthus couchii	*Thracia phaseolina*
Eunicella verrucosa	
Leucosolenia spp	
Limaria hians	
Luidia spp	
Macandrewia azorica	
Pentapora foliacea	
Pleurobranchus membranaceus	
Polyplacophora	
Polyplumaria frutescens	
Pomatoceros spp	
Raspailia spp	
Raspailia ramose	
Stelligera spp	
Terebratulina retusa	
Thyonidium drummondii	
Tonicella marmoreal	
Tubulanus annulatus	

When sampling is conducted to measure organism abundances, we should expect to recognize such community–habitat associations (though more than one community type may be associated with a given habitat type). When sampling is not conducted to measure abundance (in other words presence/absence data only is recorded – as in the Marine Recorder surveys), then we should NOT expect to be able to extract clear community–habitat associations from the data. This is because all species, including 'indicators', extend their distributions well beyond the 'key' or nuclear habitat where their populations may

Table 6.2 Estimates of sizes of areas that would be required to protect all estimated numbers of benthic species around the United Kingdom, at depths from 0 to 30m

Depth (m)	Substrate type	Estimated number of species	Estimated area (m^2)
10–20	Mixed	772	17,573.6
0–10	Mud	827	14,991.1
10–20	Rock	811	577,500.9
0–10	Sand	1210	18,472.6

Note: Estimates are maximum estimates of species numbers and areas, based on an average sample size of $10m^2$ from species-accumulation curves, and calculating the asymptote from the Chao 2 estimator (Colwell and Coddington, 1994). Substrate categories include: rock (combines bedrock and boulder categories), sand (combines cobble, gravel, pebble and sand categories), mud and mixed'

become numerically predominant. The Marine Recorder data sets deserve much closer analysis.

Community–habitat relationships: Biological facies and microcommunities

Biological facies

Even at this penultimate level of the spatial hierarchy (Table 5.1) suites of species or biological facies can be statistically and predictably related to geophysical characteristics of the marine environment, at least for the epifauna. Two examples will suffice.

In a continued examination of geophysical and biological data, photographs and grab samples from Georges Bank, Kostylev et al (2001) showed strong relationships between major taxa (groupings from classes to genera) of the epifauna and two statistical axes from a principal components analysis (PCA) (Figure 6.2). Here, several of the major taxa sorted into the statistical phase–space in relation to a combination of factors correlated with substrate particle size and the energy of water movements (depth and hydrodynamics).

Motivated by the lack of understanding of the distribution of deep-sea corals and the ecological processes associated with these assemblages, Bryan and Metaxas (2006) undertook an examination of the *Primnoidae* and *Paragorgiidae*. They qualitatively and quantitatively described the habitats of these two families of deep-water corals in relation to six oceanographic factors (depth, slope, temperature, current, chlorophyll-*a* concentration and substrate) on the Pacific and Atlantic continental margins of North America (PCM and ACM study areas, respectively). On both continental margins, coral locations were not randomly distributed, but were within specific ranges for most environmental factors. In the PCM study area, *Paragorgiidae* and *Primnoidae* locations were found in areas with slopes ranging from 0° to 10.0°, temperature from -2.0 to 11.0°C and currents from 0 to 143cm s^{-1}. In the ACM study area, *Paragorgiidae* and *Primnoidae* locations were found in areas with slopes ranging from 0° to 1.4°, temperature ranging from 0 to 11.0°C and currents ranging from 0 to 207cm s^{-1}. In both study areas, most environmental parameters in locations where corals occurred were significantly different from the average values of these parameters.

A significant problem that should be recognized is that although it may be possible to predict the potential distribution of a facies or species assemblage (in other words where a particular taxon or community should be), its actual distribution may not conform to this, because of natural or human disturbances. We should therefore expect a series of false negatives (regions where a community type is expected but not found). However, if we find a series of false positives (regions where a community is found but not expected) then our predictive model must be in error and needs to be revised. Although deep sea corals have been extensively destroyed in recent decades, these kinds of community–habitat predictions are very important as search clues to remnant populations.

Microcommunities

At present it would appear that microcommunities will only be described by direct sampling – for example of epiphytes and epizooties – or by in situ deployment of video cameras.

Conclusions and management implications

The concept of the 'community' has become firmly established in ecological/conservation terminology as a member of the hierarchy: gene, species/population, community, ecosystem. A 'community' is still generally defined as 'a set of actually or potentially interacting species sharing an environment', although the more neutral term 'assemblage' is often used where species–species interactions are unknown – which is almost universally the case!

The existence of recognizable 'community types' in the marine benthic environment has long been reported (e.g. Hedgpeth, 1957). Conventional wisdom still espouses the view that clearly-defined communities, as described by sets of species of more-or-less the same composition (biocoenoses) and associated more-or-less strongly with well definable combinations of environmental variables (the habitat), comprise identifiable biotopes. Further, it is generally accepted that the presence and geographic distributions of these 'community types' are demonstrated by the presence of indicator species (as indicators of community composition, see Chapter 9).

Communities are variable in species composition in both space and time as a result of biological, physical and human influences (see Box 6.1). It also appears likely that (just as in terrestrial environments) some species have very specific habitat requirements, while others are much more catholic. We need to understand these variable relationships, and indeed – more broadly – what drives variations of community composition within defined habitat types, and the consequences for conservation planning. Examination of relationships between distributions of biota and geophysical data is still an active field – despite decades of such research – and remains an important field for marine conservation planning.

Relationships between communities and habitats in fact lie at the heart of spatial planning for conservation and yet there are still problems in defining such relationships. However, by judicious attention to the types of communities and the issue of scale, it is possible to 'work around' many of these problems.

At the national, ecoregional and regional levels the community for which most information is likely to exist is the demersal fishes. In many studies there are well established relationships between species composition of fish assemblages and geophysical factors such as depth and water masses. At the level of geomorphological units, each one will contain a series of distinct habitats which can be well described by in situ physical survey techniques. Seascapes for both the pelagic and benthic realms can generally be mapped from available data. In a growing number of cases the distributions of even biological facies of certain taxa (e.g. deep-sea corals) can be reliably predicted. It is at the level of individual community and habitat 'types' that most problems are encountered.

Implications for a marine conservation strategy

The reality that benthic species are distributed beyond their predominant habitat (biotope) into a broader realized habitat has significant implications for marine conservation. Combined with a second reality of the heterogeneity of habitats, it means that conservation planning even at the relatively coarse level of seascapes (primary habitats – Table 5.1) should have a high probability of capturing all benthic species at some level of representation, even if the fundamental biotope of a particular species is not included at any particular location. This implies a significant relaxation of requirements for mapping and planning for what would otherwise be an onerous (perhaps currently impossible) task of ensuring representation of all possible biotopes within an ecoregion.

We may in practice never be able to completely describe the biological community structure of a region down to the lowest levels of the hierarchy such as microcommunities, except by direct surveys at a very local level. However, if we protect large enough areas (see Chapter 14 on size) of secondary or even primary biotopes, we should automatically have protected the majority of the lower levels of the hierarchy. Conservation planning can therefore follow a series of stages as will be considered in Chapter 16 (and summarized in Box 16.5).

A major consideration is the still unresolved question of whether communities are determined by either 'niche-assembly' rules or by 'dispersal assembly' rules, or (most probably) by some variable combination of both under different conditions (Hubbell, 2001).

If communities are actually assembled by niche-based requirements, then we must resign ourselves to ever-increasing and frustrating attempts at more and more precise descriptions of the geophysical characteristics of marine habitats, and the minutiae of biological-habitat relationships at the individual species level. In heterogeneous environments (however defined and measured) this also means an inevitable confounding of α and β measures of diversity (see Chapter 8).

If, however, species composition within a habitat type is subject to change (according to dispersal-assembly concepts), this reinforces the idea of habitats as being more fundamental units for conservation than communities. A practice of 'insurance' by establishing replicates of representative habitat types (which may have 'communities' of variable species composition) should then be effective in conserving species diversity. The potential consequences of confounding of α and β measures of diversity is therefore less important.

Because of the variability of species composition and the actual or potential presence of several biocoenoses within individual habitat types, we probably need to accept several things:

- We may never see a complete or definitive analysis of relationships between habitat variables and biocoenoses.
- The issue of whether communities are 'real' or not is an ecological argument that can be side-stepped for conservation purposes.
- We need to accept the habitat as the more fundamental unit of representative marine conservation, defined as cleanly as possible.
- An additional advantage of specifying habitats from geophysical characteristics, rather than communities by species, is that if regional conditions change over time (e.g. as a result of climate changes, incursions of water masses etc.) the ecological, environmental and conservation boundaries can be readily reset, without recourse to extensive biological surveys.

Particular interests

Various groups of biologists, naturalists, conservationists or environmental managers have particular interest in or responsibility for selected taxonomic groups of plants or animals. In the marine environment, the taxonomic groups of predominant economic or conservation interest are typically birds, marine mammals and fish. The first two groups at least are unlikely to be efficient indicators of marine representative areas for several reasons. They are migratory or seasonal in occurrence and therefore their use of the marine environment varies spatially and intra-annually. They are likely to aggregate or be more common in regions of high productivity (e.g. tidal mud-flats, marshes, upwelling areas, gyres, frontal systems etc.) which themselves are typically of lower than average species diversity.

The presence and abundance of these two groups (birds and marine mammals) may therefore best describe *distinctive* local regions, which would certainly become candidates for MPA status, but their distributions alone cannot form the basis for a system of marine representative areas. The tendency in marine conservation initiatives has often been to select such distinctive areas as MPAs without conducting the underlying analysis of representation which provides the necessary ecological perspective within which to evaluate a candidate MPA. The issue of conservation of distinctive areas is considered in the next chapter.

The situation with respect to fish is more complex and is considered in more detail in Chapter 13. With respect to fish, conservation efforts are generally less interested in the conservation of species diversity, and more interested in the preservation of socio-economic status by conservation of individual stocks of particular fish species. Analyses of fish populations and communities may therefore be undertaken to determine either representative or distinctive regions; the latter may include rich (i.e. productive) commercial fishing grounds. The overall distributions of fish species, or the documentation of their relationships to oceanographic and physiographic factors, may therefore become important components of the definition of marine representative areas (see Chapter 16).

Analyses of the changes in the structure of fish communities can also define how environments and habitats have become degraded. This is an extremely important area of research because of the issue of 'shifting baselines' in conservation efforts. Although Pauly (1995) defined this syndrome for fisheries scientists and managers, it is also applicable to conservation. The syndrome is based on the premise that:

> *...each generation accepts the species composition and stock sizes that they first observe as a natural baseline from which to evaluate changes. This, of course, ignores the fact that this baseline may already represent a disturbed state. The resource then continues to decline, but the next generation resets their baseline to this newly depressed state. The result is a gradual accommodation of the creeping disappearance of resource species, and inappropriate reference points for evaluating economic losses resulting from over-fishing, or for identifying targets for rehabilitation measures.* (Pauly, 1995)

Biological survey work can define whether a particular area has already been historically disturbed or degraded from its natural or 'baseline' community state. Given this perspective, subsequent management efforts can be appropriately directed at conservation of pristine areas, restoration of degraded areas and enhancement or reintroductions of selected species.

References

Auster, P. J., Michalopoulos, C., Robertson, P. C., Valentine, K. J. and Cross, V. A. (1998) 'Use of Acoustic Methods for Classification and Monitoring of Seafloor Habitat Complexity: Description of approaches', in N. W. P. Munro and J. H. M. Willison (eds) *Linking Protected Areas with Working Landscapes, Conserving Biodiversity*, Proceedings of the Third International Conference on Science and Management of Protected Areas, Wolfville, Nova Scotia, Canada

Auster, P. J. and Langton, R. W. (1999) 'The Effects of Fishing on Fish Habitat', in L. Benaka (ed) *Fish Habitat: Essential Fish Habitat and Rehabilitation*, American Fisheries Society, Bethesda, MD

Austin, M. P. (1985) 'Continuum concept, ordination methods, and niche theory', *Annual Review in Ecology and Systematics*, vol 16, pp39–61

Barnes, R. S. K. and Hughes, R. N. (1988) *An Introduction to Marine Ecology*, Blackwell Scientific Publications, Oxford

Bertnes, M. D., Gaines, S. D. and Hay, M. E. (2001) *Marine Community Ecology*, Sinauer Associates, Sunderland, MA

Bryan, T. L. and Metaxas, A. (2006) 'Distribution of deep-water corals along the North American continental margins: Relationships with environmental factors', *Deep-Sea Research*, vol 53, pp1865–1879

Caswell, H. and Cohen, J. E. (1991) 'Disturbance, interspecific interaction and diversity in metapopulations', *Biological Journal of the Linnean Society*, vol 42, pp193–218

Colwell, R. K. and Coddington, J. A. (1994) 'Estimating terrestrial biodiversity through extrapolation', *Philosophical Transactions of the Royal Society (Series B)*, vol 345, pp101–118

Connell, J. H. (1978) 'Diversity in tropical rainforests and coral reefs', *Science*, vol 199, pp1302–1310

Connor, D. W., Brazier, D. P., Hill, T. O. and Northen, K. O. (1997a) 'Marine biotope classification for Britain and Ireland. Vol 1: Littoral biotopes', *JNCC Report* no 229, Joint Nature Conservation Committee, Peterborough

Connor, D. W., Allen, J. H., Golding, N., Lieberknecht, L. M., Northen, K. O. and Reker, J. B. (2003) *The National Marine Habitat Classification for Britain and Ireland. Version 03.02*, Joint Nature Conservation Committee, Peterborough

Courtney, R. C. and Shaw J. (2000) 'Multibeam bathymetry and backscatter imaging of the Canadian continental shelf', *Geoscience Canada*, vol 27, pp31–42

Cranford, P. J. and Grant, J. (1990) 'Particle clearance and absorption of phytoplankton and detritus by the sea scallop *Placopecten magellanicus* (Gemelin)', *Journal of Experimental Marine Biology and Ecology*, vol 137, pp105–121

Diaz, R. J., Solan, M. and Valente, R. M. (2004) 'A review of approaches for classifying benthic habitats and evaluating habitat quality', *Journal of Environmental Management*, vol 73, pp165–181

Frank, K. T., Petrie, B., Choi, J. S. and Leggett, W. C. (2005) 'Trophic cascades in a formerly cod-dominated ecosystem', *Science*, vol 308, no 5728, pp1621–1623

Fraschetti, S., Terlizzi, A. and Boero, F. (2008) 'How many habitats are there in the sea (and where)?', *Journal of Experimental Marine Biology and Ecology*, vol 366, pp109–115

Gray, J. S. (1994) 'Is deep-sea species diversity really so high? Species diversity of the Norwegian continental shelf', *Marine Ecology Progress Series*, vol 112, pp205–209

Halpern, B. S., Walbridge, S., Selkoe, K. A., Kappel, C. V., Micheli, F., D'Agrosa, C., Bruno, J. F., Casey, K. S., Ebert, C., Fox, H. E., Fujita, R., Heinemann, D., Lenihan, H. S., Madin, E. M., Perry, M. T., Selig, E. R., Spalding, M., Steneck, R. and Watson, R. (2008) 'A global map of human impact on marine ecosystems', *Science*, vol 319, pp948–952

Hedgpeth, J. W. (1957) 'Treatise on marine ecology and palaeoecology', *Memorandum of the Geological Society of America*, vol 67

Hjulstrom, F. (1935) 'Studies of the morphological activity of rivers as illustrated by the River Fyris', *Bulletin, Geological Institute of Upsala*, XXV, Upsala, Sweden

Hubbell, S. P. (2001) *The Unified Neutral Theory of Biodiversity and Biogeography*, Princeton University Press, Princeton, NJ

Hutchinson, G. E. (1957) 'Concluding remarks', *Cold Spring Harbor Symposia on Quantitative Biology*, vol 22, no 2, pp415–427

Jaureguizar, A. J. J., Bava, J., Carossa, C. R. and Lasta, C. A. (2003) 'Distribution of whitemouth croaker *Micropogonias furnieri* in relation to environmental factors at the Rio de la Plata estuary, South America', *Marine Ecology Progress Series*, vol 255, pp271–282

Jumars, P. (1993) *Concepts in Biological Oceanography: An Interdisciplinary Approach*, Oxford University Press, New York

Kingsford, M. and Battershill, C. (1998) *Studying Temperate Marine Environments*, CRC Press, Boca Raton, FL

Krebs, C. J. (1972) *Ecology*, Harper and Row, New York

Kostylev, V. E., Todd, B. J., Fader, G. B. J., Courtney, R. C., Cameron, G. D. M. and Pickrill, R. A. (2001) 'Benthic habitat mapping on the Scotian Shelf based on multibeam bathymetry, surficial geology and sea floor photographs', *Marine Ecology Progress Series*, vol 219, pp121–137

Kvitek, R. G., Fukayama, A. K., Anderson, B. S. and Grimm, B. K. (1988) 'Sea otter foraging on deep-burrowing bivalves in a California coastal lagoon', *Marine Biology*, vol 98, pp157–167

Lalli, C. M. (1990) *Enclosed Experimental Marine Ecosystems: A Review and Recommendations*, Coastal and Estuarine Studies 37, Springer-Verlag, New York

Longhurst, A. (1998) *Ecological Geography of the Sea*, Academic Press, San Diego, CA

Lundquist, C. J., Thrush, S. F., Coco, G. and Hewitt, J. E. (2010) 'Interactions between disturbance and dispersal reduce persistence thresholds in a benthic community', *Marine Ecology Progress Series*, vol 413, pp217–228

MacGinitie, G. E. (1939) 'Littoral marine communities', *American Midland Naturalist*, vol 21, pp28–55

Mahon, R., Brown, S. K., Zwanenburg, K. C. T., Atkinson, D. B., Burj, K. R., Caflin, L., Howell, G. D., Monaco, M. E., O'Boyle, R. N. and Sinclair, M. (1998) 'Assemblages and biogeography of demersal fishes of the east coast of North America', *Canadian Journal of Fisheries and Aquatic Sciences*, vol 55, pp1704–1738

Mann, K. H. and Lazier, J. R. N. (1996) *Dynamics of Marine Ecosystems: Biological–Physical Interactions in the Oceans*, Blackwell Science, London

Menge, B. A., Lubchenco, J., Bracken, M. E. S., Chan, F., Foley, M. M., Freidenburg, T. L., Gaines, S. D. and Hudson, G. (2003) 'Coastal oceanography sets the pace of rocky intertidal community dynamics', *Proceedings of the National Academy of Sciences*, vol 100, no 21, pp12229–12234

Mitchell, N. C. and Somers, M. L. (1989) 'Quantitative backscatter measurements with a long-range side-scan sonar', *IEEE Journal of Ocean Engineering*, vol 14, pp368–374

Munguia, P., Osman, R. W., Hamilton, J., Whitlatch, R. B. and Zajac, R. N. (2010) 'Modeling of priority effects and species dominance in Long Island Sound benthic communities', *Marine Ecology Progress Series*, vol 413, pp229–240

Neigel, J. E. (2003) 'Species–area relationships and marine conservation', *Ecological Applications*, vol 13, pp137–145

Newell, R. C., Seiderer, L. J., Hitchcock, D. R. (1998) 'The impact of dredging works in coastal waters: A review of the sensitivity to disturbance and subsequent recovery of biological resources on the sea bed', *Oceanography and Marine Biology Annual Review*, vol 36, pp127–178

Noble, J. P. A., Logan, A. and Webb, G. R. (1976) 'The recent *Terebratulina* community in the rocky subtidal zone of the Bay of Fundy Canada', *Lethaia*, vol 9, no 1, pp1–17

Oliver, J. S. and Slattery, P. N. (1985) 'Destruction and opportunity on the sea floor: Effects of gray whale feeding', *Ecology*, vol 66, no 6, pp1965–1975

Oliver, J. S., Kvitek, R. G. and Slattery, P. N. (1985) 'Walrus feeding disturbance: Scavenging habits and recolonization of the Bering sea benthos', *Journal of Experimental Marine Biology and Ecology*, vol 91, pp233–246

Orpin, A. R. and Kostylev, V. E. (2006) 'Towards a statistically valid method of textural sea floor characterization of benthic habitats', *Marine Geology*, vol 225, pp209–222

Pauly, D. (1995) 'Anecdotes and the shifting baseline syndrome of fisheries', *Trends in Ecology and Evolution*, vol 10, no 10, p430

Petersen, C. G. J. (1913) 'Valuation of the sea. II. The animal communities of the sea bottom and their importance for marine zoogeography', *Report on Danish Biology*, vol 21, pp1–44

Pimm, S. (1991) *The Balance of Nature. Ecological Issues in the Conservation of Species and Communities*, University of Chicago Press, Chicago, IL

Post, A. L. (2008) 'The application of physical surrogates to predict the distribution of marine benthic organisms', *Ocean & Coastal Management*, vol 51, pp161–179

Rhoads, D. C. (1976) 'Organism–sediment Relationships', in I. N. McCave (ed) *The Benthic Boundary Layer*, Plenum Press, New York

Roff, J. C., Taylor, M. E. and Laughren, J. (2003) 'Geophysical approaches to the classification, delineation and monitoring of marine habitats and their communities', *Aquatic Conservation: Marine and Freshwater Ecosystems*, vol 13, pp77–90

Roff, J. C. (2004) 'Maintaining quality is primarily the role of the editor', *Marine Ecology Progress Series*, vol 270, pp281–283

Sale, P. F. (2004) 'Connectivity, recruitment variation, and the structure of reef fish communities', *Integrative and Comparative Biology*, vol 44, pp390–399

Scheibling, R. E., Hennigar, A. W. and Balch, T. (1999) 'Destructive grazing, epiphytism, and disease: The dynamics of sea urchin–kelp interactions in Nova Scotia', *Canadian Journal of Fisheries and Aquatic Sciences*, vol 56, pp2300–2314

Snelgrove, P. V. R. (1999) 'Getting to the bottom of marine biodiversity: Sedimentary habitats', *Bioscience*, vol 49, pp129–138

Snelgrove, P. V. R. and Butman, C. A. (1994) 'Animal–sediment relationships revisited: Cause versus effect', *Oceanography and Marine Biology Annual Review*, vol 32, pp111–177

Sousa, W. P. (2001) Natural Disturbance and the Dynamics of Marine Benthic Communities', in M. D. Bertness, S. D. Gaines and M. E. Hay (eds)

Marine Community Ecology, Sinauer Associates, Sunderland, MA

Spalding, M. D., Fox, H. E., Allen, G. R., Davidson, N., Ferdana, Z. A., Finlayson, M., Halpern, B. S., Jorge, M. A., Lombana, A. and Lourie, S. A. (2007) 'Marine ecoregions of the world: A bioregionalization of coastal and shelf areas', *Bioscience*, vol 57, no 7, pp573–584

Steele, J. H. (1988) 'Scale Selection for Biodynamic Theories', in B. J. Rothschild (ed) *Towards a Theory on Biological–Physical Interactions in the World Ocean*, Kluwer, Amsterdam

Tanner, J. E., Hughes, T. P. and Connell, J. H. (1994) 'Species coexistence, keystone species, and succession: A sensitivity analysis', *Ecology*, vol 75, no 8, pp2204–2219

Theroux, R. B. and Wigley, R. L. (1998) 'Quantitative composition and distribution of the macrobenthic invertebrate fauna of the continental shelf ecosystems of the northeastern United States', *NOAA Tech Rep NMFS 140*, National Oceanic and Atmospheric Administration, Silver Springs, MD

Thorsen, G. (1957) 'Bottom communities (sublittoral or shallow shelf)', *Memorandum of the Geographical Society of America*, vol 67, pp461–534

Thouzeau, G. R., Robert, G. and Ugarte, R. (1991) 'Faunal assemblages of benthic megainvertebrates inhabiting sea scallop grounds from eastern Georges Bank, in relation to environmental factors', *Marine Ecology Progress Series*, vol 74, pp61–82

Thrush, S. F. (1999) 'Complex role of predators in structuring soft-sediment macrobenthic communities: Implications of changes in spatial scale for experimental studies', *Australian Journal of Ecology*, vol 24, pp344–354

Ugland, K. I., Gray, J. S. and Ellingsen, K. E. (2003) 'The species–accumulation curve and estimation of species richness', *Journal of Animal Ecology*, vol 72, pp888–897

Urick, R. J. (1983) *Principles of Underwater Sound*, McGraw-Hill, New York

Valiela, I. (1995) *Marine Ecological Processes*, Springer, New York

Verfaillie, E., Degraer, S., Schelfaut, K., Willems, W. and Van Lancker, V. R. M. (2009) 'A protocol for classifying ecologically relevant marine zones, a statistical approach', *Estuarine, Coastal and Shelf Science*, vol 83, pp175–185

Whittaker, R. H., Levin, S. A. and Root, R. B. (1973) 'Niche, habitat, and ecotope', *The American Naturalist*, vol 107, no 955, pp321–338

Wildish, D. J. (1977) 'Factors controlling marine and estuarine sublittoral macrofauna', *Helgoland Wissenschaft Meeresuntersuchungen*, vol 30, pp445–454

Wildish, D. and Kristmanson, D. (1997) *Benthic Suspension Feeders and Flow*, Cambridge University Press, Cambridge

Wildish, D. J., Wilson, A. J. and Frost, B. (1989) 'Benthic macrofaunal production of Browns Bank, Northwest Atlantic', *Canadian Journal of Fisheries and Aquatic Sciences*, vol 46, pp584–590

Wildish, D. J., Frost, B. and Wilson, A. J. (1990) 'Stereographic analysis of the marine, sublittoral sediment–water interface', *Canadian Technical Report on Fisheries and Aquatic Sciences*, vol 1726

Woodin, S. A. (1978) 'Refuges, disturbance and community structure: A marine soft-bottom example', *Ecology*, vol 59, pp274–284

Zacharias, M. A. and Roff, J. C. (2001) 'Explanations of patterns of intertidal diversity at regional scales', *Journal of Biogeography*, vol 28, pp471–483

Zajac, R. N. (2008) 'Macrobenthic biodiversity and sea floor landscape structure', *Journal of Experimental Marine Biology and Ecology*, vol 366, nos 1–2, pp198–203

Zijlstra, J. J. (1988) 'The North Sea Ecosystem', in H. Postma and J. J. Zijlstra (eds) *Ecosystems of the World. 27: Continental Shelves*, Elsevier, Oxford

7

Distinctive areas: Species and Ecosystem Processes

Ecosystem processes – ergoclines and hotspots

All species (and spaces) are created equal, but some are more equal than others.

With apologies to George Orwell (1903–1950)

Introduction

As we have seen in Chapter 4, the term 'ecosystem' is not readily applicable to the marine environment. For purposes of representation therefore an approach based on analysis of biogeographic regions must be based on a combination of biological, ecological and geophysical data.

Nevertheless, although marine ecosystems themselves are poorly or arbitrarily defined, in contrast ecosystem-level processes – especially those involving water movements – can be well defined and recognized. Indeed, many of the larger marine species take advantage of these processes, which may produce or accumulate their required resources. Closer ecological study of such areas may help us to understand and predict when and to where migrations of some species may occur.

The significance of distinctive areas cannot be over-emphasized. The actions of water movements, of many kinds and at all scales, that accumulate or stimulate the production of resources at all trophic levels are absolutely vital for all aquatic organisms. Indeed, it has been suggested several times that in the absence of water movements, aquatic organisms could not survive in the homogeneous or average resource concentrations that would result. 'The average ocean is a lifeless ocean'. The significance of these water movements extends from the microscopic (diffusion processes) to the ocean basin scale (upwellings and divergences). Distinctive areas are therefore one class of heterogeneous marine conditions that are vital for ocean life.

Conservationists make frequent appeals for protection of an area based on assertion of its 'unique' characteristics. Distinctive areas truly are unique and 'special' in the sense that they are regionally without replicates (unlike representative areas) and essential for particular marine species. It is clearly of prime importance to understand the ecological structures and processes of such areas and their species.

The different kinds of distinctive areas

It is human nature to pay disproportionate attention to things that are different in some way. In previous chapters we have attempted to show how the various approaches to marine conservation can be founded in ecological principles, and we have concentrated on analyses of

Box 7.1 The different kinds of distinctive areas

Distinctive areas – ergoclines (this chapter)

These are areas of increased biomass or elevated production, caused by water movements of some kind. Accumulations of biomass are often confused with increased production (see text). The presence of a distinctive area is generally indicated by a geophysical anomaly of some kind and often a focal species (see Chapter 9) or an increase in species diversity. The scale of occurrence can range from regional (hundreds of kilometres) to local (tens of kilometres) or even the micro level (centimetres).

Distinctive area characteristics

- region is unique or rare in some way at some scale (from global to regional);
- region of importance for rare or endangered species;
- high primary production;
- high biomass at some trophic level: Aggregation of some food resource;
- region of importance for some species because of fitness consequences;
- presence of some focal species;
- presence of some endangered species;
- aggregation of some species (often commercial) for feeding or breeding;
- region of high species diversity.

Hotspots (Chapter 8)

The term 'hotspot' can be used to mean many things. It is frequently used to refer to regions of higher than average species diversity, however this may be caused. The scale of occurrence can range from realm to province, to regional or local.

Ecological and biologically significant areas (this chapter)

These are generally areas of significance for commercial fisheries, especially including areas of importance for seasonal feeding, spawning and recruitment. The term also includes areas important for selected species of special concern (e.g. rare or endangered species) such as deep sea corals and sponges, and regions of high species diversity. The scale of occurrence is generally regional to local.

Criteria for identifying EBSAs from the Convention on Biological Diversity (CBD)

- uniqueness or rarity;
- special importance for life history of species;
- importance for threatened, endangered or declining species and/or habitats;
- vulnerability, fragility, sensitivity, slow recovery;
- biological production;
- biological diversity;
- naturalness.

representative habitats from global to local levels. All representative areas are similar within their 'kinds', but there are many different kinds of (non-representative) distinctive areas. Because they are different, both from representative areas and from one another, they resist hierarchical classification.

Different terms have been used to describe the various kinds of distinctive areas. Those in common use include: distinctive areas (primarily ergoclines); ecologically and biologically significant areas (EBSAs, primarily of fisheries concern); hotspots (primarily of concern for areas of high species diversity). Although distinctive areas can be placed into these types, they are interrelated and overlap in various ways (see Box 7.1). In fact, each of these terms could be considered to encompass both of the others, but they are used in different ways.

The Convention on Biological Diversity (CBD) has suggested seven criteria for recognizing EBSAs (Box 7.1). The first and last of these – uniqueness or rarity, and naturalness – will be considered further in Chapter 15, where the concept of 'value' in the marine environment is considered. The remaining CBD criteria can be used to describe distinctive areas that – pragmatically – fall into two non-exclusive categories, but categories which do have different relationships between physical and biological characteristics. Distinctive habitats as ergoclines typically show clear geophysical properties and can be recognized from them. Hotspots and EBSAs – although they presumably also have distinguishing geophysical properties – may be less ecologically evident, and these habitats may be recognized more from traditional ecological knowledge (TEK) or scientific ecological knowledge (SEK).

Table 7.1 Proposed framework for conserving marine biodiversity at four levels of organization (from Zacharias and Roff, 2000), and relationships of structures and processes to representative and distinctive habitats and ecological integrity

Compositional	Structural (static)	Functional (process)
Genes	Genetic structure	Genetic processes
Species, populations	Population structure	Demographic processes, life histories
Communities	Community composition	Organism/habitat relationships
Ecosystems	Ecosystem structure	Physical and chemical processes
	REPRESENTATIVE HABITATS	
	DISTINCTIVE	HABITATS
		ECOLOGICAL INTEGRITY

Distinctive habitats: Ergoclines

Roff and Evans (2002) have provided a summary of some types of distinctive areas, and shown their relationships to environmental structures and processes. A fundamental characteristic of distinctive habitats seems to be that they are found in areas of various structural anomalies, which may define either permanent locations or temporary developments that reflect oceanographic processes – chiefly water movements of some character.

By 'anomaly' we mean any variation in a geophysical property that distinguishes an area from its surroundings. Thus anomalous topographic features would include small islands, seamounts, gulleys, elevated or depressed sea height and so on. Temperature anomalies would include areas of higher or lower values than the surroundings, and areas of rapid temperature gradient (fronts). Chlorophyll anomalies would be indicated by areas where, for example, sea colour is discernibly higher than the surroundings. Note that we may consider either that anomalous structures engender processes (e.g. local topography causes upwelling), or that processes may engender anomalous structures (e.g. local current gyres lead to anomalies in sea height).

Distinctive habitats then, often appear to be distinguished because special oceanographic processes are occurring within them on a regional or local scale, whereas representative habitats are not notable in this way (see Table 7.1). A fundamental characteristic of distinctive habitats seems to be that they are found in areas of various structural anomalies (see Tables 7.2 and 7.3).

The kinds of distinctive habitats and their properties

Several different kinds of distinctive habitats can be defined by their anomalous structures and processes. Distinctive habitats belong predominantly to a broad class of environments that Legendre et al (1986) have called 'ergoclines' (literally 'energy gradients'). Such areas – exhibiting some combination of anomalies – may typically be associated with elevated resources at some 'trophic level'. Ergoclines can in fact run the spatial spectrum from the micro-scale to the global scale of the oceans. For example, at the micro-scale, water currents may be deflected even by individual benthic organisms, causing small-scale turbulence which improves particulate food concentrations at feeding appendages

Table 7.2 Some examples of various kinds of distinctive habitats, their anomalies, processes, characteristics and focal species

	VARIOUS KINDS OF ANOMALIES						
	Resources elevated above background						Resources depleted
Resources	High in situ primary production				Resource retention	Resources advected by currents or focused	
Type of habitat	Upwellings	Island mass effect	Vents	Coral reefs	Estuaries/bays	Seamounts/shelf edge/ canyons/ gulleys/sponge and coral beds	Caves
Anomaly	Low temperature/high chlorophyll	Topography/ high chlorophyll	High temperature/ sulphur	Complex topography	High temperature/ high chlorophyll/ topography	Topography/high currents	Topography/interiority
Process/ Mechanism	Production enhanced by nutrient additions	Complex circulation/ turbulence/ wind action	Production by sulphur bacteria	Production by symbiotic and other algae	Gyres/ circulation/ physical accumulation	Suspended solids flux enhanced	Sediment by-pass/light limitation
Diversity	Low	Low	Low	High	Unchanged/low	Average/high	High
Endemicity	Low–high	??	High	High	Low	High	High
Focal species	Flagships/parasols	Indicators	Indicators	Indicators	Flagships/parasols	Corals/sponges/ indicators	Indicators

Table 7.3 Some examples of relationships between anomalies and processes in distinctive habitats and focal species

Anomaly	Location/Name/Scale of distinctive area	Physical process/Agent	Focal/Indicator species involved	Biological process	Reference
Low temperature/high or variable chlorophyll	Benguela/regional	Upwelling/ocean circulation/ wind stress	Herring species	Feeding/reproductive cell	Payne et al, 1987
Oceanic divergences	Open oceans	Ocean gyre circulation/wind stress	Commercial fisheries/marine mammals	Production/feeding	Mann and Lazier, 1996
Topography/high chlorophyll	Coral reefs everywhere	Symbiotic and other algae	Corals/fish guilds	Feeding/shelter	Jones and Endean, 1973
Topography/high temperature/ sulphur	Deep sea vents	Sea floor spreading/ hydrothermal vents	Endemic vent fauna	Sulphur metabolism	Tunnicliffe, 1988
Strong temperature gradient/ high chlorophyll	Seasonal frontal zones everywhere/regional	Boundary of stratified and non-stratified waters/tidal circulation	???	Optimal light and nutrient regime	Pingree, 1977
Low temperature/high chlorophyll	South-west Nova Scotia local	Upwelling/frontal zone between water masses	Many larval species	Recruitment cells ?	Roff et al, 1986
High temperature/high chlorophyll	Minas basin mud-flats/local	High tides/paticulate resuspension/nutrients	Migrant birds	Feeding	Daborn et al, 1993; Piccolo et al, 1993.
High temperature/high chlorophyll/reduced salinity	Estuaries everywhere	Various retention mechanisms including: estuarine null zones/ bay trapping	Many commercial species	High resource levels/ spawning/recruitment/ migrations	Roff et al, 1980
High temperature/high chlorophyll	Bays everywhere/local	Gyres	???	High resources, feeding/ recruitment	Archambeault et al, 1998
Topography/ temperature/ currents	Saguenay fjord/St Lawrence estuary/local	Complex estuarine circulation/ tides	Whales/euphausiids	Feeding	Lavoie et al, 2000
Sea surface elevation	Outer Bay of Fundy/regional	Gyre tidal circulation	Whales/euphausiids/ copepods	Feeding accumulations	Roff, 1983
Irregular topography/high currents	Passages in Bay of Fundy/ Passamoquoddy Bay/local	Upwelling/tides/currents	Migrant birds/whales/seals	Feeding	Brown et al, 1982
Topography	The Gulley/Scotian Shelf/ regional	Mixing of slope water onto the shelf/ advection of resources	High diversity of whales	Feeding	Hooker et al, 1999
Topography/tidal currents	Outer Bay of Fundy/regional	Advection of resources	High biomass of benthos	Feeding	Emerson et al, 1986
Topography/currents	Seamounts	Topographic upwelling	Sea birds/high endemicity	Feeding	Haney et al, 1995
Topography/currents	Deep-sea corals and sponge beds	Topography and high currents	Corals/sponges	Feeding	Genin et al, 1992
Topography	Caves everywhere	Resource depletion	Endemic fauna	??	Vacelet et al, 1994
Topography	Cliffs/islands everywhere/local	Geographic isolation	Birds/seals/walrus etc.	Reproduction/protection from predators	Petersen, 1982
Topography	Small islands	Island mass effect/tubulence	???	Enhanced production	Hernandez-Leon, 1991

(e.g. Wildish and Kristmanson, 1997). Examples at whole ocean scale include upwellings and divergences which enhance production, and convergences which accumulate biomass.

These elevated levels of resources may be either the product of true production (i.e. they are generated by growth in situ), or they are the product of physical circulation mechanisms that lead to the retention or accumulation (or reduction) of biomass (Tables 7.2 and 7.3). Unfortunately, much of the marine literature does not adequately distinguish between production (which is a change in biomass over time due to growth, of dimensions $M.T^{-1}$, or simply T^{-1}) and biomass (which is simply the amount or quantity of biological material, of dimensions M or L^3). These elevated resources (whether the result of production or simple increase of biomass) are of vital importance to many coastal species – especially flagship species – yet we often have poor documentation of the processes involved in their generation. These processes lie at the heart of the ecology of distinctive habitats, and are fundamental to maintenance of ecosystem health, ecological integrity, species' distributions, abundances and recruitment of species, patterns of animal migrations, and potential or actual fisheries yields.

Several kinds of processes (Table 7.2) can occur within distinctive habitats (see also Table 7.3 for some specific examples of distinctive habitats, their anomalies and processes):

- High in situ primary production due to increased supply of nutrients. Examples include: coastal upwelling areas and oceanic divergence zones (e.g. Mann and Lazier, 1996); seasonal frontal zones (e.g. Pingree, 1977); estuarine locations subject to land runoff, deep-sea vents (e.g. Tunnicliffe, 1988); coral reefs (e.g. Jones and Endean, 1973) and arctic polynyas.
- Enhanced in situ primary production due to trapping or sheer effects that allow exploitation of elevated nutrient concentrations or particulate matter. Examples would include bays in estuaries (e.g. Roff et al, 1980), mud-flats (e.g. Daborn et al, 1993), and high algal production of the epontic community at the ice-water interface in the Arctic and Antarctic.
- Accumulation or retention (either vertical or horizontal) of organisms by physical circulation mechanisms – that is, the accumulation of biomass. These mechanisms do not involve enhanced growth or production of organisms. Such physical accumulation can occur at local boundaries, eddies, estuarine null zones and in bays and fjords, or in fronts, convergences, gyres and vortices and other types of complex circulation patterns on banks. Examples would be elevated numbers of copepods in the central gyre of the outer Bay of Fundy (e.g. Roff, 1983) and accumulations of zooplankton within bays (Archambault et al, 1998).
- Circulation patterns of various types including upwelling, convergence and turbulent mixing causing displacement of deep-dwelling plankton or nekton to the surface – again a local or regional increase in biomass. This increases the availability of these resources to surface or water-column feeders such as birds, fish and whales. Examples would include the passages between Bay of Fundy and Passamoquoddy Bay and the outer Saguenay Fjord (e.g. Brown et al, 1982; Lavoie et al, 2000).
- Increased availability of resources due to advection by some combination of currents and topography. Examples would include areas of high benthic production in the outer Bay of Fundy (e.g. Emerson et al, 1986), around seamounts (e.g. Haney et al, 1995), in gulleys (e.g. Hooker et al, 1999) and around deep-sea sponges and corals (e.g. Genin et al, 1992). Even the availability of open water in the Arctic for marine mammals (in polynyas) could be considered to fall in this category.
- Depletion of resources in caves (e.g. Vacelet et al, 1994).
- Physical isolation on cliffs and islands – for example of young birds and seals – leading to protection from predators (e.g. Petersen, 1982).

Process-oriented studies are required to define the mechanisms leading to the distinctive character of such locations, although we may frequently infer such processes from the structures they engender (see Table 7.3). Thus low temperatures combined with high chlorophyll are characteristic of upwelling areas (Mann and Lazier, 1996). Intertidal mud-flats develop a temporary high temperature anomaly favouring production by microalgae and amphipods which become rich feeding grounds for migrant birds (Daborn et al, 1993). Fronts of rapid temperature change between stratified and non-stratified waters develop as areas of high production (Pingree, 1977). High plankton or nekton concentrations in gyres with an elevated sea level attract migrant whales (Roff, 1983; Woodley and Gaskin, 1996). Less visible, but still distinctive, communities also develop well below sea level; thus a combination of high currents and high slope around sea-mounts leads to distinctive communities or local areas of high production and high diversity (e.g. Boehlert and Genin, 1987). A similar combination of effects (topography and currents) appears to lead to other extraordinary and diverse communities such as deep-water coral and hexactinellid sponge beds (e.g. Genin et al, 1992; Rice and Lambshead, 1994; Bryan and Metaxas, 2006).

Many of the examples presented above, and in Table 7.3, are from the east coast of Canada. However, it should be possible to compile an inventory of distinctive habitats within any region of the world's oceans, from a combination of existing surveys, literature reviews, local knowledge and so on. In parallel, using GIS techniques based on physical data, it is possible to map areas that exhibit different kinds of anomalies. A comparison can then be made of predicted versus known distinctive habitats as a test of the distinctiveness-anomaly concept, and to ensure that all potential distinctive habitats have been discovered.

Flagships, umbrellas, parasols and distinctive habitats

Because of their anomalous properties and elevated resources, many distinctive habitats – especially in temperate waters – are disproportionately important to seasonal migrants, such as marine mammals, birds and fish, because they are important areas for feeding and/or reproduction.

Populations of larger marine species (chiefly mammals, birds and fish) make directed migrations to distinctive habitats for several reasons, including the avoidance of predators. Thus, especially at times of reproduction, many species seek isolation and safety for vulnerable young progeny – for example the topographic advantages afforded by cliffs and islands for birds and seals (e.g. Petersen, 1982). To reproduce, many species of fish select spawning grounds so that their larvae can find their way back to appropriate locations by navigating in ocean currents. The passive drift back to a home range by larval forms is compensated by the counter-migration of adults against prevailing currents (see examples in Mann and Lazier, 1996). To feed and exploit rich feeding grounds, usually associated with ergoclines (which are areas of high production or accumulation of resources), many species will make seasonal migrations often of great distances. In combination, it is frequently both the structures (e.g. topography) and the processes (e.g. high production of resources) in these distinctive habitats that support these larger and more visible 'flagship' species.

The role of flagship species in marine conservation is significant – especially because of disproportionate public interest – and will be considered in more detail in Chapter 9. These are the 'charismatic megafauna' with which the public identifies and whose conservation they are prepared to fund. However, in ecological terms (e.g. trophodynamic effects, control of other prey groups by 'top-down' processes etc.) such species may (e.g. Nybakken, 1997) or may not (e.g. Katona and Whitehead, 1988; Bowen, 1997) be important. We therefore face the peculiar situation where the public identifies with a highly visible group of animals whose role in marine dynamics is largely unknown – or at best may be ill-defined. Nevertheless, even though such species may or may not be ecologically important, they can still be useful if they contribute in practical terms to our conservation strategies – specifically if flagship species act as 'umbrellas' (see Zacharias and Roff, 2001a, and Chapter 9). By this is meant that if we protect 'more visible' species with broader habitat requirements, then many smaller species occupying the same habitat will also be protected. The term umbrella is, however, generally applied to non-migratory terrestrial species, whereas in temperate waters many marine mammals and birds are seasonal (spring to autumn) migrants. They are therefore perhaps more appropriately termed 'parasols' rather than umbrellas.

Distinctive areas: Hotspots and EBSAs

Areas of high species diversity are of major concern for the overall important strategy of biodiversity conservation. Unfortunately, we still have a very incomplete understanding of where and when and why species diversity is higher than average in the oceans, or how it is related to the geophysical environment. Species diversity is generally assumed to be associated with habitat heterogeneity, but may also be associated with ergoclines and energy availability. Species diversity is considered further in this chapter below, and species diversity hotspots will be considered in the overall context of species and habitat diversity, from global to local levels, in Chapter 8.

According to the Convention on Biological Diversity (CBD, Conference of the Parties 2008, decision IX/20), EBSAs are defined as follows:

> *Ecologically and biologically significant areas are geographically or oceanographically discrete areas that provide important services to one or more species/populations of an ecosystem or to the ecosystem as a whole, compared to other surrounding areas or areas of similar ecological characteristics…'* (CBD, 2008)

The seven criteria for their identification are listed in Box 7.1. EBSAs can therefore be considered to include all distinctive areas that can be recognized from geophysical anomalies and all hotspots of species diversity.

In addition, however, EBSAs include a series of areas and habitats that may not be characterized by geophysical anomalies, or areas whose geophysical characteristics are less evidently distinguishable from representative areas. Nevertheless such EBSAs are important in a variety of ecological ways. For example they include areas that:

> *contain: breeding grounds, spawning areas, nursery areas, juvenile habitat or other areas important for life history stages of species; or habitats of migratory species (feeding, wintering or resting areas, breeding, moulting, migratory routes); areas critical for threatened, endangered or declining species and/or habitats; areas that contain a relatively high proportion of sensitive habitats, biotopes or species that are functionally fragile (highly susceptible to degradation or depletion by human activity or by natural events) or with slow recovery.* (CBD, 2008)

The last category would include for example the deep sea coral and hexactinellid sponge beds of temperate continental shelf waters (e.g. Bryan and Metaxas, 2006).

Such additional EBSAs clearly represent many types that can only be documented from personal experience of the sea – for example as traditional ecological knowledge (TEK), or from survey and monitoring programmes such as those of federal government agencies – constituting scientific ecological knowledge or SEK.

Relationships between distinctive and representative habitats

Representative habitats can be mapped on the basis of their geophysical features (e.g. Roff et al, 2003; and see Chapters 5 and 6) and many types of distinctive habitats can be recognized on the basis of their various anomalies. It is now possible to examine the role of individual focal species or groups of focal species (including flagships, umbrellas, parasols, commercial fish species, closed fishing grounds, spawning or other reproductive areas and so on) in the development of conservation strategies.

For example, knowing the distribution range (of feeding grounds, breeding areas etc.) of a particular focal species, we can ask questions such as: What is the fidelity of association between flagship species and distinctive habitats? A cautionary note is warranted here, because marine mammals, for example, may shift their distribution patterns depending upon trophic interactions of their prey species (e.g. Kenney et al, 1996).

Recalling that representative habitats are contiguous and continuous, and will collectively occupy an entire region, whereas distinctive ones will be disjunct, discontinuous and dispersed throughout a region, we can ask: What do distinctive habitats capture in terms of regional habitat representation? Can we usefully designate a marine protected area (MPA) based on the protection of a particular flagship species or its required distinctive habitat – that is, does it act as an umbrella (or parasol) species? (See Simberloff, 1998; Zacharias and Roff, 2001a; and Chapter 9.) We might continue to ask: How many such umbrella species would we need to protect in order to conserve some assigned proportion of all types of representative habitats within a set of MPAs in a region? If we elect to protect all recognized distinctive habitats, what additional representative habitats should be protected within a region in order to achieve a goal of protecting some assigned proportion of representative habitats? These kinds of questions can be important contributions to the development of comprehensive regional conservation plans. In all cases, in order to evaluate conservation strategies based on any focal species, the correspondence between the distributions of individual species and habitat mapping is essential (see Chapter 16).

A conservation strategy, leading from studies on focal species, could have several advantages. It should rejuvenate the inherent appeal and significance of 'species' approaches to marine conservation, and provide a clear rationale for human interest and a new foundation for examination of marine ecological interactions. It would also require a novel synthesis of relationships between 'species' and 'spaces' approaches to marine conservation by asking how we can take the best advantage of both approaches (representativeness and distinctiveness), rather than seeing them as in conflict. These aspects of integration of habitat types and planning for marine conservation will be considered further in Chapter 16.

A critical question to ask is: Does selection of a set of MPAs based on distinctive habitats, occupied by designated focal species, provide protection for a greater diversity of species (or greater heterogeneity of habitats) than a comparable set of MPAs selected at random or on the basis of representation (see e.g. Andelman and Fagan, 2000)? This question has never been addressed in marine conservation ecology and the answer depends heavily on the patterns of distribution of biodiversity.

Species diversity and its relation to distinctive and representative habitats

The subject of how biodiversity (predominantly as species diversity) is distributed is considered in more detail in Chapter 8, along with many of the proposed explanations. Despite considerable interest in the distribution of species diversity, we still have a limited understanding of why it varies among habitats and why some habitats support a greater diversity of species than others. At regional scales, empirical evidence favours a hump-shaped relationship between diversity and productivity; as production rises, diversity first increases and then decreases (e.g. Rosenzweig and Abramsky, 1993). However, we should recognize that measures of species diversity may vary, and that productivity (the potential to produce) is often confused with production (the actual rate of increase in biomass) and with biomass itself, or other surrogate variables.

Diversity may be either higher or lower in distinctive habitats than in representative ones. We should consider several types of habitats with respect to their species diversity and production or biomass (Tables 7.4 and 7.5):

- Coastal upwelling areas and oceanic divergence zones represent two major classes of distinctive habitats. These habitats are readily recognized because of their anomalous temperature and chlorophyll signatures. Here production is increased due to nutrient additions, but species diversity is generally reduced (Margalef, 1978). This lower diversity may extend through higher trophic levels to fish communities. If diversity is reduced in these distinctive habitats, as it substantially is in the benthos below some upwelling areas (e.g. Sanders, 1969), then protecting them may in fact be less effective in conserving species diversity than in protecting

Table 7.4 Relationships between anomalies and species diversity in representative and distinctive habitats

	Low Diversity	Average Diversity	High Diversity
ANOMALY PRESENT	***DISTINCTIVE*** e.g. upwelling areas	***DISTINCTIVE*** e.g. accumulation areas	***DISTINCTIVE*** e.g. convergences, coral reefs
NO ANOMALY	REPRESENTATIVE e.g. areas disturbed by biological, physical or anthropogenic impacts	REPRESENTATIVE 'normal habitats'	DIVERSITY 'HOT-SPOTS' non-existent or unknown mechanisms?

Table 7.5 Some probable relationships between production OR biomass, and species diversity

Production or Biomass	Diversity of Producers	Diversity of Consumers
Production elevated by nutrient injections e.g. upwelling areas	LOW	LOW, AVERAGE, HIGH
Production naturally high (not stimulated by nutrients)	HIGH	HIGH
Production not elevated	AVERAGE	AVERAGE
Production low	HIGH	HIGH
Biomass elevated by accumulation	AVERAGE	AVERAGE, HIGH
Biomass not elevated	AVERAGE	AVERAGE
Biomass reduced	HIGH	HIGH

Notes: 1 Superimposed on these relationships are effects of variations in environmental parameters and disturbances.
2 Production is the rate of increase of biomass. Biomass is a measure of the mass of some biological tissue.

a representative habitat of similar area! However, at the highest trophic levels exploiting such areas we may again see higher diversity, for example in migrant birds of several species or mixed populations of marine mammals (e.g. Brown and Gaskin, 1988, and references therein). Thus the 'exploiters tend to become more diversified among themselves than the exploited' (Margalef, 1997).

- In distinctive habitats that accumulate biomass without an increase in the rate of production, species diversity should be unchanged from that of surrounding representative habitats. However, diversity at higher trophic levels (e.g. migrating birds) may again be higher because of the rich resources available (e.g. Brown and Gaskin, 1988).
- Complex frontal zones of mixing between water masses, for example the area off southwest Nova Scotia (Smith, 1989), may represent areas of high species diversity. This particular area contains many expatriate larval forms advected here by the Gulf Stream and Nova Scotia Currents (Roff et al, 1986).
- Coral reefs are distinctive habitats where diversity is high but production is high as well (Muscatine and Weis, 1992). However, nutrient addition to coral reefs, although first increasing production (as carbon fixation rate, Muscatine and Weis, 1992), also causes a replacement of corals by seaweeds (Miller and Hay, 1996) and a decline in their diversity and that of attendant species.
- Areas disturbed by biological, physical or human influences may show reduced species diversity.
- Underwater caves are habitats where biomass and production of resources are low, but where species diversity and endemicity may be high (e.g. Vacelet et al, 1994).
- Areas where production and biomass are not increased disproportionately above background levels, and where species diversity can be strongly related to geophysical features (e.g. Zacharias and Roff, 2001b), constitute the representative habitats.

Although relationships between biodiversity and geophysical factors at global or continental scales remain uncertain, at regional scales there are indications of stronger patterns. Patterns of species diversity may be captured at least in part by our analyses of representative habitats. Thus in British Columbia, Zacharias and Roff (2001b) showed that diversity in intertidal communities was a combined function of environmental and biological factors. Annual variability in temperature and salinity was strongly correlated to decreased species diversity, while periodic disturbance – as indexed by exposure to wave and storm action – increased it. The biological effect of increased predators was also related to higher species diversity (see Chapter 8 for further details). This confirms the significance and role of environmental stability combined with periodic physical disturbance in structuring marine communities (e.g. Sousa, 1984).

A further issue is whether we can account for all distinctive habitats in terms of recognizable anomalies, or whether there are other areas (where production and biomass of resources is not increased) of high species diversity or 'hotspots' (e.g. Norse, 1993) without recognizable anomalies, that are unaccounted for. Such areas would require explanation of their origins, persistence and resource supply – or revelation of their anomalies (see Chapter 8).

Species diversity of the deep-sea benthos – which is considerably higher than earlier thought – presents particular problems. Here the separation between representative and distinctive communities becomes blurred. This is because each area sampled contains many new species, and species-area (or species-richness, or species-accumulation) curves do not reach an asymptote over many samples and broad sampling areas (e.g. Gage and Tyler, 1991). Therefore, it could be argued that each sampling area (or set of samples) constitutes a distinctive habitat (according to our definition of being different from its surroundings at some scale) and is a distinctive community with (or without) distinctive habitat characteristics (see e.g. Gray, 1994). Thus the apparently 'homogeneous' environment of the deep sea, without identifiable anomalies, contains many contiguous species assemblages – each of high diversity – rather than the more clearly defined community types related to substrate type in shallower waters. Such local endemicity might be expected (and is observed) on individual widely spaced seamounts (e.g. De Forges et al, 2000), but was not expected in the flat and featureless deep-sea bottom. The explanation for this high diversity in a (supposedly) homogeneous environment is not clear, but it seems likely that periodic physical and biological disturbances, on a variety of spatial and temporal scales, open areas at random for colonization from a 'larval lottery' (e.g. Sale, 1977). This could lead to the localized distinctive species assemblages in areas lacking obvious anomalies. However, equilibrium and disequilibrium explanations for such diversity are not yet in harmony with its 'openness'. This conundrum will be considered further in the overall context of species diversity in Chapter 8.

Conclusions and management implications

In this chapter we have attempted to characterize the various types of distinctive marine habitats and suggest some relationships to their probable species diversity. It is important not only to recognize distinctive areas, but also to be able to delineate them in time and space, describe the processes occurring within them and evaluate their ecological integrity.

A comprehensive marine conservation strategy should involve protection of both representative and distinctive habitats as members of networks of MPAs. Such a strategy, leading perhaps from studies on flagship or other focal species (see Chapter 9), could have several advantages. It should rejuvenate the inherent appeal and significance of 'species' approaches to marine conservation, and provide a rationale for renewed human interest and a new foundation for examination of marine ecological processes. It would also require a novel synthesis of relationships between 'species' and 'spaces' approaches to marine conservation by asking how we can take the best advantage of both approaches, rather than seeing them as in conflict.

Following an inventory and mapping of representative and distinctive habitats in a region, a set of preferred or 'candidate' MPAs – some combination of representative and distinctive habitats – can be designated, based on ecological evaluation of their structures and processes. Chapter 16 deals with this process of how candidate MPAs can be selected according to ecological criteria, and Chapter 17 deals with the process of assessing connectivity among MPAs in order to ensure the development of true networks of protected areas.

Evaluation of the alternative possible combinations of protected areas, and final decisions, will involve additional socio-economic criteria and negotiation among stakeholder groups and establishment of priorities. However, distinctive areas should not be confused with priority conservation areas (PCAs or PACs). Distinctive areas are defined simply in terms of their ecological characteristics (either geophysical and/or biological). PACs are defined by some combination of uniqueness, plus current threat to their status, and opportunity for conservation initiative.

In selecting the final networks of MPAs, the objective will be to maximize protection of biodiversity, while minimizing economic, cultural and social costs. Design of MPA networks by taking into account fisheries and other extractive activities is therefore vital. If appropriately sited and designed, individual MPAs may serve several objectives including maintenance of fish stocks (e.g. Holland, 2000). Such areas may not only protect biodiversity, but may also act as natural fish hatcheries and nurseries that can export juveniles of many species to other areas. MPAs are increasingly being considered and evaluated as fisheries management tools in countries around the world, generally for non-migratory species of economic value that have larval recruitment in the pelagic realm (e.g. Murray et al, 1999; Sladek-Nowlis and Roberts, 1999). We see no inherent conflict between fishing activities and marine conservation; rather, areas selected as MPAs should have multiple purposes and rationales.

Analysis of 'ecological integrity' (see e.g. Müller et al, 2000) – primarily described by the natural processes occurring in an area (Table 7.1) – and the 'connectivity' (exchange of properties) between a set of proposed MPA sites becomes critical. In order to ensure this, we must consider processes as well as structures at all levels of the ecological hierarchy (e.g. Zacharias and Roff, 2000). In future chapters we shall further consider the ecological principles involved in the selection of MPA sites and their properties, and the design of networks of MPAs.

References

Andelman, S. J., and Fagan, W. F. (2000) 'Umbrellas and flagships: Efficient conservation surrogates or expensive mistakes?', *Proceedings National Academy of Sciences*, vol 97, pp5954–5959

Archambault, P., Roff, J. C. and Bourget, E. (1998) 'Nearshore abundance of zooplankton in relation to coastal topographic heterogeneity, and the mechanisms involved', *Journal of Plankton Research*, vol 20, pp671–690

Boehlert, G. W. and Genin, A. (1987) 'A Review of the Effects of Seamounts on Biological Processes', in B. H. Keating, P. Fryer, R. Batiza and G. W. Boehlert (eds) *Seamounts, Islands and Atolls*, Geophysical Monographs, vol 43, American Geophysical Union, Washington DC

Bowen, W. D. (1997) 'Role of marine mammals in aquatic ecosystems', *Marine Ecology Progress Series*, vol 158, pp267–274

Brown, R. G. B., Barker, S. P. and Gaskin, D. E. (1982) 'Daytime surface swarming by *Meganyctiphanes norvegica* (Crustacea, Euphausiacea) off Brier Island, Bay of Fundy, Canada', *Canadian Journal of Zoology*, vol 57, pp2285–2291

Brown, R. G. B. and Gaskin, D. E. G. (1988) 'The pelagic ecology of the gray and red-necked phalaropes *Phalaropus fulicarius* and *Phalaropus lobatus* in the Bay of Fundy, eastern Canada', *Ibis*, vol 130, pp234–250

Bryan, T. L. and Metaxas, A. (2006) 'Distribution of deep-water corals along the North American continental margins: Relationships with environmental factors', *Deep-Sea Research*, vol 53, pp1865–1879

Convention on Biological Diversity (2008) 'COP 9 Decision IX/20: Marine and coastal biodiversity', www.cbd.int/decision/cop/?id=11663, accessed 23 December 2010

Daborn, G. R., Amos, C. L., Brylinsky, M., Christian, H., Drapeau, G., Faas, R. W., Grant, J., Long, B., Paterson, D. M., Perillo, G. M. E. and Piccolo, M. C. (1993) 'An ecological cascade effect: Migratory birds affect stability of intertidal sediments', *Limnology and Oceanography*, vol 38, pp225–231

De Forges, R. B., Koslow, J. A. and Poore, G. C. B. (2000) 'Diversity and endemism of the benthic seamount fauna in the southwest Pacific', *Nature*, vol 405, pp944–947

Emerson, C. W., Roff, J. C. and Wildish, D. J. (1986) 'Pelagic–benthic coupling at the mouth of the Bay of Fundy, Atlantic Canada', *Ophelia*, vol 26, pp165–180

Gage, J. D. and Tyler, P. A. (1991) *Deep-sea Biology: A Natural History of Organisms at the Deep-Sea Floor*, Cambridge University Press, Cambridge

Genin, A., Paull, C. K. and Dillon, W. P. (1992) 'Anomalous abundances of deep-sea fauna on a rocky bottom exposed to strong currents', *Deep Sea Research*, vol 39, pp293–302

Gray, J. S. (1994) 'Is deep-sea species diversity really so high? Species diversity of the Norwegian continental shelf', *Marine Ecology Progress Series*, vol 112, pp205–209

Haney, J. C., Haury, L. R., Mullineaux, L. S. and Fey, C. L. (1995) 'Sea-bird aggregation at a deep North Pacific seamount', *Marine Biology*, vol 123, pp1–9

Hernandez-Leon, S. (1991) 'Accumulation of zooplankton in a wake area as a causative mechanism of the "island-mass effect"', *Marine Biology*, vol 109, pp141–148

Holland, D. S. (2000) 'A bioeconomic model of marine sanctuaries on Georges Bank', *Canadian Journal of Fisheries and Aquatic Sciences*, vol 57, pp1307–1319

Hooker, S. K., Whitehead, H. and Gowans, S. (1999) 'Marine protected area design and the spatial and temporal distribution of cetaceans in a submarine canyon', *Conservation Biology*, vol 13, pp592–602

Jones, O. A. and Endean, R. (1973) *Biology and Geology of Coral Reefs, Vols 1 and 2*, Academic Press, New York

Katona, S. and Whitehead, H. (1988) 'Are Cetacea ecologically important?', *Oceanography and Marine Biology. An Annual Review*, vol 26, pp553–568

Kenney, R. D., Payne, P. M., Heinemann, D. H. and Winn, H. E. (1996) 'Shifts in Northeast Shelf cetacean distributions relative to trends in Gulf of Maine/Georges Bank finfish abundance', in K. Sherman, N. A. Jaworski and T. J. Smayda (eds) *The Northeast Shelf Ecosystem Assessment, Sustainability and Management*, Blackwell Science, Cambridge, MA

Lavoie, D., Simard, Y. and Saucier, F. J. (2000) 'Aggregations and dispersion of krill at channel heads and shelf edges: The dynamics in the Saguenay–St Lawrence Marine Park', *Canadian Journal of Fisheries and Aquatic Sciences*, vol 57, pp1853–1869

Legendre, L., Demers, S. and LeFaivre, D. (1986) 'Biological Production at Marine Ergoclines', in J. C. J. Nihoul (ed) *Marine Interfaces Ecohydrodynamics*, Elsevier, Amsterdam

Mann, K. H. and Lazier, J. R. N. (1996) *Dynamics of Marine Ecosystems. Biological–Physical Interactions in the Oceans*, Blackwell Science, Cambridge, MA

Margalef, R. (1978) 'Phytoplankton communities in upwelling areas: The example of NW Africa', *Oecologia Aquatica*, vol 3, pp97–132

Margalef, R. (1997) *Our Biosphere*, Excellence in Ecology 10, Ecology Institute, Oldendorf, Germany, p544

Miller, M. W. and Hay, M. E. (1996) 'Coral–seaweed–grazer–nutrient interactions on temperate reefs', *Ecological Monographs*, vol 66, pp323–344

Müller, F., Hoffmann-Kroll, R. and Wiggering, H. (2000) 'Indicating ecosystem integrity: Theoretical concepts and environmental requirements', *Ecological Modelling*, vol 130, pp13–23

Murray, S. N., Ambrose, R. F., Bohnsack, J. A., Botsford, L. W., Carr, M. H., Davis, G. E., Dayton, P. K., Gotshall, D., Gunderson, D. R., Hixon, M. A., Lubchenco, J., Mangel, M., MacCall, A., McArdle, D. A., Ogden, J. C., Roughgarden, J., Starr, R. M., Tegner, M. J and Yoklavich, M. M. (1999) 'No-take reserve networks: Sustaining fishery populations and marine ecosystems', *Fisheries*, vol 24, pp11–25

Muscatine, L. and Weis, V. (1992) 'Productivity of Zooxanthellae and Biogeochemical Cycles', in P. Falkowski (ed) *Primary Productivity and Biogeochemical Cycles in the Sea*, Plenum Press, New York

Norse, E. A. (1993) *Global Marine Biological Diversity. A Strategy for Building Conservation into Decision Making,* Island Press, Washington, DC

Nybakken, J. (1997) *Marine Biology: An Ecological Approach*, Addison Wesley Longman Inc, New York

Payne, A. I. L., Gulland, J. A. and Brink, K. H. (1987) 'The Benguela and comparable ecosystems', *South African Journal of Science*, vol 5

Petersen, M. R. (1982) 'Predation on seabirds by red foxes (*Vulpes fulva*) at Sharak Island, Alaska USA', *Canadian Field Naturalist*, vol 96, pp41–45

Piccolo, M. C., Perillo, G. M. E. and Daborn, G. R. (1993) 'Soil temperature variations on a tidal flat in Minas Basin, Bay of Fundy, Canada', *Estuarine, Coastal and Shelf Science*, vol 35, pp34–357

Pingree, R. D. (1977) 'Mixing and Stabilization of Phytoplankton Distributions on the Northwest

European Continental Shelf', in J. H. Steele (ed) *Spatial Patterns in Plankton Communities*, Plenum Press, New York

Rice, A. L. and Lambshead, P. J. D. (1994) 'Patch Dynamics in the Deep-sea Benthos: The role of a heterogeneous supply of organic matter', in P. S. Giller, A. G. Hildrew and D. G. Raffaelli (eds) *Aquatic Ecology Scale Pattern and Process*, Blackwell Science, Oxford

Roff, J. C. (1983) 'The Microzooplankton of the Quoddy Region', in M. Thomas (ed) *Marine Biology of the Quoddy Region*, Natural Sciences and Engineering Research Council of Canada, Ottawa, Canada

Roff, J. C. and Evans, S. (2002) 'Frameworks for marine conservation: Non-hierarchical approaches and distinctive habitats', *Aquatic Conservation: Marine and Freshwater Ecosystems*, vol 12, pp635–648

Roff, J. C., Pett, R. J., Rogers, G. and Budgell, P. (1980) 'A Study of Plankton Ecology in Chesterfield Inlet, Northwest Territories: An arctic estuary', in V. Kennedy (ed) *Estuarine Perspectives, Vol II*, Academic Press, New York

Roff, J. C., Fanning, L. P. and Stasko, A. B. (1986) 'Distribution and association of larval crabs (Decapoda: Brachyura) on the Scotian Shelf', *Canadian Journal of Fisheries and Aquatic Sciences*, vol 43, pp587–599

Roff, J. C., Taylor, M. E. and Laughren, J. (2003) 'Geophysical approaches to the classification, delineation and monitoring of marine habitats and their communities', *Aquatic Conservation, Marine and Freshwater Ecosystems*, vol 13, pp77–90

Rosenzweig, M. L. and Abramsky, Z. (1993) 'How Are Diversity and Productivity Related?', in R. E. Ricklefs and D. Schluter (eds) *Species Diversity in Ecological Communities. Historical and Geographical Perspectives*, University of Chicago Press, Chicago, IL

Sale, P. (1977) 'Maintenance of high diversity in coral reef fish communities', *American Naturalist*, vol 111, pp337–359

Sanders, H. L. (1969) 'Benthic Marine Diversity and the Stability–Time Hypothesis', in G. M. Woodwell and H. H. Smith (eds) *Diversity and Stability in Ecological Systems*, Brookhaven National Laboratory, New York

Simberloff, D. (1998) 'Flagships, umbrellas, and keystones: Is single-species management passé in the landscape era?', *Biological Conservation*, vol 83, pp2247–257

Sladek-Nowlis, J. and Roberts, C. M. (1999) 'Fisheries benefits and optimal design of marine reserves', *Fisheries Bulletin*, vol 97, pp604–616

Smith, P. C. (1989) 'Seasonal and interannual variability of current, temperature and salinity off southwest Nova Scotia', *Canadian Journal of Fisheries and Aquatic Sciences*, vol 46, pp4–20

Sousa, W. P. (1984) 'The role of disturbance in natural communities', *Annual Review Ecology and Systematics*, vol 15, pp353–392

Tunnicliffe, V. (1988) 'Biogeography and evolution of hydrothermal-vent fauna in the eastern Pacific Ocean', *Proceedings of the Royal Society: London B*, vol 233, pp347–366

Vacelet, J., Boury-Esnault, N. and Harmelin, J. G. (1994) 'Hexactinellid cave, a unique deep-sea habitat in the scuba zone', *Deep-Sea Research*, vol 41, pp965–973

Wildish, D. and Kristmanson, D. (1997) *Benthic Suspension Feeders and Flow*, Cambridge University Press, Cambridge

Woodley, T. H. and Gaskin, D. E. G. (1996) 'Environmental characteristics of North Atlantic right and fin whale habitat in the lower Bay of Fundy, Canada', *Canadian Journal of Zoology*, vol 74, pp75-84

Zacharias, M. A. and Roff, J. C. (2000) 'A hierarchical ecological approach to conserving marine biodiversity', *Conservation Biology*, vol 13, no 5, pp1327-1334.

Zacharias, M. A. and Roff, J. C. (2001a) 'Use of focal species in marine conservation and management: a review and critique', *Aquatic Conservation: Marine and Freshwater Ecosystems*, vol 11, pp59-76

Zacharias, M. A. and Roff, J. C. (2001b) 'Explanation of patterns of intertidal diversity at regional scales', *Journal of Biogeography*, vol 28, no 4, pp471-483

8

Patterns of Biodiversity: Species Diversity

Theories and relationships – global, regional, local

Homage to Santa Rosalia, or why are there so many kinds of animals?

G. E. Hutchinson (1903–1991)

Introduction

Recent interest in global and regional biodiversity has centred on describing the biogeography of the oceans – that is, how species and community compositions change geographically – for example as considered in Chapters 4 and 5. There has also been considerable interest in the distribution of individual species, especially those of commercial fishing value and endangered species.

Investigations into the structure and species diversity of marine communities have covered several spatial and temporal scales, and efforts have often focused on three themes: (1) the identification and characterization of recurring assemblages and/or communities; (2) the examination of latitudinal species diversity gradients; and (3) the examination of the relationships between community structure and diversity with environmental (abiotic) variables. Research into the identification of recurring communities is perhaps the oldest form of marine community research, and formed the basis of subsequent work into examining patterns of species diversity. Much of this work has been directed at benthic environments – particularly in Europe, where investigators have attempted to characterize biocoenoses, biotopes, etages, facies, habitat types and other kinds of biogeographic units.

There have also been a number of efforts to characterize biological communities on a whole-ocean scale to test the hypothesis that species diversity decreases with increasing latitude. For example, Ekman (1953) examined the global distribution of fauna to identify 'faunistic regions' of the world, and Sanders (1968) investigated patterns of diversity between different types of marine environments at various latitudes. A third purpose of these efforts was to examine how community structure and species diversity change with environmental variables.

The intention in this chapter is to examine the underlying causes as to why some areas of the marine environment (with emphasis on the benthos) have more species than others. This is an area that has been rich in developing theory and is of active research interest at present. However, the present chapter will not give an exhaustive account of marine species diversity or its distribution; other texts do this well (e.g. Thorne-Miller and Catena, 1991; Norse, 1993; NRC, 1995; Ormond et al, 1997; Reaka-Kudla et al, 1997; Norse and Crowder, 2005). Rather the intention here is to first present some general or global patterns of species diversity distributions, and to evaluate some of the ideas advanced to explain them. This is followed by some examples of regional and local analyses of species diversity distributions along with explanations for the observed patterns.

The subject of species diversity is of fundamental significance to marine conservation and deserves our fullest attention. We hope to draw out the significant factors associated with species diversity, so that these ecological and environmental factors can be considered and allowed for in conservation planning. Although emphasis here is on patterns of distribution of species richness, inevitably this also involves consideration of habitats and their communities. Unfortunately, species diversity is often considered as synonymous with biodiversity; this leaves the impression that if we study only species diversity then we have an adequate understanding of biodiversity itself. This is not so.

Species diversity and biodiversity

The implicit or explicit synonymy of species diversity and biodiversity is unfortunate because we then fail to realize that the process of marine biodiversity conservation extends well beyond the study of species alone, and has much broader implications for habitat conservation and understanding and respecting natural environmental and ecological processes.

Both historically and currently, most attention has been devoted to the distributions of individual taxa, especially selected species deemed to be commercially important or endangered. Biodiversity can, however, be considered at each level of the ecological hierarchy, but no such distributional or global geographical synthesis has yet been attempted.

There is no present description or inventory of all the components of marine biodiversity, although Zacharias and Roff (2000) have made a preliminary definition of it. We have no theories to account for the distributions of the components of biodiversity above the species level. The Census of Marine Life project (CoML) is really just the beginning of a study of one component of biodiversity (the species complement) for the oceans as a whole. Much more could be done, even with existing data, to inventory and understand other components of marine biodiversity, including their structures and processes.

The global distributions of some of the components of biodiversity, at other levels of the ecological hierarchy, are in fact quite well known. For example at the ecosystem level, the locations and periodic processes of coastal and oceanic upwellings and divergence zones, and the locations and variations of major ocean currents, have been known and charted for decades. The analysis by Longhurst (1998) of epipelagic biomes represents the global distribution of primary productivity processes and regimes. Ocean charts of seamounts, mid-oceanic ridges and deep ocean vent systems are depictions of structures at the ecological/geomorphological level, and so on.

A major gap in a truly global framework for marine conservation is the lack of inventories of the components of biodiversity (other than selected geophysical and species components) at the global, realm, provincial and ecoregion levels. Currently, perhaps the most important contributions are in the biogeography of selected community types, predominantly those forming biogenic substrates (foundation species) or shaping associated community types, such as reef-building corals, mangroves, seagrasses, saltmarshes, macrophytic algae, deep sea vent communites and hexactinellid sponges. Such organisms have become of special interest because they may constitute or contribute to local or regional 'hotspots' of high species diversity (e.g. the Great Barrier Reef of northeastern Australia). There is less understanding of the various types of geomorphic units, including seamounts, banks, gullies and so on, and the processes surrounding them.

The bioregionalization reports from Australia (e.g. Lyne and Hayes, 2005; Lyne et al, 2007), are basically compilations of inventories of the structures and processes of biodiversity across the ecological hierarchy, at the provincial and ecoregional scales. They identify ecosystems, trophic systems and pathways, habitat types, water masses and their variations, key ecological features and so forth. These compilations of the components of biodiversity (though not referred to as such in the reports) are extremely valuable for any kind of environmental planning, risk assessment, resource exploration, fisheries and biodiversity conservation, evaluation of ecological goods and services, and so on. In short they are indispensible to devising strategies for sustainable development in the marine environment.

These bioregionalization reports and allied unpublished data (V. Lyne unpublished data and analyses) have led to some remarkable observations. For example, based on an analysis of the diversity of water masses of the global oceans, it was shown that diversity of water masses was highest at about 200m depth rather than – as generally assumed – at the surface. This may mean that seasonally variable signals of surface temperature and salinity are 'preserved' for long periods below the sea surface – but with what ecological consequences? Very high diversity of water masses was also evident at specific geographic locations, for example the junction of the Indian and Pacific oceans and around the Philippines. This is also a region of very high species diversity for corals and other taxa. What does this imply? At present there is no interpretation of the processes that led to this high diversity of water masses, or why there should be such an association between it and high species diversity. These observations are examples of what could be uncovered for other components of biodiversity from existing data, for example from the World Ocean database (www.nodc.noaa.gov).

There are many other possibilities to advance our understanding of marine biodiversity even from existing data, for example to assess risks to the biodiversity components of coastal waters – from cyclones (hurricanes) in terms of frequency, intensity, trajectories and extent of potential damage and disturbance regimes. Such risk assessment can make a valuable contribution – as a form of insurance – to the process of marine protected area (MPA) planning (see e.g. Allison et al, 2003). The overall implications of a fuller understanding of the structures and processes of biodiversity for ocean management are profound.

Measures of diversity at the species and habitat/ community levels

There are several ways in which biodiversity at either the species or community level can be quantified – see Box 8.1. Perhaps the simplest measure would be just a count of the number of species in an area. However, this suffers from several problems as soon as we wish to make comparisons among locations. For example, we could only compare the number of species of similar sizes in similar habitats of equal size or volume. Standard measures of species richness or diversity – α – must therefore compare diversity on a per individual basis. Strictly speaking, species diversity is the number of different species in a particular area (the species richness) weighted by some measure of total abundance. However, conservation biologists often speak of species diversity even when species richness is meant.

It is important to realize that our standardized estimates of species diversity are inevitably biased in several ways. Habitats obviously differ in their species diversity – the β diversity – see Box 8.1. Thus comparisons among habitats inevitably differ – for poorly researched reasons – by what different researchers accept as differences among habitats. However, it may not be a simple matter to decide whether two habitats we wish to compare are in fact identical or different; they may differ in subtle ways that confound separate estimates of α and β diversity. This annoying subject has been considered by several authors, (but see especially Gray, 1994), and is considered more fully in the following section on the deep sea, where the problem becomes most intractable if not tautological. Similarly, overall estimates of γ diversity depend on how we define a region and its composite habitats.

Estimates of species diversity are also biased towards larger, more mobile, colourful species of selected taxa. Rarely if ever do we have a complete species inventory of an area; the bias is generally to target macroscopic species. Nevertheless, because we know (and perhaps understand) more about species diversity than diversity at other levels of the ecological hierarchy, and

Box 8.1 Some measures of biodiversity

Various measures of the diversity (usually as species diversity) of a region have been devised. However, some standardization has been required. The following measures, proposed by Whittaker (1972), describe three terms for measuring biodiversity (as taxonomic diversity) over different spatial and organizational scales: alpha, beta and gamma diversity.

Alpha diversity

Alpha diversity (α diversity) is the taxonomic diversity within a particular area, habitat or community. It is usually expressed as the species richness of the area, but can also be measured by counting the number of other taxa within the area, for example families or genera. However, such estimates of taxonomic richness are strongly influenced by sample size and size of area sampled. In order to overcome these types of biases, a number of statistical techniques can be used to correct for sample size in order to derive comparable values. Both Simpson and Shannon indices are basically measures of species diversity per individual rather than per unit area.

Metrics α diversity

Simpson's diversity index

Where

S is the number of species,
N is the total percentage cover or total number of organisms,
n is the percentage cover of a species or number of organisms of a species:

$$D = \frac{\sum_{i=1}^{S} n_i(n_i - 1)}{N(N-1)}$$

Shannon index

Where

ni is the number of individuals in each species (the abundance of each species),
S is the number of species (also called species richness),
N is the total number of all individuals,
pi is the relative abundance of each species, calculated as the proportion of individuals of a given species to the total number of individuals in the community:

$$H' = -\sum_{i=1}^{S} p_i \ln p_i$$

Beta diversity

Beta diversity (β-diversity) is a measure of biodiversity which compares the species diversity between habitats or ecosystems or along environmental gradients (e.g. an estuary). This involves comparing the number of taxa that are unique to each of the ecosystems and shared between them. These indices can be viewed as the rate of change in species composition across habitats or among communities, or as a quantitative measure of diversity in communities that experience changing environments.

Metrics of β diversity

Sørensen's similarity index

The Sørensen index is a very simple measure of beta diversity, ranging from a value of 0 where there is no species overlap between the communities, to a value of 1 when exactly the same species are found in both communities.

Where
S1 is the total number of species recorded in the first community,
S2 is the total number of species recorded in the second community,
c is the number of species common to both communities:

$$\beta = \frac{2_c}{S_1 + S_2}$$

Whittaker's measure

Where
S is the total number of species recorded in both communities,
α is average number of species found within the communities:

$$\beta = \frac{S}{\bar{\alpha}} \quad or \quad \beta = \frac{S}{\bar{\alpha}} - 1$$

Gamma diversity

Gamma diversity (γ-diversity) is a measure of the total biodiversity (as species diversity) over a large area or region. Gamma diversity is the richness in species of a range of habitats in a geographic area (e.g. a landscape, a seascape, a geomorphic unit). It is the result of both the alpha diversity of the individual communities and the range of differences in beta diversity among them. It is the total (or product) of α and β diversity.

Metrics of γ diversity

$\beta = \gamma / \alpha$
OR $\gamma = \Sigma \alpha . \beta$

because we have more 'theories' to explain it, we shall concentrate on this subject.

Theories of species diversity

The literature is replete with 'theories' (also variably referred to as hypotheses) to explain the patterns of species diversity on our planet. These ideas span the range of spatial scales, from ecosystems to habitats, but they have not been integrated into a satisfactory whole – apart from attempts by Hubbell (2001) and challenges to his recent ideas. It is not possible to review the merits of all these theories here, but some of the major ones are summarized below, and in Table 8.1, along with a brief account of their supposed mechanism of operation. Most of these theories have their merits and significant explanatory power under some conditions, and they have been reviewed many times (see e.g. Pianka, 1966), but no single theory presently has general applicability or is sufficient to explain all observed patterns of species diversity at all scales from global to local.

There is also considerable confounding of ideas, ecological processes and suggested mechanisms among the various explanations of species diversity.

Several theories either contain ideas common to other theories, or propose ideas whose underlying mechanisms are in opposition to one another. For example, the stability-time hypothesis and the intermediate disturbance hypothesis contain ideas that can be considered as mutually contradictory.

At any location (and perhaps for any taxonomic group) some combination of across-scale factors must operate to determine the potential and actual species diversity (e.g. Witman et al, 2004). The central problems therefore appear to be: first, the reasons for increased (or reduced) species diversity differ with scale of observation – from global to regional and local; second, variations in species diversity most probably depend on a series of interacting factors rather than the action of single factors as favoured by most theories. Thus no single theory is likely to account for the observed species diversity at any given locality (see Ricklefs, 1987).

Undoubtedly there are several reasons why it has taken (and is still taking) so long to understand the reasons for variations in the distribution of species diversity, including the following:

- Terrestrial and aquatic systems behave fundamentally differently, yet theories may try to integrate them or deal with only one and expect explanations to transfer to the other.
- The mechanisms that operate may be largely scale independent (evolution, extinction?), while others are clearly scale dependant (competition, predation).
- There is widespread integration of measures of α and β diversity; in fact in practice it may be impossible to disentangle them, leaving the overall effect of habitats on variations in species richness unclear.

Patterns of species diversity can obviously be determined by different factors from global to regional scales. Satisfactory explanations will therefore need to invoke several factors. From a pragmatic point of view and for conservation purposes, we are interested in:

- The combination of physical and biological factors that best explain observed patterns of diversity at some scale.
- Whether such explanations are robust and transportable to other physically comparable environments (e.g. whether they have predictive as well as explanatory power).
- How such information should be used to make decisions on marine conservation.

In this chapter the emphasis is less on theory than it is on pragmatic explanations of relationships and 'explanations' that should guide practical efforts at marine conservation. Theories of species diversity are only of practical value if they allow us to predict where areas of higher than average species diversity may be located, and if they can contribute to understanding of an appropriate strategy to conserve them. Several of the theories for species diversity will be considered in the context of descriptions of the actual distribution of species diversity, along with consideration of their possible mechanisms of action.

Stability-time and intermediate disturbance

There have been a number of theoretical and empirical investigations into the relationship between diversity and theories based on time (Fischer, 1960; Simpson, 1964) and climatic stability (Fischer, 1960). Sanders (1968) proposed that stability over longer time periods would result in communities that were biologically accommodated. The biologically accommodated community as defined by Sanders (1968) is one where '...physical conditions are rather constant and uniform for long periods of time.' In these environments it was thought that biological stress is mitigated through biological means such as predation and competition and results in a community that is stable, buffered and species rich.

In contrast, environments experiencing considerable variability and correspondingly high biological stress were identified as physically accommodated. In a physically accommodated system, species are not closely coupled with their environments. Sanders (1968) proposed that species in physically accommodated environments must be able to adapt to variations in their environment, which do not allow biological interactions to structure the community. Johnson (1970) also examined the stability–time hypothesis, and concluded that succession moves a community through time towards one that is biologically accommodated. Therefore, the stability–time hypothesis can be viewed as long-term succession.

However, the relationship between species diversity and environmental stability (either as cause and effect or as effect and cause) can be discredited at least as an unqualified generalization. Observations show that some of the least diverse communities known, for example coastal salt marshes consisting of a single or two dominant species, are among the most ecologically stable known. The outcomes (and validity) of the stability–time hypothesis must depend on spatial and temporal scales, and interactions with disturbances. At smaller spatial scales, communities are maintained in various intermediate successional seres, where diversity tends to be highest. Therefore at local scales α species richness is maximized by disturbance. A number of communities distributed over a larger area and disturbed at different times and frequencies will maximize β and γ richness. Over short time scales, communities are again maintained in various intermediate successional seres where overall diversity tends to be highest. Over longer time scales, communities may evolve to adapt to the disturbance, resulting in biologically accommodated communities with higher diversity (Ricklefs and Schluter, 1993).

Closely related to the stability–time hypothesis is the intermediate disturbance hypothesis. Disturbance has long been believed to be responsible for maintaining a community in an intermediate successional state, where species diversity is neither exclusively associated with a physically nor biologically accommodated community. The notion that disturbance affects species diversity was postulated in the intermediate disturbance hypothesis (IDH), which states that species richness is greatest at intermediate levels of disturbance (Horn, 1975; Connell, 1978). The theory states that at low rates of disturbance systems become biologically accommodated, where competitively dominant species exclude sub-dominant species. At greater rates of disturbance (frequency, severity or a combination of both) competitive dominants are unable to fully colonize between disturbance events, providing opportunities for sub-dominant and opportunistic species to coexist with dominants. The net effect is greater γ diversity in a region because component sites are at various stages of successional seres.

These two hypotheses, stability–time and intermediate disturbance, are good examples of important concepts that can be interpreted in many different ways, and in some interpretations are mutually contradictory. The IDH does not in fact adequately specify either the disturbing agency or what constitutes 'an intermediate level'. Different kinds of disturbing agencies can act in opposition, some with positive effects on species diversity and others with negative consequences. How these contradictions might be resolved is considered further below.

A unified neutral theory

Many of the 'theories' of species diversity could more properly be considered as paradigms or concepts, in that they advance explanation but have little in the way of solid theoretical foundation. Most of these 'theories' also do not have applicability at scales or under circumstances beyond those envisaged by their proponents. The exception may be the unified neutral theory of biodiversity and biogeography (UNTBB) from Hubbell (2001).

Only selected aspects of the theory pertinent to conservation are reviewed here. The theory predicts the existence of a fundamental biodiversity constant (written as θ), that appears to govern species richness on a wide variety of spatial and temporal scales. In other words, the theory predicts that species diversity is globally understandable at a variety of scales, times, habitats and taxa. The UNTBB assumes that the differences between members of an ecological community of trophically similar species are 'neutral', or irrelevant to their success. Neutrality is defined as per capita ecological equivalence among all individuals of all species at a specified trophic level; this means that all species and their members are believed to behave demographically in the same way as one another – that is, they reproduce and die (behave) in the same way. The theory takes as a starting point the theory of island biogeography (MacArthur and Wilson, 1967) which is also a form of neutral theory. Mathematically, the theory shows remarkable goodness of fit to many data sets of species numbers and their relative abundances. However, the generality of the model has been challenged (see e.g. Dornelas et al, 2006).

Perhaps most importantly for conservation, the UNTBB is an example of a dispersal assembly theory which holds that local species assemblages are determined by immigration and emigration. In contrast, niche assembly theories are non-neutral, because they hold that different species behave in different ways from one another because of biological interactions and requirements for specific habitat needs. The significance of this difference was considered in Chapter 6.

Patterns of species diversity – global

Geological time

The changes in species diversity over geological time have already been described in general terms (see Chapter 1). All the factors involved in the periodic fluctuations of life on earth are far from clear, including the causes of periodic extinctions. However, there does appear to be a general correlation between species richness of our planet and the relative positions of the continental blocks. Broadly speaking, times of low species richness appear to coincide with geological periods when all the continents formed a single mass – presumably corresponding to low spatial diversity and environmental complexity. Conversely, times of high species diversity correspond to times when the continents are widely separated (as they are now), and environmental heterogeneity is at a maximum. However, many other factors may affect or be responsible for this general pattern, including climate changes and periodic mass extinctions. The same relationship between environmental complexity and species diversity seems to extend across spatial scales even to the local level (see below).

Ocean basin – Indo-Pacific versus the Atlantic Ocean

The Indo-Pacific Ocean has a significantly higher species diversity in many taxa than the Atlantic Ocean. For example, the number of species of shallow marine fish is three times higher than among corresponding taxa in the Atlantic Ocean. Bivalve mollusc species are more than twice as abundant in the Indo-Pacific region than in the Atlantic, and the hermatypic corals show ten times as many species in the Pacific as in the Atlantic Ocean (see e.g. Rocha et al, 2008 and references therein; Figures 8.1, 8.2).

These differences are largely explained on the basis of the much younger age of the Atlantic Ocean – which began life as a rift valley in Gondwanaland some 150 million years ago. Subsequent slow invasions from the Pacific to the Atlantic are still going on – at an increased pace because of human activities – especially in the opening waters of the Arctic Ocean. The stability–time hypothesis thus appears to be correct, in that time is an important factor on a global scale. However, we could hardly argue that the Pacific Ocean (despite its name) is any more or less stable overall that the Atlantic Ocean.

The very high species diversity of the Indo-Pacific junction, around Malaysia and Indonesia (Figure 8.1), seems to be related to the junction of floras and faunas (as originally proposed by Wallace, 1859) but also to the geological occurrence of regional heterogeneity and topographic complexity in a series of shallow seas in this region. Thus again it appears that the significance of environmental complexity to species richness may extend from the global to the regional level. The underlying reason here may be because we are in fact seeing a combination of both α and β diversity leading to higher overall γ diversity.

A further possible explanation for the greater species diversity of the Indo-Pacific Ocean region is simply that it is a correlate of its greater size. Species–area relationships have been studied for many terrestrial and aquatic groups and there is often a simple relationship in which the number of species in a region increases as a simple function of its size. A global study by Smith et al (2005) of the relationship between species

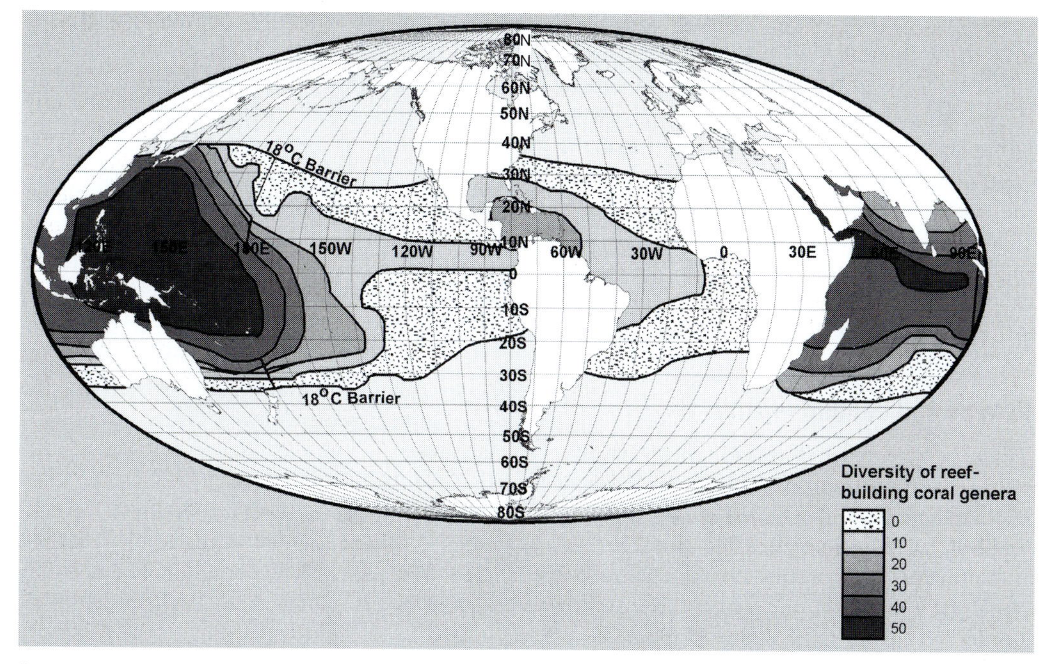

Source: Adapted from Stehli and Wells (1971)

Figure 8.1 Global distribution of diversity of reef-building coral genera

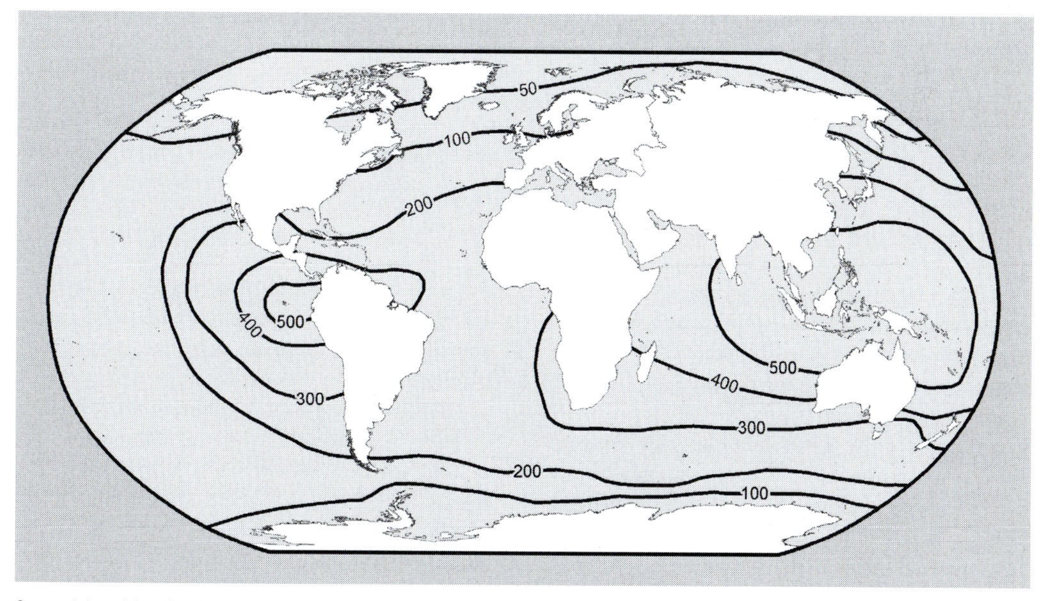

Source: Adapted from Stehli et al (1967)

Figure 8.2 Global species richness of bivalve molluscs

richness of phytoplankton and area of aquatic ecosystems showed a remarkable positive species–area relationship extending from ponds to oceans. However, an extensive study of marine prosobranch gastropods on the shelves of the western Atlantic and eastern Pacific oceans, from the tropics to the Arctic Ocean (Kaustuv et al, 1998), did not show such species–area relationships. How far we can generalize on the significance of area size as a causative factor in species richness is therefore an open question.

Pelagic and benthic realms

The greatest number of species obtainable in a single collection from the marine environment (for both plants and animals) must surely be found in an individual plankton haul. The marine plankton also contains the most numerous plants and animals on our planet; this is largely a reflection of the small size of planktonic organisms. Nevertheless, taxonomic diversity and species diversity (per individual) are higher in the benthic realm. Several major marine phyla are either rare, absent or present only as larvae (meroplankton) in the pelagic realm. For example, echinoderms, molluscs, annelids and the majority of the minor phyla are barely or sparsely represented in the pelagic realm.

Trying to substantiate the greater taxonomic diversity of the benthic realm is fraught with numerical problems. Species counts for all taxa are unreliable and under constant revision, even for the historically well known groups. For example, on average two new species of fish are described every week! Any numbers for species of various taxa should therefore be taken as very provisional. Nevertheless, some crude examples will show what is currently believed to be the general pattern. Among the divisions of the algae (some 124,000 species?), marine phytoplankton account for only about 5000 species. Among the marine crustaceans, estimates suggest about 67,000 species, although it is acknowledged that there could be more than 50,000 species of isopods alone, and the total number of marine species could be over 400,000. In contrast, the dominant group of marine zooplankton – the copepods – has only about 2500 free-living described species. Among the fish taxa, of some 33,000 known species perhaps only about 11 per cent are truly pelagic; the majority are either benthic or demersal.

The reasons for the higher diversity of the benthos involve at least in part its greater environmental complexity. The lower diversity in the pelagic realm seems to involve the greater homogeneity and mixing of the water column which over time may have facilitated gene flow, reduced isolation of populations and hindered speciation. The lack of obvious heterogeneity within the water column presumably also means that this environment presents fewer differentiated habitats and niches for exploitation than the benthos (but see Angel, 1993; Gray, 1997). In fact, even the known species diversity within the pelagic realm is hard to account for based on ecological theory, especially the expectation of competition among species. Hutchinson (1961) made this dilemma famous in the 'Paradox of the Plankton', later suggested as a form of 'Contemporaneous Disequilibrium' (Richerson et al, 1970). However, although several ideas on these subjects have been advanced, the issue of how to account for a given level of species diversity is still largely open (but see Hubbell, 2001).

Species diversity between the pelagic and benthic realms must be confounded by their global interactions, although how remains largely a mystery (see Snelgrove et al, 2000). In the meroplankton, there are distinct global gradients in larval development type, with the proportion of planktotrophic larvae increasing from polar regions to the tropics. This is expected, since the phytoplankton season in the Arctic is short, but it is long and less modulated in the tropics. Patterns are also correlated with depth; in the abyssal benthos, there are very few species with planktonic stages – as expected.

Because ocean water columns are constantly in advective motion, while in comparison the benthic environment remains largely static, this means that most conservation planning is directed to the benthic realm. This is inevitable although examples of the significance of planning for the pelagic realms will appear in later chapters.

Latitude

All descriptions of distributions of species richness should be considered as biased in various ways. For example, they almost invariably describe only a selected taxonomic group, generally the larger fish or epibenthic invertebrates. Nevertheless, certain patterns that likely apply to the majority of taxonomic groups appear to be robust (Witman et al, 2004).

Species diversity in most taxonomic groups appears to be related to latitude in the same way in all oceans, namely that it is highest in tropical waters and declines with latitude north and south, resulting in a parabolic shaped curve (Figure 8.3). The general explanation of these observations relates to physiological and energetic adaptations at higher temperatures and lower latitudes (e.g. Floeter et al, 2005; Roy et al, 2000) and/or to the more stable production regimes (and hence more predictable resources) that allow for a greatly increased number of generations per year, and hence a faster rate of speciation, among the predominant poikilotherms – invertebrates and fish (e.g. Kaustuv et al, 1998). This supposition reaffirms the importance of geological time as a global determinant of species diversity patterns.

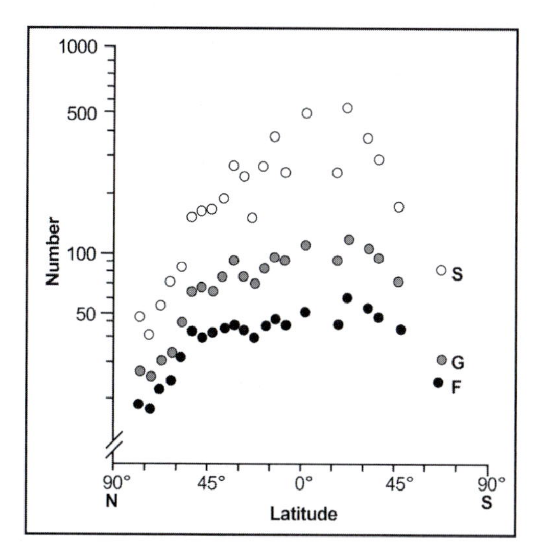

Source: Modified from Stehli et al (1967)

Figure 8.3 Latitudinal variation in the number of species (S), genera (G) and families (F) of bivalve molluscs

These patterns of species diversity and latitude seem to extend even into the deep sea. For example, Rex et al (1993) showed clear latitudinal diversity gradients in the North Atlantic deep-sea for bivalves, gastropods and isopods, and Lambshead et al (2002) showed similar patterns for nematodes related to latitudinal gradients in the seasonal sedimentation of surface production.

There may also be exceptions to these general latitudinal patterns. For example, the relationship between latitude and species diversity is either not as strong in the southern oceans, or may not be apparent at all (e.g. Gray, 2001) for reasons that are not clear. Also, such gradients may not be apparent within more restricted regions of the oceans. Thus Ellingsen and Gray (2002) failed to find such changes along a transect of 1960km (56–71°N) of the Norwegian continental shelf for 809 species of soft-sediment macrobenthos, covering a range of water depths (65–434m) and varying sediment properties. These latitudinal species diversity patterns may also not apply in freshwaters for reasons that are not clear. Other contributing factors, such as greater habitat complexity, for example in coral reef communities, may also confound these relationships (see below).

Depth

The general pattern of species diversity with depth is also parabolic in shape, first increasing (to various depths) and then decreasing (Figure 8.4). Diversity versus depth is often expressed as expected number of species per number of individuals in a random sample, so as to control for sample size and abundances. However, there is much variation and the diversity in the depths of the oceans can remain high. Even the basic pattern has been disputed (see following section on species diversity in the deep sea).

The lower diversity in shallow waters is generally explained on the basis that here biological communities experience greater disturbances. Such disturbances may be variations in temperature and salinity – causing physiological stress – or physical stress by currents, winds, waves and storms, causing disruption and extirpation of community members (e.g. Gage et al, 2000).

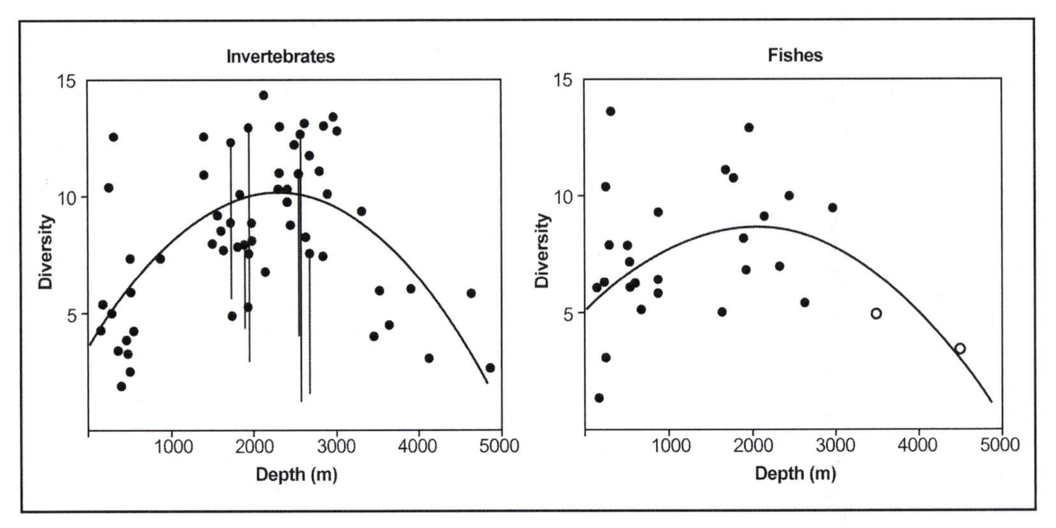

Source: Redrawn from Rex (1981)

Figure 8.4 Patterns of species diversity in the deep sea benthos based on samples from the northwest Atlantic Ocean

The increase in species diversity with depth has been explained – at first – on the basis of the increased stability of the environment, for example the stability–time hypothesis (Sanders, 1968). But at greater depths the subsequent decrease of diversity has been explained on the basis of rarefaction of food supply and increased competition for resources.

Productivity regime and energy

Relationships between species diversity, biomass and productivity are generally described as 'hump-shaped' or parabolic, with species diversity first increasing with production and then decreasing (Rosenzweig and Abramsky, 1993). Unfortunately the shape of this relationship, although often assumed, is confused by a number of factors, including: terminology (there is widespread confusion among the terms production, productivity and biomass, which will bias measurements and interpretations); how production was assessed; the type of plant community examined (terrestrial, aquatic, macrophyte or microphyte); the heterogeneity of the environment, disturbance regime and nutrient regimes (see e.g. Ptacnik et al, 2008).The subject will be considered again in Chapter 14.

Perhaps most importantly a fundamental aspect of trophodynamic ecology has been widely overlooked in both terrestrial and marine ecosystems. In both environments, a majority of production enters detrital cycles rather than being consumed directly (only in the upper pelagic world is the converse true). Based on this premise, Moore et al (2004) suggest that progress towards answering fundamental ecological questions of the distribution, abundance (and diversity) of species may therefore only be made by merging the green world of primary producers and the brown world of detritus in a new integrative ecology.

Species diversity hotspots

The subject of 'biodiversity hotspots' has become a hot topic. However, the term has been used in a variety of ways, from global to local scales, confusing subsequent discussions and arguments pro and con. In various definitions, the concept of a hotspot can encompass any of the following: sites deemed important for individual species; species richness; separate trophic levels; individual taxa above the species level; biotopes; and even representative areas. Caution in interpretation and appreciation of the term is therefore warranted (see e.g. Possingham and Wilson, 2005). There are potential problems with the identification and determination of the spatial scale and extent of hotspots (see Hurlbert and Jetz, 2007), although considerable effort has been expended in their identification for conservation purposes.

At the global level the term is generally applied to regions of high species diversity, or to regions that are known to be of major importance for selected – usually endangered – species. At the ecoregional and regional level the term is essentially synonymous with distinctive areas or ecologically and biologically significant areas (EBSAs), which are considered in Chapter 7. Distinctive areas and EBSAs are either regions of high species diversity, or they show elevated biomass at some trophic level or of some taxa – and hence have significance for feeding or reproduction for fish or focal species (see Chapter 9).

The idea of identifying biodiversity hotspots has been explicitly linked to the concept of conservation efficiency, where effort can be concentrated to get the best 'value for money' (Hiscock and Breckels, 2007). As Myers et al (2000) have noted, conservationists are far from able to assist all species under threat, if only for lack of funding. This places a premium on priorities: How can we support the most species at the least cost? One way is to identify 'biodiversity hotspots' where exceptional concentrations of endemic species are undergoing exceptional loss of habitat. In the terrestrial environment, as many as 44 per cent of all species of vascular plants and 35 per cent of all species in four vertebrate groups are confined to 25 hotspots comprising only 1.4 per cent of the land surface. This opens the way for a 'silver bullet' strategy on the part of conservation planners, focusing on these hotspots in proportion to their share of the world's species at risk. However, such a strategy concentrates only on selected larger taxa, ignores the majority of species including all of their genetic value (see Chapter 10), would protect only a tiny sample of the earth, and perhaps most importantly ignores the entire concept of ecological processes and economic value of ecosystems (see Odling-Smee, 2005). The suggestion to direct conservation funding only to the world's biodiversity hotspots may therefore be bad 'investment' advice (Kareiva and Marvier, 2003). More confusingly it has even been suggested that some hotspots and coldspots can be congruent, and thus should become candidates for conservation (Price, 2002). Nevertheless, attention to hotspots continues.

The high marine species diversity in the Indo-West Pacific is truly of global stature and significance; it has attracted attention for over a century, but the mechanisms generating such diversity remain uncertain. The region includes the so-called 'coral triangle' which even boasts its own newsletter (www.panda.org/coraltriangle). However, it should be realized that global hotspots of species richness are not necessarily congruent with degree of endemism or threat. Different mechanisms are responsible for the origin and maintenance of different aspects of diversity, and consequently the different types of hotspots also vary greatly in their utility as conservation tools (Orme et al, 2005). For example, for Indo-Pacific corals and reef fishes, the centres of high species richness and centres of high endemicity are not concordant (Hughes et al, 2002). The disparity between richness and endemicity is because many species of both corals and reef fishes are widespread, with strongly skewed range distributions. Consequently, the largest ranges overlap to generate peaks in species richness near the equator and in biodiversity hotspots, with only minor contributions from endemics. Furthermore, Hughes et al (2002) found no relationship between the number of coral vs. fish endemics at locations throughout the Indo-Pacific region, even though total richness of the two groups was strongly correlated. They suggest that the spatial separation of centres of endemicity and biodiversity hotspots in these two major taxa calls for a two-pronged management strategy to address conservation needs. The centres of endemism (e.g. Roberts et al, 2002) should perhaps therefore be the targets of conservation.

The term biodiversity hotspot should, according to our preferred usage, be defined as a region that has a higher than average set of the components of marine biodiversity, including both structures and processes across the ecological hierarchy (see Table 1.2). Stated in a different way, the term would refer to a region of high biodiversity value (see Derous et al, 2007, and Chapter 15). Unfortunately no global inventory of the components of marine biodiversity has been assembled, though such an exercise would undoubtedly point to true biodiversity hotspots.

Finally here (though by no means finalizing the discussions and arguments about marine hotspots) it might be thought impossible to produce such an inventory of all the components of marine biodiversity; for example competition and predation would seem to be processes at the community level that would defy such analysis and compilation. Yet in essence this is exactly what Worm et al (2003) have done in defining hotspots for marine predators. Using scientific-observer records from the Atlantic and Pacific oceans, they showed that oceanic predator diversity consistently peaks at intermediate latitudes (20–30° N and S), where tropical and temperate species ranges overlap. Individual hotspots were identified close to prominent habitat features such as reefs, shelf breaks or seamounts and often coincided with zooplankton and coral reef hotspots. They concluded that the seemingly monotonous landscape of the open ocean shows rich structure in predator species diversity and that these features should be used to focus future conservation efforts.

Patterns of species diversity – regional

The global distribution of marine species diversity has been established – by observation – in broad outline. However, our 'theories' to account for these patterns generally lack explanatory power especially at regional scales (hundreds to thousands of kilometres). Yet it is precisely at the regional level that nation states exert jurisdiction over their marine resources and can best effect conservation measures. Such analyses between biodiversity and geophysical factors are therefore vital to systematically implemented conservation efforts at regional scales.

Few marine organisms are globally distributed, but most are distributed well beyond the limits of small-scale research efforts; this strengthens the case for regional-scale interest and studies. Global species distributions are generally believed to be determined by the abiotic influences relating to oceanographic and physiographic properties, while local species distributions are believed to be more affected by biological processes such as competition and predation (Sanders, 1968; Ricklefs and Schluter, 1993). At regional scales, however, biotic and abiotic processes can both affect community composition and structure. The degree to which a community is biologically or physically accommodated has implications for conservation, because while humans have greatly affected marine trophic composition, many abiotic processes in marine environments (e.g. tides, circulation etc.) are largely immune to human activities.

A study of intertidal species richness

Zacharias and Roff (2001) carried out a study to examine the patterns of intertidal species richness (diversity) at regional scales (hundreds to thousands of kilometres), and to determine which abiotic or biotic variables are most strongly associated with these patterns of species richness along the coast of British Columbia, Canada. The study is reported here in some detail in order to indicate the level of effort that is required in order to reveal geophysical–biological relationships at broad regional scales.

The coast of British Columbia is composed of 29,000km of coastline and 6500 islands, and spans seven degrees of latitude. Major physiographic features of this area include the world's longest fjord (Dean Channel), an inland sea (the Georgia–Puget Basin), and a number of large rivers, archipelagos and upwelling areas. Haida Gwaii (formerly the Queen Charlotte Islands) are the most geographically isolated islands in the northeast Pacific.

Biological data were collected at 370 sites throughout the province during summer daylight low tides between 1992 and 1998. Sites were sampled using a common methodology, and carried out by a core group of biologists working from a common species list and field recording logs (Searing and Frith, 1995). Sampling consisted of a single, detailed transect from the highest water line to the boundary of the sub-tidal zone, and recorded species and the elevation relative to the highest high-water datum from which they were found. Sampling stopped when sub-tidal species were encountered. Only macrobiota were included to minimize sampling effort and disturbance to the sampled sites (in other words no rock turning). Only bedrock substrates were considered for this research. Sites comprised of mobile substrates were excluded to avoid the introduction of errors as a result of variation in sediment size between regions and the corresponding change in community composition.

For each field site, annual average and July (summer) and January (winter) temperatures were incorporated from Fisheries and Oceans Canada historical lighthouse and research datasets. In addition, fetch to the prevailing wind was also calculated using a method based on the US Army Corps of Engineers Shore Protection Manual (CERC, 1977).

A total of 205 taxa were identified, with an average of 39 species per site, a maximum of 62 and a minimum of one. Species abundance varied from the nearly ubiquitous *Fucus* and *Balanus* to 14 records of single occurrences. The ten most frequently occurring species were represented by the following seven taxa: Phaeophyta, Cirripedia, Chlorophyta, Gastropoda (Prosobranchia), Asteroidea and Bivalvia.

Salinity generally ranges from 32 to 39 parts per thousand (ppt) in most oceans, and is believed to have little effect on oceanic marine communities (Mann and Lazier, 1996). Intertidal, nearshore and neritic (continental shelf) communities, however, are affected by salinity ranges from near 0ppt (freshwater) to > 30ppt over very short distances. Overall in the data, sites with higher average salinity tended to support more species, but this relationship was weak and highly heteroscedasdic. Although there were a number of high-salinity sites with low species richness, there were no low-salinity sites with high species richness. Winter salinity exhibited a stronger relationship with species richness than summer salinity. This is probably a result of the seasonally variable pattern of salinity increasing from the inshore sheltered to the offshore exposed waters, and the corresponding increase in species richness. In summer, this trend is not as strong, as areas near the mouths of large rivers can be subject to considerable freshwater influence, while other nearshore areas can be nearly as saline as the offshore areas. A stronger inverse relationship between the number of species and increasing differences between summer and winter salinities was observed.

Temperature is thought to be the most important abiotic determinant of marine community composition at larger scales (Nybakken, 1997). Temperature's primary influence on marine biota is through its constraint on metabolic rates, as almost all intertidal plants and invertebrates are poikilotherms. Species richness tended to increase unimodally with increasing winter temperature. The increase in the number of species with winter temperature is most likely associated with the warmer temperatures located in the outer waters, which do not receive any considerable winter freshwater inputs from heavy rains, and which are less subject to the very cold influences from the continental polar airmass. The sampling sites with higher species diversity generally had lower summer temperatures and warmer winter temperatures. As for the salinity relationships, there are a number of warmer sites with low diversity, but no high-diversity sites with low winter temperatures. As with winter and summer temperature, there was a unimodal relationship between species richness and seasonal temperature differences. Again this may be a result of the different water mass properties of the protected and more exposed shorelines, which experience greater freshwater input versus continuous mixing respectively.

Wave exposure is also believed to be an important local determinant of community composition (Lewis, 1964). Increasing wave exposure results in increased mechanical stress on intertidal communities, which is manifested through a number of mechanisms. The primary effect of wave action is the physical impact on communities, which excludes those organisms with inadequate attachment mechanisms. Also of importance is the organism's ability to withstand scouring from suspended materials washing over them. Periods of large waves can also reduce attenuation depths through sediment resuspension,

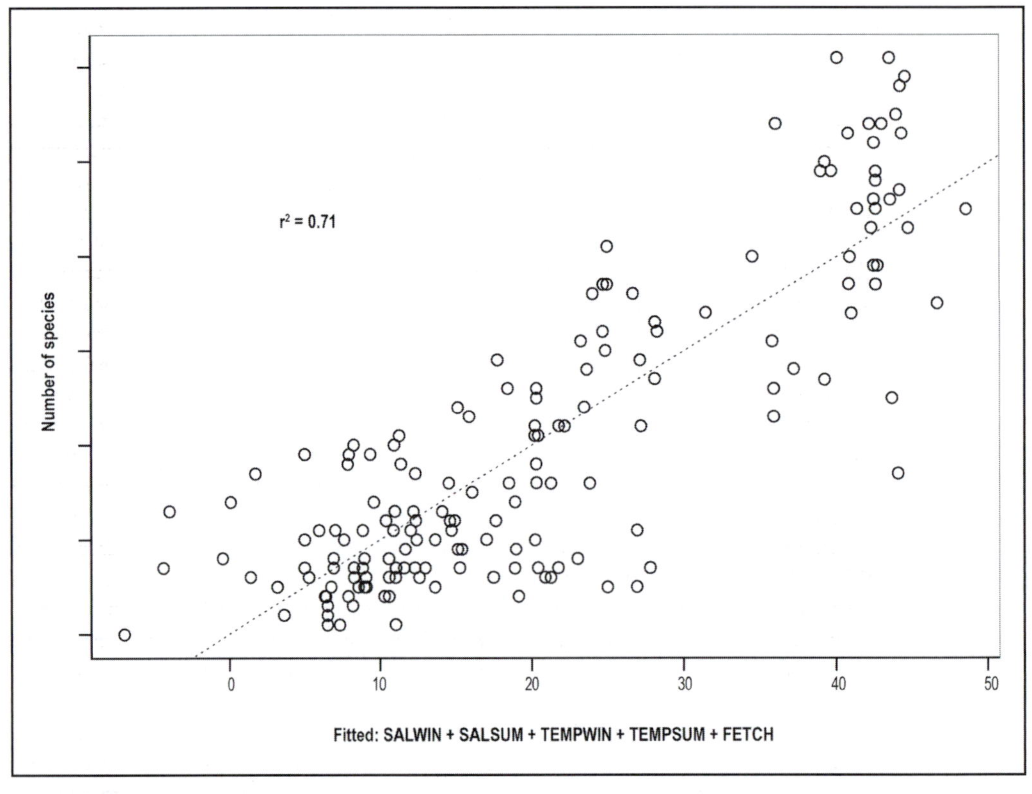

Source: After Zacharias and Roff (2001)

Figure 8.5 Fitted linear regression model of total species richness predicted using historical average winter and summer salinities (SALWIN and SALSUM) and winter and summer temperatures (TEMPWIN and TEMPSUM) and fetch

thereby limiting photosynthesis as well as burying communities in sediments.

To test the relationship between the combined environmental variables and species richness, a linear multiple regression was run. Despite the highly heteroscedasdic nature of the data, there was a highly significant relationship ($p < 0.001$, $R^2 = 0.71$) between species richness and a combination of differences between summer and winter salinities and temperatures and fetch (Figure 8.5). Adding the effects of predators to the physical variables and testing again for the 'herbivore groups' (all species minus predators), there was again a highly significant relationship to species richness relationship ($p < 0.001$, $R^2 = 0.86$).

The hypotheses to explain species richness, outlined in Table 8.1, have several similar elements in that they are all related to some combination of stability, time and disturbance. Species richness increases with increasing distance from the mainland coast, and the outer coast regions are more stable (less variable) with respect to salinity and temperature, but are more exposed to greater fetch. Therefore the greater diversity in the outer coast areas may be related to some combination of environmental homogeneity (stability) and periodic (intermediate) disturbance. These two factors are believed to interact to produce a speciose community as a result of favourable oceanographic conditions and continual succession as a result of periodic disturbance (Ricklefs and Schluter, 1993).

The stability–time and intermediate disturbance (IDH) theories are in some respects, or according to some interpretations, contradictory unless we

Table 8.1 Some 'theories' or hypotheses that seek to explain the occurrence of species richness

Major influence on species richness	Hypothesis	Basic premise	Selected references
Area	Area hypothesis	A species–area curve is a relationship between the area of a habitat and the number of species it contains. Larger areas tend to contain larger numbers of species, following systematic mathematical relationships. The species–area relationship is usually constructed for a single taxon of organisms. A wide range of factors has been invoked to explain the relationship, including: a balance between immigration and extinction; rate and magnitude of disturbance on small vs. large areas; and predator-prey dynamics.	Arrhenius, 1921; Preston, 1962; MacArthur and Wilson, 1967; Rosenzweig, 1995; Brose et al, 2004
Latitude	Latitudinal gradient	This is one of the most widely recognized patterns in ecology; lower latitudes generally have more species than higher latitudes. Many hypotheses involving latitude have been proposed; however, greater evolutionary rate and longer evolutionary time (as a function of higher temperatures and shorter generation time) are major factors.	Eckman, 1953; Sanders, 1968; Hillebrand, 2004
Isolation	Island biogeography	The theory explains species richness of islands; however, an 'island' can be any area of suitable habitat surrounded by unsuitable habitat. The theory proposes that the equilibrium number of species is determined by a balance of immigration, emigration and extinction. Immigration and emigration are related to distance from a source of colonists; extinction is a product of island size. Isolated islands foster the evolution of new species.	MacArthur and Wilson, 1967
Stability	Stability–time	The stability–time hypothesis was proposed to account for the high species diversity of the deep sea. It was supposed that older and more stable environments would support more species than younger and more disturbed ones, and that older communities would become more 'biologically accommodated' over time. However, the concept has been attacked as 'tautological' and untestable. Nevertheless, the time aspect does agree with the greater species diversity of the much older Pacific Ocean compared to the younger Atlantic.	Fischer, 1960; Simpson, 1964; Sanders, 1969
History	Historical expansion	A link is suggested between speciation and range-expansion, with genera expanding out of the tropics tending to speciate more prolifically, both globally and regionally. Alternatively, this can be viewed as essentially a composite of other ideas involving higher disturbances at higher latitude and lower rates or time for evolution.	Krug et al, 2008
Disturbance	Intermediate disturbance	The hypothesis states that local species diversity is maximized when ecological disturbance is neither too rare nor too frequent (high or low). At intermediate levels of disturbance, diversity is maximized because both competitive K-selected organisms and r-selected species can coexist. This is a result of their differing life-history strategies, which dictate a preference for either a high or low disturbance regime.	Horn, 1975; Connell, 1978

Table 8.1 Some 'theories' or hypotheses that seek to explain the occurrence of species richness (*continued*)

Major influence on species richness	Hypothesis	Basic premise	Selected references
Heterogeneity	Structural complexity	The spatial heterogeneity hypothesis predicts a positive relationship between habitat complexity and species diversity: the greater the heterogeneity of a habitat, the greater the number of species in that habitat. The hypothesis is thus really a composite of α and β diversity. Other factors such as predator–prey interactions may also be affected by habitat complexity.	Simpson, 1964
Productivity	Productivity–diversity Species–energy	The species–energy hypothesis suggests the amount of available energy sets limits to the richness of the system; higher solar energy at low latitudes causes increased net primary productivity. Higher net primary productivity supports more individuals leading to more species. However, nutrient supply and other biotic factors may alter relationships in both terrestrial and aquatic communities. Even the shape of the relationship and potential mechanisms are not decided. At different spatial scales a relationship may be linear or parabolic; a parabolic relationship indicates low species diversity at both low and high rates of production.	Rex, 1981; Grassle and Maciolek, 1992; Currie et al, 2004; Humbert and Dorigo, 2005
Succession	Succession	There are many types of ecological succession, including primary, secondary, seasonal and cyclic. In general, communities in early succession will be dominated by fast-growing species (r-selected life-histories). As succession proceeds (perhaps by colonization from a larval lottery), these species will tend to be replaced by more competitive (k-selected) species. Succession can therefore lead to changes in species diversity over time.	Connell and Slatyer, 1977; Sale, 1978
Competition	Numerous	The relationships between competition and species diversity may be complex, and several suggestions have been made as to mechanisms. However, the evidence favours the suggestions that competition and predation interact locally, because predation reduces populations of a superior competitor, thus allowing other species to exist.	Parrish and Saila, 1970; Dayton, 1972; Holt, 1977
Predation	Keystone	Predators can prevent the competitive elimination of species by superior competitors if they depress the size of the population of the dominant competitor. A predator that acts to increase the number of species in the community is called a keystone predator.	Paine, 1969; Menge et al, 1994
Neutrality	Unified neutral theory	The neutral theory is an instance of dispersal assembly theory, in contrast to niche assembly theories which are non-neutral because they hold that different species behave in different ways from one another. *Neutrality* is defined as 'per capita ecological equivalence', which means that all species and individuals are held to behave (in other words reproduce and die) in the same way as one another. The theory predicts the existence of a fundamental biodiversity constant (θ), that may govern species richness on a variety of spatial/temporal scales.	Hubbell, 2001

Table 8.2 A summary of 'theories' that could be applied to the interpretation of the study of coastal species richness at regional scales by Zacharias and Roff (2001)

Theory/hypothesis/concept	Possible interpretation/application
Stability/time	Pacific Ocean older than Atlantic – higher species diversity on the coast of British Columbia than in Atlantic provinces of Canada
Latitude	Sets the upper regional limit for species diversity within any given taxonomic group
Depth	Sets the upper regional limit for species diversity within any given taxonomic group. The intertidal area defines the tolerant species.
Stability/time/intermediate disturbance	Outer coasts are more climatically stable with respect to temperature and salinity variations, sedimentation and freshwater inputs than the inshore waters. Physiological stability/stress
Stability/time/intermediate disturbance	Wave exposure is higher on the outer coasts than on inshore waters. Physical stability/stress
Intermediate disturbance	Predator disturbances create new habitats for colonizing species from a larval lottery. Interaction of physical and biological disturbances
Competition/predation	Predator disturbances reduce competition allowing the co-existence of more species. Biological disturbances
Habitat heterogeneity	Rocky intertidal shore sets the upper regional limit for species diversity

examine their implications more closely. There are instances, however, where their mechanisms can operate in concert resulting in greater species richness. A combination of factors related to environmental stability (salinity and temperature variation) and disturbance (as indexed by fetch and predators) seems to best explain species richness in terms of a combination of physical and biological factors. We expand on these explanations as follows:

Variations in temperature and salinity lead to physiological stress – a form of disturbance to which most marine species (which are both stenohaline and stenothermal) are not well adapted, thus reducing local species diversity. Viewed conversely, a physiologically stable environment enhances species diversity. Increased wave exposure leads to physical stress, a different form of disturbance that removes individuals from substrates. These open local habitats are then colonized at random by any species producing propagules at that time (the 'larval lottery' of Sale, 1978). Because the probability is that the individual colonizing will not be the same as the one removed, local species diversity is increased. Conversely, physical stability under low exposure reduces species diversity. Predator disturbance acts by keeping populations of otherwise dominant competitors at levels below carrying capacity, ensuring open spaces for colonization, again from a larval lottery, thus increasing species diversity. Thus a series of factors, interacting at different spatial scales, is required to 'explain' species diversity at the regional level (Table 8.2). A study such as this permits the regional identification of representative sites of greater than average species diversity, as well as other outliers that could be considered distinctive. Note, however, that the problems of differentiating between α and β diversity remain. The mere fact that environmental factors are different means that the explanation for species diversity must necessarily involve a combination of both α and β diversity.

Patterns of species diversity – sub-regional to local

Biogenic substrates

Biogenic substrates are the result of the growth of species known as ecosystem engineers or

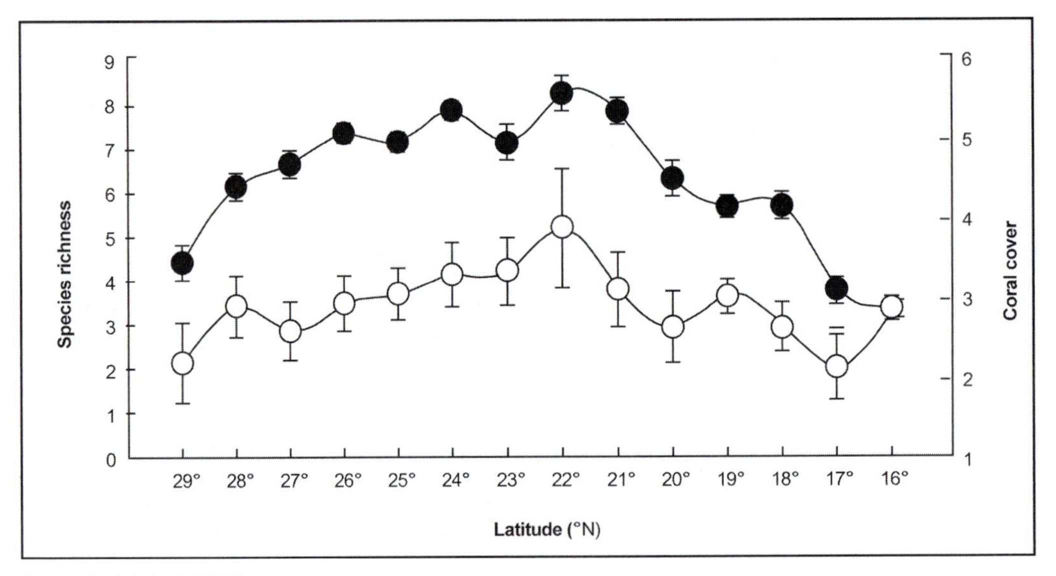

Source: After Roberts et al (1992)

Figure 8.6 Distribution of species richness may be related not to latitude but to the distribution of coral species. The distribution of species richness (number of species per ten-minute count) of butterflyfishes (Chaetodontidae) from North to South in the Red Sea (filled circles). Coral cover on a five-point scale from 0 to 100% cover.

foundation species, defined by Dayton (1972) as 'a single species that defines much of the structure of a community by creating locally stable conditions for other species, and by modulating and stabilizing fundamental ecosystem processes.' Such species, along with their attendant communities, are generally associated with areas of high regional resource productivity (especially primary production), they increase local environmental complexity, and they provide shelter, protection and reproduction sites for a diversity of attached, associated and inquiline species, including adult, larval and juvenile forms of benthic species and fish. Species or groups of species included in this important ecological category would include: scleractinian corals (reef builders), deep sea corals, sponge beds, mangroves, seagrasses, salt marshes, kelp beds and others. Such biogenic substrates can also be considered as hotspots of sub-regional or local high species diversity, comparable in significance to terrestrial rain forests (Connell, 1978). These foundation species are significant not only because of their importance to ecosystem-level processes, but also because they are under threat and in decline worldwide (Carr et al, 2002).

Among biogenic substrates, coral reefs provide the most biologically diverse of shallow water marine ecosystems, but they are being degraded worldwide by human activities and climate changes. Many coral reef taxa have highly restricted ranges, making them vulnerable to extinction. An analysis by Roberts et al (2002) argued that such restricted-range species are clustered into centres of endemism. The ten such richest centres of endemism cover 15.8 per cent of the world's coral reefs (0.012 per cent of the oceans) but include between 44.8 and 54.2 per cent of the restricted-range species which accordingly should be targeted for conservation efforts.

Although the incidence of some foundation species is higher in the tropics, they are scattered around the world primarily in shallow water habitats of the euphotic zone. However, within such regions there may be little relation between species diversity and latitude, for example in corals (Bellwood and Hughes, 2001). In fact it

Box 8.2 Some measures of substrate/habitat heterogeneity/complexity

Measure	Description and reference
Benthic complexity	Benthic complexity looks at how convoluted the bottom is, not how steep or how rough, though these both play a role. Complexity is similar to but not the same as rugosity. Rugosity can be strongly influenced by a single large change in depth whereas complexity is less so, since all changes are treated more equally. Benthic complexity is indicated by how often the slope of the sea bottom changed in a given area; that is, the density of the slope (Ardron, 2002).
Rugosity	Generally understood and defined as the roughness of the physical structure of the seafloor. Most commonly measured in an index (Dahl's surface index) characterized as the ratio of the surface area to the planimetric area (Dahl, 1973). Within a GIS the ratio of surface area to planar area can be calculated based on the surface areas and elevation grids using available tools and spatial functions, for example Jenness, 2010 (DEM surface tools), Wright et al 2005 (Bathymetric Terrain Modeler).
Vector ruggedness measure (VRM)	Quantifies terrain ruggedness by measuring the dispersion of vectors orthogonal to the terrain surface. The VRM values are low both in flat areas and in steep areas, but values are high in areas that are both steep and rugged. 'Unlike [LSRI and TRI], VRM differentiated smooth, steep hillsides from irregular terrain that varied in gradient and aspect' (Sappington et al, 2007).

is suggested that the species diversity of at least some taxa of fish is more strongly associated with the coral cover itself rather than with latitude (see Figure 8.6; and Roberts et al, 1992).

Overall the reasons for the high species diversity of communities associated with foundation species is still under debate. Among the main suggestions is that coral reefs are maintained in a non-equilibrium state, where high diversity is maintained only when the species composition is continually changing due to disturbances (Connell, 1978; Talbot et al, 1978). A corollary of this hypothesis is that species diversity depends on dispersal from a larval lottery (Sale, 1978). A second hypothesis is that high species diversity is a function of local substrate heterogeneity.

Substrate/habitat heterogeneity

There is currently great interest in the subject of habitat mapping and heterogeneity and its relationship to species diversity (see e.g. Dunn and Halpin, 2009). However, a number of problems require examination and resolution. First, it is not clear what the opposite of habitat heterogeneity – namely, habitat homogeneity – might be, nor how it can be defined or recognized. This problem becomes most acute in the deep sea (see below). Second, the heterogeneity of the environment can be described at various spatial scales by various techniques (see Box 8.2). Habitat heterogeneity appears to be important at all scales from more than thousands of metres (e.g. Manson, 2009) to centimetres (e.g. Archambault and Bourget, 1996). Third, without a clear definition of habitat homogeneneity, any relationship between habitats and species diversity has to be considered as a mix of both α and β diversity. By its very name the term 'habitat heterogeneity' must mean that it is describing a set of habitat types (however subtle the differentiation). Thus the term describes among-habitat variation, not within-habitat variation; as soon as a habitat exhibits variation (heterogeneity) it consists of a set of habitats of different types. Axiomatically then, the species diversity of heterogeneous areas must be a composite of α and β diversity, and must be higher than within a single habitat type. Fourth, habitat heterogeneity can be described in various ways, for example as slope, relief, complexity or rugosity (roughness) – see Ardron (2002) and Box 8.2. Lastly, we do not know how 'catholic' most species are with respect to variations in habitat characteristics. The expectation is that some will be very narrow and others very broad (by analogy to narrow and broad niches).

Given these essentially unresolved issues, it is hardly surprising that there is no current synthesis of the significance of habitat heterogeneity. Nevertheless, habitat heterogeneity has significant relationships to species diversity, provided that it is adequately defined. In a simple example, Abele (1974) showed that the number of different substrates was the single most important factor in determining the number of species of decapods crustaceans, probably because each species makes differential use of each substrate. In a study of species richness of reef fish, Luckhurst and Luckhurst (1978) showed a significant relationship to substrate rugosity of a coral reef rather than to the extent of coral cover. Roberts and Ormond (1987) came to similar conclusions that substrate complexity rather than coral cover was the important determinant of fish species richness.

Habitat heterogeneity may act as an important modifier of predator–prey relationships, tending to reduce foraging efficiency (e.g. Diehl, 1992; Linehan et al, 2001), which may in turn allow higher species diversity. At the finest scales of centimetres to millimetres it may also permit greater settlement of invertebrates and algae, thus increasing their biomass. In fact in the intertidal zone, a combination of scales becomes important; variability in species richness occurs at a

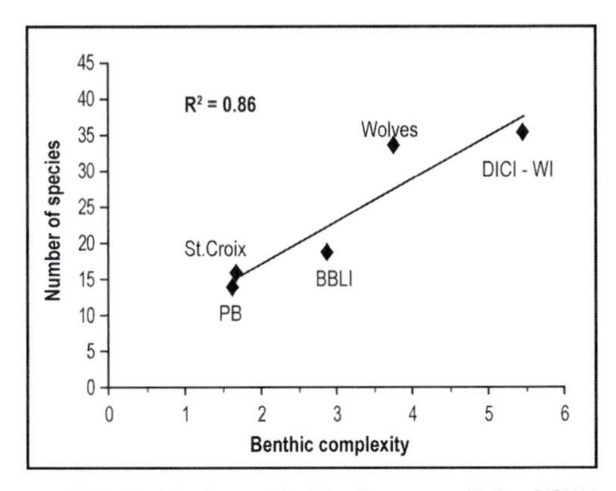

Key: BBLI = Back Bay Letang Inlet, PB = Passamoquoddy Bay, DICI-WI = Deer Island Campobello Island and Wolves Islands

Figure 8.7 Correlation between average regional substrate complexity and average species richness in the Passamoquoddy Bay, Bay of Fundy region

scale of 1km, and variation in abundances occurs at a scale of < 20cms (Archambault and Bourget, 1996). The subject of habitat heterogeneity and its relationship to species diversity, whatever the mechanism, deserves much greater attention in terms of its importance for marine conservation planning and location and characteristics of protected sites.

Patterns of species diversity - local

A study of sub-tidal species richness

A study by Buzeta and Roff (unpublished) on the east coast of Canada, between Passamoquoddy Bay and the Bay of Fundy, examined species diversity in a manner similar to that of Zacharias and Roff (2001), but for the sub-tidal rather than the intertidal zone. The study re-examined biological and oceanographic data from MacKay (1977), MacKay et al (1979), Noble et al (1976) and Robinson et al (1996), among other sources. Epibenthic species (both fixed and motile) of hard substrates (gravel to rock) were recorded by divers along transects from the low water level to 30m depth. A multibeam bathymetric survey of the region was carried out by methods essentially identical to those of Kostylev et al (2001) – see Chapter 6.

The findings were similar to those of Zacharias and Roff (2001), with significant relationships between lower species diversity and increased temperature and salinity variations. However, there was no significant relationship to the degree of exposure. This was expected for two reasons: first, the range of exposure in this study is much less in British Columbia due to shelter of land in and around the bays, and second, sub-tidal communities at depth should be less susceptible to exposure than those of the intertidal region.

The study then proceeded further to examine possible relations between species diversity and benthic substrate complexity, analysed from the multibeam data by methods adopted from Ardron (2002) – see Box 8.2 and Colour Plates 13b and 13c. However, the survey area was restricted to areas of relative homogeneity in oceanographic characteristics, in this tidally dynamic region. There was a clear and significant relationship between species diversity and substrate complexity (Figure 8.7). Although the analyses were confined to one type of substrate (hard gravel to rock bottoms), and one type of benthic community (visible epibenthos), there is obviously some confounding of α and β diversity values among sites, simply as a function of variations in substrate type as well as topographic complexity.

The relationship between higher substrate complexity and species richness is therefore not straightforward to explain. First, the increase could be a matter of preference by individual benthic species for different sizes of hard substrates. Second, the increase could be related simply to an increase of surface area due to complexity itself. Third, it could be because of an increase in the diversity of micro-habitats available (e.g. rock faces upstream, downstream, differential orientation to currents etc.) which describe changes in β diversity. Fourth, the micro-climate of food resources may have changed because of small-scale variations in turbulent advection (Wildish and Kristmanson, 1997). Which of these are the decisive factors or do they all contribute? Unfortunately it is not clear that we shall ever be able to entirely disentangle the components of α and β diversity.

Patterns of species diversity – the deep sea

The longest debate in the marine literature about interpretations of species diversity has probably concerned the deep sea. The environment here tends to (but does not attain) homogeneity, with cold temperatures, high pressure and low food supply. Consequently the numbers of animals and their individual body size are lower than in shallower waters (Rex et al, 2006), but nevertheless the species diversity (as number of species per individual) in several groups is unexpectedly high, but perhaps not hyper-diverse (see Lambshead and Boucher, 2003). Unfortunately we do not know what the number of species actually is – even to within an order of magnitude. Estimates of the number of species of benthic macrofauna range wildly from hundreds of thousands to tens of millions of species (Snelgrove, 1999). The central question was and remains: Why is species diversity of the deep sea so high? Problems of inaccessibility of the environment are confounded by lack of knowledge of the general biology of species and confusion over what is α and what is β diversity.

Following the original reporting of high species diversity in the deep sea by Hessler and Sanders (1967), Sanders (1969) proposed the stability–time hypothesis (Table 8.1). This was challenged by Dayton and Hessler (1972) who proposed instead that high species diversity was a product of continued biological disturbance rather than highly specialized competitive niche diversification. This idea was in turn challenged by Grassle and Sanders (1973) who differentiated between short-term non-equilibrium diversity (induced by disturbances), and long-term or evolutionary diversity (the product of past biological interactions in physically benign and predictable environments). They also noted that the known life-history characteristics of deep sea benthos (namely, small brood size, age classes not dominated by young stages, slow growth rates) are not consistent with predator-controlled communities.

As in other environments, species diversity in the deep sea must reflect an integration of ecological and evolutionary processes operating at different spatial and temporal scales (Rex et al, 1997). Recent deep-sea research has focused on the importance of local phenomena that permit the coexistence of species, but it is unclear how small-scale processes could account for geographic patterns of diversity. A fuller understanding of deep sea diversity needs to consider the roles of geological, biogeographic and oceanographic processes at much larger scales.

Geological origin of species diversity

Perhaps the greater mystery is of the geological and biogeographic origins of high species diversity in the ocean depths, rather than the contemporary issue of habitat patchiness and variability of species composition. Thus the deeper question is: where did all these species come from? rather than: how are they presently assorted?

It is known that deep water circulation of the oceans has been variable and has ceased at least in part at several times in the geological past (e.g. Cronin and Raymo, 1997; Horne, 1999; Hotinski et al, 2001; Friedrich et al, 2008). Indeed the prospect that this might happen again in the foreseeable future constitutes the 'nightmare scenario' of global warming. The deep and bottom waters of at least some parts of the oceans must have presumably become anoxic, eliminating all aerobic species, perhaps several times in its geological history. The deep ocean therefore had to be recolonized when deep circulation was re-established.

Recolonization likely occurred from many disparate locations, bringing different assemblages of fauna from continental margins and seamounts. Any seamounts that rose from the abyssal plains into the upper mixed layer (at depths of 600–1000m) would have retained their species complements, and these could all have acted as nucleating sites for recolonization of the deep sea benthos once deep circulation was re-established. This is one possible explanation for the high species diversity of the deep sea, namely that it is simply a consequence of periodic geological isolation and extirpation of fauna, and the high degree of endemicity and uniqueness of fauna of individual seamounts or seamount chains and continental margins.

Contemporary isolating mechanisms in the deep sea are not well known (see Chapter 12), but deep ocean circulation is very slow (of the order of centuries to a millennium). Given the general suppression of larval forms and low dispersal abilities of deep sea species, it is unlikely that wide areas of the deep ocean could effectively exchange propagules. Inevitably, regional speciation has occurred and indeed must still be occurring in this environment (see e.g. Ruzzante et al, 1998; and Chapters 10 and 17).

Contemporary origin of species diversity

Making a comparison of the benthos of the Norwegian continental shelf and the deep sea, Gray (1994) disputed that the species diversity of the deep sea really is high. He suggested that the deep-sea fauna partition the environment into habitats more finely than the fauna of the shelf, and that small-scale heterogeneity in the deep sea is produced by a variety of factors such as the activity of organisms and sedimenting phytoplankton. At more local scales, it is evident that the deep-sea environment is not at all uniform (see photographs in Heezen and Hollister, 1971), and that different habitats do support different communities of organisms. For example, the percentage of silt and clay in sediments can be highly variable and there is a significant negative correlation between percentage silt and polychaete species numbers (Abele and Walters, 1979).

However, further assessments of diversity, using species accumulation curves with randomly pooled samples, confirm the disputed claim that the deep sea supports higher diversity than the continental shelf (Levin et al, 2001).

We have considered the question of 'what constitutes a community?' (see Chapter 6), but we obviously also require more thought about what constitutes a habitat. Why for example should the species of the deep sea 'partition' habitats more finely and thereby occupy narrower niches? Although we can adequately describe species diversity (e.g. as species accumulation or species rarefaction curves), there is still considerable difficulty in deciding where to draw the line between α and β diversity. It revolves around the issue of how to define a habitat and how finely species may discriminate among them. What is a habitat, and when does it become different from another one?

The broader issue is whether communities are niche-assembled or dispersal-assembled (see e.g. Hubbell, 2001). Levin et al (2001) suggest that local communities in the deep sea may be composed of species that exist as metapopulations whose regional distribution depends on a balance among global-scale, landscape-scale and small-scale dynamics. This reasonable synthesis can be expanded as follows: A general explanation for deep sea species diversity seems to require both dispersal assembly and niche assembly. Over geological time scales, disturbances due to changes in ocean circulation may reset the entire array of deep-sea communities by recolonization from many sources (i.e. dispersal assembly). Over biological time scales, various kinds of disturbances (whose type, frequency, duration etc. are largely unknown) enhance the patchiness of substrates, allowing recolonization (dispersal assembly) from a larval lottery (e.g. Sale, 1978), which is modified by requirements of individual species for substrate types (niche assembly).

Conclusions and management implications

Species diversity has always been of paramount interest and importance to conservation and is likely to remain so, yet there is much we still have to learn about its uneven distribution in the marine environment. Studies of the distribution of species richness are vital to the proper planning of protected areas.

The global patterns of distribution of species diversity are relatively clear. The benthic realm is more speciose than the pelagic and there are also relationships to 'energy' and production regimes. There are major differences between the Pacific and Atlantic oceans, with the highest species diversity occurring at the junction of the Indian and Pacific oceans (for example in the coral triangle). There are clear gradients in several major taxa, of increasing species diversity with decreasing latitude (although in some taxa confounded with habitat complexity in coral reefs), and with increasing depth.

Superimposed on patterns at global and ecoregional levels, other factors come into play to determine species diversity at regional scales. Here, at least in shallow waters, some combination of factors is related to the physiological/ecological stability of the environment (e.g. as temperature, salinity and productivity variations) and paradoxically to environmental instability (e.g. as physical or biological disturbance regimes). Confounded with this at very local scales – from millimetres to tens of metres depending on the size of organisms themselves – is habitat heterogeneity (environmental complexity or rugosity), which really describes a transition from α to β diversity, and can also lead to enhanced species diversity.

Conservation attention has centred disproportionately on various kinds of 'hotspots' of species diversity, especially those supporting foundation (engineering) species. It should be noted that although preservation of selected hotspots may receive priority, due to their significance for ecological processes, present high rate of habitat loss, imminent threat, opportunity, public support and so on, their conservation alone is not sufficient to conserve marine biodiversity. Importantly, global hotspots of species richness are not congruent with endemism.

Concerning the numerous theories/hypotheses that seek to explain species diversity, several conclusions can be drawn. First, most of the hypotheses proposed are subject to several interpretations. Second, several of the hypotheses are mutually contradictory, depending upon the interpretation drawn. Third, no single hypothesis can currently explain all observations of variations in species richness at all spatial scales (a possible exception from Hubbell (2001) is currently under scrutiny). Fourth, the various factors affecting species diversity apparently act in concert in a spatial hierarchy from global to local scales. Fifth, there is still confusion over separating the levels of diversity, from α (within habitats) to β (among habitats) to γ (within regions). Most serious is how species diversity is related to habitat characteristics, and how habitats can be separately recognized and described – that is, the confounding of α and β diversity.

Biodiversity is not synonymous with species diversity (α diversity), and we know far less about the distribution of the other elements of biodiversity at other levels of the ecological hierarchy. Comprehensive bioregionalization studies, such as those in Australia, are in essence the first spatially explicit attempts at national inventories of all the components of biodiversity. Marine species and their habitats are being degraded and lost simultaneously. The most effective way to combat this is by planning for the conservation of marine habitats in networks of MPAs, rather than by giving disproportionate attention to individual species – even foundation species. Only such integrated planning can lead to environmentally sound management of the oceans and to true sustainable development.

The deep sea has attracted research interest because of its unexpectedly high species diversity, though how this was attained or is maintained is not clear. It is unlikely that the deep sea will achieve any significant level of conservation protection in the near future because geographically most of it lies beyond national jurisdictions (see Chapter 12). With the exception of fisheries on seamounts, the deep sea is presently also largely unexploited by humans.

The seemingly esoteric arguments about deep sea species diversity in fact have important implications for marine conservation as a whole – not just in the deep sea. If we believe that benthic communities are predominantly dispersal assembled, rather than finely niche assembled, then conservation plans need to be less concerned with the precise details of substrate type and exact composition. However, if we believe that benthic communities are incredibly 'fussy' about the precise nature of their substrate, then we require intricate surveys to determine species associations and precise distributions; this was considered further in Chapter 6.

Finally, if the current interpretation of geological origins of high species diversity of the deep sea is correct, then there is an encouraging and much broader corollary to this geological scenario. The possibility that seamounts have, perhaps repeatedly, acted as island refugia and renucleating sites for naturally disturbed areas of the deep sea during past geological times implies that MPAs in general could act in the same way (as Noah's Arks or oases) at shallower depths, as insurance against human disturbances on local, regional and global scales. The rationale for widespread establishment of MPAs can therefore be further justified in terms of 'natural' ecological recolonization.

References

Abele, L. G. (1974) 'Species diversity of decapod crustaceans in marine habitats', *Ecology*, vol 55, pp156–161

Abele, L. G. and Walters, K. (1979) 'Marine benthic diversity: A critique and alternative explanation', *Journal of Biogeography*, vol 6, pp115–126

Allison, G. W., Gaines, S. D., Lubchenco J. and Possingham, H. P. (2003) 'Ensuring persistence of marine reserves: Catastrophes require adopting an insurance factor', *Ecological Applications*, vol 13, pp8–24

Angel, M.V. (1993) 'Biodiversity of the pelagic ocean', *Conservation Biology*, vol 7, no 4, pp760–772

Archambault, P. and Bourget, E. (1996) 'Scales of coastal heterogeneity and benthic intertidal species richness, diversity and abundance', *Marine Ecology Progress Series*, vol 136, pp111–121

Ardron, J. (2002) 'A GIS Recipe for Determining Benthic Complexity: An indicator of species richness', in J. Breman (ed) *Marine Geography. GIS for the Oceans and Seas*, ESRI Press, Redlands, CA

Arrhenius, O. (1921) 'Species and area', *Journal of Ecology*, vol 9, pp95–99

Bellwood, D. R. and Hughes, T. P. (2001) 'Regional-scale assembly rules and biodiversity of coral reefs', *Science*, vol 292, pp1532–1534

Brose, U., Ostling, A., Harrison, K. and Martinez, N. D. (2004) 'Unified spatial scaling of species and their trophic interactions', *Nature*, vol 428, pp167–171

Carr, M. H., Anderson, T. W. and Hixon, M. A. (2002) 'Biodiversity, population regulation, and the stability of coral-reef fish communities', *Proceedings of the National Academy of Sciences*, vol 99, pp11241–11245

CERC (Coastal Engineering Research Center) (1977) *Shore Protection Manual*, US Army Corps of Engineers Coastal Engineering Research Center, Vicksburg, MS

Connell, J. H. (1978) 'Diversity in tropical rainforests and coral reefs', *Science*, vol 199, pp1302–1310

Connell, J. H. and Slatyer, R. O. (1977) 'Mechanisms of succession in natural communities and their role in community stability and organization', *American Naturalist*, vol 111, pp1119–1144

Cronin, T. M. and Raymo, M. E. (1997) 'Orbital forcing of deep-sea benthic species diversity', *Nature*, vol 385, pp624–627

Currie, D. J., Mittelbach, G. G., Cornell, H. V., Field, R., Guegan, J. F., Hawkins, B. A., Kaufman, D. M., Kerr, J. T., Oberdorff, T., O'Brien, E. and Turner, J. R. G. (2004) 'A critical review of species–energy theory', *Ecology Letters*, vol 7, pp1121–1134

Dahl, A. L. (1973) 'Surface area in ecological analysis: Quantification of benthic coral-reef algae', *Marine Biology*, vol 23, pp239–249

Dayton, P. K. (1972) 'Toward an Understanding of Community Resilience and the Potential Effects of Enrichments to the Benthos at McMurdo Sound Antarctica', in B. C. Parker (ed) *Proceedings of the Colloquium on Conservation Problems in Antarctica*, Allen Press, Lawrence, KA

Dayton, P. K. and Hessler, R. R. (1972) 'Role of biological disturbance in maintaining diversity in the deep sea', *Deep-Sea Research*, vol 19, pp199–208

Derous, S., Agardy, T., Hillewaert, H., Hostens, K., Jamieson, G., Lieberknecht, L., Mees, J., Moulaert, I., Olenin, S., Paelinckx, D., Rabaut, M., Rachor, E., Roff, J., Stienen, E. W. M., van der Wal, J. T., Van Lancker, V., Verfaillie, E., Vincx, M., Weslawski, J. M. and Degraer, S. (2007) 'A concept for biological valuation in the marine environment', *Oceanologia*, vol 49, pp99–128

Diehl, S. (1992) 'Fish predation and benthic community structure: The role of omnivory and habitat complexity', *Ecology*, vol 73, pp1646–1661

Dornelas, M., Connolly, S. R. and Hughes, T. P. (2006) 'Coral reef diversity refutes the neutral theory of biodiversity', *Nature*, vol 440, pp80–82

Dunn, D. C. and Halpin, P. N. (2009) 'Rugosity-based regional modeling of hard-bottom habitat', *Marine Ecology Progress Series*, vol 377, pp1–11

Ekman, S. (1953) *Zoogeography of the Sea*, Sidgwick & Jackson, London

Ellingsen, K. and Gray, J. S. (2002) 'Spatial patterns of benthic diversity: Is there a latitudinal gradient along the Norwegian continental shelf?', *Journal of Animal Ecology*, vol 71, pp373–389

Fischer, A. G. (1960) 'Latitudinal variations in organic diversity', *Evolution*, vol 14, pp64–81

Floeter, S. R., Behrens, M. D., Ferreira, C. E. L., Paddack, M. J. and Horn, M. H. (2005) 'Geographical gradients of marine herbivorous fishes: Patterns and processes', *Marine Biology*, vol 147, pp1435–1447

Friedrich, O., Erbacher, J., Moriya, K., Wilson, P. A. and Kuhnert, H. (2008) 'Warm saline intermediate waters in the Cretaceous tropical Atlantic Ocean', *Nature Geoscience*, vol 1, pp453–457

Gage, J. D., Levin, L. A. and Wolff, G. A. (2000) 'Benthic processes in the deep Arabian Sea: introduction and overview', *Deep Sea Research II*, vol 47, pp1-7

Grassle, J. F. and Maciolek, N. J. (1992) 'Deep-sea species richness: Regional and local diversity estimate from quantitative bottom samples', *American Naturalist*, vol 139, pp313–341

Grassle, J. F. and Sanders, H. L. (1973) 'Life histories and the role of disturbance', *Deep-Sea Research*, vol 20, pp643–659

Gray, J. S. (1994) 'Is deep-sea species diversity really so high? Species diversity of the Norwegian continental shelf', *Marine Ecology Progress Series*, vol 112, pp205–209

Gray, J. S. (1997) 'Marine biodiversity: Patterns, threats and conservation needs', *Biodiversity and Conservation*, vol 6, pp153–175

Gray, J. S. (2001) 'Marine diversity: The paradigms in patterns of species richness examined', *Scientia Marina*, vol 65, pp41–56

Heezen, B. C. and Holister, C. D. (1971) *The Face of the Deep*, Oxford University Press, Oxford

Hessler, R. R. and Sanders, H. L. (1967) 'Faunal diversity in the deep sea', *Deep Sea Research*, vol 14, pp65–78

Hillebrand, H. (2004) 'On the generality of the latitudinal diversity gradient', *American Naturalist*, vol 163, pp192–211

Hiscock, K. and Breckels, M. (2007) *Marine Biodiversity Hotspots in the UK: Their Identification and Protection*, World Wildlife Fund UK, Godalming, Surrey

Holt, R. D. (1977) 'Predation, apparent competition, and the structure of prey communities', *Theoretical Population Biology*, vol 12, pp197-229

Horn, H. S. (1975) 'Markovian Properties of Forest Succession', in M. L. Cody and J. M. Diamond (eds) *Ecology and Evolution of Communities*, Belknap Press, Cambridge, MA

Horne, D. J. (1999) 'Ocean circulation modes of the Phanerozoic: Implications for the antiquity of deep-sea benthonic invertebrates', *Crustaceana*, vol 72, pp999–1018

Hotinski, R. M., Bice, K. L., Kump, L. R., Najjar, R. G. and Arthur, M. A. (2001) 'Ocean stagnation and end-Permian anoxia', *Geology*, vol 29, no 1, pp7–10

Hubbell, S. P. (2001) *The Unified Neutral Theory of Biodiversity and Biogeography*, Princeton University Press, Princeton, NJ

Hughes, T. P., Bellwood, D. R. and Connolly, S. R. (2002) 'Biodiversity hotspots, centres of endemicity, and the conservation of coral reefs', *Ecology Letters*, vol 5, pp775–784

Humbert J. F. and Dorigo, U. (2005) 'Biodiversity and aquatic ecosystem functioning', *Aquatic Ecosystem Health & Management*, vol 8, no 4, pp367–374

Hurlbert, A. H. and Jetz, W. (2007) 'Species richness, hotspots, and the scale dependence of range maps in ecology and conservation', *Proceedings of the National Academy of Sciences*, vol 104, no 33, pp13384–13389

Hutchinson, G. E. (1961) 'The paradox of the plankton', *American Naturalist*, vol 95, pp137–145

Jenness, J. (2010) 'DEM Surface Tools (surface_area.exe) v. 2.0.230', www.jennessent.com/arcgis/surface_area.htm, accessed 10 October 2010

Johnson, R. G. (1970) 'Variations in diversity within benthic marine communities', *American Naturalist*, vol 104, pp285–300

Kareiva, P. and Marvier, M. (2003) 'Conserving biodiversity coldspots', *American Scientist*, vol 91, pp344–351

Kaustuv, R., Jablonski, D., Valentine, J. W. and Rosenberg, G. (1998) 'Marine latitudinal diversity gradients: Tests of causal hypotheses', *Proceedings of the National Academy of Sciences*, vol 95, no 7, pp3699–3702

Kostylev, V. E., Todd, B. J, Fader, G. B. R, Courtney, R. C., Cameron, G. D. M. and Pickrill, R. A. (2001) 'Benthic habitat mapping on the Scotian Shelf based on multibeam bathymetry, surficial geology and sea floor photographs', *Marine Ecology Progress Series*, vol 219, pp121–137

Krug, A. Z., Jablonski. D and Valentine, J. W. (2008) 'Species–genus ratios reflect a global history of diversification and range expansion in marine bivalves', *Proceedings of the Royal Society B*, vol 275, no 1639, pp1117–1123

Lambshead, P. J. D. and Boucher, G. (2003) 'Marine nematode deep-sea biodiversity: Hyperdiverse or hype?', *Journal of Biogeography*, vol 30, pp475–485

Lambshead, P. J. D., Brown, C. J., Ferrero, T. J., Mitchell, N. J., Smith, C. R., Hawkins, L. E. and Tietjen, J. (2002) 'Latitudinal diversity patterns of deep-sea marine nematodes and organic fluxes: a test from the central equatorial Pacific', *Marine Ecology Progress Series*, vol 238, pp129–135

Levin, L. A., Etter, R. J., Rex, M. A., Gooday, A. J., Smith, C. R., Pineda, J., Stuary, C. T., Hessler, R. R. and Pawson, D. (2001) 'Environmental influences on regional deep-sea diversity', *Annual Review of Ecology and Systematics*, vol 32, pp51–93

Lewis, J. R. (1964) *The Ecology of Rocky Shores*, English University Press, London

Linehan, J. E., Gregory, R. S. and Schneider, D. C. (2001) 'Predation risk of age-0 cod (*Gadus)* relative to depth and substrate in coastal waters', *Journal of Experimental Marine Biology and Ecology*, vol 263, pp25–44

Longhurst, A. (1998) *Ecological Geography of the Sea*, Academic Press, San Diego, CA

Luckhurst, B. E. and Luckhurst, K. (1978) 'Analysis of the influence of substrate variables on coral reef fish communities', *Marine Biology*, vol 49, pp317–323

Lyne, V. and Hayes, D. (2005) 'Pelagic regionalization – national marine bioregionalization – integration project', CSIRO Marine Research, Hobart, Tasmania

Lyne, V., Hayes, D. and Condie, S. (2007) 'Support tools for regional marine planning in the southwest marine region, CSIRO Marine Research, Hobart, Tasmania

MacArthur, R. H. and Wilson, E. O. (1967) *The Theory of Island Biogeography*, Princeton University Press, Princeton, NJ

MacKay, A. A. (1977) 'A biological and oceanographic study of the Brier Island region, NS', Parks Canada, Ottawa, Canada

MacKay, A. A., Bosien, R. K. and Leslie, P. (1979) 'Grand Manan archipelago', *Bay of Fundy Resource Inventory, volume 4*, Marine Research Associates Ltd., Deer Island, NB, Canada

Mann, K. H. and Lazier, J. R. N. (1996) *Dynamics of Marine Ecosystems: Biological–Physical Interactions in the Oceans*, Blackwell Science, London

Manson, M. M. (2009) 'Small scale delineation of northeast Pacific Ocean undersea features using benthic position index', *Canadian Manuscript Report of Fisheries and Aquatic Science*, vol 2864

Menge, B. A., Berlow, E. L., Blanchette, C. A., Navarrete, S. A. and Yamada S. B. (1994) 'The keystone species concept: Variation in interaction strength in a rocky intertidal habitat', *Ecological Monographs*, vol 64, no 3, pp249–286

Moore, J. C., Berlow, E. L., Coleman, D. C., Ruiter, P. C., Dong, Q., Hastings, A., Johnson, N. C., Mc-

Cann, K. S., Melville, K., Morin, P. J., Nadelhoffer, K., Rosemond, A. D., Post, D. M., Sabo, J. L., Scow, K. M., Vanni, M. J. and Wall, D. H. (2004) 'Detritus, trophic dynamics and biodiversity', *Ecology Letters*, vol 7, pp584–600

Myers, N., Mittermeier, R. A., Mittermeier, C. G., da Fonseca, G. A. B. and Kent, J. (2000) 'Biodiversity hotspots for conservation priorities', *Nature*, vol 403, pp853–858

National Research Council (1995) *Understanding Marine Biodiversity*, National Academy Press, Washington, DC

Noble, J. P. A., Logan, A. and Webb, G. R. (1976) 'The recent *Terebratulina* community in the rocky subtidal zone of the Bay of Fundy Canada', *Lethaia*, vol 9, no 1, pp1–17

Norse, E. (1993) *Global Marine Biological Diversity*, Island Press, Washington, DC

Norse, E. and Crowder, L. B. (2005) *Marine Conservation Biology. The Science of Maintaining the Seas's Biodiversity*, Island Press, Washington, DC

Nybakken, J. (1997) *Marine Biology: An Ecological Approach*, Addison Wesley Longman Inc., New York

Odling-Smee, L. (2005) 'Dollars and sense', *Nature*, vol 437, pp614–616

Orme, C. D. L., Davies, R. G., Burgess, M., Eigenbrod, F., Pickup, N., Olson, V. A., Webster, A. J., Ding, T.-S., Rasmussen, P. C., Ridgely, R. S., Stattersfield, A. J., Bennett, T. M., Blackburn, T. M., Gaston, K. J. and Owens, I. P. F. (2005) 'Global hotspots of species richness are not congruent with endemism or threat', *Nature*, vol 436, pp1016–1019

Ormond, R. F. G., Gage, J. D. and Angel, M. V. (1997) *Marine Biodiversity: Patterns and Processes*, Cambridge University Press, Cambridge

Paine, R. T. (1969) 'A note on trophic complexity and community stability', *American Naturalist*, vol 103, pp91–93

Parrish, J. D. and Saila, S. B. (1970) 'Interspecific competition, predation and species diversity', *Journal of Theoretical Biology*, vol 27, pp207–220

Pianka, E. R. (1966) 'Latitudinal gradients in species diversity: A review of concepts', *American Naturalist*, vol 100, no 910, pp33–46

Possingham, H. P. and Wilson, K. A. (2005) 'Biodiversity: Turning up the heat on hotspots', *Nature*, vol 436, pp919–920

Preston, F. W. (1962) 'The canonical distribution of commonness and rarity: Part I', *Ecology*, vol 43, pp185–215, 431–432

Price, A. R. G. (2002) 'Simultaneous 'hotspots' and 'coldspots' of marine biodiversity and implications for global conservation', *Marine Ecology Progress Series*, vol 241, pp23–27

Ptacnik, R., Solimini, A. J., Andersen, T., Tamminen, T. Brettum, P., Lepisto, L., Willen, E. and Rekolainen, S. (2008) 'Diversity predicts stability and resource use efficiency in natural phytoplankton communities', *Proceedings of the National Academy of Sciences*, vol 105, pp5134–5138

Reaka-Kudla, M. L., Wilson, D. E. and Wilson, E. O. (1997) *Biodiversity II. Understanding and Protecting our Biological Resources*, Joseph Henry Press, Washington DC

Rex, M. A. (1981) 'Community structure in the deep-sea benthos', *Annual Review of Ecology and Systematics*, vol 12, pp331–353

Rex, M. A., Etter, R. J. and Stuart, C. T. (1997) 'Large-scale Patterns of Species Diversity in the Deep-sea Benthos', in R. F. G. Ormond, J. D. Gage and M. V. Angel (eds) *Marine Biodiversity: Patterns and Processes*, Cambridge University Press, Cambridge

Rex, M. A., Etter, R. J., Morris, J. S., Crouse, J., McClain, C. R., Johson, N. A., Stuart, C. T., Deming, J. W., Thies, R. and Avery, R. (2006) 'Global bathymetric patterns of standing stock and body size in the deep-sea benthos', *Marine Ecology Progress Series*, vol 317, pp1–8

Rex, M. A., Stuart, C. T., Hessler, R. R., Allen, J. A., Sanders, H. L. and Wilson, G. D. F. (1993) 'Global-scale latitudinal patterns of species diversity in the deep-sea benthos', *Nature*, vol 365, pp636–639

Richerson, P., Armstrong, R. and Goldman, C. R. (1970) 'Contemporaneous disequilibrium, a new hypothesis to explain the "paradox of the plankton"', *Proceedings of the National Academy of Sciences*, vol 67, pp1710–1714

Ricklefs, R. E. (1987) 'Community diversity: Relative roles of local and regional processes', *Science*, vol 235, pp167–171

Ricklefs, R. E. and Schluter, D. (1993) *Species Diversity in Ecological Communities: Historical and Geographic Perspectives*, University of Chicago Press, Chicago, IL

Roberts, C. M. and Ormond, R. F. (1987) 'Habitat complexity and coral reef fish diversity and abundance on Red Sea fringing reefs', *Marine Ecology Progress Series*, vol 41, pp1–8

Roberts, C. M., McClean, C. J., Veron, J. E. N., Hawkins, J. P., Allen, G. R., McAllister, D. E., Mittermeier, C. G., Schueler, F. W., Spaulding, M., Wells, F., Vynne, C. and Werner, T. B. (2002) 'Marine biodiversity hotspots and conservation priorities for tropical reefs', *Science*, vol 295, pp1280–1284

Roberts, C. M., Shepherd, A. R. D. and Ormond, R. F. G. (1992) 'Large scale variation in assemblage structure of Red Sea butterflyfishes and angelfishes', *Journal of Biogeography*, vol 19, pp239–250

Robinson, S. M. C., Martin, J. D., Page, F. H. and Losier, R. (1996) 'Temperature and salinity of Passamoquoddy Bay and approaches between 1990 and 1995', *Canadian Technical Report of Fisheries and Aquatic Sciences*, vol 2139

Rocha, L. A., Rocha, C. R., Robertson, D. R. and Bowen, B. W. (2008) 'Comparative phylogeography of Atlantic reef fishes indicates both origin and accumulation of diversity in the Caribbean', *BMC Evolutionary Biology*, vol 8, no 157, pp1–16

Rosenzweig, M. L. (1995) *Species Diversity in Space and Time*, Cambridge University Press, Cambridge

Rosenzweig, M. L. and Abramsky, Z. (1993) 'How Are Diversity and Productivity Related?', in R. E. Ricklefs, and D. Schluter (eds) *Species Diversity in Ecological Communities, Historical and Geographical Perspectives*, University of Chicago Press, Chicago, IL

Roy, K., Jablonski, D. and Valentine, J. W. (2000) 'Dissecting latitudinal diversity gradients: functional groups and clades of marine bivalves', *Proceedings of the Royal Society of London B*, vol 267, pp293–299

Ruzzante, D. E., Taggart, C. T. and Cook, D. (1998) 'A nuclear DNA basis for shelf- and bank-scale population structure in northwest Atlantic cod (*Gadus morhua*): Labrador to Georges Bank', *Molecular Ecology*, vol 7, pp1663–1680

Sale, P. F. (1978) 'Coexistence of coral reef fishes: A lottery for living space', *Envirionmental Biology of Fishes*, vol 3, pp85–102

Sanders, H. L. (1968) 'Marine benthic diversity: A comparative study', *American Naturalist*, vol 102, pp243–282

Sanders, H. L. (1969) 'Benthic marine diversity and the stability-time hypothesis', *Brookhaven Symposium in Biology*, vol 22, pp71–80

Sappington, J. M., Longshore, K. M. and Thomson, D. B. (2007) 'Quantifiying landscape ruggedness for animal habitat anaysis: A case study using bighorn sheep in the mojave desert', *Journal of Wildlife Management*, vol 71, no 5, pp1419–1426

Searing, G. F. and Frith, H. R. (1995) *British Columbia Biological Shore Zone Mapping System*, British Columbia Resource Inventory Committee, Victoria, Canada

Simpson, G. G. (1964) 'Species density of North American recent mammals', *Systematic Zoology*, vol 13, pp57–73

Smith, V. H., Foster, B. L. Grover, J. P., Holt, R. D. and deNoyelles Jr., F. (2005) 'Phytoplankton species richness scales consistently from microcosms to the world's oceans', *Proceedings of the National Academy of Sciences*, vol 102, pp4393–4396

Snelgrove, P. V. R. (1999) 'Getting to the bottom of marine biodiversity: Sedimentary habitats', *Bioscience*, vol 49, pp129–138

Snelgrove, P. V. R., Austen, M. C., Boucher, G., Heip, C., Hutchings, P. A., King, G. M., Koike, I., Lambshead, P. J. D. and Smith, C. R. (2000) 'Linking biodiversity above and below the marine sediment–water interface', *BioScience*, vol 50, no 12, pp1076–1088

Stehli, F. G., McAlester, A. L. and Heisley, C. E. (1967) 'Taxonomic diversity of recent bivalves and some implications for geology', *Geological Society of America Bulletin*, no 78, pp455–466

Stehli, F. G. and Wells, J. W. (1970) 'Diversity and age patterns in hermatypic corals', *Systematic Biology*, vol 20, no 2, pp115–126

Talbot, F. H., Russell, B. C. and Anderson, G. R. V. (1978) 'Coral reef fish communities: Unstable high-diversity systems?', *Ecological Monographs*, vol 48, pp425–440

Thorne-Miller, B. and Catena, M. (1991) *The Living Ocean. Understanding and Protecting Marine Biodiversity*, Island Press, Washington DC

Wallace, A. R. (1859) *The Malay Archipelago*, Harper, New York

Whittaker, R. H. (1972) 'Evolution and measurement of species diversity', *Taxon*, vol 21, pp213–251

Wildish, D. and Kristmanson, D. (1997) *Benthic Suspension Feeders and Flow*, Cambridge University Press, Cambridge

Witman, J. D., Etter, R. J. and Smith, F. (2004) 'The relationship between regional and local species diversity in marine benthic communities: A global perspective', *Proceedings of the National Academy of Sciences*, vol 101, pp15664–15669

Worm, B., Lotze, H. K. and Myers, R. A. (2003) 'Predator diversity hotspots in the blue ocean', *Proceedings of the National Academy of Science*, vol 100, pp9884–9888

Wright, D. J., Lundblad, E. R., Larkin, E. M., Rinehart, R. W., Murphy, J., Cary-Kothera, L. and Draganov, K. (2005) 'ArcGIS Benthic Terrain Modeler', Oregon State University, Corvallis, OR

Zacharias, M. A. and Roff, J. C. (2000) 'A hierarchical ecological approach to conserving marine biodiversity', *Conservation Biology*, vol 13, no 5, pp1327–1334

Zacharias, M. A. and Roff, J. C. (2001) 'Explanations of patterns of intertidal diversity at regional scales', *Journal of Biogeography*, vol 28, pp471–483

9

Species and Focal Species

Keystones, umbrellas, flagships, indicators and others

The ability to focus attention on important things is a defining characteristic of intelligence.

Robert J. Shiller (1946–)

Introduction

Population biology runs the spectrum of subjects from genetics (see Chapter 10) through population ecology to management. Management – at least briefly in terms of fisheries – we shall consider in Chapter 13. For conventional population ecology, several recent texts cover the field in depth, and we refer readers elsewhere. In this chapter, our emphasis is on focal species – simply those species on which our attention is preferentially 'focused' for one reason or another (Colour plates 14a, b, c and d).

In marine conservation planning, activity seems to have swung away from individual species and towards emphasis on habitats and spaces. However, the basic ecological issue is not whether we should make efforts to conserve individual species – but rather, which ones we should pay preferential attention to, and why. Focal species are those which, for ecological or social reasons, are believed to be valuable for the understanding, management and conservation of natural environments. Many different types of focal species have been proposed, but we define them to include indicator, sentinel, keystone, umbrella, flagship, charismatic, economic and vulnerable species (Meffe and Carol, 1997).

While there are many names for the various types of focal species, the ecological concepts and societal rationale behind the nomenclature can be distilled into four distinct 'categories': indicators, keystones, umbrellas and flagships (Simberloff, 1998). Indicator, keystone and umbrella species are predicated on the expected outcomes of various ecological concepts, while the flagship species concept relies on human compassion, sense of responsibility and – to some extent – self interest. [We use the term 'concept' in the sense of a construct that may have heuristic value, rather than the terms theory or hypothesis which imply rigorous testability.] Regardless of their underlying assumptions, the expectation is that the presence or abundance of any of the four types of focal species (or in some cases guilds or taxa) is a means to understanding the composition, state or function of a more complex community.

In broad terms, the various focal species may be defined as follows: the indicator species concept has evolved into several distinct types. It suggests that either: (a) there are species whose presence or absence denotes a particular habitat, community or ecosystem; or (b) there are species that denote the condition or 'health' of a particular habitat, community or ecosystem. The keystone concept postulates that certain species are critical to the ecological function of a community or habitat, and that the importance of their role is disproportionate to their abundance or biomass. The umbrella species concept suggests that the conservation of a particular species with known (generally larger) habitat requirements will therefore also protect those species dependent upon smaller portions of the preserved habitat. The flagship species concept cannot be considered as an ecological concept; it is merely a tool to garner public support for 'charismatic megafauna'. However, similar to the umbrella concept, the ultimate goal of advocating flagships is the preservation of their habitats.

Several of the focal species concepts have been revisited by Hurlburt (1997), Menge et al (1994), Navarrete and Menge (1996), Power et al (1996) and Simberloff (1998), where all of these authors except Hurlburt (1997) suggest that there continues to be merit in the application of these concepts to conservation and management. Only recently, however, has this discussion focused on the potential value of focal species for marine conservation and management (e.g. NRC, 1995; Zacharias and Roff, 2000). However, it should be realized that approaches to marine conservation and management at the species level represent only one of an array of possible approaches ranging from the genetic to the ecosystem level of organization (see Zacharias and Roff, 2000, for further explanation).

There has been considerable debate surrounding the value of focal species to the management and conservation of terrestrial environments (e.g. Launer and Murphy, 1994; Weaver, 1995; Niemi et al, 1997; Simberloff, 1998; Brooks et al, 2004). Most criticism centres around: (a) the validity of the ecological theory behind the concepts; (b) the lack of firm definitions for the various types of focal species; (c) the lack of agreed-upon standards for their application and use; and (d) the observation that their application and popularity is often more a result of management policy and direction rather than scientific rationale (Simberloff, 1998). The major issues for conservation purposes are whether we can operationally define each of these terms and, whether we can or cannot, do focal species have utility in conservation initiatives?

The purpose of this chapter is to review the ecological and social justifications behind the use of indicators, keystones, umbrella, and flagships, and evaluate their roles in the establishment and practice of conservation strategies (see Zacharias and Roff, 2001). We see these roles as potentially including: the selection of representative and distinct areas for marine conservation (e.g. marine protected areas (MPAs)) (Roberts and Polunin, 1994; Meffe and Carol, 1997; Allison et al, 1998); integrated coastal zone management (Imperial and Hennessey, 1996); the identification and monitoring of biological communities (Paine, 1992; Kideys, 1994); habitat characterization and monitoring (Apollonio, 1994; Zacharias et al, 1999); marine ecosystem classification (Caddy and Bakun, 1994; Ray, 1996; Zacharias et al, 1998); the development of marine gap analysis (Zacharias and Howes, 1998).

We also evaluate the circumstances and conditions under which focal species may be exhibiting focal properties. Here we make a preliminary assessment of what we term the 'focal properties' or the intended roles of focal species in relation to spatial and temporal scales, degrees of disturbance, the biological hierarchy, habitat heterogeneity and species richness. This assessment is a first step in determining the ecological validity of focal species concepts so that these concepts can be tested to determine whether they should be accepted as having heuristic value or discarded as what Peters (1991) termed 'non-operational concepts'. This type of assessment could also be used to identify other kinds of focal species, document the properties of focal species, evaluate the potential utility of combining various types of focal species, and distinguish between testable and non-testable hypotheses. This assessment should be viewed as preliminary, because the utility of these concepts depends upon ecological and socio-economic considerations which are still loosely defined and may lack general consensus.

Indicator species

The indicator species concept (Table 9.1) is perhaps the broadest and most poorly defined of all the focal species, and has often been used

Table 9.1 Characteristics exhibited by the various types of focal species

Keystone species	Flagship species	Umbrella species	Condition indicator species	Composition indicator species
Hypothesis: exerts a disproportionate influence on community structure characteristics	**Concept:** garne public support to conserve habitat	**Concept:** conservation will conserve other species	**Hypothesis:** reliable surrogate for ecological integrity	**Hypothesis:** reliable surrogate of community/ habitat type
Exert a disproportionate influence on community structure relative to its abundance or biomass	Garner public support and affection	Demonstrate fidelity to a particular set of habitats	Provide an assessment over a range of stress	Exhibit a specific niche or a defined range of ecological tolerances
Substantially change the structure and/ or composition of a community or habitat upon its removal	Require large tracts of relatively natural or unaltered habitat	Limited change in community or habitat structure if removed	Differentiate between natural and anthropogenic stress	Demonstrate fidelity to community type or habitat type
Lower the number of species in a community upon its removal	Migratory or non-migratory	Non-migratory	Relevant to ecologically significant change	Relatively independent of sample size
Prevent a single species from becoming the competitive dominant	Amenable to traditional management practices (e.g. fish management)	Exhibit low inter-annual or decadal population variation	Independent of sample size	Exhibit low temporal and spatial variability
		Specialists rather than generalists	Independent of spatial scales	Cost-effectively observed and censused
		Do not thrive in disturbed or anthropogenic habitats	Exhibit low temporal and spatial variability	Compatible with national and international indicators
		Require large tracts of relatively natural or unaltered habitat	Compatible with national and international indicators	

Source: After Zacharias and Roff (2001)

as a catch-all term for other types of focal species. On one hand, authors such as Noss (1990), Landres et al (1988) and Faith and Walker (1996) view indicators as an all-encompassing term to capture approaches/techniques to monitor biodiversity, and their use of the term would include keystones, umbrellas, sentinels and charismatic species. These authors' definition of indicators is what we and others have termed focal species. The alternative view, expressed by Dufrene and Legendre (1997), Kremen (1992), Simberloff (1998) and others suggests that the indicator species concept is substantially different from other focal species and warrants separate treatment.

This confusion over this concept results from its myriad of definitions and applications; these can nevertheless be considered to fall into two categories. The first application resulted from the realization that the presence, absence or abundance of certain species can be used to identify other less easily identified species and their associated habitats (Clements, 1916). One of the first applications of the term was by Kolkwitz and Marsson (using plants in 1908 and animals in 1909) who noted that certain species could indicate such variables as soils, climate and the presence of other species. The use of indicators to indicate a particular habitat, community or ecosystem continues to be an important part of ecology, and for the purposes of this paper we have called this type of indicator a *composition indicator*. A composition indicator (Table 9.1) has also been colloquially referred to as an 'ecological' or 'environmental indicator', whose presence or abundance is used to characterize a particular habitat or biological community. This type of indicator may also be used to estimate 'biodiversity' (often 'hotspots') for the selection of candidate protected areas (Faith and Walker, 1996).

Over time the concept has been expanded to include a second application, namely to indicate the *condition* of a habitat, community or ecosystem (Meffe and Carol, 1997). These *condition indicators* (Table 9.1) form the basis of biological monitoring of environmental change as a result of anthropogenic and natural disturbances. Incorporated into condition indicators are what have been termed 'bio-indicators'. Our definition of the condition indicator is also analogous to the use of the term 'sentinel species' (Meffe and Carol, 1997).

Much of the confusion surrounding the definition and application of indicators is a result of amalgamating composition and condition indicators. This is seen in many definitions, including those from Landres et al (1988) who describe indicators as '...species that, by their response of certain environmental conditions are thought to be useful to quickly infer the effects of those conditions on other, non-indicator species.' Niemi et al (1997) imply a condition indicator when suggesting that the purpose of indicators is to monitor habitat quality that is required by other species. Block et al (1987) are referring to composition indicators when suggesting that indicator species are plants and animals that are closely associated with specific environmental factors. Meffe and Carroll (1997) see the distinction between composition and condition in their definition of an indicator as: 'A species used as a gauge for the condition of a particular habitat, community, or ecosystem. A characteristic, or surrogate species for a community or community ecosystem.'

There is also a functional differentiation between composition and condition indicators. In relation to conservation and management, composition indicators are often used in the identification of representative or distinct areas, areas of high species diversity, endemic species or critical areas that may include courting, mating, spawning, rearing, nursing, feeding, holding, staging or foraging areas. In contrast, condition indicators are only used once specific habitats or communities have been identified and there is a requirement to monitor the effectiveness of conservation and management strategies. Therefore the composition indicator is most relevant to efforts to determine areas or priorities for conservation, while the role of condition indicators in marine conservation falls under the evaluation of conservation efforts.

Here we focus on composition indicators, as these are most relevant to conservation efforts, and the canon of literature surrounding their application is considerably smaller than condition indicators. Certain characteristics of marine environments necessitate rethinking the application of composition indicator species.

Composition indicators can be further separated into what Meffe and Carroll (1997) term indicators of habitat, community and ecosystem. Community composition indicators can be used to characterize an assemblage, guild, taxon, releve, series, biocoenosis or community. Ecosystem composition indicators are used to characterize predominantly abiotic (e.g. habitat) 'structures' that may include salinity, temperature, nutrients, substrate, upwellings or productivity. Characteristics of composition and condition indicator species and some criteria to consider when evaluating them are provided in Table 9.1.

There are a number of potential benefits to using indicators in marine conservation and management. First, given the cryptic nature of most marine environments, the ability to predict community composition based on a few observable species is invaluable. For example, the presence of the giant kelp *Macrocystis spp.* indicates potential sea otter habitat, as otters are currently repopulating vast areas of habitat where they had been previously extirpated (Estes and Palmisano, 1974). Trawl data are often characterized using indicator species at small scales, such as the Canary Islands (Falcon et al, 1996), as well as oceanic scales (Pearcy et al, 1996).

Second, indicator species are often suggested as potential conservation tools because they can identify representative or distinct habitats, communities or ecosystems. They have been advocated for conservation and management purposes as being less biased than other methods such as species richness, where areas of high species diversity do not necessarily address the conservation of rare or threatened species and habitats. The concept of conserving representative and distinct areas in marine environments has gained momentum in recent years. This in turn has required cost-effective methods of identifying the presence of communities (i.e. assemblages, taxa, guilds, functional groups, biocoenoses etc.) and habitats (Webb, 1989; Cousins, 1991; Roff and Taylor, 2000; Zacharias and Roff, 2000). Consequently, measures of species richness – while still useful approaches for conservation – are not good measures of representativity, while indicator species are ideally suited to the task.

Third, there are few leaps of faith required in the use of the indicator species concept. Relative to keystones and umbrellas (see below), the notion that certain species are found in certain communities and habitats is intuitive, and there has been little disagreement that the indicator concept is valid (but see Chapter 6 for caveats concerning indicators and scale of observation). While there may be discussion surrounding which species or guilds indicate what habitats or communities, the indicator species is the most ecologically accepted of all the focal species.

Fourth, there has been nearly 30 years of effort to identify both indicator species and the communities and habitats they indicate. Numerous clustering and ordination techniques have been developed to statistically define communities and habitats, including correspondence analysis (CA), detrended correspondence analysis (DCA), principal coordinates analysis (PcoA) and non-metric multidimensional scaling (NMS). The best known site (*Q* mode) and species (*R* mode) ordination program is the TWINSPAN developed by Hill (1979) (see also Dufrene and Legendre, 1997). The program includes: (a) a simultaneous sorted data table for both sites and species; (b) indicator species at each level of the hierarchy; (c) the capability to produce dendrograms; and (d) low computational requirements. The combination of newer, more efficient ordination routines in conjunction with cost-effective computing technology is reviving the concept of indicator species (Carlton, 1996; Dufrene and Legendre, 1997; Kremen, 1992; Simberloff, 1998). To date, there are no mathematical programs to identify keystones, umbrellas or flagships.

There are, however, also a number of disadvantages to using composition indicators for marine conservation and management. First, there have been a number of studies suggesting that no one species fulfills the conservation requirements of an indicator (e.g. Landres et al, 1988). This observation is especially pertinent in marine environments, where food webs are more complex, there are more trophic levels, and where predators are more often generalist feeders than in corresponding terrestrial environments. Consequently, the ability of any single species to signal either the structure or functioning of a community may be diminished. Indicators may be more efficient when used to indicate the presence of species from a specific guild (Block et al, 1987). This concept has potential in marine environments, where – for example – a certain rockfish or reef fish could indicate the presence of less readily identified species of a guild. Functional groups have also been used as indicators of environmental conditions related, for example, to wave exposure and oceanographic conditions (Bustamante and Branch, 1996).

Second, given the fluid nature of marine environments, marine indicator species may not be as geographically or temporally stable as terrestrial ones. With the exception of birds, flying insects and some mammals, most terrestrial species are confined within a watershed, biome or other mostly impermeable boundary. Most marine boundaries are semi-permeable, therefore indicator species may be distributed over great distances and are often not endemic. This wider distribution may be part of a species natural habitat range (i.e. the distinction between its predominant habitat or biotope and its broader realized habitat – Figure 6.1), or a result of transport by storms, oceanographic events or shifts in prey or predator distributions. This instability is compounded by large-scale oceanic variations (e.g. El Nino Southern Oscillation (ENSO)) over years to decades. Terrestrial environments are also subject to these vagaries, but in marine environments entire communities may move great distances to find more favourable habitats during these events.

Third, marine species are notoriously difficult to observe and census, therefore the absence of an indicator species may be the result of incomplete observation rather than lack of a certain community type. This notion is especially critical as most of the best known species are migratory, and therefore can only be observed in an area at certain times of the year. With the exception of intertidal and nearshore sub-tidal environments, indicators will generally be comprised of vertebrates and invertebrates. Phytoplankton can and have been used as indicators, but the difficulty in their identification may be onerous.

Finally, Simberloff (1998) cautions that an indicator subject to single species management is no longer an indicator. This observation has substantial implications in marine systems, because the majority of species that are readily observable are generally harvested to some degree, and therefore make poor indicators. For instance, herring (*Clupea spp.*) has been proposed as an indicator and has also been proposed as a keystone species. Herring, however, is consumed by birds, fish, marine mammals and humans, therefore an increasing or decreasing herring abundance may be the result of a number of interconnected factors. The types of marine species that would make the best composition indicators are those not adversely affected by pollution, habitat loss, alien introductions or global climate change. Consequently, certain sea birds are potential candidates for indicators, as are seagrasses, macroalgae and certain benthic invertebrates. While many marine mammals have not been harvested for decades, these species are probably still poor composition indicators because the locations of feeding and foraging areas may be a learned behaviour passed on from mother to calf (Norse, 1993). In many areas where whales were extirpated, abundant food exists, but the utilization and genetic knowledge of these areas have disappeared with the whales.

Keystone species

The keystone species concept (Table 9.1) has received considerable attention since its designation by Paine (1969). He found that removal of *Pisaster* from an intertidal community resulted in the mussel (*Mytilus*) becoming a competitive dominant; therefore *Pisaster* appeared to exert an influence disproportionate to its abundance and biomass. He theorized that certain species are either directly or indirectly responsible for biological community structure, composition and biomass, and therefore biodiversity (Paine, 1969). The removal of a keystone has a significant impact on a community, and consequently there is an impetus to identify and conserve them. The concept holds considerable allure for managers and conservationists, as the notion of protecting and managing just a few species to the benefit of the entire community or ecosystem could make a seemingly impossible task manageable (Navarrete and Menge, 1996). A number of criteria should be met before any species can be considered a keystone. While there is debate surrounding what constitutes a keystone, their general characteristics are supplied in Table 9.1.

The keystone species concept has become an accepted and central organizing theme of population level ecology and conservation, and many species have been proposed as keystones in the marine environment (Table 9.2). The concept has, however, been ill defined, which has led to the christening of a number of species as keystones that are probably not (Hurlburt, 1997; Simberloff, 1998). The Oxford Dictionary of Ecology defines a keystone species as: 'The species, the presence or abundance of which can be used to assess the extent to which resources of an area or habitat are being exploited' (Allaby, 1996). Roughgarden's (1983) definition of a keystone species was one '...whose removal leads to a still further loss of species from the community.' Terborgh (1986) discussed keystone resources, which are those resources that comprise a small percentage of diversity or biomass, but are essential to community structure and/or diversity. Menge et al (1994) defined keystone predators '...as only one of several predators in a community that alone determines most patterns

Table 9.2 Marine species proposed as keystone species

Environment	Citation(s)	Keystone species or guild	Target of direct effect	Mechanism of effect	Evidence
Rocky intertidal	Paine, 1966	*Pisaster ochraceus* (predatory starfish)	Mussels	Consumption	Experimental, comparative
	Menge, 1976	*Nucella lapillus* (predatory snail)	Mussels	Consumption	Experimental
	Hockey and Branch, 1984	*Haematopus* spp. (black oystercatchers)	Limpets	Consumption	Comparative
	Castilla and Duran, 1985 Duran and Castilla, 1989	*Concholepas concholepas* (predatory snail)	Mussels	Consumption	Experimental
	Menge et al, 1994	*Pisaster ochraceus* (predatory starfish)	Mussels	Consumption, direct and indirect	Experimental, comparative
	Navarrete and Menge, 1996	*Pisaster ochraceus* (predatory starfish) and *Nucella* (whelks)	Mussels	Consumption, direct and indirect	Experimental, comparative
Rocky sub-tidal	Estes and Palmisano, 1974	*Enhydra lutris* (sea otter)	Sea urchins	Consumption	Comparative
	Fletcher, 1987	*Centrostephanus rodgersii* (sea urchin)	Algae	Consumption	Experimental
	Ayling, 1981	*Evechinus chloroticus* (sea urchin), herbivorous gastropods, *Parika scaber* (grazing fish)	Algae, sponges and ascidians	Consumption	Experimental
Pelagic	May et al, 1979	*Balaenoptera* spp. (baleen whales)	Krill	Consumption	Historical reconstruction
	Springer, 1992	*Theragra chalcogrammai* (walleye pollock)	Zooplankton, smaller fish	Consumption	Historical reconstruction
	Gasalla et al, 2010	*Loligo plei* (loliginid squid)	Pelagic fish	Consumption	Comparative
	Antezana, 2009	*Euphausia mucronata* (euphausiid)	Phytoplankton	Consumption	Comparative
Coral reef	Hay, 1984	Herbivorous fish, sea urchins	Seaweeds	Consumption	Experimental, comparative
	Carpenter 1988, 1990	*Diadema antillarum* (herbivorous sea urchin)	Seaweeds	Consumption	Experimental, comparative
	Hughes et al, 1987	*Diadema antillarum* (herbivorous sea urchin)	Marine plants	Consumption	Experimental, comparative
	Birkeland and Lucas, 1990	*Acanthaster planci* (coral-eating starfish)	Corals	Consumption	Experimental, comparative
	Hixon and Brostoff, 1996	*Stegastes fasciolatus* (territorial algivorous damselfish)	Schooling parrotfish and surgeonfish	Protection of seaweeds within territories from heavy grazing	Experimental
Soft sediment	VanBlaricom, 1982	*Urolophos halleri, Myliobatis californica* (carnivorous rays)	Amphipods	Consumption, disturbance	Experimental
	Oliver and Slattery, 1985	*Eschrichtius robustus* (grey whales)	Amphipod mats	Consumption, disturbance	Comparative
	Oliver et al, 1985	*E. lutris* (sea otters)	Bivalves	Consumption	Comparative
	Kvitech et al, 1992	*E. lutris* (sea otters)	Bivalves	Consumption	Experimental, comparative
Estuarine	Kerbes et al, 1990	*Chen caerulescens caerulescens* (lesser snow geese)	Salt marsh vegetation	Consumption, disturbance	Comparative
	Ray, 1996	Seagrasses and eel grasses		Substrate composition	Comparative
	Ray, 1996	*Crassostrea virginica* (eastern oyster)	Water quality	Filtration in estuary	Comparative
Other	Willson and Halupka, 1995	Anadramous fish	Terrestrial wildlife	Consumption	Comparative

Source: Adapted from Power et al (1996)

of prey community structure, including distribution, abundance, composition, size, and diversity.' This uncertainty surrounding the definition and application of the keystone concept was the impetus behind a workshop in 1994 to review and discuss the true meaning of the term. Results of this workshop were published by Power et al (1996) and suggest that keystones are 'a less abundant species that have strong effects on communities and ecosystems'. Meffe and Carroll (1997) supply probably the most comprehensive definition that still retains the intent of the concept by stating that keystone species are those species which '...play a disproportionally large role in community structure.'

Since the work of Paine (1969), a number of different types of keystone species have been proposed, some of which only vaguely resemble the spirit and intent of Paine's definition. Other types of keystones include *keystone predators* (Paine, 1966), *keystone mutualists* (Gilbert, 1980) which include plant species that support animal species which may in turn support more species, *keystone modifiers* (Naiman et al, 1986) such as the beaver (*Castor canadensis*), the African elephant (*Loxodonta africana*) or various species of seagrasses (Carlton, 1996), *keystone prey* which maintain a community through high fecundity and therefore support a population of predators (Holt, 1977, 1984), and *keystone diseases* (Sinclair and Norton-Griffiths, 1982) which may ultimately play the largest role in structuring communities. Keystones occur in all of the world's ecosystems, do not necessarily occupy higher trophic levels, and affect their communities through consumption, competition, dispersal, pollination, disease and by modifying habitats and abiotic factors. There is also growing evidence that small but critical species such as the mycorrhizal fungi and nitrogen fixing bacteria should also be termed keystones (Paine, 1995; Weaver, 1995).

There has been considerable resistance to both the keystone concept and its application towards addressing conservation objectives. There are five arguments against the use of keystones from a conceptual and empirical standpoint. These constraints are outlined and discussed in the following paragraphs:

1 Complex communities are rarely controlled by a single species.
2 All species are keystones to some degree.
3 Identifying keystones is difficult.
4 Keystone species do not always demonstrate keystone properties.
5 Conservation or management of a keystone does not guarantee that conservation objectives are met.

The first hurdle facing the keystone concept is that there is little empirical evidence that most communities are controlled by a single or relatively few predators, thus casting doubt on the universal applicability of the concept. The rocky intertidal shorelines of the Pacific Northwest USA where Paine (1969) completed his research are not representative of the more complex composition of most terrestrial and marine communities, and there is some evidence that while the keystone concept may work in simplified systems, keystones are not relevant in more complex communities. Tanner et al (1994) found that communities on the Great Barrier Reef did not have keystone species because the high species diversity of these environments reduced the chances of a single species structuring the community, and because the time required for a species to attain dominance is greater than the average period between disturbance events.

Peterson (1979) found that the exclusion of predators in estuarine and soft bottom systems did not result in a competitive dominant, thus casting doubt on the universality of the keystone concept. Peterson (1979) suggested that keystones may be present, but that predators have not been excluded long enough for a species to become dominant. He also suggested that interference competition and competitive exclusion – processes which operate on rocky shores – are absent in soft bottom systems as organisms cannot dislodge or overgrow each other in a three-dimensional sediment-dominated environment. Extrapolating these results into other pelagic and sediment-dominated marine systems suggests that the processes that permit the establishment of a keystone species may be absent in many marine communities. This hypothesis is supported in Table 9.2, where species proposed as keystones in sediment-dominated environments are all large vertebrates, with the exception of the oyster (*Crassostrea virginica*) and various seagrasses.

The second concern is that all species are keystones to some degree. The observation that most keystones have been identified through either predator exclusion experiments (e.g. starfish), or the recolonization of previously extirpated species (e.g. sea otters) suggests that the species we currently term keystones are merely products of the environments and communities under study (Mills et al, 1993; Simberloff, 1998). The fact that keystones may be artefacts of research methods and data analysis leads into the third criticism that keystones are difficult to identify, particularly in marine systems.

The difficulty in identifying keystones is the third and perhaps most damning indictment of the concept, particularly in marine systems. Most keystone predators have been identified through experimental manipulation (e.g. predator exclusions) or by observing the recovery of disturbed systems (e.g. reintroduced sea otters). Navarrete and Menge (1996) elevate the requirements for identifying keystones by correctly suggesting that a single predator should first be removed, then the other predators one by one until the entire predatory guild has been removed. Only then can the keystone properties of each species be identified. While the keystone concept evolved out of the manipulation of intertidal environments, the intertidal realm is not representative of most marine environments, where predator exclusion studies are difficult if not impossible to conduct.

However, some progress has been made in identifying 'keystoneness' of multiple species through trophic modelling. Libralato et al (2006) present an approach, based on network mixed trophic analysis, to identify keystone properties of species or functional groups by modelling community response to small biomass increases of each group. The approach can utilize existing trophic models (captured in Ecopath and Ecosim) assembled for many marine ecosystems to identify keystone functional groups (Libralato et al, 2006).

The fourth problem is that species may only act as keystones under a certain set of biotic and/or abiotic conditions. Paine's (1966, 1969) conclusion that *Pisaster* comprised a keystone species generated considerable research into testing not only whether *Pisaster* was indeed a keystone, but also the search for other potential keystone species. Paine (1969) suggested that the *Pisaster–Mytilus* interactions constituted a keystone relationship due its ubiquity from Mexico to Alaska. Foster (1990) and others, however, suggested that this relationship continuously changes over time and space, and that the spiny lobster also predates *Mytilus* in California (Fairweather et al, 1984; Foster, 1990). Paine (1980) agreed that keystones were situation specific, and speculated that in certain instances *Pisaster* was '...just another seastar.' Menge et al (1994) showed that *Pisaster* was a keystone only under higher wave exposure habitats, whereas in habitats with lower wave exposure they cease to act as keystones.

There is also considerable debate whether any species acts as a keystone throughout its life cycle, geographic range and habitats (or niches) that it occupies. Research to date has not identified any universal keystone predators, but some keystones, such as salt marshes, eelgrass beds and mangroves, probably structure communities and/or habitats in the same manner throughout their range. Such species are better regarded as belonging to the category defined as 'biogenic' or 'foundational' or 'engineering' (see Table 3.4).

As a result of attempts to define and identify keystones, Menge et al (1994) refined the keystone concept by defining *diffuse predation*, where the control of a competitively dominant species is shared by several predators. *Weak predation* occurs when predators alone do not control the abundance of a competitive dominant. Navarrete and Menge (1996) further explored these effects in intertidal environments by examining the strength of predation on mussels (*Mytilus trossulus*) and *Pisaster* and the whelks of the genus *Nucella* under various environmental conditions. They found that *Pistaster* was unaffected by the presence of the whelks, but that whelks were an important controlling influence on mussel distribution in the absence of *Pisaster*.

All the foregoing limitations lead to the fifth objection or conclusion, namely that conservation or management of keystones in the marine environment does not guarantee that objectives are met. Although the keystone concept was developed through the study of intertidal environments, the application of the concept is probably better suited to terrestrial environments for a number of reasons. Marine food webs are longer, more complex, and more apt to vary over spatial and temporal scales than most terrestrial environments. Consequently, the probability that the composition and diversity of a community rests on a single species is low.

Terrestrial habitats are generally spatially, and for the most part temporally, stable. A tropical or sub-boreal forest – for example – tends to remain the same basic habitat over geological time scales. The majority of the world's marine habitats lack this stability, and are interrupted by annual and decadal events, which include regional disturbance (e.g. hurricanes) and climatic fluctuations (e.g. ENSO) which can result in significant changes in community structure. Therefore, as demonstrated by Menge et al (1994) and Navarrete and Menge (1996), a species only exhibits keystone properties under certain conditions, which are for the most part more variable in marine systems. This is especially apparent when migratory species are identified as keystones, because they can avoid unsuitable areas as a result of changes in oceanic conditions. In contrast, many tropical environments are very stable, which suggests that keystones should be present. This very stability, however, leads to the evolution of so many species that apparently no one species controls a community (Tanner et al, 1994).

Lastly, perhaps the most important principle with keystones is that they do not change the fundamental community type in an area or community–habitat relationships. Presence or absence of a keystone simply changes the relative abundance of member species within a community. Keystones are therefore not important for the recognition of community types except if they are considered as composition indicators. However, their presence or absence may be significant in the sense that they may be condition indicators of community composition, diversity or 'health'. Where keystones are commercially exploited, management decisions can affect resultant expressions of diversity. The keystone concept is therefore currently not a globally useful concept for marine conservation, in the same manner as indicator species.

Umbrella species

The umbrella species concept (Table 9.1) hinges on the assumption that the presence of a certain species in a geographical area indicates that other species will also be present. In this sense, it could also be considered as a compositional indicator species. The umbrella concept is particularly appealing to the conservation community because the implication is that the management, conservation or protection of an identified umbrella species will protect not only the habitat and community required to support itself, but also the habitat for other (generally smaller) species as well (Roberge and Angelstam, 2004). This intuitive nature of the umbrella concept led to its use in early conservation efforts as a pragmatic and ad hoc tool to estimate optimal reserve sizes (Caro, 2003). The concept has been used in the selection of sites for preservation, where communities and habitats could be identified through the identification of umbrella species. Unfortunately, the relationships between an umbrella species and communities are usually ill-defined, especially in marine environments, where such species may span broad geographic ranges encompassing several habitat types. Umbrella species as a whole are therefore liable to be uninformative about the geographic limits of either representative or distinctive marine communities or species diversity. An umbrella species may in fact 'indicate' habitat or community types at some higher (but undefined) level of the ecological hierarchy (Roff and Taylor, 2000).

The umbrella concept is differentiated from the keystone concept in that preservation of an umbrella can lead to the preservation of communities and habitats, although these entities would continue to exist and function in the absence of the umbrella. In contrast, the removal of a keystone species may fundamentally change community composition. Certain terrestrial species could be considered as both umbrella and keystone species, but few marine species exhibit characteristics of both. Sea otters (*Enhydrus lutris*) could potentially be considered as both an umbrella and a keystone, as otters certainly modify communities-composition from invertebrate- (e.g. urchin) dominated to kelp-dominated systems, which in turn provide shelter and habitat for the juvenile life stages of a number of anadramous, meroepipelagic and demersal fish species (Bifolchi and Lode, 2005). The protection of sea otters, therefore, also protects numerous other fish species. Other marine species considered as keystones, including the grey whale (*Eschrichtius robustus*), probably fail the critical test of an umbrella species; this is because the actions of species such as grey whales – while maintaining a serial successional state through benthic foraging – are probably not 'protecting' other species as a result of their conservation. There is also the consideration that grey whales – along with most other marine mammals – are generalist feeders and migratory, and like keystones, may only act as umbrellas part of the time (Oliver and Slattery, 1985).

It is generally acknowledged that there are three types of umbrella species. The first are the classic 'single-species umbrellas' generally comprising the larger vertebrates that require large territories to survive (Wilcox, 1980; Peterson, 1988). The conservation of these species is thought to protect other species with smaller habitat requirements. The single-species umbrella concept has been applied to ungulates, terrestrial carnivores and sea otters (Noss et al, 1996).

The second type of umbrella species identifies 'mesoscale species' which are affected by the scales of human disturbance (Kitchener et al, 1980; Holling, 1992). The usefulness of this concept in marine systems is suspect, because the concept assumes that the habitat ranges of marine species are understood to an extent where a nested system of habitats can be identified. The reality, as discussed previously, is that putative marine umbrella species frequently cross both habitat types and communities to a much greater extent than their terrestrial counterparts.

A third type of umbrella species is what Ryti (1992) terms 'focal taxa' which have also been termed by Hager (1997) as 'umbrella groups'. This concept uses a speciose taxon to ensure that in the presence of these species, most of the larger community is protected. Ryti (1992) found that plant species proved to be better umbrellas than bird species on selected islands and canyons in southern California. This concept was further explored by Lambeck (1997) who identified a suite of 'focal species' within which each species is the most sensitive to a particular threat to its existence. The combination of a number of species keyed to each threat provides management direction for the conservation or management of their associated habitats and communities. Lambeck (1997) outlines a procedure to identify a number of species that must be present '...if a landscape is to meet the needs of its constituent flora and fauna.' The disadvantage of this approach is that considerable effort may be required to identify the focal species and that if numerous focal species are identified, then there is no efficiency in using the concept.

The use of umbrella species as focal taxa in marine environments is intuitively appealing, but the application of the concept is fraught with difficulties. For example, the critical question of whether the number of species protected within the range of an umbrella species is greater than would be protected within a similar range of habitat(s) selected at random has not been tested. Another primary limitation concerns the considerable unexplained spatial and temporal variation in many types of marine communities. While a terrestrial forest, for example, may be affected by occasional periods of disease, drought, infestations and other processes, the structural habitat in the form of the trees themselves is generally persistent over time (with the exception of fire and other catastrophic events). The marine analogues to trees are the kelps, which often exhibit large inter-annual and inter-decadal variation. As a result, terrestrial vegetation as a basis for focal taxa has no marine equivalent, with the notable exception perhaps of rooted nearshore seagrasses and mangroves.

In marine environments, most species – while exhibiting prey preference – are generalist feeders. There is mounting evidence that almost all species, and especially the larger vertebrates, can and do feed on numerous species within a food web. Certain species of tuna are known to feed on over 140 species, and certain species such as Pacific herring (*Clupea harengus*) are consumed by many marine mammals and sea birds in the northeast Pacific

(Nybakken, 1997). Consequently, the assumption that protecting generalist feeders will protect other species associated with the generalists may be an unwise management approach.

There is, however, some value to the approach using umbrella species as focal taxa, not necessarily for the protection of other species, but rather for the protection of the natural order and function of food webs and trophic structures. The harvest of the bowhead whale to ecological extinction in many Arctic environments resulted in the tripling of populations of sea birds and marine mammals that competed for the same krill resource (Nybakken, 1997). Consequently, the protection and subsequent (partial) recovery of bowhead stocks resulted in the decline of other vertebrate populations depending on the same resource. An application of the focal taxon concept in this instance could possibly use bowhead whales and one or more other species to gauge the relative health of the krill resource while monitoring those populations that depend on that resource. In this sense, however, the whales are being used as a condition indicator.

A further flaw in the application of the umbrella concept in marine systems is the requirement that an umbrella species should be non-migratory. The migratory nature of most marine vertebrates suggests that the concept is not as powerful an approach as for other focal species. The flagship concept may be more apropos for marine conservation strategies.

Flagship (charismatic) species

The affinity of western and other societies for marine mammals is remarkable. The ability of Greenpeace to halt certain types of commercial sealing and whaling was one of the most successful environmental campaigns in history. There is no doubt that charismatic species have been used to achieve marine conservation ends, but there are limitations to the application of the concept. In terrestrial systems, threats to charismatic species have been used to garner support for conservation efforts; this generally necessitates the conservation of the habitat required to support them. For example, the conservation of spotted owls or grizzly bears is equivalent to the conservation of habitat. Indeed, the ultimate purpose of the conservation of charismatic species is the preservation of habitat. This has resulted in recent legal challenges to the US Endangered Species Act, as certain groups charge that conservation organizations are using the Act to preserve habitat rather than the species. In contrast, the preservation of a charismatic species does not necessarily protect either the habitat or other species in marine systems. Baleen whales, for example, feed on parts of the food chain rarely extracted by humans; therefore the benefits of protecting these species are not as great as in terrestrial species (Oliver and Slattery, 1985). Terrestrial charismatics are threatened by a number of activities, including habitat loss, over-harvesting and pollution. Marine charismatics by comparison are easily conserved by not harvesting them; therefore the charismatic concept lacks several of the advantages of its use in terrestrial systems.

Another limitation to the flagship concept is that subsequent to the cessation of wholesale slaughter of marine mammals, most of the most threatened marine species have few charismatic properties. There is little public identification with taxonomic groups such as rockfish, bivalves and krill.

While the umbrella and flagship concepts share several similarities, there are important differences that affect their utility in marine systems. Flagships (Table 9.1) are better suited for marine conservation because: (a) migratory species may be considered as flagships; (b) standard management practices (e.g. fisheries management) can be used to mitigate impacts on flagships over large areas; (c) flagships may be associated with several distinctive habitats, including the feeding or breeding grounds at the extremes of their migration routes.

A major function for marine flagship species may therefore be the same as in terrestrial systems – that is, to act as surrogates for habitat protection. Habitats used by migratory marine flagship species may be relatively discrete, at least seasonally, and may define representative or distinctive areas (which can subsequently be assessed by habitat characteristics or indicator species analyses) suitable as candidates for conservation.

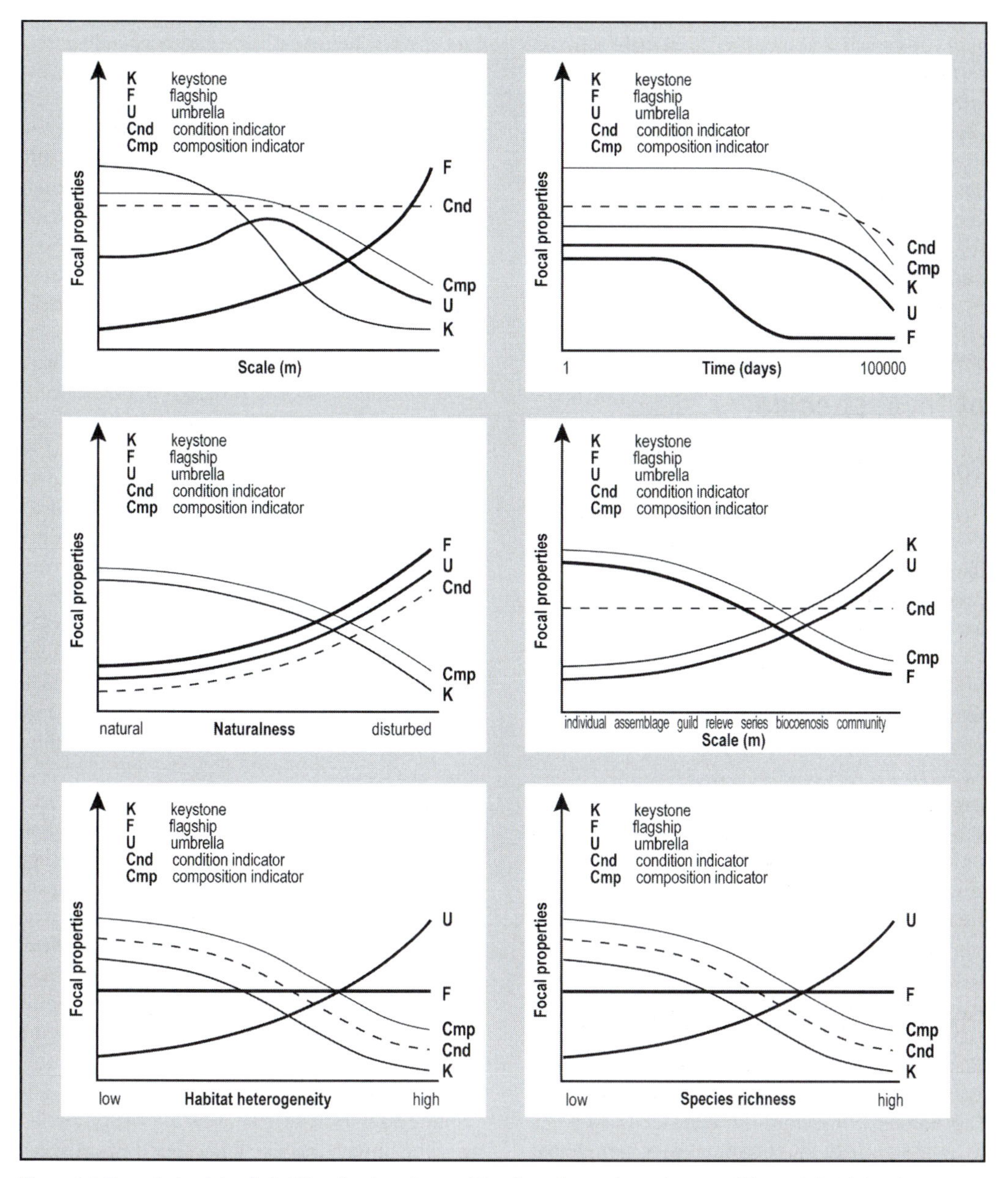

Figure 9.1 Expected or intended utility of various types of focal species under various conditions of: Spatial scale; Temporal scale; Degrees of naturalness or disturbances; The scale of the ecological hierarchy; Habitat heterogeneity; Species richness

Other focal species/ communities

The term 'focal' could also quite reasonably be applied to other levels of the ecological hierarchy, and especially to biogenic or foundational species (or their attendant communities, see Table 3.4). Thus the Great Barrier Reef of Australia could be considered a 'focal ecosystem'; coral reefs everywhere are 'focal communities'; mangroves and coastal salt marshes are 'focal habitats' and so forth. At these levels the terms begin to fuse with the concept of biological hotspots. Whatever the terminology, these regional hotspots of higher than average species diversity should be considered as 'focal entities' deserving of special attention (see Chapters 6 and 8). Commercial fish species may themselves be considered as 'focal species', as will be considered in Chapter 13.

Properties of the various types of focal species

Spatial scales

While composition indicator species have applications at all scales, their focal properties are likely reduced at larger scales for a number of reasons (Figure 9.1). First, and from a practical standpoint, to determine the effectiveness of the indicator, smaller areas are more easily observed. Second, if the composition indicator spans areas with allopatric, or partially sympatric populations, there may be different relationships between the indicator and its habitat and community. Thus a predator species may consume different prey at different locations depending on prey abundance, feeding behaviour or other reasons, and the indicator may only be valid at smaller scales. For example, the seastar *Pisaster* spp. is easily identified from aerial surveys, and is a good indicator of the presence of its primary prey (e.g. *Mytilus*) in the mid latitudes of the northeast Pacific (Paine, 1969; Searing and Frith, 1995). Farther south, however, mussel communities are consumed by additional species (e.g. lobsters); therefore the value of *Pisaster* spp. as an indicator of mussels diminishes with spatial scale (Menge et al, 1994). As another example, pacific grey whales (*Eschrichtius robustus*) generally feed on the benthos in the Bering Sea, but recent evidence shows that summer residents in British Columbia, Canada, prefer pelagic feeding behaviour. Consequently, using grey whales as indicators of benthic amphipods is applicable within the constraints of the Bering Sea, but is not universally applicable in the northeast Pacific (Dunham and Duffus, 2001). Condition indicator species, however, are scale independent and can indicate ecological state or anthropogenic disturbance at a variety of scales.

Keystone species most likely exert keystone properties at smaller scales (Figure 9.1). While keystone species may be widely distributed, their effects are localized (e.g. in the case of *Pisaster* spp.), and often context dependent (see above). There is little evidence that any organism exhibits keystone properties in the open pelagic realm, and the largest range of a keystone predator identified to date is probably the sea otter (*Enhydra lutris*), which forages on the scale of kilometres (Estes and Palmisano, 1974). In addition, most marine keystones identified to date show strong site fidelity and rarely leave a particular location. Other non-predatory keystones also demonstrate less keystone effects at larger scales. While seagrasses may be distributed over long stretches of coastline, their importance to the pelagic and benthic communities generally decreases with distance offshore (Ray, 1996). Keystone diseases or pathogens, however, which reside on or inside the host, are constrained only by the habitat range of the host, and therefore provide an exception to scale-dependency (Zacharias and Roff, 2000).

Focal properties of umbrella species may first increase and then decrease with spatial scale, because the notion behind the umbrella concept is that protecting an umbrella requires protection of its habitat. The umbrella concept may work well for species closely associated with nearshore areas, which may include sea otters, certain pinnipeds, sireneans and sea birds. The validity of the umbrella concept in offshore pelagic and benthic environments is questionable, but there are notable examples. The recent listing of the Steller sea lion under the US Endangered Species Act has forced fisheries managers to limit harvesting of groundfish stocks which are consumed by the sea lions. While this is a weak example of the umbrella concept, it does demonstrate that the concept may be valid at larger scales.

Flagship species are perhaps the only focal species whose efficacy may increase with spatial scale. The reason for this is that most flagships are vertebrates with larger ranges, and the ubiquity of a species such as the killer whale (*Orcinus orca*) results in worldwide public support. There are exceptions to this observation, however, with certain species (e.g. river dolphins) which become the focus of conservation campaigns as a result of their limited distribution. In addition, widely distributed flagships that garner considerable attention and conservation efforts fail to assist the conservation of species such as the Bowhead whale (*Balena mysticetus*), which continues to be harvested despite low populations, and is virtually ignored by the public and conservation community (Norse, 1993).

Temporal scales

Composition indicators are probably most valid at short to intermediate time scales because communities and habitats may vary considerably at decadal or longer time scales (Figure 9.1). Condition indicators, however, are generally independent of time in their ability to predict, but the prediction may be less accurate over longer time periods.

Keystone species and their resulting keystone effects may also be relatively unaffected over short to intermediate time scales, although a species may not exhibit keystone properties in a particular habitat or under certain conditions. Over longer time periods, however, influences such as habitat change or global climate change may alter the habitat or community in such a way that the species no longer exhibits keystone properties (Figure 9.1).

Umbrella species may behave in the same manner. Although the umbrella itself may remain stable over time if sufficient habitat is protected, the habitats and communities themselves are subject to change over time. The focal properties of the umbrella may therefore change at longer time scales in undefinable ways.

The flagship concept loses validity over time for two primarily non-ecological reasons. First, public interest in flagships may shift among organisms over time. Second, conservation measures applied to a flagship should hopefully lessen the threats on that species and its populations, such that conservation efforts can be directed elsewhere.

Disturbance

While disturbance relates predominantly to the degree of anthropogenic disturbance, the concept is also relevant to systems where natural changes to a system (e.g. hurricanes, volcanoes) have occurred.

The validity of composition indicators generally decreases with disturbance, because the habitats and communities that the indicator is used to predict are themselves altered (Figure 9.1). There is also no guarantee that indicators are indicating the same communities or habitat types in non-natural environments. Condition indicators, however, are increasingly valid in disturbed environments, because the purpose of the concept is to identify areas of ecological and human change.

Keystones may fail to demonstrate keystone properties in disturbed environments therefore their potential validity decreases. There are exceptions, however, where certain types of shorelines are heavily modified by infilling and the creation of seawalls, and *Pisaster* spp., for example, may still act as a keystone species in these environments.

The umbrella concept was designed to provide cues that habitats were no longer of sufficient size or natural composition to support these species. The focal properties of umbrellas are therefore generally expected to increase as environments become increasingly disturbed.

Biological hierarchy

Composition indicators would be expected to lose their focal properties at higher levels of the biological hierarchy (Figure 9.1). A composition indicator is more likely to correctly indicate the presence of an assemblage or guild of which it is a member than to indicate a more complex community type. The condition indicator, however, should be equally able to indicate an ecological or anthropogenic change at any level of the biological hierarchy.

A true keystone species (as defined by Power et al, 1996) should structure composition and abundance at the community level. Keystone effects, therefore, should become more pronounced at higher levels of the biological hierarchy. The umbrella species illustrates a similar trend with biological hierarchy. If the chosen umbrella species requires the largest habitat in an area, then conservation of this species should ideally protect the entire community, or a set of communities (Figure 9.1).

The focal properties of the flagship concept decrease with hierarchical level, as a grizzly bear, for example, may generate public interest in conserving the salmon it consumes, but not in the small prey that salmon in turn consume.

Habitat heterogeneity

The focal properties of both composition and condition indicators decrease with increasing habitat heterogeneity (Figure 9.1). An increasingly fragmented or patchy seascape results in a number of smaller community and habitat types whose composition and condition become more difficult to identify using indicators. Because species generally demonstrate keystone properties only under certain conditions, it follows that increasing habitat heterogeneity would tend to limit the effectiveness of the keystone concept as well.

The umbrella concept deals increasingly well with habitat heterogeneity and fragmentation as long as the variability of the landscape is at a smaller spatial and temporal scale than the required range of the umbrella species itself. The flagship concept is most likely not affected by habitat heterogeneity.

Species richness

Increasing species richness (number of species per site) generally diminishes the focal properties of both composition and condition indicators, because a more complex community results in greater difficulty in establishing the relationship between the indicators and what is indicated (Figure 9.1). This observation also holds true for the keystone concept, since an increasing number of species diminishes the probability that a single (or even several) species will structure a community (e.g. Tanner et al, 1994). The umbrella species concept is well suited to high species richness, as the more species that are protected as a result of conserving the umbrella, the more desirable the concept becomes.

Conclusions and management implications

Despite all the above flaws and criticisms levelled at the various focal species, they may still have useful roles in marine conservation in terms of their indicator potential. However, the appropriateness and utility of each of these focal species must be judged against specific conservation objectives. In simple terms, there are approaches that attempt to conserve spaces (e.g. marine reserves), and those that conserve species (e.g. fisheries management). How focal species might be used in the conservation of both spaces and species is indicated in Table 9.3 and outlined in the following text.

Table 9.3 Do focal species provide information for the following marine conservation strategies and initiatives?

Focal species	Identify representative community and habitat types?	Identify distinct community and habitat types (e.g. hot spots)?	Identify geographical extent of community and habitat types?	Identify condition or state as a result of ecological factors?	Identify condition or state as a result of anthropogenic factors?
Composition indicator	Yes	Yes	Yes	Unknown	No
Condition indicator	No	No	No	Yes	Yes
Keystone	No	No	No	Yes	No
Umbrella	No	No	Yes	No	Possibly
Flagship	No	No	Yes	Unknown	Possibly

A logical approach for the utilization of focal species in the conservation of spaces is as follows. First, define the objectives of the conservation strategy. For example, many countries utilize marine reserves as their primary conservation strategy. From an examination of Table 9.3, the application of composition indicators would be well suited to identifying either representative or distinctive candidate marine reserves. The additional use of condition indicators may yield information on the current state of the candidate reserve, and whether this state is a result of ecological or anthropogenic processes. This knowledge may also be required in the application of mitigation, restoration and monitoring strategies once the reserve has been established. In the identification of a geographical boundary, the use of the umbrella concept may also be applicable.

A logical approach for the utilization of focal species in the conservation of species is similar to that for the spaces approach. First, determine which species are the targets of the conservation efforts. For example, many jurisdictions are concerned with populations of harvested fish species. Second, determine which focal species can be used to design and/or implement a conservation strategy. In our example, composition indicators may be valuable in determining the habitats of the fish species under consideration. The predictive nature of these indicators may allow determination not only of where populations may occur, but where they may have occurred, or where they may occur in the future. Condition indicators may permit different populations to be evaluated (ranked) as to the degree of anthropogenic influence or successional stage. If the species of concern has the ability to garner public support for its conservation, then it becomes a flagship. The umbrella species concept may be applicable if our example fish is a prey item of a particular marine mammal or seabird, which may then act as the umbrella.

While there has been considerable criticism of the indicator species concept in terrestrial environments, and the suggestion that keystone or umbrella species may be more relevant to conservation efforts, the cryptic and fluid nature of marine environments lends greater support for the use of indicator species. By focusing on indicator species (or the indicator properties of any focal species) we can ask a number of fundamental ecological questions (which could also be phrased as testable hypotheses) which have utility in conservation planning, management, and monitoring:

1 Does a composition indicator species reliably indicate the presence or absence of a community and its affiliation with defined habitat types?
2 Does the scale (or method) of observation, or recording of data (e.g. as presence/absence versus quantitative) affect the outcome of Question 1?
3 Does the presence or absence of a condition indicator species reliably indicate the ecological state of a community (in the absence of demonstrable anthropogenic impact)?
4 Does the presence or absence of a condition indicator species reliably indicate anthropogenic stress of a defined sort and magnitude?

In terms of utility for conservation purposes, composition indicator species should indicate community types which can be related to habitat types, which can in turn be spatially mapped. This is a fundamental prerequisite for conservation initiatives based on representivity (e.g. Roff and Taylor, 2000). The composition indicator is the only focal species that can be used in this way (unless other focal species have such indicator properties). Similarly, condition indicators (whether they may be additionally considered as keystones or umbrellas) are the only biological means (using whole organisms) to infer community 'health'.

Finally, all species level concepts suffer from the same array of deficiencies for conservation planning purposes, namely:

- Any individual species may locally become extinct, thereby depriving us of our classification criterion.
- Any classification based on the species level approach requires that the distributions of those species should be known in some detail over broad geographic areas.
- Different species will be 'indicators', 'umbrellas' etc. for each region. Therefore any

classification system based on such criteria would automatically be limited in its application to that area, and hence have no general broad geographic applicability.

References

Allaby, M. (1996) *Oxford Concise Dictionary of Ecology*, Oxford University Press, New York

Allison, G. W., Lubchenco, J. and Carr, M. H. (1998) 'Marine reserves are necessary but not sufficient for marine conservation', *Ecological Applications*, vol 8, no 1, pp79–92

Antezana, T. (2009) '*Euphausia mucronata*: A keystone herbivore and prey of the Humboldt Current System', *Deep Sea Research Part II: Topical Studies in Oceanography*, vol 57, nos 7–8, pp652–662

Apollonio, S. (1994) 'The use of ecosystem characteristics in fisheries management', *Review of Fisheries Science*, vol 2, no 2, pp157–180

Ayling, A. M. (1981) 'The role of biological disturbance in temperate subtidal encrusting communities', *Ecology*, vol 62, no 3, pp830–847

Bifolchi, A. and Lode, T. (2005) 'Efficiency of conservation shortcuts: An investigation with otters as umbrella species', *Biological Conservation*, vol 126, no 4, pp523–527

Birkeland, C. and Lucas, J. S. (1990) *Acanthasterplanci: Major Management Problem of Coral Reefs*, CRC Press, Ann Arbor, MI

Block, W. M., Brennan, L. A. and Gutierrez, R. J. (1987) 'Evaluation of guild-indicator species for use in resource management', *Environmental Management*, vol 11, no 2, pp265–269

Brooks, T. M., da Fonseca, G. A. B. and Rodrigues, A. S. L. (2004) 'Protected areas and species', *Conservation Biology*, vol 18, pp616–618

Bustamante, R. H. and Branch, G. M. (1996) 'Large scale patterns and trophic structure of southern African rocky shores: The roles of geographic variation and wave exposure', *Journal of Biogeography*, vol 23, pp339–351

Caddy, J. F. and Bakun, A. (1994) 'A tentative classification of coastal marine ecosystems based on dominant processes of nutrient supply', *Ocean and Coastal Management*, vol 23, pp201–211

Carlton, J. T. (1996) 'Pattern, process, and prediction in marine invasion ecology', *Biological Conservation*, vol 78, pp97–106

Caro, T. M. (2003) 'Umbrella species: Critique and lessons from East Africa', *Animal Conservation*, vol 6, pp171–181

Carpenter, R. C. (1988) 'Mass mortality of a Caribbean sea urchin: Immediate effects on community metabolism and other herbivores', *Proceedings of the National Academy of Sciences of the United States of America*, vol 85, pp511–514

Carpenter, R. C. (1990) 'Mass mortality of *Diadema antillarum*. I. Long-term effects on sea urchin population-dynamics and coral reef algal communities', *Marine Biology*, vol 104, pp67–77

Castilla, J. C. and Duran, L. R. (1985) 'Human exclusion from the rocky intertidal zone of central Chile: The effects on *Concholepas concholepas* (Gastropoda)', *Oikos*, vol 45, pp391–399

Clements, F. E. (1916) *Plant Succession: An Analysis on the Development of Vegetation*, Carnegie Institution, Washington, DC

Cousins, S. H. (1991) 'Species diversity measurement: Choosing the right index', *Trends in Ecology and Evolution*', vol 6, pp190–192

Dufrene, M. and Legendre, P. (1997) 'Species assemblages and indicator species: The need for a flexible asymmetrical approach', *Ecological Monographs*, vol 67, no 3, pp345–366

Dunham, J. S. and Duffus, D. A. (2001) 'Foraging patterns of gray whales in central Clayoquot Sound, British Columbia, Canada', *Marine Ecology Progress Series*, vol 223, pp299–310

Duran, L. R. and Castilla, J. C. (1989) 'Variation and persistence of the middle rocky intertidal community of central Chile, with and without human harvesting', *Marine Biology*, vol 103, pp555–562

Estes, J. A. and Palmisano, L. R. (1974) 'Sea otters: Their role and structuring nearshore communities', *Science*, vol 185, pp1058–1060

Fairweather, P. G., Underwood, A. J. and Morgan, M. J. (1984) 'Preliminary investigations of predation by the whelk *Morula marginalba*', *Marine Ecology Progress Series*, vol 17, pp143–156

Faith, D. P. and Walker, P. A. (1996) 'How do indicator groups provide information about the relative biodiversity of different sets of areas?: On hotspots, complementarily and pattern based approaches', *Biodiversity Research*, vol 3, pp18–25

Falcon, J. M., Bortone, S. A., Brito, A. and Bundrick, C. M. (1996) 'Structure of and relationships within and between the littoral, rock-substrate fish communities off four islands in the Canarian Archipelago', *Marine Biology*, vol 125, no 2, pp215–231

Fletcher, W. J. (1987) 'Interactions among subtidal Australian sea urchins, gastropods and algae: Effects of experimental removals', *Ecological Monographs*, vol 57, no 1, pp89–109

Foster, M. (1990) 'Organization of macroalgal assemblages in the North-East Pacific: The assumption of homogeneity and the illusion of generality', *Hydrobiologia*, vol 192, pp21–33

Gasalla, M. A., Rodrigues, A. R. and Postuma, F. A. (2010) 'The trophic role of the squid *Loligo plei* as a keystone species in the South Brazil Bight ecosystem', *ICES Journal of Marine Science*, vol 67, pp1413–1424

Gilbert, L. E. (1980) 'Food web organization and conservation of neotropical diversity', in M. E. Soule and B. A. Wilcox (eds) *Conservation Biology: An Evolutionary-Ecological Perspective*, Sinauer Associates, Sunderland, MA

Hager, H. A. (1997) *Conservation of Species Richness: Are all Umbrella Species of Similar Quality?* University of Guelph, Ontario, Canada

Hay, M. E. (1984) 'Patterns of fish and urchin grazing on Caribbean coral reefs: Are previous results typical?', *Ecology*, vol 65, pp446–454

Hill, M. O. (1979) *TWINSPAN – A FORTRAN Program for Arranging Multivariate Data in an Ordinated Two-Way Table by Classification of the Individuals and Attributes*, Cornell University, Ithaca, NY

Hixon, M. A. and Brostoff, W. N. (1996) 'Succession and herbivory: Effects of differential fish grazing on Hawaiian coral-reef algae', *Ecological Monographs*, vol 66, pp67–90

Hockey, P. A. R. and Branch, G. M. (1984) 'Oystercatchers and limpets: Impacts and implications', *Ardea*, vol 72, pp199–200

Holling, C. S. (1992) 'Cross-scale morphology, geometry, and dynamics of ecosystems', *Ecological Monographs*, vol 62, pp447–502

Holt, R. D. (1977) 'Predation, apparent competition, and the structure of prey communities', *Theoretical Population Biology*, vol 12, pp197–229

Holt, R. D. (1984) 'Spatial heterogeneity, indirect interactions, and the coexistence of prey species', *American Naturalist*, vol 124, pp337–406

Hughes, T. P., Reed, D. C. and Boyle, M. L. (1987) 'Herbivory on coral reefs: Community structure following mass mortality of sea urchins', *Journal of Experimental Marine Biology and Ecology*, vol 113, pp39–59

Hurlburt, S. H. (1997) 'Functional importance vs keystoneness: Reformulating some questions in theoretical biocenology', *Australian Journal of Ecology*, vol 22, pp369–382

Imperial, M. T. and Hennessey, T. M. (1996) 'An ecosystem based approach to managing estuaries: An assessment to the National Estuary Program', *Coastal Management*, vol 24, pp115–139

Kerbes, R. H., Kotanen, P. M. and Jefferies, R. L. (1990) 'Destruction of wetland habitats by lesser snow geese: A keystone species on the west coast of Hudson Bay', *Journal of Applied Ecology*, vol 27, pp242–258

Kideys, A. E. (1994) 'Recent dramatic changes in the Black Sea ecosystem, the reason for the sharp decline in Turkish Anchovy fisheries', *Journal of Marine Systems*, vol 14, pp171–181

Kitchener, D. J., Chapman, A., Muir, B. G. and Palmer, M. (1980) 'The conservation value for mammals of reserves in the Western Australian wheatbelt', *Biological Conservation*, vol 18, pp179–207

Kremen, C. (1992) 'Assessing the indicator properties of species assemblages for natural areas monitoring', *Ecological Applications*, vol 2, no 2, pp203–217

Kvitek, R. G., Oliver, J. S., DeGange, A. R. and Anderson, B. S. (1992) 'Changes in Alaskan soft bottom prey communities along a gradient in sea otter predation', *Ecology*, vol 72, no 2, pp413–428

Lambeck, R. J. (1997) 'Focal species: A multi-species umbrella for nature conservation', *Conservation Biology*, vol 11, no 4, pp849–856

Landres, P. B., Verner, J. and Thomas, J. W. (1988) 'Ecological uses of vertebrate indicator species: A critique', *Conservation Biology*, vol 2, pp316–328

Launer, A. E. and Murphy, D. D. (1994) 'Umbrella species and the conservation of habitat fragments: A case of a threatened butterfly and a vanishing grassland ecosystem', *Biological Conservation*, vol 69, pp145–153

Libralato, S., Christensen, V. and Pauly, D. (2006) 'A method for identifying keystone species in food web models', *Ecological Modelling*, vol 195, nos 3–4, pp153–171

May, R. M., Beddington, J. R., Clark, C. W., Holt, S. J. and Laws, R. M. (1979) 'Management of multispecies fisheries', *Science*, vol 205, pp267–277

Meffe, G. K. and Carroll, C. R. (1997) *Principles of Conservation Biology*, Sinauer Associates, Sunderland, MA

Menge, B. A. (1976) 'Organization of the New England rocky intertidal community role of predation competition and environmental heterogeneity', *Ecological Monographs*, vol 46, pp355–393

Menge, B. A., Berlow, E. L., Blanchette, C. A., Navarrete, S. A. and Yamada S. B. (1994) 'The keystone species concept: Variation in interaction strength in a rocky intertidal habitat', *Ecological Monographs*, vol 64, no 3, pp249–286

Mills, J. S., Soule, M. E. and Doak, D. F. (1993) 'The keystone-species concept in ecology and conservation', *BioScience*, vol 43, no 4, pp219–224

Naiman, R. J., Melillo, J. M. and Hobbie, J. M. (1986) 'Ecosystem alteration of boreal forest streams by beaver (*Castor canadensis*)', *Ecology*, vol 67, pp1254–1269

Navarrete, S. A. and Menge, B. A. (1996) 'Keystone predation and interaction strength: Interactive effects of predators on their main prey', *Ecological Monographs*, vol 66, no 4, pp409–429

Niemi, G. A., Hanowski, J. M., Lima, A. R., Nichols, T. and Weiland, N. (1997) 'A critical analysis on the use of indicator species in management', *Journal of Wildlife Management*, vol 61, no 4, pp1240–1251

Norse, E. A. (1993) *Global Marine Biological Diversity. A Strategy for Building Conservation into Decision Making*, Island Press, Washington, DC

Noss, R. (1990) 'Indicators for monitoring biodiversity: A hierarchical approach', *Conservation Biology*, vol 4, no 4, pp355–364

Noss, R. F., Quigley, H. B., Hornocker, M. G., Merrill, T. and Paquet, P. C. (1996) 'Conservation biology and carnivore conservation in the Rocky Mountains', *Conservation Biology*, vol 10, pp949–963

NRC (National Research Council) (1995) *Understanding Marine Biodiversity*, National Academy Press, Washington, DC

Nybakken, J. (1997) *Marine Biology: An Ecological Approach*, Addison Wesley Longman Inc, New York

Oliver, J. S. and Slattery, P. N. (1985) 'Destruction and opportunity on the sea floor: Effects of gray whale feeding', *Ecology*, vol 66, no 6, pp1965–1975

Oliver, J. S., Kvitek, R. G. and Slattery, P. N. (1985) 'Walrus feeding disturbance: Scavenging habits and recolonization of the Bering sea benthos', *Journal of Experimental Marine Biology and Ecology*, vol 91, pp233–246

Paine, R. T. (1966) 'Food web complexity and species diversity', *American Naturalist*, vol 100, pp65–75

Paine, R. T. (1969) 'A note on trophic complexity and community stability', *American Naturalist*, vol 103, pp91–93

Paine, R. T. (1980) 'Food webs, linkage interaction strength, and community infrastructure', *Journal of Animal Ecology*, vol 49, pp667–685

Paine, R. T. (1992) 'Food-web analysis through field measurements of per capita interaction strength', *Nature*, vol 355, pp73–75

Paine, R. T. (1995) 'A conversation on refining the concept of keystone species', *Conservation Biology*, vol 9, no 4, pp962–968

Pearcy, W. G., Fisher, J. P., Anma, G. and Meguro, T. (1996) 'Species associations of epipelagic nekton of the North Pacific Ocean, 1978–1993', *Fisheries Oceanography*, vol 5, no 1, pp1–20

Peters, R. H. (1991) *A Critique for Ecology*, Cambridge University Press, New York

Peterson, C. H. (1979) 'Predation, Competitive Exclusion and Diversity in the Soft Bottom Benthic Communities of Estuaries and Lagoons', in R. J. Livingston (ed) *Ecological Processes and Coastal and Marine Systems*, Plenum Press, New York

Peterson, R. O. (1988) 'The Pit or the Pendulum: Issues in large carnivore management in natural ecosystems', in J. K. Agee and D. R. Johnson (eds) *Ecosystem Management for Parks and Wilderness*, University of Washington Press, Seattle, WA

Power, M. E., Tilman, D., Estes, J. A., Menge, B. A., Bond, W. J., Scott Mills, L., Daily, G., Castilla, J. C., Lubchenko, J. and Paine, R. T. (1996) 'Challenges in the quest for keystones', *BioScience*, vol 46, no 8, pp609–621

Ray, G. C. (1996) 'Coastal-marine discontinuities and synergisms: Implications for biodiversity conservation', *Biodiversity and Conservation*, vol 5, pp1095–1108

Roberge, J. M. and Angelstam, P. (2004) 'Usefulness of the umbrella species concept as a conservation tool', *Conservation Biology*, vol 18, pp76–85

Roberts, C. M. and Polunin, N. V. C. (1994) 'Hol Chan: Demonstrating that marine reserves can be remarkably effective', *Coral Reefs*, vol 13, p90

Roff, J. C. and Taylor, M. (2000) 'A geophysical classification system for marine conservation', *Journal of Aquatic Conservation: Marine and Freshwater Ecosystems*, vol 10, pp209–223

Roughgarden, J. (1983) 'The Theory of Coevolution', in D. J. Futuyma and M. Slatkin (eds) *Coevolution*, Sinauer Associates, Sunderland, MA

Ryti, R. T. (1992) 'Effect of the focal taxon on the selection of nature reserves', *Ecological Applications*, vol 2, pp404–410

Searing, G. F. and Frith, H. R. (1995) *British Columbia Biological Shore Zone Mapping System*, Resource Inventory Committee, Victoria, Canada

Simberloff, D. (1998) 'Flagships, umbrellas, and keystones: Is single-species management passé in the landscape era?', *Biological Conservation*, vol 83, pp247–257

Sinclair, A. R. E. and Norton-Griffiths, M. (1982) 'Does competition or facilitation regulate migrant ungulate populations in the Serengeti? A test of hypotheses', *Oecologia*, vol 53, pp364–369

Springer, A. M. (1992) 'A review: Walleye pollock in the North Pacific – how much difference do they really make?', *Fisheries Oceanography*, vol 1, pp80–96

Tanner, J. E., Hughes, T. P. and Connell, J. H. (1994) 'Species coexistence, keystone species, and succession: A sensitivity analysis', *Ecology*, vol 75, no 8, pp2204–2219

Terborgh, J. (1986) 'Keystone Plant Resources in the Tropical Forest', in M. E. Soule (ed) *Conservation Biology: The Science of Scarcity and Diversity*, Sinauer Associates, Sunderland, MA

VanBlaricom, G. R. (1982) 'Experimental analysis of structural regulation in a marine sand community exposed to oceanic swell', *Ecological Monographs*, vol 52, pp283–305

Weaver, J. C. (1995) 'Indicator species and scale of observation', *Conservation Biology*, vol 9, no 4, pp939–942

Webb, N. R. (1989) 'Studies on the invertebrate fauna of fragmented heathland in Dorset, UK, and the implications for conservation', *Biological Conservation*, vol 47, pp153–165

Wilcox, B. A. (1980) 'Insular Ecology and Conservation', in M. E. Soule and B. A. Wilcox (eds) *Conservation Biology: An Ecological–Evolutionary Perspective*, Sinauer Associates, Sunderland, MA

Willson, M. F. and Halupka, K. C. (1995) 'Anadromous fish as keystone species in vertebrate communities', *Conservation Biology*, vol 9, no 3, pp489–497

Zacharias, M. A. and Howes, D. E. (1998) 'An analysis of marine protected areas in British Columbia using a marine ecological classification', *Natural Areas Journal*, vol 18, no 1, pp4–13

Zacharias, M. A. and Roff, J. C. (2000) 'A hierarchical ecological approach to conserving marine biodiversity', *Conservation Biology*, vol 13, no 5, pp1327–1334

Zacharias, M. A. and Roff, J. C. (2001) 'Use of focal species in marine conservation and management: A review and critique', *Aquatic Conservation: Marine and Freshwater Ecosystems*, vol 11, pp59–76

Zacharias, M. A., Morris, M. C. and Howes, D. E. (1999) 'Large scale characterization of intertidal communities using a predictive model', *Journal of Experimental Marine Biology and Ecology*, vol 239, no 2, pp223–241

Zacharias, M. A., Howes, D. E., Harper, J. R. and Wainwright, P. (1998) 'The British Columbia marine ecosystem classification: Rationale, development, and verification', *Coastal Management*, vol 26, pp105–124

10

Genetic Diversity

Significance of genetics: From genes to ecosystems

They have come a long way, those replicators. Now they go by the name of genes, and we are their survival machines.

Richard Dawkins (1941–)

Introduction

Historically a driving interest in the biological sciences has been to describe the numbers and distributions of species on our planet. Starting even before Linnaeus, this led to the sciences of taxonomy and systematics giving us inventories and classifications of species. In turn, taxonomy and systematics led to the formal recognition and organization of species by global biogeography. All this historical understanding is now being revised and in fact revolutionized by modern genetic techniques that allow objective quantification of patterns that were previously rather subjective.

Genetic techniques, and the information they can provide, have now become indispensible to conservation planning. Genetics can provide information at all the other levels of the ecological hierarchy, ranging from the identity of individual organisms and assessment of fish stocks and other migratory species, to regional patterns of biological and oceanographic connectivity (see Chapter 17) or isolation, and global patterns of biogeography (see Chapters 4 and 5). It is probably fair to say that the discipline of genetics has not yet contributed anything like its actual potential to the field of marine conservation. In fact, viewed dispassionately, modern genetics will not only become indispensible, but may actually come to dominate the field of marine conservation in the near future. Although this prediction may currently seem overblown, it could be argued that most of our future conservation decisions should be based largely on a combination of genetic data and oceanographic mapping. In fact it can be readily appreciated that genetic information lies at the heart of the fundamental question: 'What should we conserve?' (see Allendorf and Luikart, 2007). The discipline of genetics can drive the conservation of all the identities and processes of life, not just genes and species, but ecosystems as well. Genetics in fact has become the common denominator between conservation ecology and evolution.

In a single short chapter we cannot hope to explore the full scope of the power, utility and applicability of genetic techniques to marine conservation. To the uninitiated, the field of genetics presents a bewildering array of modern techniques and terminology (see Box 10.1). Collectively their power is that they can statistically define relationships, not just taxonomic ones, but also relationships in time and space. They can therefore inform both structures and processes over the whole spectrum of biodiversity

Box 10.1 Definitions of some frequently used genetic terminology

AFLP (amplified fragment length polymorphism): A combination of random amplified polymorphic DNA (RAPD) and restriction fragment length polymorphism (RFLP). DNA is cut with restriction enzymes to produce many fragments, some of which vary in length from individual to individual (polymorphic).
Allele: One of two or more forms of the DNA sequence of a particular gene. Each gene can have different alleles.
Allele phylogeny: Seeks to explain the evolutionary origins and relationships among alleles.
Allozymes: Variant forms of an enzyme that are coded by different alleles at the same locus. These are distinguished from isozymes, which are enzymes that perform the same function, but which are coded by genes located at different loci.
Chromosome: An organized structure of DNA and protein that is found in cells. It is a single piece of coiled DNA containing many genes.
Clade: A group of individuals consisting of an organism and all its descendants. It is a single 'branch' of the 'tree of life'.
Deme: A local population of organisms of one species that actively interbreed and share a distinct gene pool. When demes are isolated they can become distinct subspecies or species.
Directional selection: Occurs when natural selection favours a single phenotype and therefore allele frequency continuously shifts in one direction.
DNA fingerprinting: A technique used in the identification of individual organisms on the basis of their respective DNA profiles – that is, the DNA nucleotide sequences.
DNA sequence: The specific sequence of nucleotides along a DNA strand.
Gene: A unit of heredity in a living organism. It is normally a stretch of DNA that codes for a type of protein or for an RNA chain that has a function in the organism.
Genealogy: The study of families and tracing their lineages and history.
Gene banks: Established to preserve genetic material of plants and animals.
Genetic bottleneck: An evolutionary event in which a significant percentage of a population or species is killed or otherwise prevented from reproducing.
Genome: The entirety of an organism's hereditary information. It is encoded either in DNA or, for many types of virus, in RNA. The genome includes both the genes and the non-coding sequences of the DNA.
Genetic diversity: The proportion of polymorphic loci across the genome.
Genetic drift: An important evolutionary process, which leads to changes in allele frequencies over time. It may cause gene variants to disappear completely, and thereby reduce genetic variability. Changes due to genetic drift are not driven by environmental or adaptive pressures, and may be beneficial, neutral or detrimental. Effects of genetic drift are larger in small populations, and smaller in large populations.
Genetic lineages: A series of mutations which connect an ancestral genetic type to derivative type.
Genetic stochasticity: Changes in the genetic composition of a population unrelated to systematic forces (selection, inbreeding or migration) – that is, effectively synonymous with genetic drift.
Genotype: The genetic constitution of a cell, an organism or an individual, that is the specific allele complement of the individual.
Haplotype: A combination of alleles at different loci on a chromosome that are transmitted together.
Heritability: The proportion of phenotypic variation in a population that is attributable to genetic variation among individuals.
Heterozygosity: The mean number of individuals with polymorphic loci.
Hypervariable minisatellite DNA: Short, tandemly repeated DNA sequences present at numerous loci in eukaryotes.
Introgressive hybridization: Also known simply as introgression, it is the movement of a gene from one species into the gene pool of another by repeated backcrossing of an interspecific hybrid with one of its parent species.

Microsatellites or STRs (short tandem repeats): A simple DNA sequence that is repeated several times at various points in the organism's DNA. Such repeats are highly variable enabling that location (polymorphic locus or loci) to be tagged or used as a marker.
Mini-satellite: A section of DNA that consists of a short series of bases (10–60 base pairs), that may occur in hundreds of locations within a genome.
Mitochondrial DNA (mtDNA): The DNA located within the mitochondria.
Molecular clock: A technique that uses rates of molecular change to deduce the time in geologic history when two species or other taxa diverged. The molecular data used is usually nucleotide sequences for DNA, or amino acid sequences for proteins.
Molecular markers: Used in molecular biology to identify a particular sequence of DNA.
mtCOI gene: The gene for the enzyme mitochondrial cytochrome oxidase I, now extensively used in the Barcode of Life project for the identification of species.
Multilocus: Multiple loci.
Mutation: A change in a genomic sequence of DNA. Mutations are caused by radiation, viruses, transposons and mutagenic chemicals, as well as errors that occur during DNA replication.
Nucleotide: DNA consists of nucleic acid which is a chain of linked units (nucleotides) of four types: adenine, cytosine, guanine, thymine.
Nucleotypes: A set of nucleotide sequences that may distinguish between clades.
Polymorphic loci/polymorphisms: Regions or points in the genetic structure that may vary among individuals, basically differences in the number of repeats (of nucleotides) in the same location.
Phenotype: Any observable characteristic or trait of an organism such as its morphology, development, biochemical or physiological properties and behaviour. Phenotypes result from the expression of an organism's genes as well as the environmental influences.
RAPD: A random amplification of anonymous loci by PCR (polymerase chain reaction) – amplification of DNA fragments that may be unique to a locus or gene.
RFLP: Polymorphism represented by the presence or absence of 'restriction' sites, which are short sequences along the DNA that can be cut by commercially available 'restriction enzymes'. The length of the cut fragment depends on whether particular restriction sites are present or not (polymorphic). The presence and absence of fragments resulting from changes in recognition sites are used to identify species or populations.
Reciprocal monophyly: Means that all DNA lineages within an ESU (evolutionary significant unit) must share a more recent common ancestor with each other than with lineages from other ESUs.

components. Our goal here is simply: to indicate what genetics can tell us about the identity and distributions of marine organisms; to review the role that modern genetics can play in the conservation of marine biodiversity throughout its ecological hierarchy; to show how various techniques can be used in decision-making in marine conservation. Finally we shall summarize what we consider to be the major contributions that genetics can presently play in conservation of marine biodiversity.

We have barely begun to describe the genetic richness of the oceans, although we have already significantly exploited it. There is a vast potential heritage for future generations in pharmacy, genetic engineering and other purposes that we cannot yet imagine. Loss of marine genetic diversity must compromise the future potential for its uses.

Genetic level

Significance of genetic diversity

Genetic diversity is the level of biodiversity that describes the total number of genetic characteristics in the make-up of a species; it differs from genetic variability, which is the tendency of genetic characteristics to vary. Genetic diversity can be considered in three components: within an individual (e.g. the proportion of polymorphic loci across the genome); within a subpopulation (i.e. the types and frequencies of alleles present, or mean individual heterozygosity within a population); and divergence among populations (i.e. the mean genetic difference among population locations). For any management purposes it is crucial to allocate these proportions of variation to determine how genetic diversity is spatially distributed, so that regions of conservation interest can be identified.

Genetic diversity plays a major role in the survival and adaptability of a species. When a species' environment changes, slight genetic variations can result in changes in an organism's biology that enables it to adapt and survive. A species with a high degree of genetic diversity among its populations will have more variations from which the alleles that are most fitted to a habitat can be selected. A high level of genetic diversity is also essential for species to evolve. Species that have low genetic variation are at a greater risk of extinction. With low gene variation within a species, reproduction leads to inbreeding and its associated problems. For example, the vulnerability of a population to certain types of diseases can increase with reduction in genetic diversity. The modern causes for the loss of genetic diversity among organisms include selective breeding for agriculture and aquaculture, and habitat destruction leading to the loss of whole species and reduction in the numbers of separate species' populations (meta-population erosion). Genetic diversity and biodiversity as a whole are dependent upon each other; genetic diversity within a species is necessary to maintain diversity among species, and vice versa.

Genetic 'value'

At the genetic level itself the concept of 'value' has been applied to describe the genetic uniqueness of individual taxa. The best known example is perhaps that of the New Zealand lizard – the Tuatara (*Sphenodon* species) – which occupies a genetic lineage quite distinct and genetically isolated from all other extant reptiles. In fact it has been estimated that this single species (or species group) represents up to 7 per cent of all genetic information among reptiles (May, 1990). The management implications of this sort of analysis are immediately apparent, and raise important questions. Where should conservation efforts be preferentially directed: towards genetically unique taxa, or to regions of greatest genetic diversity? A major argument and logical perspective is that our concerns about extinction should focus more on the preservation of distinct genetic lineages rather than on the preservation of individual species (e.g. Hobbs and Mooney, 1998). This concept of genetic value can therefore obviously radically alter our views of priorities for conservation efforts.

Systematists would undoubtedly find similarly distinct genetic lineages in each of the major and minor marine animal phyla and plant divisions. For example, among the agnathan fishes, the hagfishes (*Cyclostomata* – which may not even be true fish) represent a small group of species (~ 60) that have been distinct for over 400 million years. In the crustaceans the small (some nine species) obscure group of *Cephalocarida* is distinct from the remainder of the phylum. In the *Bryozoa*, the *Ctenostomata* have also been distinct from other members of this composite phylum for some 400 million years. Each of these small 'living fossil' groups thus retains a distinct and valuable genetic heritage.

Following the logic of this concept of genetic value, then the organisms of the marine environment as a whole take on a new and dramatically increased importance. This is because the oceans as a whole contain a far greater heritage of genetic diversity and are collectively far more valuable than the organisms of the terrestrial environment. The oceans contain not only many genetically unique individual species (as does the terrestrial environment) but – more importantly – whole divisions of plants (as phytoplankton) and phyla of animals that are completely absent on land (see Table 3.1), and likely numerous bacteria and viruses as well. Genetic distinctiveness and diversity are by definition greatest among divisions and phyla. The combined genetic significance of these unique taxa (from species to phyla and divisions) must therefore be enormous, and their phylogenetic and potentially practical value is incalculable (or at least has not been calculated in a manner analogous to that done for the Tuatara).

Species level

Species identities and inventories

A major problem for conservation biologists is that we have little idea of the actual number of species on our planet. Estimates have been growing significantly over the years but still vary widely. As described by classical taxonomy, many species are still denoted as, for example, cosmopolitan or endemic – that is, with either very wide or very restricted geographic ranges. In fact, until examined more closely with genetic techniques, we often cannot be certain that these 'cosmopolitan' species are not in fact cryptic species complexes, and that the 'endemics' are simply local phenotypic variants. Perhaps we can never be sure that we have either the correct species identification or that we have enumerated all the species in a given taxonomic group within a region, unless we have confirmation of identity from genetic analyses.

Worldwide initiatives such as the Census of Marine Life (www.coml.org/) and the Barcode of Life (www.boldsystems.org/) projects are now beginning to greatly expand our species inventories and to disentangle the problems that cannot be resolved by classical taxonomy and systematics.

The Barcode of Life Initiative began in 2003 with a proposal that species could be discriminated by using a very short gene sequence from a standardized position in the genome (see e.g. Hebert et al, 2004). Selected gene sequences can be viewed as a genetic 'barcode', which is enclosed in every cell, and barcoding has become a standardized approach for characterizing species. DNA barcodes (short DNA sequences that discriminate species and aid in recognition of unknown species) are based on the mitochondrial cytochrome oxidase I gene (mtCOI). DNA barcoding has now emerged as a global standard for assigning biological specimens to the correct species, although the technique has been challenged from several perspectives. Research projects on many taxonomic groups are under way, and many more are being planned. Some are global research campaigns involving dozens to hundreds of contributors, and others are the work of a small team focusing on a small taxonomic group.

All these barcoding projects share the goal of building an open-access database of reference barcodes that will improve our understanding of biodiversity and will allow non-taxonomists to identify species. Morphological identification of organisms can be difficult. Many species look so much alike that only a few experts can distinguish them. Even these experts cannot identify many juvenile forms or specimens that have been damaged. This becomes critically important when it comes to identifying pests or invasive species, or detecting products made from endangered species. Giving non-experts a way to identify species would open a goldmine of biological knowledge to students, teachers, government officials and the general public, and would transform our ability to understand and protect biological diversity

Testing relationships between classical taxonomic species and barcode species assignments and their phylogenetic relationships continues in many groups. For example Ward et al (2005) examined 207 species of mostly Australian marine fish, and sequenced (barcoded) their mitochondrial cytochrome oxidase subunit I gene (cox1). Most species were represented by multiple specimens, and 754 sequences were generated. All species could be differentiated by their cox1 sequences. Although DNA barcoding is primarily for species identifications, phylogenetic signals were also apparent in the data. Four major clusters were apparent: chimaerids, rays, sharks and teleosts. Species within genera invariably clustered, and generally so did genera within families. Three taxonomic groups – dogfishes of the genus *Squalus*, flatheads of the family *Platycephalidae*, and tunas of the genus *Thunnus* – were examined more closely. The clades revealed generally corresponded well with expectations. Individuals from operational taxonomic units designated as *Squalus* species B through F formed individual clades, supporting morphological evidence for each of these as belonging to separate species. The authors therefore concluded that the cox1 sequencing, or 'barcoding', can be used to identify fish species.

A major asset of genetic techniques is that they can also be applied to any developmental life stages (e.g. planktonic larvae) which may be very different in morphology from their adult parents and contain very little tissue. For example Pegg et al (2006) captured planktonic fish larvae around coral reef study sites on the Great Barrier Reef. These larvae could be identified to the genus or species level by comparison with a phylogenetic tree of adult tropical marine fish species using mtDNA HVR1 sequence data or mtDNA gene marker (cox1).

The species concept

Trying to define a species is a trying task! The populations of many species may be divided into several units, separated in some way by ecological barriers (of greater or lesser effectiveness), and whose component units consequently experience reduced gene flow among themselves. It has proven just as difficult to define these units as it has to define the species itself. Various terms have been used to describe these sub-specific units: subspecies, morphs, phenotypes, varieties, races and so on. These terms were all used before it became possible to distinguish phenotypic and genotypic effects and to estimate genetic relatedness and distance.

The biological species concept (i.e. that species are reproductively isolated from one another, e.g. Mayr, 1963) is something learned by all beginning university biology students, only to be told that it is quite inadequate. Despite their conviction of the fundamental principle of evolution, biologists have been reluctant to acknowledge the inherent variability of species, and still cling to it as the 'fundamental' taxonomic unit. This has given rise to problems. Reality is that members of a 'species' simply experience a greater rate of gene flow within – rather than among – other such units, and that species can become infinitely subdivided as gene flow is restricted for any reason. In a complex and heterogeneous world, following the principle of evolution, we should therefore expect the complex genetic patterns that we see.

Over two dozen definitions of 'a species' have been proposed within the last 20 years (see Box 10.2 for a sample). These include an ecological species concept, based on distinctness of ecological niche (e.g. Van Valen, 1976) and an evolutionary species concept based on analysis of lineages over time (e.g. Simpson, 1961). The phylogenetic species concept (e.g. Cracraft, 1989) suggests that all members of a species must share a single common ancestor. However, many populations around the globe have become split off from one another although they are clearly still members of the same species (i.e. they constitute a 'population of populations' – a metapopulation). Where genetic analysis reveals fixed DNA differences (polymorphisms), then individual populations can be interpreted as separate 'species'; such analyses can therefore lead to over-splitting and misplaced conservation effort (see e.g. Avise, 2000).

So what is a species? A species is essentially what a taxonomist says it is, as long as she has consulted a geneticist and an ecologist! The combination of morphological, anatomical, genetic and ecological data becomes important, although data types do not necessarily support one another. An example is the black sea turtle (putatively *Chelonia agassizii*) which shows morphological differences from other *Chelonia mydas* populations, yet is not genetically or reproductively isolated from them (Karl and Bowen, 1999).

Evolutionary significant units (ESUs) and 'units' of conservation

In order to address these problems and recognize that species and their populations are in a constant state of flux (isolation and reconnection) and evolution, Ryder (1986) proposed the concept of ESUs. ESUs can be defined as populations that warrant separate management or conservation priority because of genetic and ecological distinctiveness. The debate about what constitutes 'distinctiveness' began immediately! The conservation problem becomes how to recognize these 'sub-specific' units, and to decide which of them are the most significant ecologically, genetically and socio-economically. Unfortunately there is now considerable confusion in the literature about the term ESU and the competing terms: distinct population segment (DPS) and management units (MUs) – see Box 10.2.

Much of the debate about ESUs centres on the implications under the US Endangered

Box 10.2 Some definitions of species and subspecies units

Species concepts

Biological species concept: A species is a group of interbreeding natural populations that is reproductively isolated from other such groups.
Evolutionary species concept: An entity composed of organisms that maintains its identity from other such lineages and has its own independent evolutionary tendencies and historical fate.
Cohesion species concept: A species is the most inclusive population of individuals having the potential for phenotypic cohesion through intrinsic cohesion mechanisms.
Phylogenetic species concept: The smallest diagnosable cluster of individual organisms with which there is a parental pattern of ancestry and descent.
General lineage concept of species: Species are segments of population-level lineages

Evolutionarily significant units – ESUs

Subsets of the more inclusive entity species, which possess genetic attributes significant for the present and future generations of the species in question.

A population or group of populations that:

- is substantially reproductively isolated from other conspecific population units;
- represents an important component of the evolutionary legacy of the species.

Populations or groups of populations demonstrating significant divergence in allele frequencies.
Sets of populations derived from consistently congruent gene phylogenies.

Populations that:

- are reciprocal monophyletic for mtDNA alleles;
- demonstrate significant divergence of allele frequencies at nuclear loci.

Groups diagnosed by characters which cluster individuals or populations to the exclusion of other such clusters.

A lineage demonstrating highly restricted gene flow from other such lineages within the higher organizational level (lineage) of the species.

Distinct population segment – DPS

The smallest division of a taxonomic species permitted to be protected under the US Endangered Species Act.

Management unit – MU

Demographically independent sets of populations identified to aid short-term management, delimited by differences in frequencies of mtDNA or nuclear alleles, irrespective of allele phylogeny.

Source: Adapted from Fraser and Bernatchez (2001), Ryder (1986), Allendorf and Luikart (2007) among others

Species Act (ESA) which grants protection to species, subspecies and 'distinct population segments' of vertebrate species. Historically, the power to list species or populations as distinct population segments has been used to tailor management practices to unique circumstances, grant varied levels of protection in different parts of a species' range, protect species from extinction in significant portions of their ranges, as well as to protect populations that are unique evolutionary entities.

The fear is that a strict redefinition of distinct population segments as evolutionarily significant units will compromise management efforts, and that strictly cultural, economic or geographic justifications for listing populations as threatened or endangered will be greatly curtailed (see Pennock and Dimmick, 1997). The objectives and intent of the US ESA considers 'distinct' populations of vertebrates to be 'species' – and hence eligible for legal protection – but does not explain how distinctness should be evaluated (Waples, 1998). However, the unifying theme of the policy is the desire to identify and conserve important genetic resources in nature, thus allowing the dynamic process of evolution to continue largely unaffected by human factors.

The recognition of appropriate population units is an important step in managing threatened taxa but has thus been plagued by uncertainty about criteria and conservation goals. Moritz et al (1995) suggested that some of the conflicts which have arisen over ESUs in practice can be resolved by recognizing two types of conservation unit, each type being important for practical conservation of natural populations. ESUs can be defined as consisting of historically isolated sets of populations, for which a stringent and qualitative criterion is reciprocal monophyly for mitochondrial DNA (mtDNA) combined with significant divergence in frequencies of nuclear alleles. Such ESUs complement described species and are identified in order to contribute to the setting of conservation priorities. In contrast, management units (MUs) are demographically independent sets of populations identified to aid short-term management of the larger entities and are delimited by differences in frequencies of mtDNA or nuclear alleles, irrespective of allele phylogeny. Moritz et al (1995) suggest that both ESUs and the MUs that constitute ESUs or described species should be eligible for listing under the ESA.

Whatever the resolution to the problems, it appears that some flexibility in criteria for ESUs is necessary in order to reconcile opposing views. Fraser and Bernatchez (2001) suggest that, like species concepts, conflicting ESU concepts are all essentially aiming to define the same thing: segments of species whose divergence can be measured or evaluated by putting differential emphasis on the role of evolutionary forces at varied temporal scales. Thus, differences between ESU concepts lie more in the criteria used to define the ESUs themselves rather than in their fundamental essence. A review of the major approaches to discerning ESUs suggests that no single approach will work best in all situations, but that each has its strengths and weaknesses under different circumstances. Yet they all aim towards preserving the adaptive genetic variance within species.

Therefore, maintaining evolutionary potential in the face of uncertainty may be better served by using a more malleable system to delineating conservation units, like adaptive evolutionary conservation (AEC), which is able to incorporate the positive attributes of each approach (see Figure 10.1). Just as there is wide consensus that different evolutionary processes give rise to similar entities that we call species, there is also consensus that entities which we define as ESUs may arise by the accumulation of genetic differences through the various roles of evolutionary forces (e.g. novel mutations vs. directional selection) through time. The point to consider on the evolutionary continuum will vary with the organism at hand and so too will the criteria used. Differing ESU approaches should therefore be considered as only tools in the AEC tool box; they need not conflict with one another, but can operate in a complementary and adaptive fashion.

Migration and dispersal of individual species

The technique of DNA fingerprinting, which first gained notoriety in the field of forensic science, has now been widely applied for marine conservation to delineate the migrations and dispersals of individual members of a species. These techniques can be highly revealing about some fundamental aspects of the biology and ecology of individual species and their members.

Most species of whale spend the majority of their lives well away from land, are capable of migrating over large distances and are difficult to identify individually. However, conservation

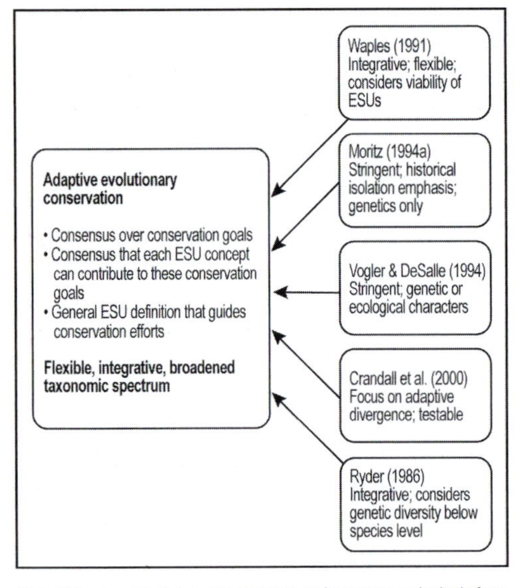

Note: This conceptual view encompasses various proposed criteria from other ESU concepts and definitions

Source: After Fraser and Bernatchez (2001)

Figure 10.1 Conceptual view of adaptive evolutionary conservation.

measures require a detailed understanding of their social structure, breeding behaviour and migration patterns. The advent of DNA fingerprinting permits a systematic investigation of such parameters (e.g. Baker et al, 1993). For example, humpback whales exhibit a remarkable social organization that is characterized by seasonal long-distance migration (> 10,000km/year) between summer feeding grounds in high latitudes, and winter calving and breeding grounds in tropical or near-tropical waters. All populations are currently considered endangered as a result of intensive commercial exploitation during the last 200 years. Using three hypervariable mini-satellite DNA probes originally developed for studies of human genetic variation, Bowen et al (1997) examined genetic variation within and among three regional subpopulations of humpback whales from the North Pacific and one from the North Atlantic oceans. Analysis of DNA extracted from skin tissues collected by small biopsy darting from free-ranging whales revealed considerable variation in each subpopulation. The extent of this variation argues against a recent history of inbreeding among humpback whales as a result of 19th- and 20th-century hunting. Significant categorical differences were found between the two oceanic populations, and the relationship between DNA fingerprint similarities and geographic distance suggests that nuclear gene flow between regional subpopulations within the North Pacific is restricted by relatively low rates of migratory interchange between breeding grounds or assortative mating on common wintering grounds.

DNA fingerprinting can be used not only to establish relationships among populations but also to disentangle their origins and geological dispersions. The Kemp's ridley sea turtle (*Lepidochelys kempi*) is restricted to the warm temperate zone of the North Atlantic Ocean, whereas the olive ridley turtle (*L. olivacea*) is globally distributed in warm-temperate and tropical seas, including nesting colonies in the North Atlantic that nearly overlap the range of *L. kempi*. To assess their biogeographic distributions, mtDNA was compared among 89 ridley turtles (Bowen et al, 1997). These data confirm a fundamental partition between *L. olivacea* and *L. kempi*, shallow separations within *L. olivacea*, and strong geographic partitioning of mtDNA lineages. The most divergent *L. olivacea* haplotype is observed in the Indo-West Pacific region, as are the central haplotypes in a parsimony network, implicating this region as the source of the most recent radiation of olive ridley lineages. The most common olive ridley haplotype in Atlantic samples is distinguished from an Indo-West Pacific haplotype by a single nucleotide substitution, and East Pacific samples are distinguished from the same haplotype by two nucleotide substitutions. These shallow separations are consistent with the recent invasion of the Atlantic, and indicate that the East Pacific nesting colonies were also recently colonized from the Indo-West Pacific region. Molecular clock estimates place these invasions within the last 300,000 years.

At more local levels, Peare and Parker (1996) used multilocus mini-satellite DNA fingerprinting to examine the genetic structure within nesting populations of green turtles (*Chelonia mydas*) in Tortuguero, Costa Rica and Melbourne, Florida,

USA. In the Tortuguero population, there was a significant negative correlation between genetic similarity of pairs of nesting females and the distance between their nest sites both within years and between years. In the Melbourne population, however, no relationship between genetic similarity and distance was found. The distance-related genetic structure of the Tortuguero population indicates that these females exhibit low levels of dispersal from natal sites, and that nest-mates return independently to nest near their natal sites. The lack of a similar structure in the Melbourne population suggests that females from this population may not return to natal sites with comparable precision. High levels of mortality among nests, hatchlings or maturing turtles produced in the Melbourne rookery may also be responsible for the absence of distance-related local genetic structure.

These techniques can also define dispersal or retention behaviours of invertebrate species and macrophytic plants. Clonal reproduction is a common life history strategy among sessile marine invertebrates, and can lead to high local abundances of one to a few genotypes in a population. Analysis of the clonal structure of such populations can provide insight into the ecological and evolutionary history of the population, and DNA fingerprinting can provide markers that are unique for an individual genotype. Coffroth et al (1992) generated DNA fingerprints for over 70 colonies of the clonal gorgonian, *Plexaura A* (*Plexaura sp. A*) in the San Blas Islands, Panama. DNA fingerprints within single individuals were identical, and fingerprinting resolved multiple genotypes within and among reefs. On one reef in the San Blas Islands, Panama, 59 per cent of the colonies sampled were of one genotype which was not found on any other sampled reefs, thus indicating a high degree of local clonal development and a low degree of dispersal of the species.

Populations of the temperate seagrass, *Zostera marina*, a widespread and ecologically important species, often exist as discontinuous beds in estuaries, harbours and bays where they can reproduce either sexually or vegetatively. Alberte et al (1994) examined the genetic structure of three geographically and morphologically distinct populations from central California (Elkhorn Slough, Tomales Bay and Del Monte Beach), using DNA fingerprints of multilocus restriction fragment length polymorphisms. The three eelgrass populations showed significantly less genetic similarity between locales than was found within populations, indicating that gene flow is restricted between locales even though two of the populations were separated by only 30km. The study demonstrates that: (a) natural populations of *Z. marina* from both disturbed and undisturbed habitats possess high genetic diversity and are not primarily clonal; (b) gene flow is restricted even between populations in close proximity; and (c) an intertidal population from a highly disturbed habitat showed much lower genetic diversity than an intertidal population from an undisturbed site.

Invasive species

The existence and transfer of non-indigenous (invasive) species (NIS) has been documented for many years (e.g. Ruiz et al, 1997). They are increasingly conspicuous in marine and estuarine habitats throughout the world, and the number, variety and effects of these species continue to accrue. Most of these NIS invasions result from anthropogenic dispersal. Although the relative importance of different dispersal mechanisms varies both spatially and temporally, the global movement of ballast water by ships appears to be the largest single vector for NIS transfer today, and many recent invasions have resulted from this mechanism of transfer. The rate of new invasions may have increased in recent decades, perhaps due to changes in ballast water transport. Estuaries have been especially common sites of invasions, accumulating from tens to hundreds of NIS per estuary including most major taxonomic and trophic groups. We now know of approximately 400 NIS along the Pacific, Atlantic and Gulf coasts of the US, and hundreds of marine and estuarine NIS are reported from other regions of the world. Although available information about invasions is limited to a few regions and likely underestimates the actual number of NIS invasions, there are apparent differences in the frequency of NIS among sites. Mechanisms responsible for observed patterns among sites likely include variation in supply of NIS, and perhaps variation in properties of recipient or donor communities, but the role of these mechanisms has not been tested. Although our present knowledge about the extent, patterns and mechanisms of marine invasions is still in its infancy, it is clear that NIS are a significant force of change in marine and especially estuarine communities globally. Taxonomically diverse NIS are having significant effects on many, if not most, estuaries that fundamentally alter population, community and ecosystem processes. The impacts of most NIS remain unknown, and the predictability of their direct and indirect effects remains uncertain. Nonetheless, based upon the documented extent of NIS invasions and scope of their effects, studies of marine communities that do not include NIS must be considered incomplete. Many of these species may be closely related to indigenous species, may act as cryptic species, and may only be distinguished by genetic analyses.

Species complexes and cryptic species complexes

Cryptic species complexes comprise two or more genetically distinct species which live in similar habitats, which were previously classified as a single species, and whose populations may not be morphologically distinguishable from one another. Cryptic species complexes are not the same as populations undergoing speciation; rather they indicate that speciation has already broken gene flow between populations, but that evolution has not progressed to a point where recognizable phenotypic adaptations have taken place. The progressive recognition of cryptic species (and species complexes) is a major source of the rapidly increasing species tally of our planet.

High gene flow, especially by larval dispersal, had been viewed as limiting geographic isolation and hence limiting population differentiation among marine species. However, it now appears that many marine species with high dispersal are more genetically subdivided than originally thought, even within the pelagic realm.

Since the advent of molecular phylogenetics, there is increasing evidence that many small aquatic and marine invertebrates – once believed to be single, cosmopolitan species – are in fact cryptic species complexes. The record for the number of cryptic species contained in a single classical species seems to be for *Brachionus plicatilis*. Suatoni et al (2006) explored different methods of empirically delimiting species boundaries in the salt water rotifer *B. plicatilis* by comparing reproductive data (i.e. the traditional biological species concept) to phylogenetic data (the genealogical species concept). Based on a high degree of molecular sequence divergence, the genealogical species hypothesis indicates the existence of at least 14 species. Although the diversity of the group is higher than previously understood, geographic distributions remain broad. Efficient passive dispersal has resulted in global distributions for many species with some evidence of isolation by distance over large geographic scales where sympatry of genetically distant strains is common.

Even among the larger and well known nektonic organisms many questions of species identification and distributions are still unsettled. For example, many species of euphausiids (Euphausiacea, Crustacea) are distinguished by subtle or geographically variable morphological characters, and erroneous identification of euphausiid species may be more frequent than currently acknowledged. DNA barcodes are of use for this group. Bucklin et al (2007) sequenced regions of mtCOI for 40 species of 10 euphausiid genera: *Bentheuphausia*, *Euphausia*, *Meganyctiphanes*, *Nematobrachion*, *Nematoscelis*, *Nyctiphanes*, *Stylocheiron*, *Tessarabrachion*, *Thyssanoessa* and *Thysanopoda*. MtCOI sequence variation reliably discriminated all species including a separation within Atlantic and Pacific Ocean populations of *Euphausia brevis* which may deserve status as distinct species. The mtCOI gene tree for 20 species of *Euphausia* reproduced one of three morphologically defined species groups, and resolved relationships among closely related species of most genera, usually in accord with morphological groupings. The technique could therefore help ensure accurate species identification, recognition of cryptic species and evaluation of taxonomically meaningful geographic variation and distributions.

Unfortunately, a lack of systematic studies leaves many questions open, such as whether cryptic species complexes are more common in particular habitats, latitudes or taxonomic groups. The discovery of cryptic species is likely to be non-random with regard to taxon and biome and, hence, could have profound implications for evolutionary theory, biogeography and conservation planning (Bickford et al, 2007).

Genetic markers: Identifying poaching and products from endangered species

Identification of individual animals is now possible by application of the same techniques for DNA fingerprinting that have become notorious in cases of human forensics. These techniques supplement existing and perhaps less certain techniques, for example for identification of individual whales and porpoises from photographs of flukes and fins. Identification of individual animals by genetic techniques is now being combined with other tracking techniques such as direct tagging. The advantage of modern tags on individual marine mammals, large fish species, turtles and so on is that they can return data on environmental and oceanographic conditions, in addition to tracking the movements of single animals (see the OTN website http://oceantrackingnetwork.org/ for further review).

The use of gene sequences has now become indispensible for investigation of biosecurity, food authentication, investigation against poaching or illegal trade of endangered species, and wildlife enforcement. A single example from Roman and Bowen (2000) will serve as illustration. Much of the demand for turtle meat in North America and Europe has been met using green turtle (*Chelonia mydas*) and other marine turtles. As stocks of marine turtles dwindled, harvest of the alligator snapping turtle (*Macroclemys temminckii*), the largest freshwater turtle in North America, increased in the southeastern USA. As a result, this species has declined and is now protected in every state of the USA except Louisiana. There is concern that the remaining legal trade in turtle products may serve as a cover for illegally harvested species. To assess the composition of species in commerce, 36 putative turtle meat products were purchased in Louisiana and Florida. Using cytochrome b and control region sequences of the mitochondrial genome, 19 samples were identified as common snapping turtle (*Chelydra serpentina*), three as Florida softshell (*Apalone ferox*), one provisionally as softshell turtle (*Apalone sp.*), one as alligator snapping turtle and eight as American alligator (*Alligator mississippiensis*). It appears that *M. temminckii* is no longer the predominant species in markets of Louisiana, and the presence of alligator meat in a quarter of the samples indicates that the trade in turtle products is not entirely legitimate. As is often the case for unsustainable wildlife harvests, large esteemed species, such as green turtle and alligator snapper, have been replaced by smaller, more abundant or mislabelled species, a phenomenon that Roman and Bowen (2000) refer to as the 'mock turtle syndrome'.

Aquaculture escapement

Marine aquaculture has been growing exponentially around the world, even as conventional capture fisheries decline. More than a decade ago 94 per cent of all adult Atlantic salmon were derived from aquaculture, and presently over 80 per cent of coho salmon on the west coast of Canada are hatchery reared (see DFO ESTR report www.dfo-mpo.gc.ca/csas/Csas/Publications/SAR-AS/2010/2010_030_E.pdf). Such gross imbalances between nature and culture lead to considerable disturbances both genetic and environmental, causing conflicts among the fishing industry, conservationists, aquaculturists and environmentalists.

Adult escapement of reared salmon from impounds has become common throughout the northern hemisphere. Escapees seem to behave like homeless fish, and enter rivers at random for spawning (Egidius et al, 1991). Diseases common to wild and cultured populations such as the fluke *Gyrodactylus salaris* have been spread to many rivers, probably by stocking fish from infected hatcheries. The salmon lice, which are normally considered harmless to wild salmon, also affect salmon reared in net pens. Bacterial and fungal diseases are found among free-living as well as among cultured salmon; wild populations may act as reservoirs for the disease agents. Even more seriously, escaped salmon cause gene flow

Box 10.3 Relationships between genetic variation and population size proposed by Soulé (1987)

Theory predicts that levels of genetic variation should increase with effective population size. Soule (1976) compiled the first evidence that levels of genetic variation in wildlife were related to population size, but the issue remains controversial.

The hypothesis that genetic variation is related to population size leads to the following predictions:

- Genetic variation within species should be related to population size.
- Genetic variation within species should be related to island size.
- Genetic variation should be related to population size within taxonomic groups.
- Widespread species should have more genetic variation than restricted species.
- Genetic variation in animals should be negatively correlated with body size.
- Genetic variation should be negatively correlated with rate of chromosome evolution.
- Genetic variation across species should be related to population size.
- Vertebrates should have less genetic variation than invertebrates or plants.
- Island populations should have less genetic variation than mainland populations.
- Endangered species should have less genetic variation than non-endangered species.

between cultured and wild populations through introgressive hybridization, thus reducing the variation among natural populations and consequent fitness and reproductive success (e.g. Utter and Epifanio, 2002).

In a study of rivers in northwest Ireland where escapes of adult Atlantic salmon (*Salmo salar*) were known to have occurred from sea cages, farmed populations showed a significant reduction in mean heterozygosity over wild populations. The proportion of juveniles of maternal farm parentage in two rivers ranged from 18 per cent to a maximum 70 per cent in an individual sample. Only a small proportion of adult farm salmon that escaped appear to have bred successfully in the rivers studied (Clifford et al, 1998). Survival of the progeny of farmed salmon to the smolt stage was significantly lower than that of wild salmon, with increased mortality being greatest in the period from the eyed egg to the first summer. However, progeny of farmed salmon grew faster and competitively displaced the smaller native fish downstream. Growth of hybrids was generally either intermediate or not significantly different from the wild fish. The demonstration that farmed and hybrid progeny can survive in the wild to the smolt stage, taken together with unpublished data that show that these smolts can survive at sea and home to their river of origin, indicates that escaped farmed salmon can produce long-term genetic changes in natural populations. These changes affect both single-locus and high-heritability quantitative traits, for example growth and sea age of maturity. While some of these changes may be advantageous from an angling management perspective, they are likely, in specific circumstances, to reduce population fitness and productivity.

Initiatives to protect natural gene pools are now being developed, including the technical improvement of farming facilities, the establishment of gene banks, restrictions on the transfer of living material, and the use of indigenous fish for enhancement and establishment of areas protected from fish farming.

Population level

This is probably the level of the ecological hierarchy where genetics has been most informative for conservation purposes. The three most important metrics of populations for conservation are: population structure (e.g. subpopulations and fragmentation); population size (and other demographics); and population health (in terms of physiological or – for this chapter – genetic condition). It has been hypothesized that population size and health (as genetic variation) are related (e.g. Soulé, 1987; and Box 10.3), but relationships are perhaps not as simple as once assumed.

Population structure

Most marine species consist of several subpopulations that are connected or fragmented on various space and time scales – that is, they consist of meta-populations. Population fragmentation is a term most often used in terrestrial ecology, where corridors – linking subpopulations isolated by habitat fragmentation and destruction – become of great importance. However, marine populations can also become fragmented in various ways. A knowledge of population structure is therefore of broad significance both for the preservation of individual species and for the sustainable exploitation of resource species. Especially for migratory, widespread and commercially exploited species it is vital to know whether we are dealing with a single pan-mictic population or a series of subpopulations (demes or a meta-population).

Appropriate conservation and management measures (whether for preservation or exploitation) may need to be applied to each of the genetically separate (and separately breeding) subpopulations rather than treating them as a combined 'stock'. Various genetic techniques can be used to: assign individual members to a defined population; assign groups of individuals to a population of defined origin; define the composition of a stock of mixed populations. For example, an examination of red king crab (*Paralithodes camtschaticus*) purportedly caught in one location in fact identified them as belonging to a subpopulation from another location 1500km away, on the basis of differences in 14 polymorphic loci (Allendorf and Luikart, 2007).

Population size

The concept of the size of natural populations has become of extreme importance in recent years as human impacts progressively dominate the natural world. Historically, census population levels (Nc) have been assessed by some form of statistical procedure, for example by mark and recapture, or in the case of exploited populations by analysis of catch and effort data. Many exploited populations or those in reduced habitats have undergone a population bottleneck (see Box 10.4) of greatly reduced numbers. For conservation purposes this has lead to population viability analysis (PVA) or estimates of minimum viable populations (MVP) – the number of individuals required to ensure survival of a species for a specified time period. However, this concept has recently become less popular for two prime reasons: first it is perceived as setting targets for species' conservation too low; and second, census population numbers mask the potential genetic problems associated with effective population size (Ne). See Box 10.4 and Allendorf and Luikart (2007) for definitions of Ne and how it can be calculated.

Genetic stochasticity (i.e. genetic drift unrelated to systematic forces such as natural selection, inbreeding and migrations) due to small population size contributes to risk of extinction, especially when population fragmentation disrupts gene flow. Estimates of Ne can therefore be more informative and more critical than estimates of Nc. Comprehensive estimates of Ne/Nc (that included the effects of fluctuation in population size, variance in family size and unequal sex-ratio) made by Frankham (1995) averaged only 0·10–0·11, showing that wildlife populations have much smaller effective population sizes than previously recognized. Palstra and Ruzzante (2008) confirmed the role of gene flow in countering genetic stochasticity and underlined the importance of gene flow for the estimation of Ne, and of population connectivity for conservation in general. They argue that reductions in contemporary gene flow due to ongoing habitat fragmentation will likely increase the prevalence of genetic stochasticity, which should therefore remain a focal point in the conservation of biodiversity.

Effective population size (Ne) is a central evolutionary concept, but its genetic estimation can be significantly complicated by age structure. An evaluation of Ne estimates in a demographic context suggests that life history diversity,

Box 10.4 Some definitions of population size

Census population size (Nc) is the numbers of members of a population as determined by conventional census techniques, such as mark-recapture and stratified random sampling.

Effective population size (Ne) is the number of breeding individuals in an idealized population that would show the same dispersion of allele frequencies under random genetic drift or the same degree of inbreeding as the population under consideration. The effective population size is usually smaller than the census population size Nc.

Minimum viable population (MVP) is a lower bound on the population numbers of a species, such that it can survive in the wild (in other words excluding domesticated or captive populations). MVP is usually estimated (by computer simulations of population viability using demographic and environmental information) as the population size necessary to ensure between 90 and 95 per cent probability of survival for 100 to 1000 years into the future.

Population viability analysis (PVA) is a species-specific method of risk assessment, defined as the process that determines the probability that a population will go extinct within a specified given number of years. It combines species characteristics including genetic parameters, environmental variability to forecast population health and extinction risk. Each PVA is uniquely developed for a target population or species.

Population bottleneck or **genetic bottleneck** is an historical or geological event in which a significant percentage of a population or species dies or is otherwise prevented from reproducing. Population bottlenecks increase genetic drift and inbreeding due to the reduced population size. A genetic bottleneck (called a founder effect) can occur if a small group becomes reproductively separated from the main population.

For further explanation and statistical methods, see Allendorf and Luikart (2007).

density-dependent factors, metapopulation dynamics and life history characteristics may all affect the genetic stability of these populations (Palstra et al, 2009).

Differences between Ne and Nc can be not just significant, but dramatic. Using microsatellite loci and a variety of analytical methods, Turnera et al (2002) estimated genetic effective size (Ne) of an abundant and long-lived marine fish species, the red drum (*Sciaenops ocellatus*), in the northern Gulf of Mexico (Gulf). The ratio Ne/Nc was ~ 0.001, whereas in an idealized population this ratio should approximate unity. The extraordinarily low value of Ne/Nc appeared to be related to high variance in individual reproductive success, and to variance in productivity of critical spawning and nursery habitats located in spatially discrete bays and estuaries throughout the northern Gulf. The authors caution that models that predict Ne/Nc exclusively from demographic and life-history features will seriously overestimate Ne if variance in reproductive success and variance in productivity among spatially discrete demes is underestimated. The study also shows that vertebrate populations with enormous adult census numbers may still be at risk relative to decline and extinction from genetic factors.

Genetic techniques can also be used to estimate historical population sizes and estimate numbers during bottleneck phases. Hunting during the past 200 years has reduced the populations of many marine mammals to near extinction, but with differing genetic consequences.

The northern elephant seal was hunted extensively in the 19th century and forced through a bottleneck of approximately 10–20 seals. All measures of molecular genetic variation (including mtDNA and allozymes) show current levels for the northern elephant seal to be low (Hoelzel, 2008). The killer whale (*Orcinus orca*) is an abundant, highly social species which also shows reduced genetic variation. Hoelzel et al (2002) found no consistent geographical pattern of global diversity and no mtDNA variation within some regional populations. The worldwide pattern and paucity of diversity may again indicate a historical population bottleneck.

However, genetic patterns in other marine mammals can be very different. Populations of humpback whales are presently 6–20 times lower than genetic estimates of historical population size. Despite this reduction, there is an abundance of genetic variation in all but one of the world-wide oceanic subpopulations in humpback whales (Baker et al, 1993). Phylogenetic reconstruction of nucleotypes and analysis of maternal gene flow showed that current genetic variation is not due to post-exploitation migration between oceans, but is a relic of past population variability. Preservation of pre-exploitation variation in humpback whales may be attributed to their long life span and overlapping generations and to an effective, though perhaps not timely, international prohibition against hunting.

Population 'health'

Assessment of 'health' or fitness (naturalness?) at any level of the ecological hierarchy really comes down to some measure of diversity. The level of genetic diversity found in a population highly depends on the mating system, the evolutionary history of a species, the population history (which is usually unknown) and the level of environmental heterogeneity.

Genetic diversity is one of the three forms of biodiversity recognized by the International Union for the Conservation of Nature (IUCN) as deserving conservation. The need to conserve genetic diversity within populations is based on two arguments: the necessity of genetic diversity for evolution to occur, and the expected relationship between heterozygosity and population fitness. Because loss of genetic diversity is related to inbreeding, and inbreeding reduces reproductive fitness, a correlation is expected between heterozygosity and population fitness. Long-term Ne, which determines rates of inbreeding, should also be correlated with fitness. However, other theoretical considerations and empirical observations suggest that the correlation between fitness and heterozygosity may be weak or nonexistent. Reed and Frankham (2003) used 34 data sets to perform a meta-analysis in an attempt to resolve the issue. Data sets were included in the study provided that fitness – or a component of fitness – was measured for three or more populations, along with heterozygosity, heritability and/or population size. The mean weighted correlation between measures of genetic diversity, at the population level, and population fitness was highly significant and explained 19 per cent of the variation in fitness. The study strengthens concerns that the loss of heterozygosity has a deleterious effect on population fitness and supports the IUCN designation of genetic diversity as worthy of conservation.

In studies of individual populations, strong relationships between genetic diversity and fitness have been demonstrated. For example, in a study of populations of the eelgrass *Zostera marina*, there were significant positive associations between genetic diversity and sexual reproduction, with a similar trend for vegetative propagation and development of flowering shoots. In addition, more seeds germinated from a genetically diverse, untransplanted population than from a transplanted population with low genetic diversity (Williams, 2001). Genetic diversity clearly contributes to eelgrass population viability even over the short term.

Similar relationships apply in the mussel *Mytilus edulis* in natural populations. Koehn and Gaffney (1984) showed that growth rates in this species were positively correlated with individual heterozygosity. The authors generalized that the relationship between multiple locus heterozygosity and growth rate is one that is general to a diversity of outbreeding plant and animal populations. However, this relationship does not emerge from experimental designs in which there has been limited genetic sampling of the natural genetic variation.

Community level

This may be the level of the ecological hierarchy at which genetics (presently) has the least to contribute. Compared to the importance of genetic knowledge at the species and population levels, an understanding of the significance of genetic data at the community level is still developing. Relationships between environmental stress (presumed to be a major force in evolution), inbreeding and genetic variance have been investigated predominantly in terrestrial species, whereas in marine species stress is typically assessed as physiological or biochemical responses.

In marine systems, potentially interesting genetic questions arise especially with respect to focal species and their 'behaviour' within communities and among locations. For example: do community composition indicators (see Chapter 9) actually 'indicate' membership of the same community type among locations, or – because of genetic differences and consequent differences in adaptations to different environments – do they actually indicate different relationships, or indeed anything at all? The same type of question could be asked of the reliability of community condition indicator species and keystone species. Keystone species show keystone effects in some locations but apparently not in others. Are these differences simply the result of undefined ecological or environmental interactions, or are there regional differences in genetics?

Success of invasive species

It has been suggested that management and control of non-indigenous species is perhaps the greatest challenge that will face conservation biologists in the coming decades. Despite the severe economic and ecological damage already caused by introduced species, the factors that permit them to become successful are paradoxical. Allendorf and Luikart (2007) suggest the factors involved are the same as those facing populations threatened with extinction, namely: genetic drift and effects of small population size; gene flow and hybridization; natural selection and adaptation. As a rule of thumb only one species in a hundred introduced species actually becomes invasive – that is, becomes a serious economic/ecological pest.

Invasive species must pass through an introduction/colonizing phase consisting of few members; that is – it must pass through a population bottleneck. Why then, if population bottlenecks are harmful, are invasive species that have gone through a colonizing bottleneck so successful? One answer to this paradox is that introduced species often have greater genetic variation than native species because they are a mixture of source populations. In addition – at least for plants – asexual reproduction avoids the effects of inbreeding depression, and the progeny are identical to the parents. However, a second paradox is now immediately apparent. If adaptation to local conditions is important, why are introduced species so successful at replacing native species? Three suggestions are: that invasive species are inherently better competitors; that their natural enemies are absent; that introduced species may outperform native ones only in the short-term because native species are constrained by long-term adaptations.

Molecular genetic techniques can provide valuable information about origins, life histories and mode of reproduction of invasive species. For example, most strains of the green alga *Caulerpa taxifolia* are non-invasive. The invasive strains differ from native ones because they reproduce asexually, grow more rapidly and are resistant to lower temperatures. The common cordgrass –*Spartina anglica* – is one of the world's worst invasive species, whose spread leads to the exclusion of native species and reduction of habitat for shorebirds. The species originated by chromosome doubling of hybrids between Old World *S. maritima* and the New World *S. alterniflora*, leading to fixed heterozygosity and an almost total lack of genetic difference among individual plants.

Genetic diversity and community disturbance

Much empirical and theoretical research suggests that more species-rich systems exhibit enhanced productivity, nutrient cycling or resistance to disturbance or invasion relative to systems with fewer species. In contrast, few data are available to assess the potential community and ecosystem-level importance of genetic diversity within species known to play a major functional role. In a manipulative field experiment Hughes and Stachowicz (2004) showed that increasing genotypic diversity in the seagrass *Zostera marina* actually enhances community resistance to disturbance by grazing geese. The time required for recovery to near pre-disturbance densities also decreased with increasing eelgrass genotypic diversity. Genotypic diversity did not affect ecosystem processes in the absence of disturbance. These results suggest that genetic diversity, like species diversity, may be most important for enhancing the consistency and reliability of communities and ecosystems by providing biological insurance against environmental change.

Genotypic diversity in populations of the Hawaiian reef coral, *Porites compressa*, is also directly related to habitat-disturbance history (Hunter, 1993). The highest diversity (lowest amount of clonal proliferation) was found in populations that had been intensely or recently disturbed – as indicated by small mean colony size – suggesting early stages of recolonization. In an undisturbed, protected habitat, lower genotypic diversity was a result of a significant degree of clonal replication of established genotypes and larger colony size. Populations in intermediately disturbed habitats showed intermediate levels of diversity and clonal structure as a result of the combined contributions of sexual and asexual reproduction. This study also emphasized the importance to conservation measures of knowing which corals are reproducing sexually versus asexually.

Ecosystem level

Global

Ultimately all the waters of the oceans are connected over some time and space scale, but with variable consequences for organisms as geological forces periodically open and close dispersal and migration routes. The relevant time and space scales for organisms are those of their life cycles and dispersal distances. The longest circulation time scale in the oceans is that of the deep and bottom waters of the oceans, which circulate around the globe in about 1600 years. This is more than an order of magnitude greater than even the longest lived marine species. Even in the much faster circulating surface waters, the various regions of the oceans exhibit a whole series of physical and physiological barriers that can effectively isolate members of all taxonomic groups, both geographically and genetically, leading to the evolution of new species.

Recent advances in paleoclimate research have prompted a re-examination of oceanographic processes as a fundamental influence on genetic diversity. Evidence from ice cores and anaerobic marine sediments document strong regime shifts in the world's oceans, in concert with periodic climatic changes. These changes in sea surface temperatures, current paths, upwelling intensities, and retention eddies were likely associated with severe fluctuations in population sizes or regional extinctions. Grant and Bowen (1998) assessed the consequences of such oceanographic processes on marine fish intrageneric gene genealogies of sardines (*Sardina*, *Sardinops*) and anchovies (*Engraulis*). Representatives of these two groups occur in temperate boundary currents on a global scale, and these regional populations are known to fluctuate markedly. Biogeographic and genetic data indicate that *Sardinops* has persisted for at least 20 million years, yet the mtDNA genealogy for this group coalesces in less than half a million years and points to a recent founding of populations around the rim of the Indian-Pacific Ocean. Phylogeographic analysis of Old World anchovies reveals a Pleistocene dispersal from the Pacific to the Atlantic, almost certainly via southern Africa, followed by a very recent recolonization from Europe to southern Africa. These results demonstrate that regional populations of sardines and anchovies are subject to periodic extinctions and recolonizations. Such climate-associated dynamics may explain the low levels of their nucleotide diversity and the shallow coalescence of their mtDNA genealogies. If these findings apply generally to marine fishes, management strategies should incorporate the idea that even extremely abundant populations may be relatively fragile on ecological and evolutionary time scales. In fact, concordant phylogeographic patterns among independently evolving species provide evidence of other vicariant histories of population separation, which can also be related tentatively to episodic changes in environmental conditions during the Pleistocene (Avise, 1992).

The fastest contemporary changes in the oceans seem to be occurring in the Arctic, where passage of species from Pacific to Atlantic oceans has been facilitated by climate change in the past (e.g. Vermeij and Roopnarine, 2008).

Accumulating evidence suggests that such passage is already occurring again (see DFO www.dfo-mpo.gc.ca/csas/Csas/Publications/SAR-AS/2010/2010_030_E.pdf).

Our understanding of the patterns and processes that control global biogeography are in fact constantly and increasingly undergoing revision as genetic techniques reveal both connections and barriers to gene flow that were previously not imagined. It is the tension between species' abilities to disperse and their subsequent tolerance of environmental conditions (both geophysical and biological), played out over geological time scales, that has resulted in the biogeographic ecoregions that we recognize today. Fortunately there appears to be an evolving concordance between palaeoclimate research and genetic data. A major role for the discipline of genetics is to examine, revise and sharpen the presently proposed ecoregion boundaries, to show how they evolved and how they may change under various scenarios of future climate change.

Genetic isolation in peripheral marine ecosystems

Ecosystems that are geographically and ecologically isolated or marginal are under high selection pressures, resulting in an increase in genetically atypical populations. Johannesson and André (2006) analysed genetic data from 29 species inhabiting the low salinity Baltic Sea, a geographically and ecologically marginal ecosystem. On average Baltic populations had lost genetic diversity compared to Atlantic populations. This pattern was unrelated to dispersal capacity, generation time of species or to the taxonomic group of organism, but was strongly related to the type of genetic marker (mitochondrial DNA loci had lost about 50 per cent diversity and nuclear loci 10 per cent). Genetic isolation showed clinal patterns of differentiation between Baltic and Atlantic regions, with a sharp slope around the Baltic Sea entrance, indicating impeded gene flows between Baltic and Atlantic populations. Despite the short geological history of the Baltic Sea (about 8000 years), populations inhabiting the Baltic have evolved substantially differently from Atlantic populations, probably as a consequence of isolation and bottlenecks. The Baltic Sea thus acts as a refuge for unique evolutionary lineages that constitute an important genetic resource for management and conservation.

Similar concerns apply – on a more local scale – to marine reserves. Many current marine reserves are small in size and isolated to some degree (e.g. sea loughs and offshore islands). While such features may enable easier management, they may have important implications for the genetic structure of protected populations, the ability of populations to recover from local catastrophes, and the potential for marine reserves to act as sources of propagules for surrounding areas. Bell and Okamura (2005) demonstrated genetic differentiation, isolation, inbreeding and reduced genetic diversity in populations of the dogwhelk *Nucella lapillus* in Lough Hyne Marine Nature Reserve (an isolated sea lough in southern Ireland), compared with populations on the local adjacent open coast and populations in England, Wales and France. This sea lough is isolated from open coast populations, and it indicates that there may be long-term genetic consequences of selecting reserves on the basis of isolation and ease of protection. This is one reason for the significance of connectivity among marine protected areas (MPAs) (see Chapter 17).

However, isolation can apparently also have the opposite effect. The Hawaiian Islands represent one of the most geographically remote locations in the Indo-Pacific, and are a refuge for many rare and endemic taxa. LaJeunesse et al (2004) surveyed the diversity of symbiotic dinoflagellates (*Symbiodinium sp.*) inhabiting zooxanthellate corals and other symbiotic cnidarians from the High Islands region of Hawaii. From the 18 host genera examined, there were 20 genetically distinct symbiont types. Most types were found to be associated with a particular host genus or species, and nearly half of them have not been identified in surveys of western and eastern Pacific hosts. A clear dominant generalist symbiont is lacking among Hawaiian cnidarians. This is in marked contrast with the symbiont community structures of the western Pacific and Caribbean, which are dominated by a few prevalent generalist symbionts inhabiting numerous host taxa. Geographic isolation, low host diversity and a high proportion of coral species that directly transmit their symbionts from generation to generation are implicated in the formation of a coral reef community exhibiting high symbiont diversity and specificity.

The ecoregional level

The important subject of connectivity among populations at the regional (ecoregional) level will be considered in Chapter 17. This is because genetics is only one of the contributing disciplines to the vital subject of connectivity, which also requires knowledge of life histories and physical oceanography.

Summary and management implications

Genetic information has become indispensible for marine conservation at all levels of the ecological hierarchy. The significance of the genetic diversity of marine species and their 'value' has yet to be fully appreciated. Genetic techniques allow verification of the identity of known species and the demographics of their populations, the discovery and systematic affinities of new species, and the movements and invasions of species into new habitats.

Many marine species of conservation concern are harvested commercially or illegally and thus economic, social, jurisdictional and forensic matters often arise in population management, in addition to biological considerations. For a diversity of marine taxa, molecular markers have uncovered previously unknown aspects of behaviour, natural history and population demography that can inform conservation and management decisions.

The information that genetic techniques has provided about species and their populations has dramatically changed our views about the fundamental units of conservation, including ESUs and MVPs and estimates of effective population size and health.

Arguments should not be concerned with whether we need to conserve species or spaces or genes, but rather how to combine these needs and approaches together. However, the most fundamental level for conservation is the genetic level. Genetic diversity is the net expression of the variability of life forms on the planet. Any reduction represents a loss of scope for future evolution, and a diminution of the potential of human resources. In order to be successful, conservation efforts must preserve the processes of life. This task requires the identification and protection of diverse branches in the tree of life (phylogenetics), the maintenance of life-support systems for organisms (ecology and the environment), and the continued adaptation of organisms to changing environments (evolution). None of these objectives alone is sufficient to preserve the threads of life across time.

The discipline of conservation biology has already incorporated many technologies to speed up and increase the accuracy of conservation decision-making, for example to characterize endangered species and their populations, or areas that contain endangered species. In the field of fisheries management, genetic aspects must be considered in management programmes in order to maximize probability of the long-term survival and continued adaptability of stocks. Technical advances will also allow for more precise conservation decisions to be made and, more importantly, will allow conservation genetics to contribute to biogeography, seascape level and ecosystem-based decision-making and planning of MPA networks.

There are perhaps few emergent generalizations that genetics can yet provide for marine conservation but perhaps two things can be said. First, loss of variation by genetic drift becomes a major consideration in populations below a critical size (e.g. Lacy, 1987). Second, at the ecosystem level we still have much to learn (in both genetic and ecological terms) about the balance between isolation of populations, which leads to speciation, and its 'opposite' – connectivity – which leads to the colonization of new habitats. In this second respect, the field of molecular genetics is now making major contributions to our understanding of global marine biogeography – how plants and animals move around the world (Riddle et al, 2008).

References

Alberte, R. S., Suba, G. K., Procaccini, G., Zimmerman, R. C. and Fain, S. R. (1994) 'Assessment of genetic diversity of seagrass populations using DNA fingerprinting: Implications for population stability and management', *Proceedings of the National Academy of Science*, vol 91, pp1049–1053

Allendorf, F. W. and Luikart G. (2007) *Conservation and the Genetics of Populations*, Blackwell, Malden, MA

Avise, J. C. (1992) 'Molecular population structure and the biogeographic history of a regional fauna: A case history with lessons for conservation biology', *Oikos*, vol 63, pp62–76

Avise, J. C. (2000) 'Cladists in wonderland', *Evolution*, vol 54, pp1828–1832

Baker, C. S., Gilbert, D. A., Weinrich, M. T., Lambertsen, R., Calambokidis, J., McArdle, B., Chambers, G. K. and O'Brien, S. J. (1993) 'Population characteristics of DNA fingerprints in Humpback whales (*Megaptera novaeangliae*)', *The Journal of Heredity*, vol 84, no 4, pp281–290

Bell, J. J. and Okamura, B. (2005) 'Low genetic diversity in a marine nature reserve: Re-evaluating diversity criteria in reserve design', *Proceedings of the Royal Society B*, vol 272, pp1067–1074

Bickford, D., Lohman, D. J., Sodhi, N. S., Ng, P. K. L., Meier, R., Winker, K., Ingram, K. K. and Das, I. (2007) 'Cryptic species as a window on diversity and conservation', *Trends in Ecology & Evolution*, vol 22, no 3, pp148–155

Bowen, B. W., Clark, A. M., Abreu-Grobois, F. A., Chaves, A., Reichart, H. A. and Ferl, R. J. (1997) 'Global phylogeography of the ridley sea turtles (*Lepidochelus spp.*) as inferred from mitochondrial DNA sequences', *Genetica*, vol 101, pp179–189

Bucklin, A., Wiebe, P. H., Smolenack, S. B., Copley, N. J., Beaudet, J. K., Bonner, K. G., Färber-Lorda, J. and Pierson, J. J. (2007) 'DNA barcodes for species identification of euphausiids', *Journal of Plankton Research*, vol 29, no 6, pp483–493

Clifford, S. L., McGinnity, P. and Ferguson, A. (1998) 'Genetic changes in Atlantic salmon (*Salmo salar*) populations of northwest Irish rivers resulting from escapes of adult farm salmon', *Canadian Journal of Fisheries and Aquatic Sciences*, vol 55, pp 358–363

Coffroth, M. A., Lasker, H. A., Diamond, M. E., Bruenn, J. A. and Bermingham, E. (1992) 'DNA fingerprints of a gorgonian coral: A method for detecting clonal structure in a vegetative species', *Marine Biology*, vol 114, pp317–325

Cracraft, J. (1989) 'Speciation and its Ontology: The empirical consequences of alternative species concepts for understanding patterns and processes of differentiation', in D. Otte and J. A. Endler (eds) *Speciation and Its Consequences*, Sinauer Associates, Sunderland, MA

Egidius, E., Hansen, L. P., Jonsson, B. and Nævdal, G. (1991) 'Mutual impact of wild and cultured Atlantic salmon in Norway', *ICES Journal of Marine Science*, vol 47, no 3, pp404–410

Frankham, R. (1995) 'Effective population size/adult population size ratios in wildlife: A review', *Genetical Research*, vol 66, pp95–107

Fraser, D. J. and Bernatchez, L. (2001) 'Adaptive evolutionary conservation: Towards a unified concept for defining conservation units', *Molecular Ecology*, vol 10, pp2741–2752

Grant, W. A. S. and Bowen, B. W. (1998) 'Shallow population histories in deep evolutionary lineages of marine fishes: Insights from sardines and anchovies and lessons for conservation', *Journal of Heredity*, vol 89, pp415–426

Hebert, P. D., Penton, E. H., Burns, J. M., Janzen, D. H. and Hallwachs, W. (2004) 'Ten species in one: DNA barcoding reveals cryptic species in the neotropical skipper butterfly *Astraptes fulgerator*', *Proceedings of the National Academy of Sciences*, vol 101, no 41, pp14812–14817

Hobbs, R. J. and Mooney, H. A. (1998) 'Broadening the extinction debate: Population deletions and additions in California and Western Australia', *Conservation Biology*, vol 12, pp271–283

Hoelzel, A. R. (2008) 'Impact of population bottlenecks on genetic variation and the importance of life-history: A case study of the northern elephant seal', *Biological Journal of the Linnaean Society*, vol 68, no 12, pp23–39

Hoelzel, A. R., Natoli, A., Dahlheim, M. E., Olavarria, C., Baird, R. W. and Black, N. A. (2002) 'Low worldwide genetic diversity in the killer whale (*Orcinus orca*): Implications for demographic history', *Proceedings of the Royal Society of London B*, vol, 269, no 1499, pp1467–1473

Hughes, A. R. and Stachowicz. J. J. (2004) 'Genetic diversity enhances the resistance of a seagrass ecosystem to disturbance', *Proceedings of the National Academy of Sciences*, vol 101, pp8998–9002

Hunter, C. L. (1993) 'Genotypic variation and clonal structure in coral populations with different disturbance histories', *Evolution*, vol 47, no 4, pp1213–1228

Johannesson, K. and André, C. (2006) 'Life on the margin: Genetic isolation and diversity loss in a peripheral marine ecosystem, the Baltic Sea', *Molecular Ecology*, vol 15, pp2013–2029

Karl, S. A. and Bowen, B. W. (1999) 'Evolutionary significant units versus geopolitical taxonomy: Molecular systematics of an endangered sea turtle (Genus *Chelonia*)', *Conservation Biology*, vol 13, pp990–999

Koehn, R. K. and Gaffney, P. M. (1984) 'Genetic heterozygosity and growth rate in *Mytilus edulis*', *Marine Biology*, vol 82, no 1, pp1–7

Lacy, R. C. (1987) 'Loss of genetic diversity from managed populations: Interacting effects of drift, mutation, immigration, selection, and population subdivision', *Conservation Biology*, vol 1, pp143–158

LaJeunesse, T. C., Thornhill, D. J., Cox, E. F., Stanton, F. G., Fitt, W. K., Schmidt, G. W. (2004) 'High diversity and host specificity observed among symbiotic dinoflagellates in reef coral communities of Hawaii', *Coral Reefs*, vol 23, pp596–603

May, R. M. (1990) 'Taxonomy as destiny?', *Nature*, vol 347, pp129–130

Mayr, E. (1963) *Animal species and evolution*, Belknap Press, Cambridge, MA

Moritz, C., Lavery, S. and Slade, R. (1995) 'Using allele frequency and phylogeny to define units for conservation and management', *American Fisheries Society Symposium*, vol 17, pp249–262

Palstra, F. P. and Ruzzante, D. E. (2008) 'Genetic estimates of contemporary effective population size: What can they tell us about the importance of genetic stochasticity for wild population persistence?', *Molecular Ecology*, vol 17, pp3428–3447

Palstra, F. P., O'Connell, M. F. and Ruzzante, D. E. (2009) 'Age structure, changing demography and effective population size in Atlantic salmon (*Salmo salar*)', *Genetics*, vol 182, pp1233–1249

Peare, T. and Parker, P. G. (1996) 'Local genetic structure within two rookeries of *Chelonia mydas* (the green turtle)', *Heredity*, vol 77, pp619–628

Pegg, G. G., Sinclair, B., Briskey, L., and Aspden, W. J. (2006) 'MtDNA barcode identification of fish larvae in the southern Great Barrier Reef, Australia', *Scientia Marina*, vol 70, pp7–12

Pennock, D. S. and Dimmick, W. W. (1997) 'Critique of the evolutionarily significant unit as a definition for "Distinct Population Segments" under the U.S. Endangered Species Act', *Conservation Biology*, vol 11, pp611–619

Reed, D. H. and Frankham, R. (2003) 'Correlation between fitness and genetic diversity', *Conservation Biology*, vol 17, pp230–237

Riddle, B. R., Dawson, M. N., Hadly, E. A., Hafner, D. J., Hickerson, M. J., Mantooth, S. J. and Yoder, A. D. (2008) 'The role of molecular genetics in sculpting the future of integrative biogeography', *Progress in Physical Geography*, vol 32, pp173–202

Roman, J. and Bowen, B. W. (2000) 'The mock turtle syndrome: Genetic identification of turtle meat purchased in the south-eastern United States of America', *Animal Conservation*, vol 3, pp61–65

Ruiz, G. M, Carlton, J. T., Grosholz, E. D. and Hines, A. H. (1997) 'Global invasions of marine and estuarine habitats by non-indigenous species: Mechanisms, extent, and consequences', *American Zoologist*, vol 37, no 6, pp621–632

Ryder, O. A. (1986) 'Species conservation and systematics: The dilemma of subspecies', *Trends in Ecology & Evolution*, vol 1, pp9–10

Simpson, G. G. (1961) *Principles of Animal Taxonomy*, Columbia University Press, New York

Soulé, M. E. (1976) 'Allozyme variation: its determinants in space and time', in F. Ayala (ed) *Molecular Evolution*, Sinauer Associates, Sunderland, MA

Soulé, M. (1987) *Viable Populations for Conservation*, Cambridge University Press, Cambridge

Suatoni, E., Vicario, S., Rice, S., Snell, T. and Caccone, A. (2006) 'An analysis of species boundaries and biogeographic patterns in a cryptic species complex: The rotifer – *Brachionus plicatilis*', *Molecular Phylogenetics and Evolution*, vol 41, pp86–98

Turnera, T. F., Waresa, J. P. and Gold, J. R. (2002) 'Genetic effective size is three orders of magnitude smaller than adult census size in an abundant, estuarine-dependent marine fish (*Sciaenops ocellatus*)', *Genetics*, vol 162, pp1329–1339

Utter, F. and Epifanio, J. (2002) 'Marine aquaculture: Genetic potentialities and pitfalls', *Reviews in Fish Biology and Fisheries*, vol 12, pp59–77

Van Valen, L. (1976) 'Ecological species, multi-species and oaks', *Taxon*, vol 25, pp233–239

Vermeij, G. J. and Roopnarine, P. D. (2008) 'In a future warmer climate, mollusks and other species are likely to migrate from the Pacific to the Atlantic via the Bering Strait', *Science*, vol 321, pp780–781

Waples, R. S. (1998) 'Evolutionary significant units, distinct population segments, and the Endangered Species Act: Reply to Pennock and Dimmick', *Conservation Biology*, vol 12, pp718–721

Ward, R. D., Zemlak, T. S., Innes, B. H., Last, P. R. and Hebert, P. D. N. (2005) 'DNA barcoding Australia's fish species', *Philosophical Transactions of the Royal Society of London. Series B, Biological Sciences*, vol 360, pp1847–1857

Williams, S. L. (2001) 'Reduced genetic diversity in eelgrass transplantations affects both population growth and individual fitness', *Ecological Applications*, vol 11, pp1472–1488

11

Coastal Zones

Components, complexities and classifications

Thanks to the Interstate Highway System, it is now possible to travel from coast to coast without seeing anything.

Charles Kuralt (1943–1997)

Introduction

The coastal zone is the part of the ocean most familiar to us. It is the most complex of all marine environments, and certainly the most heavily utilized by humans. Yet the coastal zone is perhaps more difficult to actually define and study than any other part of the oceans. There are now several recent texts on the coastal zone, with emphasis on some mix of science (ecology and environment) and management or conservation. Among others, readers are recommended to see Mann (2000), Salm et al (2000), Ray and McCormick-Ray (2004), Beatley et al (2002) and Barnabé and Barnabé-Quet (2000). Integrated coastal zone management (ICZM) or integrated coastal management (ICM) is a process that tries to combine all disciplines and approaches to the coastal zone, in attempts to achieve sustainability. Unfortunately, such efforts have not met with notable success around the globe. However, in approaching the process of planning for conservation of the coastal zone, again a prime prerequisite is a recognition and classification of its biodiversity components, and the natural processes that shape and define them. This is the prime subject of this chapter.

Definition of the coastal zone

A variety of terms and definitions have been used to describe the areas of the coastal realm or coastal zone, and some of these are illustrated in Figure 11.1. However, there is still little agreement as to where the coastal zone even begins or ends. Some definitions of the coast: 'The general area between land and sea', or coastal zone: 'The interface between land and water', are too vague. Other definitions are too inclusive: 'The regions of the continental shelves up to and including the watersheds whose drainage influences the region.' The last definition would encompass most of the neritic zone and indeed much of the terrestrial environment.

There is an obvious need for clear operational definitions, because the coastal zone is where humans most strongly interact with the oceans. The *Encyclopedia of Earth* gives the following definition:

> *The coastal ocean is the portion of the global ocean where physical, biological and biogeochemical processes are directly affected by land. It is either defined as the part of the global ocean covering the continental shelf or the continental margin. The coastal zone usually includes the coastal ocean as well as*

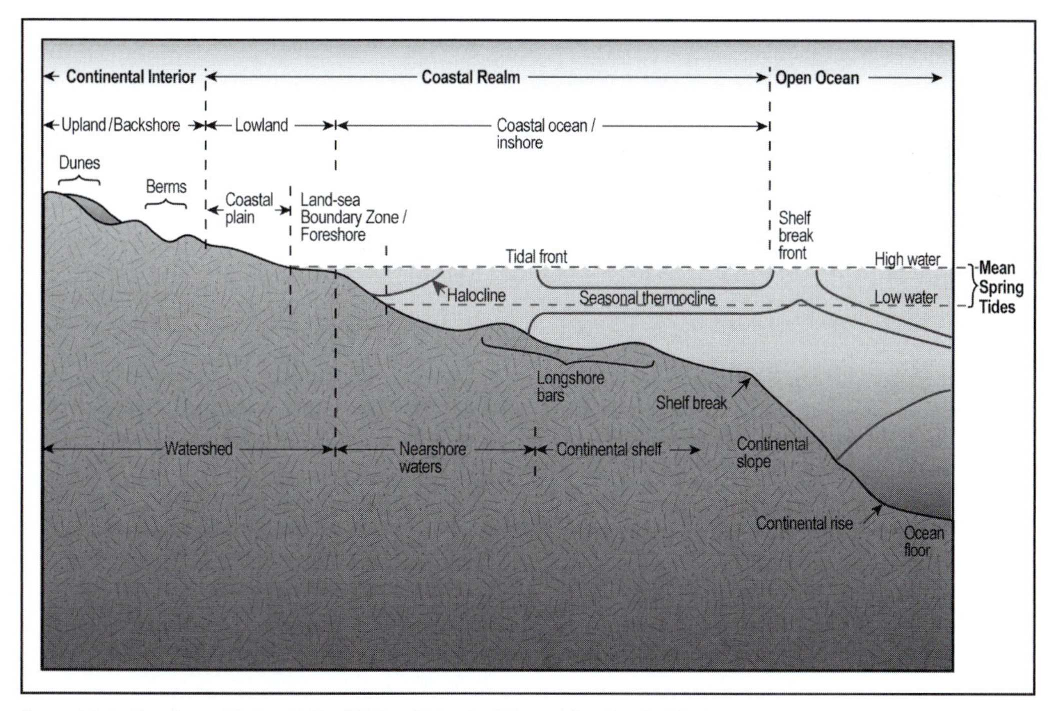

Source: Adapted from Ray and McCormick-Ray (2004) and University of Liverpool *Encyclopedia of Earth*

Figure 11.1 Showing various definitions and extents of the coastal realm and coastal zone

> *the portion of the land adjacent to the coast that influences coastal waters. It can readily be appreciated that none of these concepts has a clear operational definition.*

Again, by following this definition, the coastal zone would seem to include the entire continental shelf and any land (all of any island, and most of any continental land mass) that drains into the oceans.

More realistically from an ecological perspective, and more restrictively, the coastal zone might be defined by: seaward – the extent of the fringing communities (i.e. the depth limit of the benthic euphotic zone, some 30–50m in temperate waters); and landward – by the extent of beaches, cliffs and headlands (i.e. a terrestrial morphological limit), and by wetlands associated with bays and estuaries to the null zone or 'zero' salinity point (i.e. a marine influence limit). It is perhaps easier to define the limits of the sea on the land than vice versa, but these suggested limits would define the major types of influences (especially trophic and energetic) within the zone. Seaward of this zone lies the neritic province and then the open ocean. Landward of this zone lies the terrestrial environment and freshwater ecosystems. In this chapter, we shall adopt this approximate definition of the coastal zone.

Why the coastal zone should be treated separately

The coastal zone (however defined) deserves to be considered separately from other parts of the marine environment for several reasons, first because of its inherent complexity. In addition to its identity problem, the coastal zone also suffers from a form of multiple personality disorder because of the many influences that impinge upon it. The coastal zone consists of a greater array

Table 11.1 Complexities of the coastal zone that can be defined and measured

STRUCTURES
Latitude and seasonal cycles
Depth, area, volume, shape, slope, rugosity
Shoreline length, shoreline slope and profile, complexity
Watershed catchment properties
Topographic variability and environmental heterogeneity:
vertical
horizontal
Substrate, sediment type, particle size
Geological formation type
Temperature
Salinity
Water turbidity and chlorophyll concentrations
Nature and distributions of primary producer communities
Degree of exposure to:
ocean waves and currents
atmospheric storms
Geomorphological units, biotopes and habitat types: see Table 11.3
PROCESSES
Tidal regime
Local currents, entrainment and retention mechanisms
Wind stress and direction
Storms
Erosion/deposition regimes
Upwelling locations
Temperature and salinity variations
Seasonal stratification
Land drainage
Annual, seasonal productivity regime, dominance of pelagic or benthic realm
HUMAN
Fishing gears
Engineering construction/shoreline stabilization
Eutrophication nutrients
Water supply/elimination
Waste disposal
Resource use:
sand gravel extraction
fisheries
mangrove use
coral use
Drainage modification, saline intrusion into aquifers
Tourism
Wetland degradation and reclamation
Wildlife exploitation
Climate change effects
Erosion/deposition

Table 11.2 Types of coasts

Coastal types:
collision coasts
trailing edge coasts
marginal coasts
Mountain coast
Narrow shelf:
headlands and bays
coastal plain
Wide shelf:
headlands and bays
coastal plain
Deltaic coast
Reef coast
Glaciated coast
Ria coast
Fjord coast
Unglaciated lowland coast

Note: The type of coast significantly determines sources, sinks and types of sediments and their dynamics and distributions. This in turn largely determines the types and distributions of biological community types. Note that many other terms are used to describe the geology and geomorphology and evolution of coasts

Source: After Carter (1988) and Guilcher (1958)

of biodiversity structures (both abiotic and biotic), and is subject to a greater array of processes (both abiotic and biotic) than any other part of the marine environment. It also demonstrates greater variability of its structures and processes (for example, temperature, salinity, productivity regimes), in both time and space, than anywhere else in the oceans. Tables 11.1, 11.2, 11.3 and 11.4 summarize some aspects of these components of the complexities and variability of the coastal zone.

Coastal zones are of various types (Table 11.2); they have attained their characteristics by a variety of geological processes and interactions, and undergo various stages in their evolution (see Carter, 1988). Regions of the coastal zone may be dominated by one or more of a set of physical processes, on a range of time scales, including wave activity, tidal range and currents, and river inputs (Figures 11.2 and 11.3). The coastal zone is in a constant state of flux and modification over time (see Figure 11.3), and is subject to a variety of influences. Periodic or aperiodic catastrophic events, such as storms and hurricanes, can have particularly profound and unpredictable

Table 11.3 Selected geomorphic and biogenic units and habitats of the coastal zone

Zone	Geomorphic unit		Primary biotope or biogenic unit	Habitat type
Supra-tidal (backshore)	Rocky headlands/cliffs			Rocky shore
	Beaches			
	Coastal dunes		Coastal dune	
	Lagoon			
	Marsh			
	Sand bars/spits			
	Barrier islands			
Supra/intertidal	Salt marsh		Salt marsh	Salt marsh
	Delta		Mangroves	
Intertidal (foreshore)	Rocky shore			Tidal pools
	Sandy beach			
	Mud-silt flats			
Intertidal/sub-tidal	Inlets	Coves	Rocky, sandy shores	Rocky, sandy shores
		Bays	Sand, mud	Sand, mud
		Estuaries	Sand, mud silt	Sand, mud silt
	Sandy banks			
Sub-tidal (offshore)	Ridges		Coral reef	Fringing
	Reefs			Barrier
	Banks			Atoll
	Gullies		Kelp forest	
	Canyons		Seagrass beds	
	Sea Caves			

Note: Note that the identical terminology may be used to describe a geomorphic unit, a biogenic unit or a habitat type
Source: Compiled from a variety of sources and not intended to be comprehensive

effects, of much greater significance than seasonal or annual events, that may locally or regionally change the entire character of the coastal zone. In this respect, any conservation initiative in the coastal zone should consider the time-frame, or longevity, of the geomorphological structures being evaluated. Questions related to how coasts originate, how likely they are to evolve and how long they will last are vital to considerations of regional ecological significance.

The second reason to consider the coastal zone separately is because it is, uniquely, the interface between the marine and terrestrial parts of the biosphere (Figure 11.2). Interfaces are always of much greater ecological and biological significance than bulk properties, and the land–sea interface is one of the most significant of our planet. The coastal zone is influenced by both marine and terrestrial processes, and it is here that exchanges important to each part of the biosphere occur. Such exchanges include the important fluxes of nutrients in various dissolved and particulate forms and the dispersals and migrations of organisms, mediated by the unique organisms of the fringing communities. The combination of physical processes leads to a generally convoluted shoreline consisting of headlands and inlets of varying substrate type and size, which in turn determine the type of biological community of an area (Figure 11.4).

A third (but perhaps most important) reason is that the coastal zone contains ALL the types of marine primary producers in its fringing communities, substrates and open waters. Here are found the following communities of producers: phytoplankton; benthic microphytes of mud-flats; benthic microphytes of rocky shores; macrophytic algae of rocky shores; seagrasses; salt-marsh plants; mangrove forests; and coral reefs. In combination, these communities of producers make the coastal zones the most productive regions of the oceans in terms of carbon fixation and fisheries (see Table 11.4).

A fourth reason to consider the coastal zone separately is that it is the major site of human

Table 11.4 Annual rates of production for communities of the coastal zone

Location/community type	Production rate (net) gC m^{-2} $year^{-1}$
Open waters	
ocean waters	50–150
upwelling zones	200–2000
shelf waters	100–300
coastal bays	50–4000
surf zone	30–500
Sub-tidal	
seaweeds	1000–1800
coral reefs	500–5000
seagrasses	120–800
Intertidal	
rockweeds	300–600
sandy beaches	20–30
estuarine mud-flats	25–500
Supra-tidal	
salt marshes	500 - 3700
mangroves	200 - 1200
sand dunes	150 - 175

Source: Adapted from Carter (1988), Mann (2000) and Barnabé and Barnabé-Quet (2000)

interactions with the oceans, and the place where our activities have their greatest impacts. This is a region of major physical and biological resources for humans.

General descriptions and conventional classifications of the coastal zone

Accurate description and mapping of the coastal zone has always depended heavily on aerial photography; this technique enabled recognition of coastal types, description of the variety of coastal structures, and interpretation of the geological and oceanographic forces shaping them (e.g. McCurdy, 1947). For conservation purposes, there followed various types of descriptions of coastal areas, making use of a combination of topographic land maps and coastal oceanographic charts.

Because of the nature of its complexity, the coastal zone can be described in many ways. For example, it can be described and analysed in terms of its geological formations and geomorphology (e.g. Carter, 1988), or the ecology of its component ecosystems' habitats and communities (e.g. Mann, 2000; Knox, 2001; Madden et al, 2005), or its environmental characteristics and their conservation (e.g. Ray and McCormick-Ray, 2004). Specific types of geomorphological structures (e.g. estuaries) can be described in quantitative terms from a combination of tidal activity and freshwater runoff (i.e. the opposing influences of sea and land, see Hansen and Rattray, 1966). From an ecological perspective, for analytical and conservation purposes and in terms of broad-scale availability of data, the most useful descriptions are made in terms of its geophysical components, in order to describe and map the types of coastal landscapes and seascapes, or geomorphological units.

As for offshore regions there are two ways to proceed with analysis – by direct mapping of geomorphological units and their habitats, or by using geophysical data to reconstruct coastal landscapes or seascapes. Which approach is used depends on the size of the ecoregions under study, the availability of data and the nature of the fringing communities. However, any classification of the coastal zone structures should recognize the self-evident and natural geomorphological features themselves. Unfortunately, there is little standardized terminology of structures in the coastal zone, and terminologies at the various 'levels' of the ecological hierarchy begin to converge or overlap. Terms used depend partly upon the spatial scale of investigation, and partly on the preference of individual authors. Thus a salt marsh could be considered as an ecosystem, a geomorphic unit or biogenic unit, a primary biotope or a habitat (see Table 11.3).

Direct mapping of coastal habitats can mean extensive aerial or ground surveys, and these have been the conventional approaches. This is quite feasible for shorelines of a few kilometres, but it involves excessive work for hundreds or thousands of kilometres, where more indirect techniques must be used. In tropical and subtropical regions, the extent and nature of biogenic communities (or foundation species or biogeomorphological units)

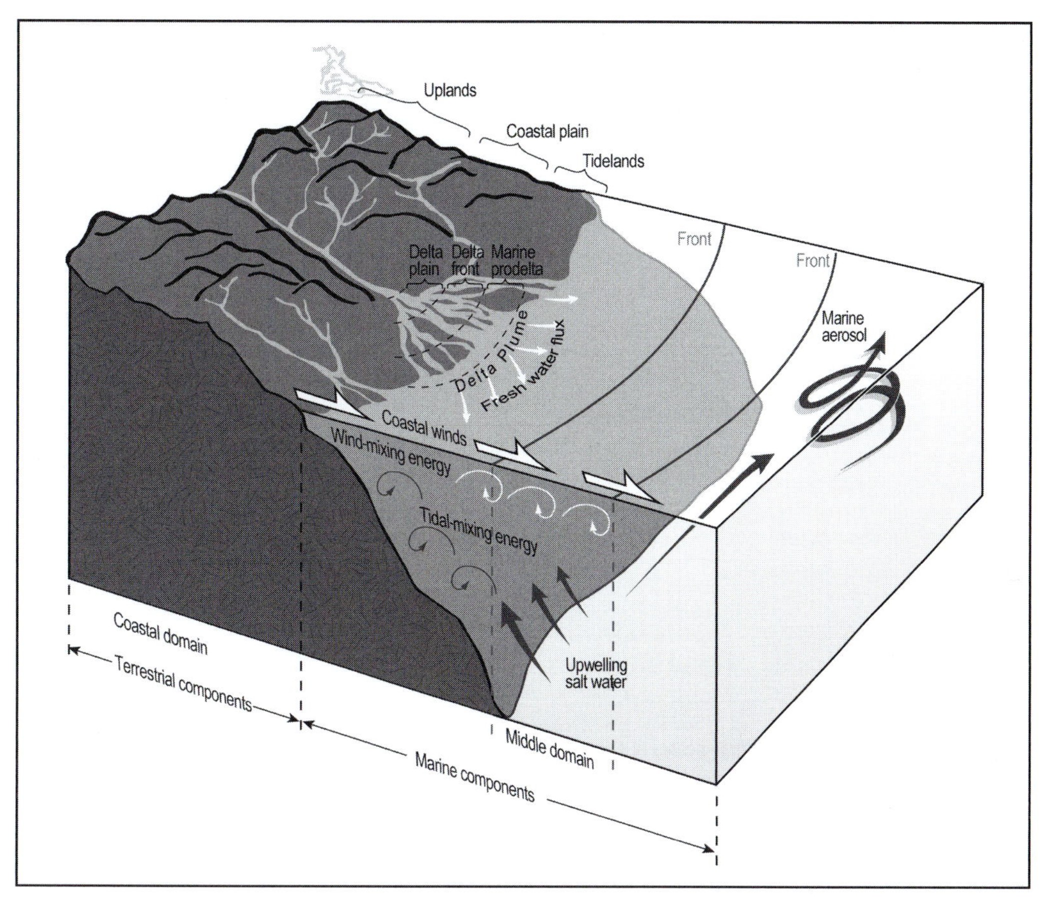

Source: Adapted from Ray and McCormick-Ray (2004)

Figure 11.2 Some features (structures and processes) of the coastal realm indicating its complexity

such as coral reefs, seagrass beds and mangrove forests will generally be well mapped. In temperate regions, direct survey of sub-tidal habitats is much more demanding, requiring in situ surveys using either diver effort or submarine photography and video camera survey, or multibeam and sonar techniques.

However, despite considerable recent attention to the coastal zone, especially in terms of attempts at integrated management, there is still little indication of any comprehensive analyses of available ecological and environmental information. Nor has there been progress towards defining how to develop coastal networks of marine protected areas (MPAs) through inclusion of representative or distinctive coastal areas. Even where protected areas currently exist, analyses of their contribution to a network of protected areas has rarely been undertaken, and where this has been done, results indicate that we are far from achieving such networks (see e.g. Johnson et al, 2008).

Geographic information system (GIS) studies of coastal inlets and new methods of coastal zone classification

The level of conventional descriptive analysis summarized above is not sufficient for coastal zone planning or to define representative and

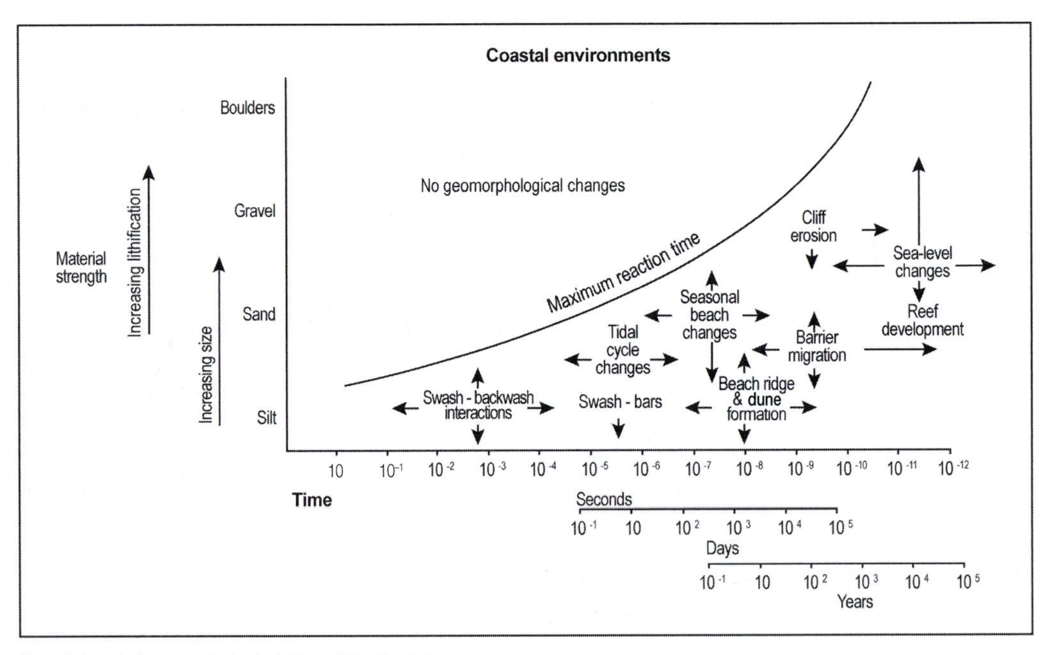

Note: Ordinate indicates nominal material 'strength' and particle sizes

Source: Adapted from Carter (1988)

Figure 11.3 Schematic view of the time scales of coastal processes

distinctive areas. Where any given geomorphological unit or habitat type is unique within an ecoregion (e.g. a spatially delimited mangrove forest, estuary etc. without replicate), it may be considered distinctive (*sensu* Roff and Evans, 2002). However, where there are multiple such units with an ecoregion (e.g. bays, wetlands, coral reefs etc.), without further spatial analysis we do not know how they may vary in habitat and community composition, nor do we know the proportions of different habitats they may contain. Thus we have no idea of which ones may be considered as representative of the set of units. This requires two levels of analysis: one of the units themselves and a second of individual units for calibration and definition of habitat types in relation to coastal processes (levels 5 and 6 in Table 5.1). Without such analysis of the ecological components of a region, we cannot be sure of developing an effective conservation strategy that will capture the maximum number of components of biodiversity within an ecoregion.

Even when classifications are based on some combination of geomorphological/environmental and ecological features (e.g. the ENCORA system, Madden et al, 2005), we have still only assembled lists of types of structures and units. There is still no fundamental understanding of the quantitative representation of habitat types within those units. Thus for example, although we may know the numbers of estuaries, we do not know that any kind of estuary contains the same, similar or same extent of any particular type of habitat. Only quantitative categorization of the geomorphological types within the coastal zone can lead to quantitative inventories of their component habitat types.

Development of GIS techniques and their increasing applications to oceanography and fisheries (e.g. Valavanis, 2002) have now permitted the quantitative geographic description of the coastal zone, and assessment of its ecological and environmental structures. This has facilitated the development of many resource inventories

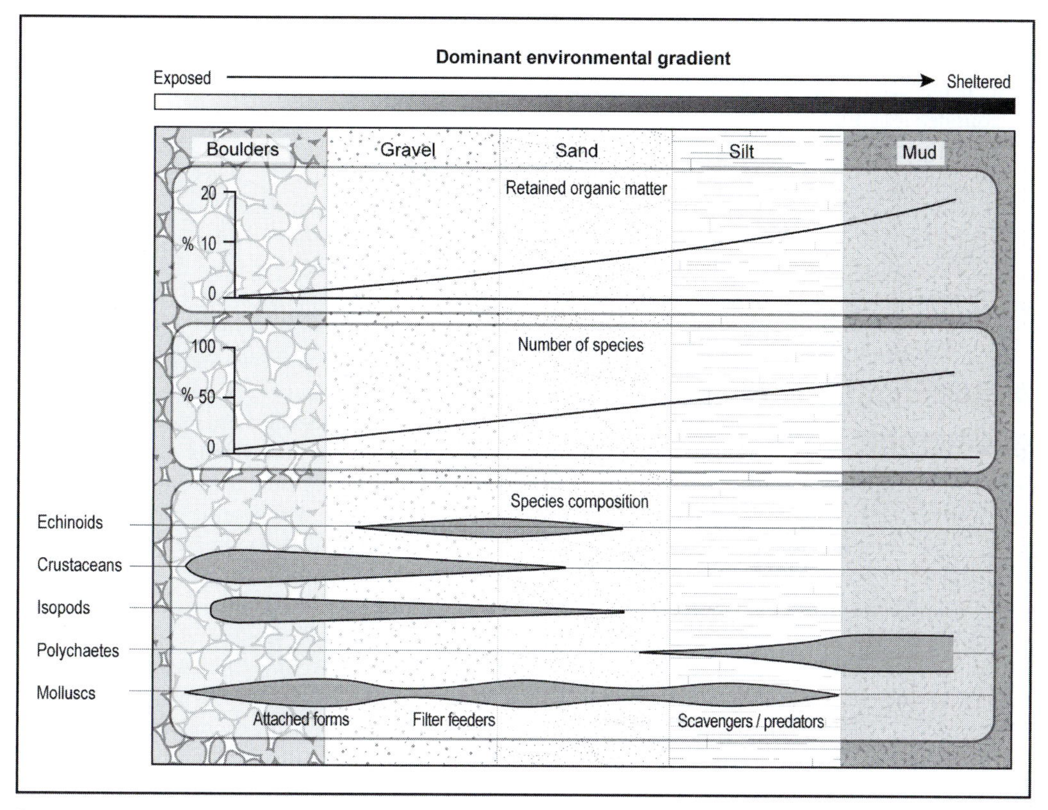

Source: Adapted from Carter (1988)

Figure 11.4 The dominant structural features and environmental gradients of beaches, from exposure to shelter, showing particle sizes, retained organic matter, and species diversity and composition

and conservation initiatives. For example, extensive coastal mapping and planning for conservation at the national and regional levels has been carried out in the National Marine Bioregionalization study of Australia (Commonwealth of Australia, 2005), and in the analysis of protected sites in the European NATURA 2000 inventory (see e.g. Boedeker and von Nordheim, 2002). Quantitative descriptions of specific types of coastal geomorphological structures have also been undertaken at the national or regional level, for example in the inventory and description of sea lochs in west Scotland (Dipper et al, 2007). Habitat suitability models have also been developed in coastal waters for a variety of species, generally those of commercial interest (e.g. Brown et al, 1997).

Spatial analysis and planning, using GIS techniques, has now become indispensible to all initiatives in the coastal zone, whether for development or conservation. GIS techniques have generally been used to map geomorphological units or 'ecosystems' within the coastal zone and the component habitat types within individual such units. However, further analysis or complete surveys of geomorphological units and their habitat types within an entire ecoregion have generally not been attempted. Yet this is precisely what is required if we are to define representative habitat types.

In order to do this, we need to define the habitat structures (characteristics) and analyse the processes acting upon them. Within an ecoregion, the outcome of such an analysis and classification can now tell us which geomorphic units may be representative of a region, and which – because of their unusual or anomalous structures – may be considered as outliers of a group and therefore be considered as potentially 'distinctive' areas, perhaps with a unique array of habitat types. In fact, such analysis actually affords a means of defining distinctive coastal regions – which may harbour components of biodiversity not found in the more average representative members.

A classification of coastal geomorphic units

In this section, the coastal nearshore zone is defined as the region between the highest high water mark and a line joining coastal headlands. This region therefore includes all types of inlets, and the headlands themselves. Within the coastal zone, geomorphic units are those repeating units of similar type and with comparable environmental characteristics and array of habitats. Here, landscape/seascape types are the geomorphic units, comprising various types of inlets (e.g. bays, coves and estuaries) and headlands. Headlands in temperate regions tend to be rather ecologically uniform – although modified in their community composition as a function of exposure (see Chapter 8).

In a single chapter on the coastal zone, we cannot hope to deal with all its complexities and all possible approaches to conservation. Here we present new techniques for quantitative analysis of the inlets of a coastal zone using available geophysical information (Greenlaw, Roff and Redden, unpublished). From this analysis we can begin to assess the proportions of different types of habitats in each inlet type, although at the level of the entire geomorphic unit we may not know how to describe each of them precisely. After this, we present techniques for further analyses of selected inlets that analyse the distributions of the component habitats themselves, following Valesini et al (2010).

Previous classifications of coastal inlets (see e.g. Hansen and Rattray, 1966; Heath, 1976; Cowardin et al, 1979; Hume, 1988; Gregory et al, 1993; Cooper, 2001; Ryan et al, 2003; Engle et al, 2007; Hume et al, 2007) have rarely focused on developing a system that is representative of their biological patterns. A planning technique of representative area mapping, similar to that of Roff and Taylor (2000) can be applied to inlets by mapping quantitative geophysical variables that shape the nearshore environment to produce inlet types. The representative area approach (Chapter 5) focuses on ecosystem biodiversity through the conservation of ordinary habitats occupied by the majority of communities. The approach uses physical environmental variables as descriptors of habitat types and as surrogates for biological variables. This approach is taken not only because community level data is normally very limited, but because it is truly impossible to sample all species within an ecosystem. Areas of similar geophysical character are then delineated to produce a classification that is expected to predict biological communities, and which may then be validated by subsequent surveys.

The definition of inlet adopted was the shoreline, benthic environment and associated marine environment within any stretch of shoreline that is measurably concave, from mean high water to the straight line distance passing between two headlands at least 500m apart. Each inlet will contain an array of primary habitats (Tables 5.1 and 11.3) and community types. The proportion of each of these primary habitats within each inlet should be related to its type, as designated by the geophysical classification created. For example, an inlet with low average exposure will have a higher proportion of primary habitats with soft substrates and benthic community types with large proportions of infauna.

The geophysical classification created is also expected to encompass both α-diversity (species richness) and β-diversity (change in species composition along a spatial dimension) (Izsak and Price, 2001). For instance, an inlet with a large watershed area to tidal volume ratio that is largely sheltered from the open ocean is likely to contain typical estuarine habitat with large amounts of sandy to silty sediment, with decreasing salinity towards the head of the estuary and low physical disturbance from wave energy. These conditions, altered from the typical offshore environment, should also predictably alter diversity and abundance patterns within the inlet. In such a highly estuarine environment, high β-diversity is expected due to the additional presence of species with the ability to tolerate lowered salinity. Species diversity patterns will also be altered due to the sheltered environment where low physical disturbance will create conditions where species diversity may be low but species abundances may be high (Wilkinson, 1999). Such an environment, primarily composed of fine sediments, would accommodate benthic invertebrates primarily of deposit feeder functional groups.

Study area

The study area where the classification was applied has one of the most folded and indented coasts anywhere in the world. It is located almost completely within the nearshore Scotian Shelf marine ecoregion between 43 and 47 degrees N, and includes over 4000km of heterogeneous shoreline of mainland Nova Scotia (see Figure 16.1a for location).

Physical conditions in the study area are a sharp contrast to the surrounding bodies of water such as the Gulf of St. Lawrence and the Bay of Fundy which are sheltered from ocean swells. Most of the study area has practically unlimited fetch at the mouth of the numerous embayments and estuaries which contain many types of sediment. This high exposure, along with the continuing geological sea-level rise, contributes to the highly crenulated shoreline and high number of inlets.

River flow into the Atlantic Nova Scotian estuaries is not strong enough at any time of year to develop salt wedge estuaries. Mostly, well mixed estuaries are found on the coast except during periods of spring runoff and heavy rainfall (Gregory et al, 1993). Nova Scotia's estuaries are influenced by tide and wind-driven currents with salinity progressively decreasing towards the head of the estuary (Davis and Browne, 1996).

The nearshore stretch has heterogeneous tidal influences with semi-diurnal tides. Winds are primarily from the southwest in summer and from the west/northwest in winter, although circulation in the nearshore is dominated by the southwestward Nova Scotia Current. The Nova Scotia Current, tidal mixing, topographic upwelling and wind-driven nearshore upwelling all contribute to persistent nearshore temperature and salinity patterns.

Selection of geophysical variables that structure nearshore communities

What is required for the coastal zone is a classification that is based on well defined criteria and quantified characteristics that will lead to separation of coastal geomorphological units which possess clearly differentiated arrays of primary habitats. In order to achieve this it is necessary to ask two questions. First: what are the interactions between structures and processes that determine the essential character (i.e. the set of primary habitats) of coastal geomorphological units? Second: what data is available to characterize these units so as to differentiate them according to question one? Inevitably there will be some compromises or requirements for surrogate variables because of lack of available data.

Important geophysical variables that are generally believed to determine nearshore habitat types were identified by literature review (Table 11.1). One or two physical variables from each category were chosen to be included in the classification, in order to eliminate redundancy and auto-correlation among variables, according to three criteria: those that have been shown to be most influential to community patterns; those that best discriminated between inlet types (using a principal components analysis); those for which good data was available over the entire study area.

Some of the physical variables chosen have not previously been used in a classification of coastal inlets, even though it has been shown that these factors influence community patterns in nearshore areas. These factors include: the measurement of fetch at each 20m grid square throughout the inlet, shoreline sinuosity, benthic complexity and the diversity of physical categories of depth and fetch.

There were also factors that could not be included in the classification because of data-gaps; these included nature of the sub-tidal substrate, temperature at the inlet scale, local currents, nutrient concentrations, turbidity (or water transparency), location and regime of upwelling areas and the presence of a sill at the mouth of an inlet. Satellite turbidity measurements are available, but are only currently appropriate for offshore analyses (Selvavinayagam et al, 2003). In many cases the geological history has a large effect on the coastal substrate type, especially of estuaries. However, it is to be expected that very fine sediments will rarely be found in any of the cove inlet types, and mud will have a higher frequency of occurrence in estuaries and bays. Upwelling areas and the presence of a sill were not included in the classification as it was too difficult to include them as a quantitative factor. However, the presence of a sill may limit the exchange of water, and can result in anoxic conditions. This factor could be mapped and overlaid on to the classification as a distinctive factor, along with others that could not be included quantitatively – that is, upwelling zones and structural anomalies. The classification also excluded one inlet type along the shoreline – lagoons. Lagoons have very different properties that are influenced by small tidal exchanges (Gonenc and Wolflin, 2004). There are lagoon-like estuaries in the classification at the far end of the estuarine spectrum, but lagoons that are currently closed off from tidal exchange were not included.

It should be appreciated that the selection of variables used here was made for a particular region and was limited by data availability. In other regions, other factors may or may not be available, but basic hydrographic chart information forms the basis of the classification proposed here, and this is available worldwide.

Inlet classification framework

The physical variables chosen to be included in the classification were separated into three categories based on how they influence habitat patterns, and based on underlying concepts of the 'natural' divisions of the coastal zone. They should also be intuitively 'obvious' to environmental managers and even the general public. In contrast, the classification of representative areas in offshore regions (Roff and Taylor, 2000; Roff et al, 2003; Chapter 5; and see also CLF/WWF, 2006; Chapter 16), resulted in defining seascapes which, although also ecologically and environmentally 'natural', have no directly obvious meaning and perhaps have little intuitive appeal for management purposes. The three primary categories for coastal inlets were therefore selected so as to make the resultant classification easier to understand for management purposes.

The three categories chosen to describe inlets were: the relative influence of land versus sea, as morphological/hydrographic type; the dominant primary productivity regime – pelagic versus benthic; the physical complexity of the inlet – from simple to complex – which acts as a predictor of habitat diversity (Figure 11.5).

Level 1: Morphological/hydrographic inlet type

The first of the three categories, the morphological/hydrographic inlet type, describes the major geomorphological and hydrographic characteristics of inlets resulting from a combination of environmental structures and processes that are established at the land/sea interface. It separates them into the three inlet archetypes: bays, estuaries and coves (Figure 11.5 and Colour plate 15a). The dominant character of the coastal zone is set

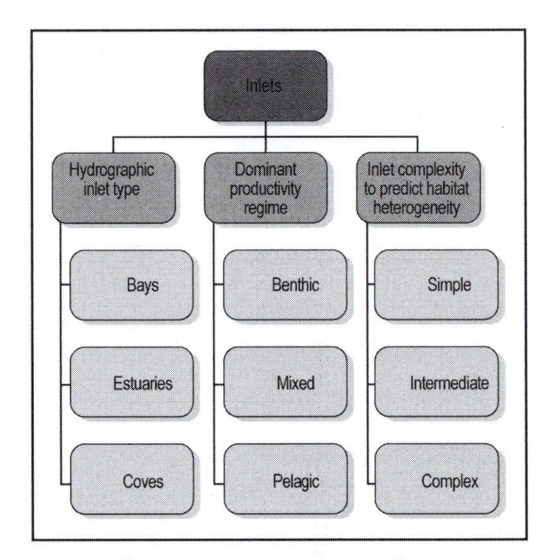

Figure 11.5 Overview of the hierarchy of inlet types according to Greenlaw and Roff (unpublished), classified according to: hydrography and topography; dominant productivity regime; inlet complexity

by the fact that it is here that sea and land meet and interact. The local character of the coastal zone must therefore be determined by which of these major environments predominates at any given location. Inlets are where shelter from open ocean conditions begins to have an effect in the nearshore, providing a physical environment that is altered from the open ocean. This is where the inclusion of factors such as wave exposure and freshwater input becomes necessary.

Morphological/hydrographic inlet types were defined by two variables: freshwater runoff from a defined watershed, and shelter – the proportion of the inlet that is sheltered from open ocean conditions (see Box 11.1). Bays were defined by their variable degree of shelter, and a freshwater per tidal cycle to tidal prism value of close to zero. Coves were defined by their low proportion of shelter and mid-range of freshwater per tidal cycle to tidal prism value. Estuaries were defined by their large proportions of shelter and by high values of freshwater per tidal cycle to tidal prism.

This first level, describing the major inlet types, should separate them into categories with major differences in γ and β diversity levels between the three archetypic inlet types.

Level 2: Dominant productivity regime type

The dominant productivity regime category is designed to differentiate between inlets that have predominantly pelagic productivity (generally deeper) versus those that are predominantly driven by benthic productivity (generally shallower). Phytoplankton production will dominate in environments that are predominantly pelagic, while in environments dominated by shallow benthic habitats and intertidal habitats, benthic microalgae, macrophytic algae, seagrasses and salt marshes will contribute up to or over 50 per cent of total primary production (Table 11.4; Valiela, 1995). Productivity regime types were chosen to represent the three archetypic productivity regimes, those dominated by benthic or pelagic productivity regimes, or of mixed type. The three archetypic productivity regimes were distinguished by two factors: mean depth and the proportion of intertidal zone (Box 11.1 and Colour plate 15b).

Level 3: Inlet complexity type

It is suggested that the inlet complexity category may be used to predict a mix of α and β diversity in each inlet, by assessing habitat heterogeneity, with increasing diversity expected from low to high inlet complexity. Habitat heterogeneity is thought to influence species diversity by creating microhabitat types, a greater total niche space, and providing refuge (in the form of structural complexity). Three archetypic inlet types (Figure 11.5) were chosen to represent the inlet complexity types: simple, intermediate and complex.

Six measures were used to index overall environmental complexity at the geomorphic unit level. Topographic complexity and shoreline complexity (shoreline sinuosity) were measured as proxies for structural complexity. Depth (Colour plate 15c), exposure diversity (Colour plate 15d) and evenness were calculated as measures of physical diversity (Box 11.1).

Classification of inlet types

Inlets were assigned to classes using a supervised clustering algorithm which organizes data into groups based on the similarity of one or more variables. The specific clustering algorithm used was possibilistic C-means (PCM) that weights class membership values on their distance from each cluster centre while ignoring their distance to other cluster centroids (Bezdek et al, 1984; Krishnapuram and Keller, 1993). Clustering algorithms are widely used in data mining, bioinformatics, GIS and remote sensing-based analyses, which have proliferated due to their utility in sorting out complex interactions between variables in high dimensional data (Bezdek et al, 1984). A supervised cluster analysis is used when the user has some previous knowledge of the system and therefore either chooses samples they know to belong to a certain class and 'trains' the algorithm to choose similar samples, or defines the cluster centroid. In this analysis, the cluster centroid was chosen to represent the archetypic inlets. The centroid values chosen are input to the supervised clustering algorithm which then assigns a membership value to each inlet based on how far away (in Euclidean distance) each

Box 11.1 Categories of parameters to describe coastal inlets

Three categories of parameters in three levels were selected to describe coastal inlets. They were: the relative influence of land versus sea, as morphological/ hydrographic type; the dominant primary productivity regime – pelagic versus benthic; the physical complexity of the inlet – from simple to complex - which acts as a predictor of habitat diversity. The process of quantifying these parameters is described here.

Level 1. Morphological/hydrographic inlet category

The first level, the hydrographic inlet type category, indicates the major hydrographic and geomorphological characteristics of an inlet resulting from a combination of environmental factors at the land/sea interface. Under stable tectonic and sea-level conditions this can be determined principally by the influence of waves, tide and river power on an inlet and it's morphometry (Penthick, 1984).
Hydrographic inlets types were defined by two variables: the ratio of predicted freshwater runoff per ½ tidal cycle to the total volume of an inlet, and the proportion of the inlet that is sheltered from open ocean conditions. **Bays** were defined by their variable amount of shelter and a R12/V (freshwater per ½ tidal cycle/ tidal prism) value of close to 0. **Coves** were defined by their low proportion of shelter and mid-range R12/V. **Estuaries** were defined by their large proportions of shelter and by high R12/V ratios.

Freshwater runoff per ½ tidal cycles to total inlet volume (R12/V)

A large R12/V indicates the inlet is dominated by river forcing thus experiencing decreased salinity.
The factor freshwater per ½ tidal cycles (R12/V) was measured by taking the ratio of predicted freshwater entering the system during ½ tidal cycles to the volume of each inlet. The predicted freshwater runoff per ½ tidal cycles was calculated by determining relationships between gauged river outflow and the area of the watershed for 75 of the 141 of the inlets presented in Gregory et al (1993), that coincided with the study area.

Gauged river outflow and watershed areas are highly positively correlated (R^2= 0.93, $p < 0.5$).

Inlet volume was calculated using the digital elevation model created to measure depth characteristics (see below). Volume was measured from a Triangular Irregular Network (TIN) surface using the ArcGIS Desktop TIN Polygon Volume function where the volume of each area outlined as a polygon boundary was determined.

Proportion of Sheltered Area

The Proportion of sheltered area was derived by the sum of the inlet area in two categories of fetch: 0-700 and 700-5000 m. Fetch is the metric widely used to predict wave exposure in the absence of wave power information. Wave exposure, although not precisely defined, is typically taken as an index of the severity of the hydrodynamic environment to which nearshore plants and animals are exposed (Denny, 1995).

Modified effective fetch was calculated as the arithmetic average of two fetch grids, one referenced to the normal (perpendicular direction) to the headland direction of the inlet to account for waves entering the system from the offshore environment, and the other to the average wind direction during the year 2005 to account for the development of wind waves that can effect shallow areas of inlets. Wind direction data were downloaded from Environment Canada Archives and averaged over 2005 at five wind stations located along the study area. For these areas the frequency at which wind blew in each of 16 binned directions was calculated based on the wind rose.

The fetch algorithm calculated a fetch measurement for each pixel below 0 in the digital elevation model. The algorithm took the arithmetic average of 9 fetch radials at each point in a 20 m grid using an algorithm created for ArcGIS 9.0 (Finlayson, 2005). The average, min, max and standard deviation of fetch for each inlet were calculated using functions within ArcView 9.1. Proportions were calculated for each of five categories modified from Howes et al (1994), and categories of fetch were chosen based on the formula (Mardia et al, 1979):

Modified Effective Fetch *(Fm)* $= (0.5 * (\sum i_{wn})/ n) + (0.5 * (\sum i_{an})/ n)$

Where:

i = Fetch length which is the open water distance from the point to a maximum value of 25 km
n = radial number from 1 to 9
w = average wind direction
a = azimuth to the mouth direction

Level 2. Dominant productivity regime type

The dominant productivity regime type category is designed to differentiate between inlets that have predominantly pelagic productivity processes or predominantly benthic productivity processes.

Depth and proportion of intertidal zone were used to distinguish three main classes of inlet types; benthic, mixed and pelagic dominant productivity regime types. Inlets with a large intertidal zone and shallow depth will have primarily benthic productivity while large average depths and a narrow intertidal zone will have primarily pelagic productivity.

Proportion of intertidal zone

The proportion of intertidal zone was delineated based on Canadian Hydrographic Service maps at an approximate 1:50000 scale and area was measured using ArcView 9.1 zonal area function. Tidal Volume was calculated as the volume between LNT and MHW of the inlet using a Triangular Irregular Network of the nearshore (See Depth Characteristics). Volume at Low Low Water Large Tide (LLWLT) and MHW were calculated using the ArcDesktop 9.1 TIN Polygon Volume function. Total inlet volume was calculated as the average of the MHW and LLWLT volumes i.e. value at mid-tide.

Depth characteristics

Depth characteristics were measured by creating a digital elevation model of the nearshore. The digital elevation model was created using Digital Coastal Series Map Data provided by the Nova Scotia Geomatics Centre (NSGC) for the area surrounding the study area land/water boundary and out to a depth of 50 m from LLWLT (the average of the lowest low waters, based on 19 years of predictions).

The Coastal Map Series was created as two layers, a land layer and water layer. The land features were derived by the NSGC from the Nova Scotia Topographic Database 1:10000 Map Series, and include 10 m contours and individual spot heights. Clearly defined features have a horizontal accuracy of 20 m or better.

The water layer is derived from the Canadian Hydrographic Service digital bathymetric charts at various scales and digitizing conventions. Point depths from the Canadian Hydrographic service are collected using various methods, more recently multibeam – with nearshore accuracies between 0.15 – 1 m (Canadian Hydrographic Service, 2005), although there are also still areas where lead-line measurements taken in the early 1900s are used (Canadian Hydrographic Service, 2005). The maps have been digitized to the 1:50 000 level and extend to the 12-mile limit.

Traditional (MHW) shorelines represented on topographic maps are derived from remote sensing photogrammetry.

Once bathymetric and terrestrial elevation data were reconciled (to MHW), a continuous surface was created by interpolating between contours and sounding using a Triangular Irregular Network (TIN). This TIN was then transformed to a raster dataset for subsequent analysis. Depth properties of each inlet were extracted from the digital elevation model (DEM) using the spatial zonal functions included in ArcView 9.1.

Level 3. Inlet complexity to predict habitat heterogeneity

Six measures were used to predict habitat heterogeneity: topographic complexity and shoreline complexity (shoreline sinuosity) were measured as proxies for structural complexity, diversity and evenness of depth and fetch were calculated as measures of physical diversity.

Habitat heterogeneity is a widely used criterion in conservation planning used to establish protected areas (Mumby, 2001) as habitat heterogeneity is expected to be positively correlated with species diversity (Kallimanis et al, 2008).

Shoreline sinuosity

Shoreline sinuosity was calculated as an index of the complexity of the shoreline (with and without island shoreline lengths), where the length of the shoreline is divided by the circumference of a circle with equal surface area. The index is analogous to the limnological index of shoreline development. Shoreline sinuosity is not a proxy for shoreline length.

$$\textit{Shoreline Sinuosity (SS)} = L/2\sqrt{a\pi}$$

Where: L = length of shoreline

a = inlet surface area

Benthic complexity

The benthic complexity surface was created from the nearshore digital elevation map by taking the maximum change in slope in the eight surrounding grid cells on the raster slope surface and is adapted from Ardron and Sointula (2002). It can be calculated using the formula

$$\textit{Complexity} = ATAN\left(\sqrt{\left([ds/dx]^2 + [ds/dy]^2\right)}\right) * 57.29578$$

Where: s = slope

Evenness and diversity indices

Depth and Fetch category evenness and diversity indices were calculated based on the Shannon (1948) Diversity index. Evenness is a diversity index that quantifies how equal the physical categories are and is calculated by Pielou's (1975) evenness index:

$$\textit{Pielow's Evenness (E)} = \frac{H^{I}}{H^{I}_{max}}$$

Where: H^1 = Shannon's Diversity Index

$$\textit{Shannon's Diversity Index } (H^{I}) = -\sum_{i=1}^{s} p_i \ln p_i$$

$$p_i = \frac{n_i}{N}$$

$$H^{I} \max = \ln S$$

Where: n_i = proportion of depth in each category and N = 100 percent;

s = total number of depth categories

The measurements of fetch and depth are described above.

References

Ardron, J. A., and Sointula, B. C. (2002) 'A GIS Recipe for Determining Benthic Complexity: An Indicator of Species Richness', in J. Breman (ed.), *Marine Geography GIS for the Oceans and Seas*, ESRI, Redlands, CA

Canadian Hydrographic Service (2005) *Standards for the Hydrographic Survey*, Canadian Hydrographic Service, Fisheries and Oceans Canada

Denny, M. (1995) 'Predicting Physical Disturbance: Mechanisitic Appraoches to the Study of Survivorship on Wave-Swept Shores', *Ecological Monographs*, vol 65, no4

Finlayson, D. (2005) 'UW Waves Toolbox for Arc9.0', accessed from http://david.p.finlayson.googlepages.com/gisscripts

Gregory, D., Petrie, B., Jordan, F., and Languille, P. (1993) *Oceanographic, Geographic and Hydrological Parameters of Scotia-Fundy and Southern Gulf of St. Lawrence Intets*: Bedford Institute of Oceanography, Dartmouth, NS

Howes, D. E., Harper, J. R., and Owens, E. (1994) 'BC Physical Shore-Zone Mapping System' [Electronic Version], from www.ilmb.gov.bc.ca/risc/pubs/coastal/pysshore/index.htm

Kallimanis, A. S., Maxaris, A. D., Tzanopoulos, J., Halley, J. M., Pantis, J. D., and Sgardelis, S. P. (2008) 'How Does Habitat Diversity Affect The Species-Area Relationship?', *Global Ecology and Biogeography*, vol 17, no 4, pp532–538

Mardia, K., Kent, J., and Bibby, J. (1979) *Multivariate Analysis*, Academic Press, London

Mumby, P. J. (2001) 'Beta and Habitat Diversity in Marine Systems: A New Approach to Measurement, Scaling and Interpretation', *Oecologia*, vol 128, pp274–280

Penthick, J. (1984) *An Introduction to Coastal Geomorphology*, Edward Arnold, London

Shannon, C.E. (1948) 'A mathematical theory of communication', *Bell System Technical Journal*, vol 27, pp379–423 and 623–656

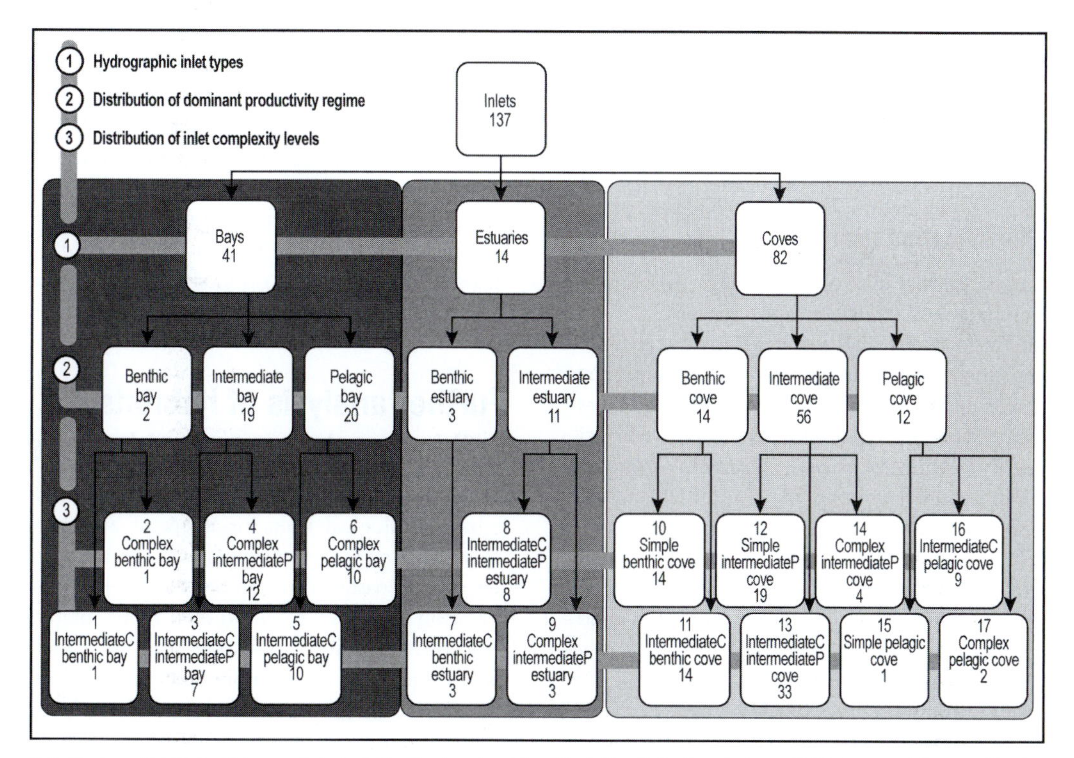

Note: Numbers of inlets of each type are indicated below the class description. The first level is the three hydrographic inlet types, and the second level shows the distribution of the dominant productivity regime types, within each hydrographic inlet type. The final level shows the distribution of the inlet complexity levels within the dominant productivity regime types and hydrographic inlet types.

Figure 11.6 The resulting full classification of Nova Scotia's inlet types according to category

inlet is from each archetypic cluster centre. More detailed discussion of procedures can be found in Jain et al (1999), Clarke and Warwick (2001) and Bezdek (1981).

The full inlet classification resulted in 17 classes (Figure 11.6). The coves with intermediate complexity and a mixed dominant productivity regime were most frequent in the classification. There were three classes which were represented by one member only: intermediate complexity benthic bays, complex benthic bays and simple pelagic coves. Many possible classes did not have any members: simple benthic bays; simple mixed productivity bays; simple pelagic bays; simple benthic estuaries; complex benthic estuaries; simple mixed productivity regime estuaries; simple/intermediate/complex pelagic productivity regime estuaries; and complex benthic productivity regime coves. The first level of classification resulted in 41 bays, 14 estuaries and 82 coves, the second level resulted in 19 inlets with benthic productivity regimes, 86 with mixed productivity regimes and 32 with pelagic productivity regimes, and the third level of classification resulted in 28 simple inlets, 77 intermediate inlets and 32 complex inlets.

Classification uses

The 17 distinct inlet types should predict varying combinations of habitat and community types, and α and β-diversity patterns. The first level, the morphological/hydrographic inlet type predicts major differences in β-diversity between bay, estuary and cove types. Differences between primarily benthic and pelagic community types can be determined from where an inlet falls in level two – the dominant productivity regime type. The final level three – the inlet complexity level – is designed to predict a mix of β (habitat types) and α (species level) diversity, with both increasing from simple to complex inlets.

As an example, from this classification, the highest species richness would be expected in an inlet that has a broad suite of characteristics such as in a complex mixed bay. In such an inlet, a mixed productivity regime would result in a range of intertidal and pelagic habitats (and benthic and pelagic community types), while its physical complexity results from variability in depths, fetches, microhabitats and salinity gradients. In contrast, simple coves would present habitats and conditions little different from the background marine environment, but will house species that can withstand wave pressure. Estuaries, with salinity and temperature values that differ the most from open ocean conditions, will have species that can tolerate physiologically stressful environments. The type of quantitative classification presented here can be validated by actual biological surveys of selected locations (see below).

Along with predicting biodiversity patterns of each inlet type, there is considerable value in providing a standard classification system. The method outlined here could be applied globally to shorelines with similar conditions – a highly crenulated shoreline with a large number of inlets experiencing highly variable physical characteristics – although some modification of the variables used may be required.

There are important questions to which this classification can provide answers, such as: what are the critical inlets along the coast? Are any of them ecologically unique or distinctive? Which ones should be protected in order to ensure representation? Which inlets are currently protected, and are they representative of the coastal inlets of a region as a whole? Critical inlets would be those that are not only distinctive, but ones that also have a significant level of anthropogenic stress.

Further analysis of habitats within coastal geomorphic units

Following the kind of quantitative analyses of geomorphological units or inlets of an entire coastline as suggested above, specific areas or inlets of the coastal zone can be selected for conservation initiatives, based on their representative or distinctive features. However, from such analyses of the set of units within ecoregions, we do not yet have explicit inventories or knowledge of the quantitative distributions of the various habitat types. Clearly, further analysis of selected or candidate units is required in order to establish the kinds and extents of the different component habitat types within them.

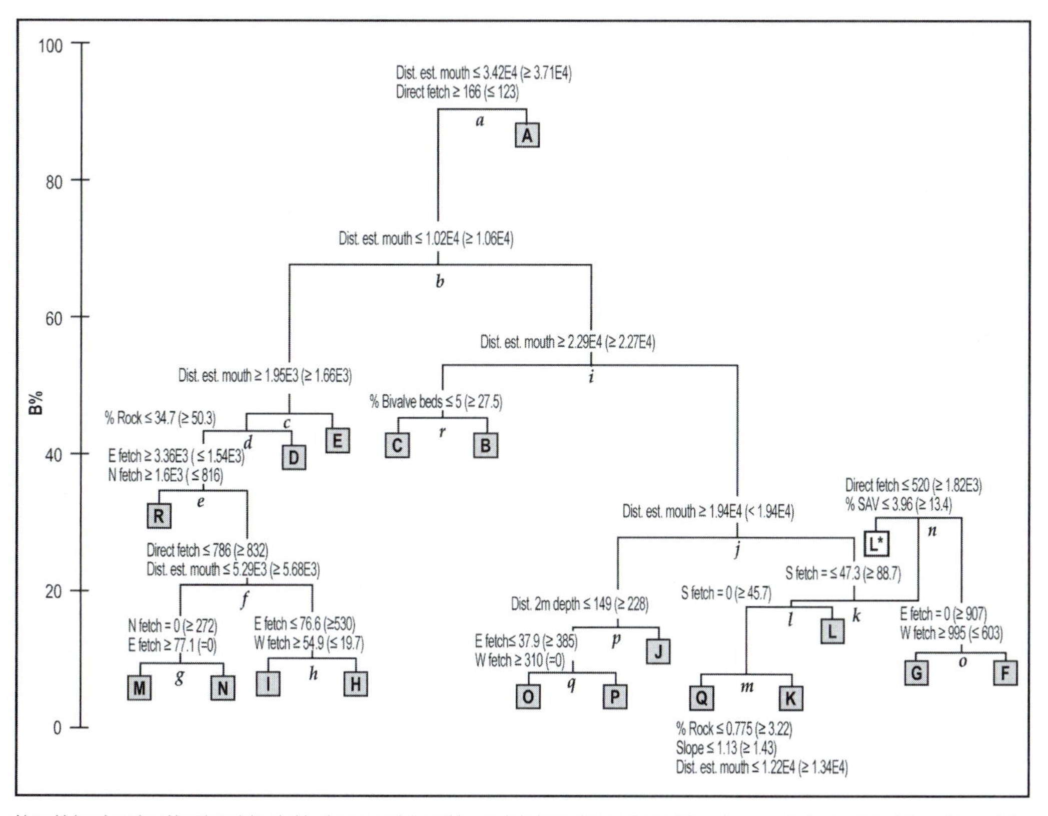

Note: Unbracketed and bracketed thresholds given at each branching node indicate that a left and right path, respectively, should be followed through the tree. B% reflects the extent of inter-habitat differences as a proportion of that between the most dissimilar habitats. E = exponential. For further details see Valesini et al (2010)

Figure 11.7 Linkage tree and associated enduring environmental variable thresholds for assigning new nearshore sites in the Swan Estuary Australia to their appropriate habitat type (terminal nodes in grey boxes)

In a recent paper, Valesini et al (2010) have shown how habitats within entire single coastal geomorphic units (in their study a set of estuaries in Australia) can be analysed based on available geophysical data and limited field work. This part of the chapter draws heavily on that study and is described here in some detail as an example of coastal conservation planning. It also demonstrates how careful analysis of even limited data can provide a quantitative environmental classification of habitats within inlets of a complex coastal zone.

It is important to distinguish between habitat maps derived from a classification scheme and those created from seabed mapping techniques that delineate geomorphological features associated with the substrate. The former are produced from a framework that collates information on spatial differences in the environment and can systematically assign sites to a group using specified differences in a suite of criteria (Diaz et al, 2004; Valentine et al, 2005; Snelder et al, 2007). However, in many cases, the latter do not employ such decision rules and represent only benthic features, such as seabed topography and texture, different substrates or broad groups of sessile biota (Diaz et al, 2004; Valentine et al, 2005). While such information can provide an important component of coastal habitat classification schemes (e.g. Kenny et al, 2003), it does not provide a systematic framework for defining or predicting habitat types, and has no application beyond the area for which it was created.

In addition, simple seabed mapping techniques fail to capture other biologically relevant habitat attributes, such as differences in wave exposure or water quality.

The most useful classification schemes are those that:

- Are based on quantitative data and decision rules.
- Employ a suite of enduring environmental criteria that can be easily and accurately measured from readily available mapped data and that either directly influence the distribution of biota or provide good surrogates for influential variables.
- Are flexible in their ability to incorporate new data and can be applied to areas beyond those for which they were developed.
- Are applicable at the spatial scales at which most ecologists and resource managers operate, i.e. local to regional scales.
- Are easy to use.
- Are biologically valid, i.e. the characteristics of biotic assemblages differ significantly among habitats.
- Are predictive, both in their ability to identify the habitat of any new site and the species likely to typify its faunal assemblages (e.g. Zacharias et al, 1999; Roff and Taylor, 2000; Banks and Skilleter, 2002; Roff et al, 2003; Valesini et al, 2003; Madden et al, 2005; Hume et al, 2007; Snelder et al, 2007).

However, many existing habitat classification schemes are deficient in one or more of the above criteria.

The main aims of the study by Valesini et al (2010) were as follows:

- Devise an approach for classifying nearshore habitats within a range of estuaries which (i) is fully quantitative; (ii) ascertains statistically that derived habitat types are significantly different; (iii) is based on measurements of a suite of enduring, biologically relevant and easily obtainable environmental criteria; (iv) is able to accommodate new environmental criteria; and (v) is directly pertinent to ecologists and managers working at local scales.
- Develop a quantitative and readily usable method for predicting the habitat type to which any new nearshore site in an estuary should be assigned.
- Determine, for each estuary and season, the extent to which the spatial differences among habitats in a suite of non-enduring water quality variables reflects the enduring environmental characteristics used to identify those habitats.

Five estuaries of varying characteristics were chosen as study systems by Valesini et al (2010), but analyses concentrated on two: the Swan and Peel-Harvey estuaries in south-western Australia. The systems varied mainly in: (i) the frequency with which they are open to the sea; (ii) their overall morphology; and (iii) the extent to which they have been anthropogenically modified.

Data sources and data processing

The main data sources employed for identifying the various habitat types in the nearshore shallow waters (< 2m deep) of each estuary were: (a) a high resolution, digitally georeferenced and remotely-sensed image, that is the red, green and blue bands of digital aerial photos (1 pixel = 40cm) or Quickbird satellite images (1 pixel = 2.4m); and (b) high resolution bathymetric data (one depth sounding per 10–50m). All preliminary processing of these mapped data and the subsequent measurement of the suite of enduring environmental variables employed in the habitat classification and prediction approaches for each estuary were carried out using GIS software Idrisi Kilimanjaro v14 or ArcGIS v9.1.

The above data for each system were prepared for measuring the suite of enduring variables by (a) tracing the shoreline, including the outline of any islands or larger structures such as marinas, from the remotely-sensed image and (b) constructing a digital elevation model (DEM) of the bathymetric data. The outline and DEM of each estuary were then used in combination to mask out all unwanted areas for each band of the remotely sensed images, namely all waters greater than 2m deep and land. The masked bands for each image were then further prepared for measuring the areas of the various substrate/submerged vegetation types they contained by enhancing differences in the spectral reflectance of pixels belonging to different benthic categories.

Measurement of enduring environmental variables

A large number of environmentally diverse nearshore sites were initially selected throughout each estuary, which were considered likely to capture the full extent of the nearshore habitat diversity. These sites were chosen from a visual assessment of the high resolution images of each system and several reconnaissance trips in the field. Three broad categories of enduring environmental variables were measured at each site in each estuary, thus providing the data required to allocate those sites to their respective habitat types. The three categories were: location with respect to marine and riverine water sources; exposure to wave activity; and substrate and submerged vegetation types.

Location with respect to marine and riverine water sources was intended as a group of variables mainly as a surrogate for the range of water quality parameters that typically vary spatially throughout an estuary due to differences in the extent of mixing between marine and riverine waters, such as salinity, water temperature, dissolved oxygen concentration, water colour, turbidity and ion composition. Note that, while these water quality variables are not necessarily expected to vary in accordance with a simple gradient from estuary mouth to the riverine extent of tidal influence, they are likely to exhibit spatial differences throughout those systems.

Exposure to wave activity was a group of variables reflecting the exposure of each site to waves generated by local winds (i.e. fetch in each cardinal direction and that along the bearing perpendicular to the aspect of each site) and the impact of local bathymetry on waves as they approach the shoreline (i.e. average slope of the substrate and distance to the wave shoaling margin). Northerly, southerly, easterly, westerly and direct fetch were measured at each site using the formula for modified effective fetch (MEF) (Coastal Engineering Research Centre, 1977). This method encompasses a range of fetches within a limited arc of a given bearing, and thus provides a robust reflection of wave exposure.

For substrate and submerged vegetation types, the pretreated images for each estuary were subjected to a non-hierarchical unsupervised cluster analysis to assign each pixel to one of ten nominal benthic classes on the basis of differences in their spectral signatures. They were then assigned to one of three broad and more distinct groups – bare unconsolidated substrate, rock or submerged aquatic vegetation. The last group represented both seagrass and macroalgae, which could not be reliably discriminated from each other on the images, due either to the fact that they grew in close association with each other and/or their spectral signatures were not sufficiently different. The overall accuracies of the benthic classification maps for the Swan and Peel-Harvey estuaries were 74 and 76 per cent, respectively. The area occupied by each substrate/submerged vegetation type within the boundary of each site was then calculated and converted to a percentage of the total site area.

Measurement of non-enduring environmental variables

Measurement of a suite of water quality characteristics, namely salinity (practical salinity scale), water temperature (°C) and dissolved oxygen (mg L^{-1}), each of which was considered likely to influence the spatial distribution of fish and benthic invertebrate faunas, was undertaken at two sites representing particular habitat types in the Swan and Peel-Harvey estuaries during the last month of each of six seasons between autumn 2005 and summer 2007. The measurements at each site in each season were collected on two occasions separated by at least a week to reduce the chances of the resultant data being unduly affected by an atypical sample.

Statistical analyses

Data analyses were carried out using the PRIMER v6 multivariate statistics package (Clarke and Gorley, 2006) with the PERMANOVA+ for PRIMER add-on module (Anderson et al, 2008).

To ensure that each of the three broad categories of enduring variables – location, exposure to wave activity and substrate/submerged vegetation type – contributed equally to the habitat classification procedure, each variable was weighted on the basis of the total number of variables that comprised the broad category to which it was assigned. Thus, each of the three categories was assumed to contribute an equal and arbitrary proportion to the overall data matrix. For each estuary, the enduring environmental data were then used to construct a Manhattan distance matrix containing the resemblances between each pair of sites.

To identify those groups of sites within each estuary that did not differ significantly in their suite of enduring environmental characteristics and thus together represented distinct habitat types, the Manhattan distance matrix was subjected to hierarchical agglomerative clustering with group average linking (CLUSTER) and an associated similarity profiles (SIMPROF) test (Clarke et al, 2008). The latter routine is a permutation test that determines whether any significant group structure exists within a set of samples for which there is no priori grouping hypothesis.

For each estuary, any new nearshore site (i.e. one not used in the habitat classification procedure) could be quantitatively assigned to its appropriate habitat type using a novel application of the linkage tree (LINKTREE) routine. This approach was used to ascertain which enduring environmental variables, and their true quantitative thresholds, were most tightly linked with the progressive separation of sites into the habitats identified by the above classification procedure. These variables and their thresholds were then used as the quantitative criteria to predict the habitat of any new site. LINKTREE (Clarke et al, 2008) is a non-metric modification of the multivariate regression tree technique published by De'ath (2002). Thus, a binary 'linkage tree' is constructed that reflects how samples are most naturally split into successively smaller groups.

For those sites representing habitats at which the suite of water quality variables were recorded, the RELATE routine was used to determine whether the pattern of relative differences among sites, as defined by their water quality, was significantly correlated with that defined by their suite of enduring environmental characteristics. The biota and environment-matching routine (BIOENV) (Clarke and Ainsworth, 1993; Clarke et al, 2008) was then used to ascertain whether a greater correlation with the second of these matrices could be obtained by using only a particular subset of water quality parameters, rather than the full suite.

Habitat classification and prediction results

The CLUSTER and SIMPROF procedures performed on the data for the suite of enduring environmental variables recorded at the 101 nearshore study sites throughout the Swan Estuary yielded 18 habitat types, while those carried out on the 102 sites in the Peel-Harvey Estuary produced 17 habitat types. Habitat clusters were distinguished by a variety of combinations of characteristics. Some were distinguished by distance from the mouth of the estuary and limited fetch, while others nearby were separated by a combination of empty bivalve shells, or lack of submerged vegetation, shallow slopes and narrow wave shoaling margins. Middle estuary habitat types were distinguished mainly by differences in their exposure to winds from different directions, the widths of their wave shoaling margins and the proportions of submerged vegetation and rock comprising the substrate. Other habitat types, located in the upper channel and lowermost reaches of the main basin or located in the middle channel, were distinguished largely by differences in their exposure to prevailing winds, quantities of rock and submerged vegetation and the slope of their substrates.

The linkage trees representing the separation of study sites into the habitat types identified by the CLUSTER and SIMPROF procedures, as well as the quantitative thresholds of the enduring environmental variable(s) that best reflect the division at each branching node of the tree, are shown for the Swan Estuary in Figure 11.7. Such linkage trees provide a set of quantitative decision rules for assigning any new nearshore site in these systems (in other words ones not used in the habitat classification procedure) to the appropriate habitat on the basis of measurements of its enduring environmental characteristics. They also provide a way of detecting which of the enduring variables employed in the classification procedure were most important for defining the habitat types in any given system.

The path at several nodes in the linkage tree was defined by thresholds for only one environmental variable – distance from estuary mouth, contributions of rock or bivalve shells to the substrate, width of the wave shoaling margin or southerly fetch in the case of the Swan estuary. However, in both estuaries, most of the remaining enduring variables were selected in some combinations at other nodes of the linkage trees (Figure 11.7).

Relationships between habitat types and non-enduring environmental variables

The non-enduring characteristics of the estuaries (temperature, salinity, dissolved oxygen concentration) frequently displayed seasonal and location variations as expected. In some seasons, there was a lack of good agreement in spatial pattern between the non-enduring water quality variables and the matrices of enduring environmental parameters, which was due to some water quality variables exhibiting relatively little change throughout the estuary, for example salinity in autumn and summer, and temperature in autumn. However, in other cases when there were marked spatial differences in the magnitude of the water quality variables, the pattern of those differences was still not well reflected by that of the enduring environmental data.

Although the resident biota will be exposed to these variations, those in the benthos will necessarily need to tolerate them in situ. These non-enduring factors, although conventionally used by physical oceanographers to typify and classify estuaries, their processes and circulation regimes, are not particularly useful for conservation planning. In addition, salinity and dissolved oxygen cannot be remotely sensed and must be measured in situ.

Overall evaluation of the approach to habitat classification

The present approach to classifying nearshore habitat types within estuaries (Valesini et al, 2010), which was developed for a range of systems in southwestern Australia, has produced a logical and intuitive separation of an environmentally diverse range of sites throughout the study estuaries. Furthermore, the study developed a quantitative method for subsequently predicting the habitat type to which any new nearshore site in each of these systems should be assigned. These approaches to estuarine habitat classification and prediction are among the first of their kind to be developed.

The quantitative nature of the approach is two-fold. First, it is based on fully quantitative measurements for each of the enduring environmental criteria. Second, the decision rules for assigning sites to habitat types are entirely quantitative, objective and derived from rigorous statistical tests, and ensured that each habitat type was significantly distinct from all others in the system – that is, there were no redundant classes. These features remove any ambiguity in the use of the classification, ensure that the results are both reliable and repeatable, and provide a sound foundation for statistically ascertaining the extent to which the spatial differences among habitats are reflected by those of various faunal assemblages.

Furthermore, the habitat prediction approach provided, at each node of the resultant linkage trees, quantitatively defined thresholds for those environmental variables that were most important in separating sites into their respective habitats. These thresholds thus provide sound and easily interpretable decision rules for assigning any new nearshore site to its most appropriate habitat, on the basis of measurements for its enduring environmental characteristics.

Although several other habitat classification approaches have adopted a hierarchical clustering technique to determine the patterns of environmental similarity among sites, they have typically chosen an arbitrary resemblance level as a 'cut-off point', below which those groups of sites that have formed in the clustering process are considered to represent different habitats (e.g. Edgar et al, 2000; Connor et al, 2004; Snelder et al, 2007). Such approaches do not demonstrate statistically that the resultant groups actually represent distinct habitat types, or whether any such group may contain more than one habitat.

The use of environmental rather than biological criteria in habitat classification schemes, and particularly those that are enduring, has several advantages which have been recognized by numerous workers (e.g. Roff and Taylor, 2000; Banks and Skilleter, 2002; Roff et al, 2003; Valesini et al, 2003; Hume et al, 2007; Snelder et al, 2007).

First, the resultant habitats are applicable to a range of fauna, while the 'biotopes' (i.e. communities and their habitats) (Connor et al, 2004; Olenin and Ducrotoy, 2006) that are often derived from biological schemes are applicable only to the biota on which they are based and the area for which they were devised (e.g. Zacharias et al, 1999; Connor et al, 2004; Stevens and Connolly, 2005). Second, enduring environmental criteria are easy to measure directly from mapped sources, whereas the costs of acquiring quantitative biotic data over appropriate spatio-temporal scales and levels of replication are often prohibitive (e.g. Edgar et al, 2000; Roff and Taylor, 2000; Banks and Skilleter, 2002). Third, enduring criteria often represent good surrogates for complex suites of non-enduring environmental variables that may be difficult and/or costly to measure. The magnitude of non-enduring environmental characteristics (e.g. salinity, temperature, oxygen concentration, nutrients, wave height etc.) at each habitat will of course vary over a range of temporal scales, particularly in dynamic environments such as estuaries. However, habitats defined on the basis of enduring characteristics are still expected to remain distinct and display largely similar spatial patterns over time.

The enduring environmental criteria employed in this study are also likely to be useful for classifying nearshore habitats within other estuaries in other areas of the world. However, the current approach is flexible in that the particular enduring criteria employed can be easily tailored to suit the regional conditions in any estuary, or indeed, any other type of environment, so long as they can be easily measured from available data sources.

Conclusions and management implications

The coastal zone is the region of greatest variety of components of biodiversity. It is also the region where humans interact most strongly with the marine environment, and where they have the greatest impact upon it. Consequently this should be the region where we pay closest attention to the oceans, and pay the greatest attention to environmental management. Yet unfortunately, CZM has not been notably successful around the globe.

Good management requires an understanding of the structures and processes being managed. This applies to all forms of management for all purposes, not just for conservation. Unfortunately management in the coastal zone is complicated by overlapping or competing political and governance jurisdictions, lack of knowledge of environmental conditions and even lack of definition of what the coastal zone really comprises (Mercer Clarke et al, 2008).

There are many ways to define and describe the coastal zone and its biological communities. However, for conservation planning, the coastal zone can be considered to encompass: seaward – the fringing communities to the depth of the euphotic zone; and landward to the extent of the sea – either physically as encountered by land, or chemically as limited by the null zone in estuaries. Conservation planning in the coastal zone has not received adequate attention due partly to the high level of complexity and lack of analysis of data required for the appropriate mapping of geophysical factors.

The intention in this chapter was to show how the coastal zone and its component inlets can be quantitatively described and analyzed, largely from existing or readily obtainable data, as a basis for decision-making for conservation and other management purposes.

In the nearshore, coastal inlets define naturally repeating geomorphic units at a scale above that of habitats, which can be classified using an ecosystem-based conservation approach (Greenlaw, Roff and Redden, unpublished). A three-level classification is suggested that recognizes and quantifies both marine and freshwater influences on the coastal zone and its inherent physical and ecological complexity. The categories created are based on three sets of factors: geomorphological/hydrographic inlet type determined by marine versus freshwater influence (as quantified by exposure and freshwater drainage); productivity regime – whether dominated by the pelagic water column or benthic communities; inlet complexity – assessed in several dimensions that in combination act as surrogate measures of α and β diversity.

The resultant classification allows recognition of inlets as replicate representative units, or draws attention to outliers which require further examination as potentially distinctive. The method could be applied globally to any type of shoreline, but is especially applicable to complex highly crenulated shorelines with large numbers of inlets of variable character, and experiencing highly variable physical influences. The categories created – hydrographic inlet type, dominant productivity regime type and inlet complexity level – can be modified to include physical variables appropriate to other regime types.

A study by Valesini et al (2010) has developed fully quantitative approaches for firstly classifying local-scale nearshore habitats within individual coastal inlets (in this case – estuaries), and secondly predicting the habitat of any nearshore site within those systems. Both of these methods have employed measurements of enduring and biologically relevant environmental factors that can be easily obtained from digitally mapped or remotely sensed data sources. Furthermore, the classification approach demonstrates statistically that the characteristics of derived habitats differ significantly. Although these approaches have been developed for selected estuaries in south-western Australia, they can be easily tailored to any estuary by the inclusion of new enduring environmental factors. These approaches represent advances on other published methods for classifying and predicting habitats at local scales in estuarine and coastal waters. They also provide a reliable framework for, firstly, investigating whether spatial differences in habitats are correlated with those in fish and benthic invertebrate assemblages and, if so, predicting the faunal species that are most likely to occupy any estuarine site on the basis of the habitat type to which it belongs.

Some comparisons with other nested habitat classification schemes

As Valesini et al (2010) have noted, many habitat classification schemes developed for coastal waters have adopted a hierarchical approach, in which finer spatial units of the classification are nested within successively broader groups. In these classifications the user is guided through a series of interconnected decision rules to reach a final classification unit. In several cases, the broadest level of these schemes incorporates all marine and/or estuarine waters within a national economic exclusion zone, and the lowest levels can represent highly localized habitats at the scale of metres (e.g. Allee et al, 2000; Connor et al, 2004; Madden et al, 2005).

Necessarily, the finest levels of these hierarchies must be tailored by the individual user to accommodate the particular features of their local environment. This kind of approach facilitates the growing trend towards the development of standardized habitat classification systems at national and continental scales, which has been motivated by a requirement for consistency in habitat definitions between one part of a country and another (e.g. Diaz et al, 2004; Madden et al, 2005; Mount et al, 2007).

Given that ecological studies and resource management of coastal waters typically occur at regional to local scales, the lower levels of hierarchical classification schemes are usually the most critical. However, while several of these schemes provide clear decision rules at the broader levels of the hierarchy, those at lower levels (e.g. the local habitat, biotope or eco-unit levels) are often less clear, either because they are more qualitative or because they present users with a myriad of ways of ultimately defining their local unit of interest. Consequently, choices made at the lower levels of such schemes may differ among users, which to a significant extent contradicts the very purpose of these standardized hierarchical methods. Furthermore, the number of potential habitats/biotopes/eco-units that can be derived from such schemes is often extraordinarily large, particularly when they are designed to be adaptable across temporal scales and to different-sized biota and each of their activities, for example feeding or spawning. These issues highlight the fact that the outcomes of such classification schemes depend heavily on the objectives of the study, and may differ among users due to differences in interpretation. They also demonstrate that a substantial amount of quantitative data for a suite of abiotic factors at a diversity of spatio-temporal scales needs to be acquired in the field before such schemes can be used with confidence at lower levels of a hierarchy.

Several classification schemes for estuaries have focused on categorizing whole systems and/or their catchments, or making very broad distinctions among environmental zones within estuaries (e.g. Digby et al, 1998; Edgar et al, 2000; Roy et al, 2001; Engle et al, 2007; Hume et al, 2007). While these classifications are useful at a national level for summarizing broad differences in estuarine function, identifying their susceptibility to particular environmental impacts and/or qualifying their environmental or cultural value, they are of limited use to local resource managers and may not provide a reliable basis for predicting the distribution of biota at finer taxonomic levels, particularly for small benthic fauna.

Future steps

As estuarine ecologists and managers work at regional to local scales, the results of the current approaches, in combination with associated studies of biota, will provide them with highly useful tools to plan for coastal zone conservation.

A next and important step in developing the current approaches to habitat classification and prediction is to test their ability to reliably reflect spatial differences in coastal biological community types – the plants, fish and benthic invertebrate assemblages – at appropriate temporal scales.

Another obvious development of the current approaches is to produce digital, spatially continuous habitat maps of the coastal zone in a GIS, in which all inlets display their habitat types. This could be achieved by automating the habitat prediction technique for every site along a coastline. Thus, users of the scheme would simply need the geographic coordinates of their site of interest in order to ascertain its habitat type, without the need to undertake any measurements of its enduring environmental characteristics.

Coastal zone classification and habitat mapping also allows the evaluation of human impacts. Human impact layers (e.g. Halpern et al, 2008) should also be mapped to manage human activities and include them in protection planning schemes. Human impact layers can be mapped as discrete activities (shipping, fishing, organic input, habitat modification etc.) and their impact or compatibility with habitat and biological community types can be assessed.

Finally, the kinds of classifications of the coastal zone described here are simply the beginning of the process of conservation planning. The two stage process of classification can lead to the definition of representative and distinctive areas of the coastal zone and its inlets, followed by further investigation of selected inlets to define their array of habitat types. The subsequent process of selecting sites as protected areas, determination of their location, size and boundaries will be considered in Chapters 14 and 16, and the process of integrating sites into networks of protected areas will be considered in Chapter 17. To this point the processes of the coastal zone have barely been considered, or only considered from surrogate measures (e.g. as estimates of shoreline exposure). One of the most important processes in the coastal zone concerns the ocean circulation in coastal cells, which define the connectivity among areas, and the retention or local advection of water masses including any larvae or other reproductive propagules they contain; this will be considered in Chapter 17.

References

Allee, R. J., Dethier, M., Brown, D., Deegan, L., Ford, R. G., Hourigan, T. F., Maragos, J., Schoch, C., Sealey, K., Twilley, R., Weinstein, M. P. and Yoklavich, M. (2000) 'Marine and estuarine ecosystem and habitat classification', *NOAA Technical Memorandum NMFS-F/SPO-43*, National Oceanic and Atmospheric Administration, Silver Spring, MD

Anderson, M. J., Gorley, R. N. and Clarke, K. R. (2008) 'PERMANOVA+ for PRIMER: Guide to software and statistical methods', PRIMER-E, Plymouth

Ardron, J. (2002) A GIS Recipe for Determining Benthic Complexity: An indicator of species richness', in J. Breman (ed.) *Marine Geography. GIS for the Oceans and Seas*, ESRI Press, Redlands, CA

Banks, S. A. and Skilleter, G. A. (2002) 'Mapping intertidal habitats and an evaluation of their conservation status in Queensland, Australia', *Ocean & Coastal Management*, vol 45, pp485–509

Barnabé, G. and Barnabé-Quet, R. (2000) *Ecology and Management of Coastal Waters: The Aquatic Environment*, Springer-Praxis, Chichester

Beatley, T., Brower, D. J. and Schwab, A. K. (2002) *An Introduction to Coastal Zone Management*, Island Press, Washington, DC

Bezdek, J. C. (1981) *Pattern Recognition with Fuzzy Objective Function Algorithms*, Plenum, New York

Bezdek, J., Ehrlich, R. and Full, W. (1984) 'Fcm: The fuzzy C-Means clustering algorithm', *Computers and Geosciences*, vol 10, no 2, pp191–203

Boedeker, D. and Von Nordheim, H. (2002) 'Application of NATURA 2000 in the marine environment', report of a workshop at the International Academy for Nature Conservation (INA) on the Isle of Vilm (Germany), from 27 June to 1 July 2001, German Federal Agency for Nature Conservation, Bonn, Germany

Brown, S. K., Buja, K. R., Jury, S. H., Monaco, M. E. and Banner, A. (1997) 'Habitat suitability index models in Casco and Sheepscot Bays, Maine', National Oceanic and Atmospheric Administration, Silver Spring, MD

Canadian Hydrographic Service (2005) *Standards for the Hydrographic Survey*, Fisheries and Oceans Canada, Ottawa

Carter, R. W. G. (1988) *Coastal Environments. An Introduction to the Physical, Ecological and Cultural Systems of Coastlines*, Academic Press, London

Clarke, K. R. and Ainsworth, M. (1993) 'A method of linking multivariate community structure to environmental variables', *Marine Ecology Progress Series*, vol 92, pp205–219

Clarke, K. R. and Gorley, R. N. (2006) 'PRIMER v6: User Manual/Tutorial', PRIMER-E, Plymouth

Clarke, K. R. and Warwick, R. M. (2001) *Change in Marine Communities: An Approach to Statistical Analysis and Interpretation*, Primer-E, Plymouth

Clarke, K. R., Somerfield, P. J. and Gorley, R. N. (2008) 'Testing of null hypotheses in exploratory community analyses: Similarity profiles and biota-environment linkage', *Journal of Experimental Marine Biology and Ecology*, vol 366, pp56–69

Coastal Engineering Research Centre (1977) *Shore Protection Manual*, U.S. Army Corp of Engineers Coastal Engineering Center, Vicksburg, MS

Commonwealth of Australia (2005) 'National marine bioregionalization of Australia', Department of Environment and Heritage, Canberra, Australia

Connor, D. W., Allen, J. H., Golding, N., Howell, K. L., Lieberknecht, L. M., Northen, K. O. and Reker, J. B. (2004) *The Marine Habitat Classification for Britain and Ireland Version 04.05*, Joint Nature Conservation Committee, Peterborough

Cooper, J. G. (2001) 'Geomorphological variability among microtidal estuaries from the wave-dominated South African coast', *Geomorphology*, vol 40, pp99–112

Cowardin, L. M., Carter, V., Francis, C. G. and LaRoe, E. T. (1979) *Classification of Wetlands and Deepwater Habitats of the United States. No. 79/31*, U.S. Department of the Interior, Washington, DC

Davis, D. S. and Browne, S. (1996) *The Natural History of Nova Scotia: Theme Regions*, The Nova Scotia Museum, Halifax, Canada

De'ath, G. (2002) 'Multivariate regression trees: A new technique for modeling species–environment relationships', *Ecology*, vol 83, pp1105–1117

Denny, M. (1995) 'Predicting physical disturbance: Mechanistic approaches to the study of survivorship on wave-swept shores', *Ecological Monographs*, vol 65, no 4, pp371–418

Diaz, R. J., Solan, M. and Valente, R. M. (2004) 'A review of approaches for classifying benthic habitats and evaluating habitat quality', *Journal of Environmental Management*, vol 73, pp165–181

Digby, M. J., Saenger, P., Whelan, M. B., McConchie, D., Eyre, B., Holmes, N. and Bucher, D. (1998) 'A physical classification of Australian estuaries', Southern Cross University, Lismore, NSW, Australia

Dipper, F. A., Howson, C. M. and Steele, D. (2007) 'Marine Nature Conservation Review Sector 13. Sealochs in West Scotland: Area summaries', Joint Nature Conservation Committee, Peterborough

Edgar, G. J., Barrett, N. S., Graddon, D. J. and Last, P. R. (2000) 'The conservation significance of estuaries: A classification of Tasmanian estuaries using ecological, physical and demographic attributes as a case study', *Biological Conservation*, vol 92, pp383–397

Engle, V. D., Kurtz, J. C., Smith, L. M., Chancy, C. and Bourgeois, P. (2007) 'A classification of U.S. estuaries based on physical and hydrologic attributes', *Environmental Monitoring and Assessment*, vol 129, pp397–412

Finlayson, D. (2005) 'UW Waves Toolbox for Arc9.0', http://david.p.finlayson.googlepages.com/gisscripts, accessed 10 September 2010

Gonenc, I. E. and Wolflin, J. P. (2004) *Coastal Lagoons*, CRC Press, Boca Raton, FL

Gregory, D., Petrie, B., Jordan, F. and Languille, P. (1993) *Oceanographic, Geographic and Hydrological Parameters of Scotia-Fundy and Southern Gulf of St. Lawrence Inlets, Canadian Technical Reports of Hydrographic and Ocean Sciences*, Bedford Institute of Oceanography, Nova Scotia, Canada

Guilcher, A. (1958) *Coastal and Submarine Morphology*, Methuen, London

Halpern, B. S., Walbridge, S., Selkoe, K. A., Kappel, C. V., Micheli, F., D'Agrosa, C., Bruno, J. F., Casey, K. S., Ebert, C., Fox, H. E., Fujita, R., Heinemann, D., Lenihan, H. S., Madin, E. M., Perry, M. T., Selig, E. R., Spalding, M., Steneck, R. and Watson, R. (2008) 'A global map of human impact on marine ecosystems', *Science*, vol 319, pp948–952

Hansen, D. V., and Rattray, M. (1966) 'New dimensions in estuary classification', *Limnology and Oceanography*, vol 11, no 3, pp319–326

Heath, R. A. (1976) 'Broad classification of New Zealand inlets with emphasis on residence times', *New Zealand Journal of Marine and Freshwater Research*, vol 10, no 3, pp429–444

Howes, D. E., Harper, J. R. and Owens, E. (1994) 'BC Physical Shore-Zone Mapping System', www.ilmb.gov.bc.ca/risc/pubs/coastal/pysshore/index.htm, accessed 13 July 2010

Hume, T. M. (1988) 'A geomorphic classification of estuaries and its application to coastal resource management: A New Zealand example', *Ocean and Shoreline Management*, vol 11, pp249–274

Hume, T. M., Snelder, T., Wetherhead, M. and Liefting, R. (2007) 'A controlling factor approach to estuary classification', *Ocean & Coastal Management*, vol 50, nos 11–12, pp905–929

Izsak, C. and Price, A. R. G. (2001) 'Measuring beta diversity using a taxonomic similarity index, and its relation to spatial scale', *Marine Ecology Progress Series*, vol 215, pp69–77

Jain, A. K., Murty, M. N. and Flynn, P. J. (1999) 'Data clustering: A review', *ACM Computing Surveys*, vol 31, no 3, pp232–264

Johnson, M. P., Crowe, T. P., Mcallen, R. and Alcock, A. L. (2008) 'Characterizing the marine NATURA 2000 network for the Atlantic region', *Aquatic Conservation: Marine and Freshwater Ecosystems*, vol 18, pp86–97

Kallimanis, A. S., Maxaris, A. D., Tzanopoulos, J., Halley, J. M., Pantis, J. D. and Sgardelis, S. P. (2008) 'How does habitat diversity affect the species-area relationship?', *Global Ecology and Biogeography*, vol 17, no 4, pp532–538

Kenny, A. J., Cato, I., Desprez, M., Fader, G., Schüttenhelm, R. T. E. and Side, J. (2003) 'An overview of seabed-mapping technologies in the context of marine habitat classification', *ICES Journal of Marine Science*, vol 60, pp411–418

Knox, G. A. (2001) *The Ecology of Seashores*, CRC Press, Boca Raton, FL

Krishnapuram, R. and Keller, J. M. (1993) 'A possibilistic approach to clustering', *Transactions on Fuzzy Systems*, vol 1, no 2, pp98–110

Madden, C. L., Grossman, D. H. and Goodin, K. L. (2005) 'Coastal and marine systems of North America: Framework for an ecological classification standard:Version II', *Natureserve*, Arlington, VA

Mann, K. H. (2000) *Ecology of Coastal Waters: With Implications for Management*, Blackwell Science, Oxford

Mardia, K., Kent, J., and Bibby, J. (1979) *Multivariate Analysis*, Academic Press, London

McCurdy, P. G. (1947) *Manual of Coastal Delineation from Aerial Photographs*, Hydrographic Office, Washington, DC

Mercer Clarke, C. S. L., Roff, J. C. and Bard, S. M. (2008) 'Back to the future: Using landscape ecology to understand changing patterns of land use in Canada, and its effects on the sustainability of coastal ecosystems', *ICES Journal of Marine Science*, vol 65, no 8, pp1534–1539

Mount, R., Bricher, P. and Newton, J. (2007) 'National intertidal/subtidal benthic (NISB) habitat classification scheme', Australian Coastal Vulnerability Project, Hobart, Tasmania

Mumby, P. J. (2001) 'Beta and habitat diversity in marine systems: A new approach to measurement, scaling and interpretation', *Oecologia*, vol 128, pp274–280

Olenin, S. and Ducrotoy, J. P. (2006) 'The concept of biotope in marine ecology and coastal management', *Marine Pollution Bulletin*, vol 53, pp20–29

Penthick, J. (1984) *An Introduction to Coastal Geomorphology*, Edward Arnold, London

Ray, G. C. and McCormick-Ray, J. (2004) *Coastal Marine Conservation: Science and Policy*, Blackwell, Malden, MA

Roff, J. C. and Evans, S. (2002) 'Frameworks for marine conservation: Non-hierarchical approaches and distinctive habitats', *Aquatic Conservation: Marine and Freshwater Ecosystems*, vol 12, pp635–648

Roff, J. C. and Taylor, M. E. (2000) 'National frameworks for marine conservation: A hierarchical geophysical approach', *Aquatic Conservation: Marine and Freshwater Ecosystems*, vol 10, pp209–223

Roff, J. C., Taylor, M. E. and Laughren, J. (2003) 'Geophysical approaches to the classification, delineation and monitoring of marine habitats and their communities', *Aquatic Conservation: Marine and Freshwater Ecosystems*, vol 13, pp77–90

Roy, P. S., Williams, R. J., Jones, A. R., Yassini, I., Gibbs, P.J., Coates, B., West, R. J., Scanes, P. R., Hudson, J. P. and Nichol, S. (2001) 'Structure and function of south-east Australian estuaries', *Estuarine, Coastal and Shelf Science*, vol 53, pp351–384

Ryan, D. A., Andrew, D. H., Radke, L. and Heggie, D. T. (2003) 'Conceptual models of Australia's estuaries and coastal waterways: Applications for coastal resource management', *Geoscience Australia*, Canberra, Australia

Salm, R.V., Clark, J. R. and Siirila, E. (2000) 'Marine and coastal protected areas: A guide for planners and managers', IUCN, Gland, Switzerland

Selvavinayagam, K., Surendran, A. and Ramachandran, S. (2003) 'Quantitative study on chlorophyll Using Irs-P4 Ocm data of Tuticorin coastal waters', *Journal of the Indian Society of Remote Sensing*, vol 31, no 3, pp227–235

Shannon, C. E. (1948). 'A mathematical theory of communication', *Bell System Technical Journal*, vol 27, pp379–423 and 623–656

Snelder, T. H., Leathwick, J. R., Dey, K. L., Rowden, A. A., Weatherhead, M. A., Fenwick, G. D., Francis, M. P., Gorman, R. M., Grieve, J. M., Hadfield, M. G., Hewitt, J. E., Richardson, K. M., Uddstrom, M. J. and Zeldis, J. R. (2007) 'Development of an ecologic marine classification in the New Zealand region', *Environmental Management*, vol 39, pp12–29

Stevens, T. and Connolly, R. M. (2005) 'Local scale mapping of benthic habitats to assess representation in a marine protected area', *Marine and Freshwater Research*, vol 56, pp111–123

Valavanis, V. D. (2002) *Geographic Information Systems in Oceanography and Fisheries*, Taylor and Francis, London

Valentine, P. C., Todd, B. J. and Kostylev, E.V. (2005) 'Classification of marine sublittoral habitats, with application to the northeastern North America region', *American Fisheries Society Symposium*, vol 41, pp183–200

Valesini, F. J., Clarke, K. R., Eliot, I. and Potter, I. C. (2003) 'A user-friendly quantitative approach to classifying nearshore marine habitats along a heterogeneous coast', *Estuarine, Coastal and Shelf Science*, vol 57, pp163–177

Valesini, F. J., Hourston, M., Wildsmith, M. D. and Cohen, N. J. (2010) 'New quantitative approaches for classifying and predicting local-scale habitats in estuaries', *Aquatic Conservation: Marine and Freshwater Ecosystems*, in press

Valiela, I. (1995) *Marine Ecological Processes (Vol. 2)*, Springer, New York

Wilkinson, D. (1999) 'The disturbing history of intermediate disturbance', *Oikos*, vol 84, pp145–147

Zacharias, M. A., Morris, M. C. and Howes, D. E. (1999) 'Large scale characterization of intertidal communities using a predictive model', *Journal of Experimental Marine Biology and Ecology*, vol 239, pp223–242

12

High Seas and Deep Seas

Pelagic and benthic, hydrography and biogeography

A smooth sea never made a skilled mariner.

English proverb

Introduction

The open oceans – or high seas, a quasi-legal term – are generally recognized as those lying beyond the coastal jurisdictions of sovereign states – presently therefore beyond the 200nm limit from any coastline. The deep seas – a topographic rather than a legal term – are generally recognized as those areas (primarily applicable to the seabed) lying at depths greater than about 200 to 300m – that is, areas of the oceans beyond the continental shelf. Because the 200nm limit and the edge of the continental shelf rarely coincide spatially, it is immediately apparent that some high seas may not be deep, and that some deep seas are not high.

Our knowledge of the deep seas and open oceans beyond the limits of national jurisdiction is limited, although even four decades ago a number of remarkable photographs of the deep sea had been taken (see Heezen and Hollister, 1971). To date, no comprehensive and agreed upon biogeographic classification exists for all of the world's open ocean and deep seabed areas, although some work has been undertaken for certain ecosystems. Previous global marine biogeographic classifications (see Box 12.1) have generally been limited to coastal waters, including the recent marine ecoregions of the world (MEoW) classification of Spalding et al (2007), which extends only to the 200m isobaths or 200nm offshore.

Because of these limitations for the progress of global marine conservation, an international workshop was convened in Mexico City in order to initiate work towards biogeographic classifications for the open oceans and deep seas. This chapter draws heavily on the results of that workshop presented in the GOODS report (Global Open Oceans and Deep Seabed) and its classifications (UNESCO, 2009).

The workshop first defined a set of basic principles and a framework for the recognition and classification of coherent biogeographic regions in deep and open oceans (Box 12.2). These basic principles should allow scientists to spatially delineate biogeographic provinces that have recognizably different biodiversity components. Available information was processed using geographic information systems (GIS) in order to present the geophysical and hydrographic features that can help delineate biogeographic regions and explain species distributions. The primary focus of the report from the workshop was to delineate major representative 'ecosystems' in the open ocean and deep seabed area outside national exclusive economic zones (EEZ) or comparable zones, and oceanward of continental

Box 12.1 A selection of classifications of global marine biogeography

Zoogeography of the Sea (Ekman, 1953)
One of the first classic volumes, originally published in German in 1935, Ekman recognized but did not clearly map a number of 'faunas', 'zoogeographic regions' and 'subregions' of the oceans.

Treatise on marine ecology and palaeoecology (Hedgpeth, 1957)
This work relied on Ekman, but also reviewed other sources, and produced a first global map showing the distribution of the highest level 'littoral provinces'.

Marine Zoogeography (Briggs, 1974)
A thorough taxonomic-based classification that still forms the basis for much biogeographic work. The focus is on shelf areas and it does not provide a biogeographic framework for the high seas. It presents a system of regions and provinces, with broad-scale provinces (53 in total) defined as areas having at least 10 per cent endemism.

Classification of coastal and marine environments (Hayden et al, 1984)
An important attempt to devise a system of spatial units to inform conservation planning. The coastal units are closely allied to those proposed by Briggs.

Biomass Yields and Geography of Large Marine Ecosystems (Sherman and Alexander, 1989)
One of the mostly widely used classifications, LMEs (64 globally) are 'relatively large regions on the order of 200,000km^2 or greater, characterized by distinct properties of: bathymetry, hydrography, productivity and trophically dependent populations'. LMEs were devised through expert consultation, taking account of governance and management practicalities. The system is restricted to shelf areas and some adjacent major current systems but does not include all island systems. LMEs are not defined by their biotas.

A Global Representative System of Marine Protected Areas (Kelleher et al, 1995)
Although not strictly a classification, this is one of the few global efforts to consider global marine protected areas coverage. Contributing authors were asked to consider biogeographic representation in each of 18 areas and this work provides important guides to the biogeographic literature and potential spatial units.

Ecological Geography of the Sea (Longhurst, 1998)
This system of broad biomes and 'biogeochemical provinces' is based on abiotic measures and surrogates for productivity. The classification consists of four biomes and 57 provinces. They are largely determined by satellite-derived surrogate measures of surface productivity (ocean colour as chlorophyll) and temperature regimes, and inferred locations of changes in other parameters (including mixing and the location of the nutricline). Some of the proposed divisions lie close to lines suggested by taxonomic biogeographers, but others cut right across major ocean gyres, splitting some of the most reliable units of taxonomic integrity.

Ecoregions: The Ecosystem Geography of the Oceans and Continents (Bailey, 1998)
Bailey has provided critical input into the development of terrestrial biogeographic classification, but his work also provides a tiered scheme for the high seas. The higher level 'domains' are based on latitudinal belts similar to Longhurst, while the finer-scale divisions are based on patterns of ocean circulation.

Marine Ecoregions of the World: A bioregionalization of coastal and shelf areas (Spalding et al, 2007)
This latest classification system is based on review and synthesis of existing biogeographic boundaries as well as expert consultation. It covers coastal areas and continental shelves, but not the deep and open oceans beyond national jurisdictions. The classification system includes 12 realms, 58 provinces and 232 ecoregions.

Source: Modified from Spalding et al (2007) and UNESCO (2009)

shelves in those regions where continuity of the same ecosystem exists.

Current biogeographic knowledge suffers from limited data of open ocean and deep sea ecosystems, as well as from a lack of understanding of their vulnerability, resilience or the functioning of marine biodiversity. Most marine research has been conducted in shallow coastal waters where the components of biodiversity are far more accessible than in remote deep sea environments which require specialized technology and equipment to access. The greater costs of research in these areas has meant that deep and open ocean research has been given a far lower priority than issues closer to shore, which are generally seen as being of more direct relevance to day-to-day uses of the ocean.

Our knowledge about deep and open ocean areas beyond the limits of national jurisdiction is limited both in terms of numbers of samples, and in their uneven spread around the globe. Many of the existing samples from the deep seas have now been documented by the Census of Marine Life (CoML, www.coml.org) project on the diversity of abyssal marine life (CeDaMar, www.cedarmar.org). These samples have provided for the description of patterns of species distribution in areas beyond national jurisdiction, and will, in the future, help our understanding of the composition and richness of species through ongoing programmes such as CoML, and the associated ocean biogeographic information system (OBIS, www.iobis.org). It is with the help of OBIS programmes and other databases worldwide that the GOODS study has provided the first preliminary attempt at classifying the open oceans and deep seas into distinct biogeographic classes.

Existing approaches to marine biogeographic classifications

In the deep and open ocean areas, biogeographic classification is far less developed than in terrestrial, coastal and continental shelf areas, although there have been substantial efforts at marine biogeographic classification at the local, national and regional scales in coastal waters. There have been fewer such attempts to delineate marine bioregions globally, due mainly to the difficulties in acquiring data on this scale. In the pelagic environment, the only purely data-driven global marine biogeographic classification is that of Longhurst (1998) – see Chapter 4.

Another widely used, although not strictly biogeographic, classification is that of large marine ecosystems (LMEs), which are perhaps the most widely used for management purposes (see Chapter 4). Regional classifications exist for many coastal and shelf waters, although most are only described in the 'grey literature'. The OSPAR (Oslo and Paris Convention) maritime area is an example of a well-developed regional classification (Dinter, 2001).

However, open ocean and deep sea areas beyond national jurisdiction are not covered, nor are many island systems, and the boundaries of LMEs have been set by a combination of biological and geopolitical considerations. The more recent Marine Ecoregions of the World (MEOW) (Spalding et al, 2007; see Chapter 5) provides a more comprehensive classification based solely on biodiversity criteria, but does not extend to the open ocean and deep sea areas.

The preferred system of biogeographic classification should be consistent with available knowledge from a variety of disciplines, including taxonomy, physiognomy, palaeontology, oceanographic processes and geomorphology. A summary of some present approaches to classification of marine environments was presented in Table 5.3, indicating that coastal, shelf and deep and open ocean areas can all be viewed from a variety of perspectives, and classified according to a variety of attributes, for a variety of purposes.

Taxonomic methods

There is a long history of biogeography based on species ranges, and the broad global patterns of taxonomic distributions are relatively well known, though continuously subject to revision as new genetic methods are applied (see Chapter 10) and bio-exploration of the seas continues. Taxonomic methods and surveys alone are, however, not presently sufficient to fully classify the biodiversity of the oceans. For the vast majority of the oceans insufficient information is

Box 12.2 The set of principles decided by GOODS workshop participants to guide designation of biogeographic regions

1. Pelagic and benthic environments should be considered separately.
To a first approximation the pelagic realm is fully three-dimensional, whereas the benthic realm features two-dimensional properties. The ecological scales and processes operating in the two systems are also fundamentally different (see Chapters 2 and 3). The pelagic system is dominated by oceanographic processes operating on large spatial scales but relatively shorter time scales. These processes are reflected strongly in the patterns of occurrence of pelagic species. In contrast, the patterns of benthic species occurrences are strongly influenced by processes reflecting the depth, topography and substrates of the seafloor, processes that often have much finer spatial scales but persist on longer temporal scales. Although the two realms exchange energy and organisms, and are coupled, their complements of taxa, size-spectra of species, life spans of species, and communities of organisms are significantly different. Thus it is reasonable to expect that different combinations of factors will need to be used to classify these two environments.

2. A classification of biogeographic regions for the selection of representative areas cannot be based upon unique characteristics of distinctive areas or upon individual focal species.
Conservation efforts may legitimately be directed towards protection of distinctive areas or species because of their unique value to biodiversity, but attention to such areas alone will not address the overall patterns of species distribution in the great majority of the oceans.

3. The classification system needs to reflect taxonomic identity, which is not addressed by ecological classification systems that focus on biomes.
Although geographically widely separated biomes may have similar physical environments, functions and types of communities, their community species compositions and hence biogeographies are different. The benefits of protecting representative portions of one biome will therefore not accrue to the different species found in other similar functional biomes. A consequence of principles 1–3 is that biogeographic classification of deep and open ocean areas must use the taxa themselves to delineate biogeographic provinces, and this inevitably becomes the first level (see number 6 below) of a classification for broad-scale biogeographic boundaries.

Next, within such biogeographic areas – where the faunal and floral assemblages are already defined at some scale – physiognomic and other factors can be used to achieve finer scale classifications.

4. The biogeographic classification system should emphasize generally recognizable communities of species, and not require presence of either a single diagnostic species or abrupt changes in the whole species composition between regions.
Both endemic species and discontinuities in the ranges of many species may occur within properly delimited biogeographic zones, but there will always be anomalies in distributions of individual species, and some species are cosmopolitan. What is important is that the community structure should change in some definable and consistent way, such that the dominant species determining ecosystem structure and regulating ecosystem function have changed, whether the types of ecosystem characteristics of the zone or lists of species have changed greatly or not.

5. A biogeographic classification should recognize the influences of both ecological structures and processes in defining habitats and their arrays of species, although the operative factors will be different in the pelagic and benthic worlds.
In the pelagic world, processes of ocean circulation dominate. These broadly correspond to biogeographic provinces and biomes, but their boundaries are dynamic and influenced by water motions in both vertical and horizontal planes. In the benthic world, geomorphological structures (seamounts, ridges, vents etc.), topography and physiography (scales of rugosity and complexity, and substrate

composition) determine the type of benthic community and its characteristic species assemblages, and these structures are comparatively less dynamic than circulation features, resulting in more static biogeographical boundaries.

6. A meaningful classification system should be hierarchical, based on appropriate scales of features, although the number of divisions required in a hierarchy is less clear.
Any factor used in a biogeographic classification system should enter the hierarchy at the scale at which it is judged to most significantly affect distributions (local, regional, global) – or to have done so historically. To do otherwise will produce neither a comprehensive hierarchy nor clear and inclusive categories within any level of the hierarchy. Thus, for example, in the pelagic environment water masses of the ocean gyres and depth categories delimit species assemblages, while smaller scale features such as convergences and other frontal systems may serve to mark their boundaries or transitions. Such large-scale oceanographic features that strongly influence the species assemblages are inherently dynamic, with boundaries whose positions change over time. In the benthic environment, the largest scale biogeographic provinces will be determined by evolutionary history and plate tectonic movements of the basin. In addition, the local-scale units would be determined by topography, geochemistry of the sediment–water interface and substrate characteristics which are more persistent over time.

Source: Modified from UNESCO (2009)

available, and at regional scales it is impossible to directly conduct comprehensive biological surveys. Instead, it is necessary to rely on extrapolations of relationships between biota and their habitats – that is, on geophysical (physiognomic) data of the physical environment itself (see Chapter 6).

Physiognomic methods

The term physiognomic is derived from terrestrial biogeographic work where habitats could be broadly defined by the structural or physiognomic characteristics of a region and its vegetation. Ensuing classifications across a broad range of scales were then shown to be closely allied to driving abiotic influences, and that such influences could be used to map patterns of vegetation. In the marine environment, the potential for predicting patterns using abiotic drivers is potentially extremely valuable given the poor state of knowledge of biotic distributions.

Environmental factors can adequately define habitat characteristics and associated biological community types from regional to seascape scales (Chapters 5 and 6), although aliasing of physical and biological data may be problematic.

Within ecoregions, where the array of community types is already biogeographically defined, geophysical factors can predict at least the major community types fairly accurately (OSPAR, 2003; Kostylev et al, 2005). Physiognomic data can therefore provide a level of calibration for mapping representative areas, and this general approach is now in widespread use in coastal and shelf waters.

Ecological geography

Longhurst (1998) described regions of the epipelagic oceans, based primarily on remotely observed temperature and ocean colour, and added additional data to infer oceanographic and trophodynamic processes. However epipelagic boundaries and productivity regimes are only one part of the patterns of marine biodiversity, they do not ensure taxonomic identity, and cannot alone form a general basis for delineating marine ecozones.

The concept of large marine ecoystems (Sherman and Alexander, 1989) also has several drawbacks when considered as a global marine biogeographic classification. The boundaries of LMEs reflect a set of compromises among a variety of considerations. Boundaries are at least partially determined by geopolitical considerations, and with a few exceptions the concept has been restricted to shelf areas. Also, the concept of LMEs does not consistently incorporate physiognomy or global ecological geography, and the boundaries do not consistently demonstrate a greater degree of homogeneity of biodiversity components within LMEs than across adjacent ones.

Political or governance management regions

The boundaries used to delineate regional fisheries or oceans management organizations are generally based on the distributions of fish stocks managed by the jurisdictions of the participating states. Although they may be somewhat internally homogeneous in fauna, their boundaries cannot be counted on to coincide with any major discontinuities in species composition. Rather, boundaries reflect the limits of legal agreements and historic patterns of fisheries or other ocean uses, and coverage of deep and open ocean areas beyond the limits of national jurisdiction is far from complete.

Hydrography and biogeography of the world ocean

Hydrography

The GOODS report provides a summary of salient features of the hydrography of the world's oceans (UNESCO, 2009). There are several such summaries available in oceanographic works (e.g. Tomczak and Godfrey, 1994). Unfortunately, common to most such presentations is the fact that variables important to our understanding of biogeography (e.g. temperature, salinity and dissolved oxygen) are given broadly only for surface and abyssal waters with single profiles characterizing entire ocean basins.

Over the last several decades, most of the hydrographic data taken during research cruises has been compiled by the National Oceanic and Atmospheric Administration (NOAA) National Oceanographic Data Center, and is available online (www.nodc.noaa.gov). From this data, UNESCO (2009) generated maps for visualization of patterns and developed GIS layers for temperature, salinity and dissolved oxygen. Hydrographic patterns were summarized at depth intervals of 800, 2000, 3500 and 5500m, and then plotted on the bathymetric maps in a manner that emphasized contact of the water with the benthos at the relevant biogeographic depths of 800, 2000, 3500 and 5500m. This is a far more instructive approach than the conventional presentation of hydrographic profiles, and readers are referred to the GOODS report (UNESCO, 2009) for these important figures.

Temperature

At 800m water temperatures still differ significantly among the major ocean basins. The Arctic is very cold, below 0°C, as is the Southern Ocean. A steep front exists along the northern border of the Southern Ocean with temperatures rising by 3 to 6°C over a distance as short as 5 degrees of latitude. At 40° S the Atlantic is the coldest ocean with water about 4°C; the Pacific is slightly warmer at 4°C in the east and 7°C in the west. The Indian Ocean overall is warmer (6–10°C) than the Pacific (3.5–6°C). The Atlantic Ocean is cold in the south, but due to the effects of the Gulf Stream and Mediterranean outflow warms to more than 10°C between 20 and 40° N.

At 2000m the water has cooled considerably in the Indian Ocean, being about 2.5 to 3°C everywhere north of 40–45° S. The Pacific over most of its area at this depth is about 0.5 degrees cooler. The Atlantic at this depth is for the most part between 3 and 4°C. The Southern Ocean is coldest to the east of the Weddell Sea, the latter being the locus of formation of Antarctic Bottom Water, and warmest south of the eastern Pacific. At 3500m the ocean basins become more subdivided by topography. The effects of Antarctic Bottom Water (from the Atlantic sector) are still clearly seen in both the Indian and Pacific oceans, where temperatures are between 1.25 and 1.5°C over most of the area, but can reach 2°C. The Atlantic remains the warmest of the major basins, being about 2.5°C over most of the basins. At 5500m, the deepest parts of the ocean basins reflect the temperature pattern seen at 3500m, the major exception being the deep water in the Weddell Sea, where bottom temperatures are below 0°C.

Temperature gradients can also indicate the location of frontal zones, where water masses meet and mix. The major surface water convergence areas (e.g. the Subtropical Convergence, the Antarctic Convergence) signify large changes in water characteristics, such as between Antarctic, temperate and tropical waters. Many species do not cross such boundaries, either because of physiological limitations to adults or their early life stages, or because of physical limits to dispersal. These convergence zones may not extend below the upper bathyal depths, but the resultant effects of increased productivity and other ecological processes may well influence benthic composition and abundances.

Salinity

Salinities of the world's open and deep oceans do not vary by much more than 1psu (practical salinity unit) over most of the area and at all depths. However, salinity ranges and salinity gradients (in combination with temperature) are indicators of different water masses that often determine species and community distributions, and indicate origins and sources of recruitment. One of these water masses, the Antarctic Intermediate Water, is characterized by a salinity minimum at around 1000m in the South Pacific. This water mass does not extend northwards into the North Pacific, and many deepwater fish species associated with such water do not occur in the northern Pacific.

Other areas where salinity is very different are at 800m in the NW Indian Ocean where the salinity may be over 36psu, and in the North Atlantic where the salinity is influenced by the Gulf Stream and Mediterranean outflow. Because of the Gulf Stream, high salinity water extends as far north as the Iceland–Faeroes Ridge on the eastern side of the Atlantic. In deeper waters, the salinity becomes more uniform.

Oxygen

As with temperature, oxygen is an important correlate of the presence of species in various parts of the ocean. Oxygen values vary over a wide range; the highest values are generally associated with the colder, deeper and younger waters, and the lowest values with older deeper waters and the oxygen minimum layer of intermediate waters.

At 800m the waters with highest oxygen concentrations are in the Arctic with values about 7mL.L^{-1}, and the Antarctic Intermediate Water in all three major basins where values are between 5 and 5.5mL.L^{-1}.

At 2000m the influence of the upper Antarctic Bottom Water can be seen in both the Indian and Pacific oceans where dissolved oxygen values are between 3 and 4mL.L^{-1} over most of the southern portions of both basins, but fall to below 2 mL.L^{-1} in the northern parts of these oceans. In contrast, Atlantic waters at this depth are very highly oxygenated at 6.5 to 5.5mL.L^{-1} north to south due to the southward flowing North Atlantic Deep Water. From 3500m to the deepest parts of all the basins the pattern of dissolved oxygen follows that seen at 2000m. However, in the Indian and Pacific basins, the better oxygenated Antarctic Bottom Water has spread all the way to the northern reaches, so that dissolved oxygen values are always more than 3mL.L^{-1}.

Summary

In summary, according to the GOODS report (UNESCO, 2009), from a benthic biogeographical perspective the important hydrographic variables are temperature and dissolved oxygen. Although temperature and salinity become progressively less variable with depth in all oceans, nevertheless, along with oxygen (which retains its variability with depth) these three factors still differ considerably in various parts of all ocean basins. However, temperature and salinity in combination are most useful to characterize individual water masses – as indicators of biological origins and patterns of connectivity.

The general water mass circulation and topography of the deep and bottom waters of the oceans is becoming known, but what this means in terms of the origins and maintenance of biogeographic patterns is still far from clear. We still have little idea of how the deeper parts of the oceans are connected on time scales relevant to evolutionary processes such as generation and dispersal times of the biota, or how they may be physiographically isolated. Nevertheless, hydrographic factors and bathymetry may provide clues to potential biogeographic distributions, but these need to be tested against species distributional data.

Biogeography

The GOODS report presents a comprehensive review of the history and state-of-the-art deep sea biogeography, and summarizes the biogeographic provinces suggested by various authors; these will not be repeated in detail here (see UNESCO, 2009). Ideas on deep sea biogeography have undergone significant evolution, but are still very speculative and many conceptual problems remain to be clarified. Any suggestions on the biogeography of the deep sea depend significantly on the spatial coverage, and on the group of fauna studied, particularly on: the taxonomic level (family, genus, species, genetic); the taxonomic group (e.g. molluscs, crustaceans, fish); dispersal abilities of the group. Only some of the major issues will be summarized in this part of the chapter.

Vinogradova (1997) summarized the literature on deep sea fauna studies up to that time. From that analysis the studies of deep sea benthic fauna were categorized into three major schools of thought regarding deep sea zoogeographic patterns:

1 Those who think that the bottom fauna should be very widespread because of the lack of ecological barriers and relative homogeneity of conditions on the deep sea floor.
2 Those who think that the deep sea fauna is fractionated by the presence of topographic features that divides the sea floor into about 50 separate ocean basins.
3 Those who subscribe to the idea that species have more extensive ranges at greater depth.

Based on earlier work, Vinogradova (1979) noted that the ranges of species in fact tended to contract, rather than expand with depth (although the opposite view has been proposed), and believed that species' ranges were constricted due to the presence of deep sea ridges, causing a delimitation of basins with their own distinct faunas. Such a view would be consistent with differences in temperature and oxygen ranges as a result of deep ocean circulation patterns and topography. For the entire World Ocean, Vinogradova (1979) found that 85 per cent of the species examined occurred in one ocean only, and only 4 per cent were common to the Atlantic, Indian and Pacific oceans. Overall, Vinogradova characterized the fauna of the deep sea regions as highly endemic with a large number of endemic genera and families.

Reviewing species distributions in the Pacific, Vinogradova (1979) concluded that there was an apparent bipolarity of bottom fauna distribution in certain groups. Most seemed to be eurybathic species following the deep abyssal cold waters, from the Antarctic to the northern Pacific. This bipolarity is possibly related to the existence of cold shallow waters at the poles connected by deep cold waters, and to the higher sedimentation of organic matter to the deep sea at high latitudes. Zezina (1997) noted the bathyal zone is a place where relict species, 'living fossils', have often been found. Such organisms are prevalent among crustaceans and fish, but also among crinoids, gastropods and others. Patterns of endemism vary greatly among taxa, taxonomic level and topographic feature.

The deep-sea soft-sediment environment hosts a diverse and highly endemic fauna of uncertain origin, but genetic studies are now shedding light on the processes of its evolution. As an example, Zardus et al (2006) quantified patterns of genetic variation in the protobranch bivalve *Deminucula atacellana*, a species widespread throughout the Atlantic Ocean at bathyal and abyssal depths. They showed that genetic divergence was much greater among populations at different depths within the same basin than among those at similar depths but separated by thousands of kilometres, and they suggested that isolation by distance probably explains much of the inter-basin variation. It is clear that broadly distributed deep-sea organisms can possess highly genetically divergent populations, despite the lack of any morphological divergence.

In sharp contrast to the cryptic diversity often revealed by molecular studies of benthic marine organisms, however, are the studies by Lecroq et al (2009) and Pawlowski et al (2007). Their studies revealed the existence of exceptionally widespread benthic foraminifera, which constitute the major part of deep-sea meiofauna. Analyses of nuclear ribosomal RNA revealed high genetic similarity between Arctic and Antarctic populations of three common deep-sea foraminiferal species even though they were separated by distances of up to 17,000km. These very broad ranges of the deep-sea foraminifera support the hypothesis of global distribution of small eukaryotes, and suggest that deep-sea species diversity may be more modest at global scales than present estimates suggest (see Chapter 8).

Principles and practices for a classification system for deep and open ocean areas

The development of a science-based biogeographic classification requires a framework and definition of a set of basic principles (see Box 12.2). These basic principles should permit the spatial delineation of separate areas that have recognizably different and predictable taxonomic compositions. Confidence in the delineation of such areas would be greater if they can be linked to oceanographic processes or geophysical structures that contribute to making them definably separate, and which suggest evolutionary mechanisms whereby their relative homogeneity could have arisen and their taxonomic diversity could be maintained.

The GOODS biogeographic classification (UNESCO, 2009) considered and rejected a number of alternative approaches, including distinctive areas, hotspots (of whatever kind including areas of high species diversity), ecologically and biologically significant areas, or the 'naturalness' of an area. Such considerations, while important for overall marine planning, do not lie within the scope of representativity.

Practical issues and open questions

A number of practical issues must be addressed in order to derive any meaningful biogeographic classification, and it is still not clear exactly how to derive such a framework for the open ocean and deep seas. A series of questions that are currently 'unresolved' are summarized in Box 12.3. Nevertheless, however a classification is finally derived for the open ocean and deep seas, a general considerations is clear: to define and map biogeographic regions and select representative areas will necessitate dealing with a 'mixed' system that combines taxonomic, ecological and physiographic approaches and factors.

The present-day distributions of organisms have resulted from a series of interacting processes at different time scales, including evolution, regional oceanographic processes of production, dispersal or retention and local adaptation to oceanographic and substrate factors. It is therefore to be expected that large-scale patterns in taxonomic occurrences, ecology and physiognomy should all have some coherence. This may provide the foundation of a synthesis of factors needed to describe the planet-wide patterns of representative marine faunas and floras. However, the extent, nature and causal basis for the concordance of these patterns has not yet been well explored.

As the data and patterns from classification systems are explored and consistencies are identified, it should be possible to synthesize them into coherent descriptions of global biogeography. In the pelagic realm this appears to be an attainable goal in the near future, but in the benthic environment, with a multiplicity of finer scale features, finding a consistent classification framework will require longer.

Data sources

The data used to inform the GOODS biogeographic classification process is presented in that report (UNESCO, 2009). The data were sourced from a number of publicly available databases and from researchers working in deep and open ocean environments. Because the biogeographic classification covers large oceanic areas around the world, the data needed to have consistent global coverage. The geographical coverage of biological data is often insufficient, and physical data such as bathymetry, temperature and substratum have commonly been used as surrogates of the

Box 12.3 How to develop a biogeographic classification for the open ocean and deep seas? A series of questions and issues that are currently unresolved

- Information is not equally available on community taxonomic composition around the globe, such that different groups of experts, each using the best information available in their area and discipline, may not draw the same maps. How can differences among biogeographic schemes be reconciled, where they are based on community taxonomic composition?
- Much of the taxonomy of deep sea species is still unknown to the species level, and for some animal groups many genera are widespread. What level of taxonomy should be used (species, genera, families)? Is there a biological reason to justify any one as more suitable than the others, and are there problems with using mixed levels in one classification?
- Regardless of level, which taxonomic groups should be used (e.g. zooplankton, macrobenthos, fish)? Is there a better strategy than just using whatever is available?
- How should we deal with transition zones, faunal breaks and other discontinuities, given that ocean processes are dynamic and that abrupt community discontinuities will be rare?
- Marine boundaries and conditions, particularly in the upper water column, are variable in both space and time, and any mapping can only be one 'snapshot' of current and recent historical knowledge; thus it will only describe the biogeography of a quiescent ocean. How should we deal with variability in general, especially seasonal and inter-annual, given that the same dynamic oceanographic processes suggest that boundaries of biogeographic zones are unlikely to be spatially very stable?
- Regardless of the classification used, any scheme should clearly and explicitly state the principles and strategies upon which it is based, so that subsequent communications have an identifiable and unambiguous starting point.
- A final consideration is clear: to define and map biogeographic regions and select representative areas will necessitate dealing with a 'mixed' system that combines taxonomic, ecological and physiographic approaches and factors.

ecological and biological characteristics of habitats and their associated species and communities.

Pelagic realm and biogeography

The key purpose of networks of marine protected areas (MPAs) on the high seas is a universally acknowledged need to ensure the conservation of the characteristic composition, structure and functioning of ecosystems. Although conceptually different, taxonomic and physiognomic classification systems are highly interdependent. Composition is best reflected in biogeographic classification systems based on taxonomic similarity, whereas structure and function require consideration of systems also based on physiognomic classifications.

One of the desired features of a network of MPAs is the inclusion of representative areas. This objective requires considering a taxonomically based system, because marine biomes with the same physiognomic features in different parts of the sea will have different species compositions. Hence even a well-positioned MPA in one zone would not be representative of the species in a similar biome elsewhere, even if the main physical features and processes were very similar.

Characteristics of the pelagic realm and their importance to biogeographic classification

The physical and biological characteristics of the pelagic realm have been reviewed in Chapters 2 and 3. After reviewing a variety of options, UNESCO (2009) concluded that the main large-scale physical features that an appropriate classification system should capture should include: core areas of gyres; equatorial upwellings;

upwelling zones at basin edges; important transitional areas – including convergence and divergence areas.

Each ocean basin has a large gyre located at approximately 30° North and South latitude in the subtropical regions. The currents in these gyres are driven by the atmospheric flow produced by the subtropical high pressure systems. Smaller gyres occur in the North Atlantic and Pacific oceans centred at approximately 50° N latitude. Currents in these systems are propelled by the circulation produced by polar low pressure centres. In the southern hemisphere, these gyral systems do not develop as clearly because of the lack of constraining land masses and the existence of the circum-global West Wind Drift.

Upwelling regions tend to have very high levels of primary production compared to the rest of the ocean. Equatorial upwelling occurs in the Atlantic and Pacific oceans where the southern hemisphere trade winds reach into the northern hemisphere, giving uniform wind direction on either side of the equator. Surface water is drawn away from the equator, causing the colder water from deeper layers to upwell. The equatorial region, as a result, has high phytoplankton concentrations and high production.

Areas of convergence and divergence represent transition regions. For example, the Antarctic Polar Front, which fluctuates seasonally, is considered to separate the Southern Ocean from other oceans. This ocean zone is formed by the convergence of two circumpolar currents, one easterly flowing and one westerly flowing. These oceanographic features are readily differentiated, and generally have distinct assemblages of species and some distinct species.

Starting with these main geophysical features, finer-scaled biographic units nested within the large-scale features were then considered by UNESCO (2009), such as basin-specific boundary current upwelling centres, and core areas of gyres. Such nested areas, although functionally defined, were considered to generally reflect distinctive taxonomic biogeography. Information on species ranges is available for validation of the taxonomic meaningfulness of the candidate boundaries in enough of those nested cases to allow a tentative acceptance of the patterns. A further level of nesting is often ecologically reasonable, in order to reflect functional ecologically holistic regions at finer scales. These have been defined for the coast and shelf areas by Spalding et al (2007), but there are currently insufficient data to apply this nested scale on a global basis.

Classifying the largest scale units into a set of ecological biomes can produce useful ecological insights. These would recognize the commonalities between, for example, eastern boundary currents, equatorial upwellings and so on that may be repeated in different oceans. The need for consistent use of terms, many of which may have broad or variable interpretations in the wider scientific and technical community, was recognized in the GOODS report. For example, the concepts of 'core' versus 'edge' are particularly important. The term 'core areas' is used to represent areas of stability in critical ecosystem processes and functions, whereas at 'edges' important ecosystem processes are often in transition and may display sharp gradients. Although the central importance of ecological processes – most notably productivity – is recognized, such processes are not the basis for delineating biogeographic units.

Further challenges

The pelagic realm contains several features which present specific challenges for biogeographic classification. For example, little information is available that could be used to indicate biogeographic patterns of the deeper pelagic biota. The current proposal of UNESCO (2009) was focused on observations in the photic zone, down to 200m; it is to be expected that patterns will diverge from those of surface waters with increasing depth. Available information on taxonomic patterns or even of the abiotic drivers below this depth remains so poor that it is unlikely that global-scale classification of deep-pelagic biogeography is possible at the present time.

It is presently not known whether all biodiversity 'hotspots' were captured in ecologically appropriate ways by the proposed GOODS system. However, the UNESCO (2009) group agreed that centres of species richness probably are well captured, sometimes by transition/

convergence areas which are rich through the mix of different communities, and sometimes by core areas of features that capture stable conditions for community maintenance, and major productivity processes.

Three types of migratory patterns of pelagic species were identified in the GOODS report (see also Chapters 7, 9, 16 and 17):

1 Those shifting consistently between two locations or general areas; for example humpback whales. A good classification system should ensure that each general area was within a clearly defined unit, but the classification would not have to show any particular relationship between the two locations.
2 Those aggregated at one location and then moving widely; for example species with fixed breeding grounds and wide feeding ranges. A good classification system should ensure that the consistent location was within a clearly defined unit, but on a case-by-case basis the distribution of the species otherwise might or might not be informative about boundaries of other units, depending on what affected the migration.
3 Those showing more constant movements. The species of this class most appropriate for delineating biogeographic regions were species of limited motility, whose pelagic life history stages depend on oceanographic processes. Their distributions can be informative about the effects of water-mass, gyres and boundary/transitional zones on ranges and distributions of other species in the assemblages.

Pelagic biogeographic units differ from benthic, shelf and terrestrial ones in showing far greater temporal and spatial variability in the location of their boundaries. Although some boundaries are clean and fairly abrupt (spanning only a few tens of kilometres), others are broader gradients with mixing of species from different zones across an area sometimes hundreds of kilometres in width. Some of these transitions zones are relatively permanent features of biodiversity and were considered sufficiently distinct to be separately classified in the GOODS study. In most cases, however, the sharp lines of boundaries portrayed on maps must be regarded only as general indicators of a zone of change which is broad, and which is often moving through time.

The pelagic classification system and data sources

Many data sources are available for the pelagic realm, and these are cited in the GOODS report (UNESCO, 2009). The GOODS group initially used a Delphic (expert-driven) approach to prepare a preliminary global mapping of biogeographic zones for open ocean pelagic systems. Next a series of biogeographic publications was consulted, and additional expert knowledge was applied. The Atlantic map was influenced particularly strongly by White (1994), the Pacific map by Olson and Hood (1994), and the map of the Southern Ocean by Grant et al (2006).

UNESCO (2009) produced a map of pelagic biogeographic classes, which is presented in Figure 12.1. The biogeographic classification included 30 provinces. These provinces each have unique environmental characteristics in terms of variables such as temperature, depth and primary productivity (see UNESCO, 2009 for details).

In the regions of the world's oceans that have better inventories of pelagic biodiversity, some major oceanographic features like central gyres and boundary currents consistently coincided with provinces delineated on taxonomic grounds, leading to confidence in the proposed classification. However, the exact boundaries on the pelagic biogeographic map remain a work in progress. Notwithstanding the need for additional refinements, the major zones indicated by UNESCO (2009) are considered reasonable for use in planning and management for conservation and sustainable use of the components of pelagic marine biodiversity.

Benthic realm and biogeography

The benthic realm is inherently more complex and divided than the pelagic realm (see Chapters 2 and 3); available data tends to be sparse and

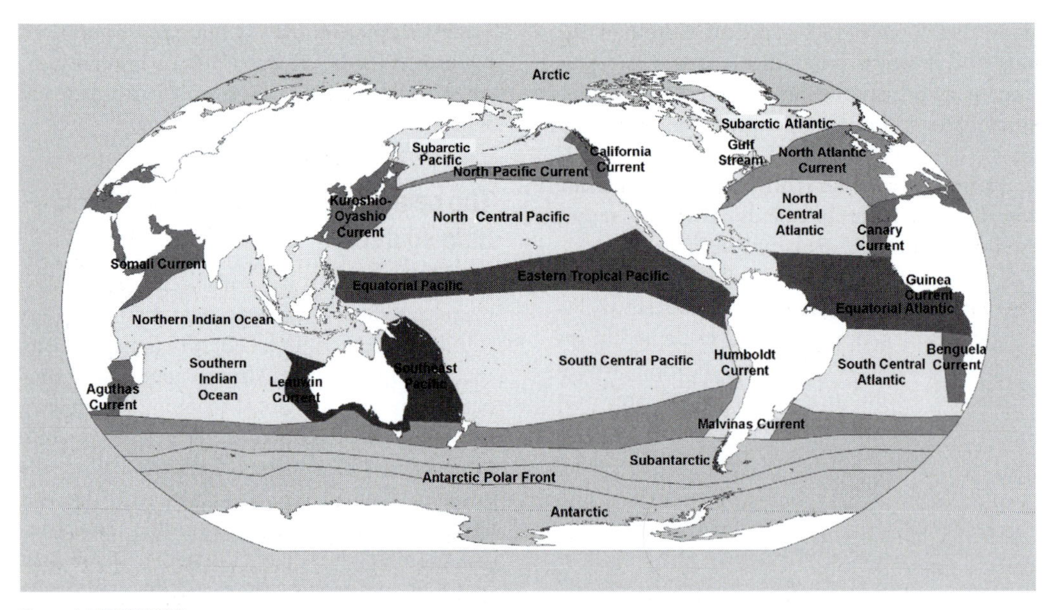

Source: UNESCO (2009)

Figure 12.1 Global map of pelagic provinces

heterogeneous. For much of the deep sea there is little information that can be used to delineate biogeographic units at the level of either province or region. This lack of information is partly due to lack of sampling, but is also due to a lack of synthesis of existing data. Available physical and chemical data taken over the past century have all been compiled by the US National Oceanographic Data Center (NODC, www.nodc.noaa.gov).

Despite these limitations, the GOODS report (UNESCO, 2009) provides preliminary maps containing the locations of 'the centers of distributions' of deep sea provinces at bathyal and abyssal depths. Available biological information was compiled, and the distribution patterns of this data, along with as much of the hydrographic data as possible, were then examined for correlations among biota and geophysical variables.

These efforts are predicated on the premise that benthic species are influenced in their distributions by the major water masses of the ocean. Unfortunately, while the surface water mass distributions are well known, at depths below about 800m, water masses and currents have not been comprehensively mapped. The objective of the GOODS report (UNESCO, 2009) was therefore to produce maps of the bathymetry, bottom temperature, salinity, oxygen and organic matter flux for discrete depth layers and to assess the relationship between known organism distributions and these water mass characteristics. These factors, although limited in number, are generally accepted as being key determinants of distributions of the benthic biota.

Bathymetry

Benthic biogeographic provinces are distributed vertically as well as horizontally, reflecting the fundamental importance of depth as a surrogate for several physical and biological structures and processes (see Chapters 2 and 3). The GOODS report (UNESCO, 2009) provides maps of ocean characteristics within the depth zones 300–800m (upper bathyal), 800–2000 and 2000–3500m (upper and lower portions of the lower bathyal), 3500–6500m (abyssal), and > 6500m (ultra-abyssal and hadal). The 0–300 and 300–800m layers were not generally considered in the GOODS report because these areas lie almost exclusively within the EEZs of various nations, and are therefore largely included in Spalding et al (2007). Although these depth categories were carefully chosen after examination of existing data, there is still significant discussion about the appropriateness of the remaining depth zones below 800m.

Lower bathyal (800–3500m)

The lower bathyal provinces (Colour Plate 16a) consist almost entirely of three physiographic categories: lower continental margins, isolated seamounts and oceanic island slopes, and mid-ocean ridges. The lower bathyal of the continental margins is for the most part sedimentary and is really part of the extended continental shelves of coastal nations. In contrast, seamounts, island flanks (and often the summits) and mid-ocean ridges, although they may have some sediment cover, also offer expanses of hard substrate for invertebrates, and bathyal fishes.

Seamounts

Seamounts (see also further section below) and ridges provide areas of lower bathyal depth in offshore areas otherwise dominated by the abyssal plains. These elevated topographic features have different faunas from the surrounding seafloor because they are 'islands' of shallower habitat providing a wide range of depths for different communities. Bare rock surfaces can be common on seamounts because of current flows that scour the often steep flanks. Even though the area covered by ridges and seamounts may be small in relation to the surrounding seafloor, their geographical location may be very important in determining the distribution of bathyal species across the wider ocean basins.

The distribution of seamounts is shown in Colour Plate 16b which summarizes their estimated summit depths (excluding seamounts on the abyssal plains with summit depths > 3500m) based on satellite altimetry (Kitchingman and Lai, 2004). This figure illustrates the extent to which seamounts extend the distribution of bathyal habitat throughout the world ocean. Most of the seamounts at less than 800m of depth are at least partially within national EEZs, as are a large number of those seamounts with summits between 800 and 2000m depth.

Abyssal (3500–6500m)

The abyssal provinces (Colour Plate 16c) cover the bulk of the deep ocean floor. Most of the abyss is characterized by deep, muddy sediments, although hard substrate in the form of metalliferous nodules may also be present. The abyssal provinces were designated based on the deep basin(s) in which they occur. The scheme modifies that of Menzies et al (1973) and Vinogradova (1997) based on newer data (see UNESCO, 2009).

With the exception of the Central Pacific, the ocean basins are separated by the mid-ocean ridge system. There are, however, gaps in nearly all the ridges, allowing some water flow from one basin to another. In the Indo-West Pacific region there are a few small basins that are completely isolated from the rest of the abyssal ocean, but these are mostly within the EEZs of various nations.

Ultra-abyssal and hadal areas (> 6500m)

The GOODS report adopted the scheme presented by Belyaev (1989). These regions (see UNESCO, 2009) are primarily restricted to lithosphere plate boundaries where subduction occurs. Most of the trenches are in the western Pacific, between the Aleutians and Japan, the Philippines, Indonesia, the Marianas, and the Kermadec trench around New Zealand. The eastern Pacific has only the Peru–Chile trench and the Atlantic has the Puerto Rico and Romanche trenches. All but two of the trenches are within the EEZs or management areas of various countries.

Proposed benthic biogeographic provinces – summary

The benthic biogeographic units adopted in the GOODS report started with the concepts regarding regions and provinces promoted by Menzies et al (1973) and Vinogradova (1979) for the abyssal areas, Belyaev (1989) for the hadal (ultra-abyssal) areas, and Zezina (1973, 1997) for the bathyal. Boundaries were adjusted on the basis of more recent data (see UNESCO, 2009).

The proposed deep sea benthic biogeographic classification encompasses the three large depth zones outlined above: the lower bathyal, 800–3500m, the abyssal, 3500–6500m, and the hadal, which is found only at depths greater than 6500m, primarily in the trenches. Little consideration was given to the upper bathyal, depth range (300–800m), because almost the entire bottom at that depth is within the EEZ of one country or another.

The GOODS report acknowledged that the lower bathyal probably covers too broad a depth range, and may warrant further splitting at around 2000m where there are marked changes in species composition or diversity for a number of taxa (e.g. demersal fish). The hadal is also for the most part encompassed by the EEZs of various countries, and the biogeographic provinces for that realm are well established by the work of Belyaev (1989).

All of the proposed provinces should be considered as provisional, especially for the lower bathyal where data are sparser. The proposed abyssal classification for the Atlantic basins is, however, probably robust, where the pattern has been tested using the distributions of deep sea protobranch bivalves (Allen and Sanders, 1996). In contrast, the GOODS report cautions that the Indian and Pacific ocean basins are much less well studied and the patterns there were deduced using proxies including temperature and organic matter input. For the present, all boundaries between provinces should be considered as transition areas of unknown extent.

Compatibility between GOODS and other biogeographic classifications

In order to be of maximum usefulness, the GOODS biogeographic classification (UNESCO, 2009) should be compatible with existing global and regional classification systems. Particular attention was paid to the compatibility between the GOODS biogeographic classification and the marine ecoregions of the world classification (MEOW; Spalding et al, 2007), which is the newest classification system covering coastal areas and continental shelves (see Chapter 5).

Because the MEOW classification has already defined key biogeographic boundaries at the national and regional level in coastal and shelf waters, compatibility between MEOW and GOODS should ultimately allow for a fully globally integrated classification system that incorporates the finer-scale classifications in coastal waters with the larger spatial units in the open ocean and deep sea area. There will always be some overlaps, mismatches and ill-defined boundaries in any biogeographic system, but continued research and new data will permit this important subject of complementarity of classification systems to be further refined.

Seamounts and other geomorphic units of deep seas

Various kinds of anomalous structures and geomorphic units are found in the deep oceans; all warrant conservation in one form or another, and all could be considered to occupy a position intermediate between representative and distinctive areas. Chief among these areas are seamounts and deep sea vents.

The location of ocean ridge systems containing vent systems is relatively well known, but how many of these areas may be active at any time is not known. These areas require special attention for several reasons. First, any individual vent only has a 'life span' of some 40 to 150 years, so that a strategy for conservation needs to be mobile in some way because the locations of active systems will change over time. Second, on the basis of their genetic uniqueness alone the faunas (and bacterial floras) of these areas deserve close attention and preservation (see Chapter 10). Because hydrothermal vent communities are governed by processes that differ from those determining the locations of broad bathyal provinces, a separate hydrothermal vent biogeographic map was also produced by the GOODS group. The GOODS scheme follows that of Van Dover et al (2002) and their unpublished data (UNESCO, 2009).

Here, however, attention centres on seamounts. Unfortunately seamounts were not considered at any length in the GOODS report and they are still poorly known components of the deep sea. However, they are of vital interest both from the perspective of their unique contributions to marine biodiversity and because of growing resource use and impacts by the fishing community. Because of their hard substrates and often distant location offshore, seamounts and mid-ocean ridges have only recently been investigated using modern oceanographic tools such as submersibles, moorings and remotely operated vehicles.

A seamount is a mountain arising from the ocean seafloor but that does not break through to the sea surface. They are typically formed from extinct volcanoes and rise steeply as independent features (or sub-marine islands) from the seafloor, at 1000–4000 metres depth, to at least 1000 metres above the seafloor. The peaks are therefore still hundreds to thousands of metres below the sea surface. Seamounts are common throughout the world's oceans though they are believed to be more abundant in the Pacific Ocean. They may number over 100,000 globally, although some estimates suggest as many as one million (depending upon 'seamount' definition).

Their unique characteristics – including hard substrates (as opposed to the typical deep sea soft substrate), steep slopes, interaction with currents leading to upwelling of both nutrients and plankton, retention of biota in associated gyres, geographic isolation and high degree of endemism – make them unique structures of the ocean floor. The physical presence of the seamount intercepts lateral currents, and creates hydrographic eddies and flows that can restrict the dispersal of larvae and plankton and keep species and production processes concentrated over the seamount.

Seamounts may play a key role in the maintenance of open ocean biodiversity by acting as 'stepping-stones' for trans-oceanic dispersal of biota, refugia for relict populations and areas of local speciation. They possess highly individualistic and endemic biotas. For example, Richter de Forges et al (2000) report the discovery of more than 850 macro- and mega-faunal species from seamounts in the Tasman Sea and southeast Coral Sea, of which 29–34 per cent are new to science, and potential seamount endemics. Low species overlap between seamounts in different portions of the region indicates that the seamounts in clusters or along ridge systems function as 'island groups' or 'chains', leading to highly localized species distributions and apparent speciation between groups or ridge systems that is exceptional for the deep sea.

The current inventory of 535 species of seamount fishes (Froese and Sampang, 2004) represents only about 2 per cent of the recent fish species on earth. However, seamount fishes belong to 130 (25 per cent) of 515 families and 29 (47 per cent) of 62 orders. They are thus mostly not closely related to each other – that is, their genetic diversity at higher taxa is higher than suggested by the number of species (see also Chapter 10). Many seamount families are small, with 13 families consisting only of seamount species and 12 families with half or more of their members living on seamounts. According to Marshall (1979) and Haedrich (1997), deep-sea fishes are representatives of groups that appeared early in the evolution of modern fishes. Many of them are highly adapted to the environment and ecological conditions of the deep sea, with specialized eyes, complex bioluminescent organs, elaborate gas glands and swim bladder constructions, and often remarkable jaws and teeth.

The high level of endemicity at seamounts can be explained in terms of island biogeography theory. The theory was developed by MacArthur and Wilson (1967) to explain species richness of actual newly created islands, but has since been applied to any ecosystem surrounded by unlike ecosystems (here – seamounts which act as underwater 'islands'). The theory proposes that the number of species found on an undisturbed island is determined by rates of immigration, emigration and extinction. Immigration and emigration are affected by the distance of an island from a colonizing source. The rate of extinction is affected by island size as larger habitat size reduces the probability of extinction due to chance events. After immigration, habitat heterogeneity increases the number of species that will be successful. Over time, these opposing forces determine an equilibrium number of species. The factors that influence the number of species on any 'island' can include degree of isolation, time since isolation, size of island, climate/temperature, productivity and human activity.

The degree of endemicity of seamounts is still the subject of investigations and differences. The seamount endemicity hypothesis (SMEH) states that seamounts possess a set of isolating mechanisms that produce highly endemic faunas. However, McClain et al (2009) found little support for the SMEH among megafauna of a northeast Pacific seamount, instead finding an assemblage of species that also occurs on adjacent continental margins. A large percentage of these species were also cosmopolitan with ranges extending over much of the Pacific Ocean basin. Conversely, in support of the SMEH, the genetic structure of Patagonian toothfish populations in the Atlantic indicated that populations from around the Falkland Islands were genetically distinct from those at South Georgia (Rogers et al, 2006). Genetic differentiation between these populations is thought to result from hydrographic isolation, because these areas are separated by large geographic distance and water in excess of 3000m deep, below the distributional range of toothfish (< 2200m).

An understanding of the significance of the biodiversity of seamounts is especially urgent because of their increasing exploitation by deep-sea fisheries, and serial depletion of long-lived, late-maturing fishes. Migratory fish and cetaceans rely on visiting seamount food webs, so that the impact of overfishing raises serious concerns. As pelagic fisheries have declined, fisheries have increasingly turned to deeper waters to maintain catches, progressively exploiting deep-sea species of seamounts (e.g. Koslow et al, 2000). Deep-sea fishing techniques have been shown to destroy benthic habitats and in particular the complex coral habitats upon which many invertebrates depend. Controlling these activities is of major concern, especially since about half of all seamounts lie in international waters.

The GOODS report (UNESCO, 2009) did not contain a biogeographic classification of seamounts, but this is a necessary prerequisite for any systematic conservation approach. A very preliminary classification of seamounts is given in Box 12.4 which attempts to take into account the basic features of their biogeography, geomorphology and ecology, with the intention of summarizing their probable biodiversity and ecological significance.

Conclusions and management implications

The pelagic and benthic biogeographic classifications in the GOODS report (UNESCO, 2009) are the first global attempts to comprehensively classify the open ocean and deep seafloor into distinct biogeographic regions. The classifications use geophysical and environmental characteristics of the benthic and pelagic environments to identify regions of similar habitat and their associated biological communities. Such biogeographic classifications have as their prime purpose the definition of distributions of species and habitats for the purposes of scientific research, conservation and management. Future refinements of these classifications should eventually provide a basis for describing global patterns of representative marine fauna and flora.

The present GOODS report is still preliminary although even in its present format it provides a basis for discussions that can assist policy development and implementation. However, further basic research on 'what lives where' and what affects the patchy nature of deep sea biotic distributions is needed to advance our understanding of this vast reservoir of largely still unexplored marine diversity and its associated biogeographic classifications.

The policy processes and classification of deep seas and open oceans

Recent policy discussions on the conservation and sustainable use of biodiversity, including genetic resources, in marine areas that lie beyond national jurisdictions have pointed out the need for more information on the various components of biodiversity to be found in those areas, and for a classification of those areas to be developed based on scientific criteria. Such discussions have

Box 12.4 A suggested classification of seamounts

A hierarchical classification system, starting from the GOODS report (UNESCO, 2009) and based on concepts of Roff et al (2003) and Roff and Taylor (2000), suggests proceeding through a series of factors each of which is assumed to act at a larger scale than the ones below it. See also Rowden et al (2005).

Level	Factor	Description
1	Biogeography	Provinces from GOODS report
2	Depth	Zones from GOODS report
3	Height	Height of seamount summit below sea level
4	Size	Surface area of summit
5	Productivity regime	Flux of organics from surface Taylor columns and eddies
6	Degree of isolation	Distance to continental shelf or next 'upstream' seamount
7	Substrate complexity	Complexity of topography/substrate

NOTES

1. Biogeography. The first level is defined from deep sea provinces of GOODS. (Note that province and depth may need to be reversed in relative importance.)
2. Depth. Seamounts may have different characters and species arrays depending upon their location on: shelf, slope, upper or lower bathyal, abyssal.
3. Height of seamount summit above substrate, whether it penetrates into the upper mixed layer of the ocean or thermocline or not.
4. Size of seamount as surface area of summit. Size will be related to several factors, including productivity regime, habitat diversity, species and community diversity. Island biogeography theory can be applied to seamounts to predict rates of species colonization and extinctions. This is appropriate since seamounts have often been referred to as 'underwater islands'. A size of 50km may be one critical divide, between seamounts having associated eddies and those without. Another, pragmatic, consideration may be a critical size of seamounts that are 'worth' fishing and those too small to attract such attention.
5. Productivity regime (essentially organic matter flux at depth) is a complex factor dependant on several variables, including:

   - flux at depth which may be predicted from Suess's (1980) relationship between surface production (indexed from surface chlorophyll a [e.g. Longhurst, 1998 categories]) and depth;
   - actual flux at depth, related to presence or absence of the Taylor column;
   - stratification of water column;
   - current speed and direction;
   - latitude.

   Productivity regime may be indexed at the surface by evidence of:

   - chlorophyll plumes or other index of water clarity;
   - sea surface temperature anomaly (local upwelling effect);
   - sea height anomaly and gyres.

   However, any of these anomalies may also be sub-surface (100+ metres).

   Taylor columns (and local eddies) may or may not be formed depending upon a series of critical parameters, including the blocking parameter, and the Burger number (a class of stratification parameter). Depth is also a critical factor in stratification, which may determine the development of downstream eddies. This has consequences for energy flux and dispersasl of propagules.

- Blocking parameter: B = h.f.L/DU (see Roden 1987 for theory). Taylor columns only form when B > 1.

Where:

h = seamount height, f = Coriolis parameter, L = seamount diameter [note that < or > 50km may be a critical value for development of eddies around a seamount], D = water depth, U = local current speed

- Burger number: $S = g.\Delta\rho.D/\rho.f^2.L^2$ (see Roden 1987 for complete theory). The Burger number interacts with the blocking parameter, the β parameter [important for seamounts > 50km diameter], and directions of flow in upper and lower layers.

Where: g = gravity, $\Delta\rho$ = density difference between water layers, ρ = water density

6. Degree of isolation (see Grigg et al, 1987) may be related to uniqueness and degree of endemism of fauna. Two elements may be important for the source of colonizing propagules and probability of species extinctions and colonizations: 'upstream' distance (direction and current speed) to nearest continental shelf and to nearest seamount neighbour. The nearest continental shelf area is likely to be the main source of recruitment of species.
7. Topographic/substrate complexity is strongly related to changes in species composition (community type) and abundances (see Chapters 6 and 8). However, this level of data may be redundant if entire seamounts are to be protected.

all recognized that biogeographic classifications can contribute importantly to policy-setting and implementation. Biogeographic classification enhances our knowledge and global understanding of marine life by integrating and centralizing information on its taxonomy, distribution and the biogeophysical characteristics that influence it (see Box 12.5).

Marine biogeographic classifications such as UNESCO (2009) are vital for the implementation of ecosystem-based management measures, and for the development of spatial management tools including networks of representative MPAs. By identifying the range and distribution of marine species, habitats and ecosystem processes, the spatial analysis of biogeographic data provides visual information that can be viewed in conjunction with information on human impacts in order to set boundaries for management actions. It can also: (i) serve as a basis to identify areas representative of major marine ecosystems and habitat types to be included in networks of representative MPAs; (ii) help to assess gaps in existing MPA programmes; (iii) help to set priorities for management action in areas of high human use; and (iv) guide further marine scientific research into areas where significant information gaps exist.

Given these applications, biogeographic information, especially when combined with ecological information, can assist the implementation of the provisions of a number of international and regional conventions, such as the Convention on Biological Diversity (CBD, www.cbd.int), which relate to the conservation and sustainable use of biodiversity. In addition, the CBD addresses deep seabed genetic resources beyond the limits of national jurisdictions.

Collecting further biogeographic information is therefore crucial to consolidating current knowledge about the status of, trends in, and possible threats to deep seabed resources of all kinds, beyond national jurisdiction, and for providing information relevant to the identification and implementation of technical options for their conservation and sustainable use (UNESCO, 2009).

Box 12.5 Some applications of biogeographic classifications to the conservation and sustainable use of deep sea and open ocean areas and their biodiversity

Sound biogeographic information has many possible applications. Two examples of practical applications of biogeographic classification are summarized here.

Applying biogeographic classification for marine protected areas

To date it has been difficult to undertake strategic action towards the development of a comprehensive, effectively managed and ecologically representative system of protected areas in deep and open ocean areas due to our limited knowledge about how and where species and their habitats are geographically distributed. Such a system of protected areas should incorporate the full range of all components of marine biodiversity in protected sites, including all habitat types. The amount of each habitat type should be sufficient to cover the variability within it, and to provide duplicates (as a minimum) so as to maximize potential connectivity and minimize the risk of impact from large-scale effects (CBD, 2004).

By informing governments about the large-scale distribution of the elements of marine biodiversity within a science-based framework for biogeographic classification, the results of the GOODS report (along with recommendations from the Azores Workshop, see: http://cmsdata.iucn.org/downloads/iucn_information_paper.pdf) provide tools that can assist governments in making significant progress towards the 2012 target for establishing representative networks of MPAs.

The following three initial steps recommended by the Azores expert meeting are now feasible:

Scientific identification of an initial set of ecologically or biologically significant areas

The criteria as proposed by the workshop should be used, considering the best scientific information available, and applying the precautionary approach. This identification should focus on developing an initial set of sites already recognized for their ecological values, with the understanding that other sites could be added as new and/or better information comes available.

Develop/choose a biogeographic habitat and/or community classification system

This system should reflect the scale of the application, and address the key ecological features within the area. Usually, this will entail a separation of at least two realms – pelagic and benthic.

Drawing upon steps 1 and 2 above, iteratively use qualitative and/or quantitative techniques to identify sites to include in a network

Their selection for consideration of enhanced management should reflect their recognized ecological importance and vulnerability, and address the requirements of ecological coherence through: representativity; connectivity; replication; and assessing the adequacy and viability of the selected sites. Consideration should be given to their size, shape, boundaries, buffering and appropriateness of the site management regime.

Applying biogeographic classification to marine spatial planning

In the context of marine spatial planning, biogeographic scientific information is combined with information on uses, impacts and opportunities for synergy among stakeholders to identify specific areas for protection or for specific uses over different time scales. This approach has been successfully used in the marine coastal areas of many countries around the world (Ehler and Douvere, 2007).

In a policy setting, normally stakeholders' aspirations, expectations and interests are analysed against biogeographic and other similar scientific information such as knowledge of ecological processes and biodiversity impact assessments, so as to agree on possible common agendas. In this way, the resulting policies represent the combination of scientific knowledge, stakeholders' interests and political decisions for actions such as the identification of areas to be subjected to restricted management measures or areas to conduct further investigations. An example in this regard is given by the regional units identified in the context of the UN's Regular Process for the Global Reporting and Assessment of the State of the Marine Environment including Socioeconomic Aspects, as these regions represent a combination of ecological, legal, policy and political criteria that serve well the purpose of assessing the state of the marine environment from a combined ecological and human use perspective (see also Chapters 16 and 17).

Source: From UNESCO (2009)

Unfortunately, the value and contribution of biogeographic knowledge to the policy-making process is still not widely understood. The overarching international legal framework governing human activities in marine areas beyond national jurisdiction is set forth in the 1982 United Nations Convention on the Law of the Sea (UNCLOS) and other sector-based and environmental agreements. In recent years, the CBD, the United Nations Informal Consultative Process on Oceans and the Law of the Sea (UNICPOLOS) and other UN working groups have devoted significant attention to the need to enhance international cooperation and action in areas beyond national jurisdictions.

Future links between biogeographic classifications and policy-making

There is increasing recognition of the significance of sound biogeographic classification to policy development and implementation, and also an increasing demand for biogeographic information on open ocean and deep sea areas beyond national jurisdiction. Consequently, as in many fields of applied ecology, there is a need to bridge the gap between policy demand and scientific research which generates biogeographic knowledge (UNESCO, 2009).

Biogeographic research programmes will benefit from the political support needed to build international scientific cooperation at a global scale, as well as adequate funding. An example is provided by the Census of Marine Life (CoML, www.coml.org) and its ocean biogeographic information system (OBIS, www.iobis.org). The Census and OBIS have existed for almost ten years and have provided a body of scientific knowledge that is comprehensive, globally unique in content, and vital for policy and applications for both conservation and development. Yet the future of these and of similar programmes is unclear.

A further challenge facing the scientific community is the transfer of biogeographic information to policy makers in a manner that is accurate, timely and relevant. The GOODS report (UNESCO, 2009) demonstrates that the scientific community involved in the biogeography of the oceans is increasingly aware of this responsibility to address policy needs, so that the conservation and sustainable use of biodiversity in marine areas beyond national jurisdiction at all levels – genetic, species, ecosystems and seascapes – can be achieved in the years to come.

References

Allen, J. A. and Sanders, H. L. (1996) 'The zoogeography, diversity and origin of the deep-sea protobranch bivalves of the Atlantic: The epilogue', *Progress in Oceanography*, vol 38, pp95–153

Bailey, R. G. (1998) *Ecoregions: The Ecosystem Geography of the Oceans and Continents*, Springer-Verlag, New York

Belyaev, G. M. (1989) *Deep Sea Ocean Trenches and Their Fauna*, Nauka Publishing House, Moscow, Russia

Briggs, J. C. (1974) *Marine Zoogeography*, McGraw-Hill, New York

CBD (2004) 'Technical advice on the establishment and management of a national system of marine and coastal protected areas', *CBD Technical Series*, vol 13

Dinter, W. P. (2001) *Biogeography of the OSPAR Maritime Area*', German Federal Agency for Nature Conservation, Bonn

Ehler, C. and Douvere, F. (2007) 'Visions for a sea change. Report of the first international workshop on marine spatial planning', *IOC*

Manual and Guides no. 48, ICAM Dossier no. 4, Intergovernmental Oceanographic Commission and Man and the Biosphere Programme, UNESCO, Paris

Ekman, S. (1953) *Zoogeography of the Sea*, Sidgwick & Jackson, London

Froese, R. and Sampang, A. (2004) 'Taxonomy and Biology of Seamount Fishes', in T. Morato and D. Pauly (eds) *Seamounts: Biodiversity and Fisheries*, Fisheries Centre Research Report, vol 12, no 5, Vancouver, Canada

Grant, S., Constable, A., Raymond, B. and Doust, S. (2006) 'Bioregionalisation of the Southern Ocean: Report of experts workshop, Hobart, September 2006', WWF-Australia and Antarctic Climate and Ecosystems Cooperative Research Centre, Hobart, Australia

Grigg, R. W., Malahoff, A., Chave, E. H. and Landahl, J. (1987) 'Seamount Benthic Ecology and Potential Environmental Impact from Manganese Crust Mining in Hawaii', in B. H. Keating, P. Fryer, R. Batiza and W. Boehlert (eds) *Seamounts, Islands, and Atolls*, Geophysical Monograph, vol 43, American Geophysical Union, Washington, DC

Haedrich, L. H. (1997) 'Distribution and Population Ecology', in D. J. Randall and A. P. Farrell (eds) *Deep-sea Fishes*, Academic Press, San Diego, CA

Hayden, B. P., Ray, G. C. and Dolan, R. (1984) 'Classification of coastal and marine environments', *Environmental Conservation*, vol 11, pp199–207

Hedgpeth, J. W. (1957) 'Treatise on marine ecology and palaeoecology', *Memorandum of the Geological Society of America*, vol 67

Heezen, B. C. and Hollister, C. D. (1971) *The Face of the Deep*, Oxford University Press, New York

Kelleher, G., Bleakley, C. and Wells, S. (1995) *A Global Representative System of Marine Protected Areas*, World Conservation Union, Washington, DC

Kitchingman, A. and Lai, S. (2004) 'Inferences on Potential Seamount Locations from Mid-resolution Bathymetric Data', in T. Morato and D. Pauly (eds) *Seamounts: Biodiversity and Fisheries*, Fisheries Centre Research Reports, University of British Columbia, Canada

Koslow, J. A., Boehlert, G. W., Gordon, J. D. M., Haedrich, R. L., Lorance, P. and Parin, N. (2000) 'Continental slope and deep-sea fisheries: Implications for a fragile ecosystem', *ICES Journal of Marine Science*, vol 57, pp548–557

Kostylev, V. E., Todd, B. J., Longva, O. and Valentine, P.C. (2005) 'Characterization of Benthic Habitat on Northeastern Georges Bank, Canada', in P. W. Barnes and J. P. Thomas (eds) *Benthic Habitats and the Effects of Fishing*, American Fisheries Society Symposium, vol 41

Lecroq, B., Gooday, A. J., and Pawlowski, J. (2009) 'Global genetic homogeneity in the deep-sea foraminiferan *Epistominella exigua* (Rotaliida: Pseudoparrellidae)', in W. Brökeland and K. H. George (eds) 'Deep-Sea Taxonomy: A Contribution to our Knowledge of Biodiversity', *Zootaxa*, vol 2096

Longhurst, A. (1998) *Ecological Geography of the Sea*, Academic Press, San Diego, CA

MacArthur, R. H. and Wilson, E. O. (1967) *The Theory of Island Biogeography*, Princeton University Press, Princeton, NJ

Marshall, N. B. (1979) *Developments in Deep-sea Biology*, Blandford, Poole

McClain, C. R., Lundsten, L., Ream, M., Barry, J. and DeVogelaere, A. (2009) 'Endemicity, biogeography, composition, and community structure on a northeast Pacific seamount', *Public Library of Science One*, vol 4, no 1, e4141

Menzies, R. J., George, R. Y. and Rowe, G. T. (1973) *Abyssal Environment and Ecology of the World Oceans*, John Wiley and Sons, New York

Olson, D. B. and Hood, R. R. (1994) 'Modelling pelagic biogeography', *Progress in Oceanography*, vol 34, pp161–205

OSPAR (2003) 'Criteria for the Identification of Species and Habitats in need of Protection and their Method of Application', Annex 5 to the OSPAR Convention for the Protection of the Marine Environment of the North-East Atlantic, OSPAR, London

Pawlowski, J. J., Fahrni, B., Lecroq, D., Longet, N., Cornelius, L., Excoffier, T., Cedhagen T. and Gooday, A. J. (2007) 'Bipolar gene flow in deep-sea benthic foraminifera', *Molecular Ecology*, vol 16, pp4089–4096

Richter de Forges, B., Koslov, J. A. and Poore, G. C. B. (2000) 'Diversity and endemism of the benthic fauna in the southwest Pacific', *Nature*, vol 405, pp944–947

Roden, G. (1987) 'Effect of Seamounts and Seamount Chains on Ocean Circulation and Thermohaline Structure', in B. H. Keating, P. Fryer, R. Batiza and W. Boehlert (eds) *Seamounts, Islands, and Atolls*, Geophysical Monograph, vol 43, American Geophysical Union, Washington, DC

Roff, J. C. and Taylor, M. (2000) 'A geophysical classification system for marine conservation', *Journal of Aquatic Conservation: Marine and Freshwater Ecosystems*, vol 10, pp209–223

Roff, J. C., Taylor, M. E. and Laughren, J. (2003) 'Geophysical approaches to the classification, delineation and monitoring of marine habitats and their communities', *Aquatic Conservation, Marine and Freshwater Ecosystems*, vol 13, pp77–90

Rogers, A. D., Morley, S. A., Fitzcharles, E., Jarvis, K. and Belchier, M. (2006) 'Genetic structure of Patagonian toothfish (*Dissostichus eleginoides*) populations on the Patagonian Shelf and Atlantic and western Indian Ocean sectors of the Southern Ocean', *Marine Biology*, vol 149, no 4, pp915–924

Rowden, A. A., Clark, M. R. and Wright, I. C. (2005) 'Physical characterisation and a biologically focused classification of "seamounts" in the New Zealand region', *New Zealand Journal of Marine and Freshwater Research*, vol 39, pp1039–1059

Sherman, K. and Alexander, L. M. (1989) *Biomass Yields and Geography of Large Marine Ecosystems*, Westview Press, Boulder, CO

Spalding, M. D., Fox, H. E., Allen, G. R., Davidson, N., Ferdana, Z. A., Finlayson, M., Halpern, B.,S., Jorge, M. A., Lombana, A. and Lourie, S. A. (2007) 'Marine ecoregions of the world: A bioregionalization of coastal and shelf areas', *Bioscience*, vol 57, no 7, pp573–584

Suess, E. (1980) 'Particulate organic carbon flux in the oceans: Surface productivity and oxygen utilization', *Nature*, vol 288, pp260–263

Tomczak, M. and Godfrey, J. S. (1994) *Regional Oceanography: An Introduction*, Pegamon, Oxford

UNESCO (2009) *Global Open Oceans and Deep Seabed (GOODS) – Biogeographic Classification*, UNESCO, Paris

Van Dover, C. L., German, C. R., Speer, K. G., Parson, L. M. and Vrijenhoek, R. C. (2002) 'Evolution and biogeography of deepsea vent and seep invertebrates', *Science*, vol 295, pp1253–1257

Vinogradova, N. G. (1979) 'The geographical distribution of the abyssal and hadal (ultra-abyssal), fauna in relation to the vertical zonation of the ocean', *Sarsia*, vol 64, nos 1–2, pp41–49

Vinogradova, N. G. (1997) 'Zoogeography of the abyssal and hadal zones', in A. V. Gebruk, E. C. Southward and P. A. Tyler (eds) 'The Biogeography of the Oceans', *Advances in Marine Biology*, vol 32

White, B. N. (1994) 'Vicariance biogeography of the open-ocean Pacfic', *Progress in Oceanography*, vol 34, pp257–284

Zardus, J. D., Etter, R. J., Chase, M. R., Rex, M. A. and Boyle, E. E. (2006) 'Bathymetric and geographic population structure in the pan-Atlantic deep-sea bivalve *Deminucula atacellana* (Schenck, 1939)', *Molecular Ecology*, vol 15, pp639–651

Zezina, O. N. (1973) 'Benthic biogeographic zonation of the world ocean using brachiopods', *Proceedings of the Russian Scientific Research Institute of Marine Fishery and Oceanography*, vol 84, pp166–180

Zezina, O. N. (1997) 'Biogeography of the bathyal zone', in A. V. Gebruk, E. C. Southward and P. A. Tyler (eds) 'The Biogeography of the Oceans', *Advances in Marine Biology*, vol 32

13

Linking Fisheries Management with Marine Conservation Objectives through Ecosystem Approaches

Compatibility of exploitation and preservation

The best fishermen I know try not to make the same mistakes over and over again; instead they strive to make new and interesting mistakes and to remember what they learned from them.

John Gierach (1946–)

Introduction

A text on marine conservation ecology would not be complete without exploring the ecological relationships between marine capture fisheries and efforts to conserve marine biodiversity. While other parts of this book examine specific conservation issues and applications at the genetic, population, community and ecosystem levels, this chapter specifically focuses on marine capture fisheries and commercial fisheries in particular given their potential to impact biodiversity and ecosystem function. This chapter begins with a summary of the current global status of marine fisheries followed by a discussion on how fishing affects marine biodiversity and ecosystems. Traditional (current) approaches to fisheries management are then discussed for the purposes of highlighting limitations of current management paradigms, including single-species management approaches, application of maximum sustained yield (MSY) and its variants, and limitations of current management controls. In light of recent national and international laws and conventions requiring that fisheries be managed in an 'ecosystem' context, this chapter focuses on improvements to fisheries ecology and management under the guise of 'ecosystem approaches'. Ecosystem approaches will be broadly explored in terms of their definition, purpose and application to conserving marine biodiversity while maintaining goods and services to humanity. In particular, the 'ecosystem approach to fisheries' (EAF) will be explored as fisheries are the most challenging aspect of implementing ecosystem approaches in marine environments. Lastly, current and successful applications of the EAF will be discussed and future directions of the EAF will be indicated.

Current status and trends of marine fisheries

The importance of marine fisheries to humanity is often overlooked and, from a Western perspective, seafood is just another source of protein and consumer choice. However, currently over one billion people depend on seafood as their primary source of protein and 2.6 billion people rely on seafood for at least 20 per cent of their protein (Davies and Rangely, 2010). Fishing has been estimated to employ 22 million in full- or part-time jobs and the first-sale value of the world's fisheries was estimated at US$84.9 billion in 2004 (Sinclair et al, 2002; Davies and Rangely, 2010).

However, the sustainability of marine capture fisheries has recently been shown to be at risk.

The United Nations Food and Agriculture Organization (FAO) currently reports that 28 per cent of the world's fisheries are now overexploited, depleted or recovering, 52 per cent are fully exploited and only 20 per cent underexploited or moderately exploited (FAO, 2009). For context, in 1974, the corresponding percentages were 10, 50 and 40 per cent respectively. The Commission of the European Communities has determined that, for the 43 per cent of European fish stocks that can be determined, less than 35 per cent are currently within safe biological limits (Commission of the European Communities (CEC), 2008). In the US, 47 per cent of the 188 stocks or stock complexes with sufficient information to determine whether overfishing is occurring are either overfished or currently have overfishing taking place (National Marine Fisheries Service (NMFS), 2009). In New Zealand, 29 per cent of the 101 stocks or stock complexes in which status can be determined are below their management targets (NZ Ministry of Fisheries, 2009). Status of high-seas stocks are even less well known. Of the 17 Atlantic fish stocks managed under the North Atlantic Fisheries Organization (NAFO) where sufficient status can be determined, six are collapsed and three are considered to be sustainable (NAFO, 2008). Atlantic cod biomass is estimated at 6 per cent of historical levels and North Sea cod stocks are depressed to such a degree that for years fishers have been unable to harvest up to their allowable catches.

While the above numbers are sobering, a closer look at certain aspects of marine capture fisheries shows that 50–70 per cent of pelagic predators have been removed by fisheries (Myers and Worm, 2003) and fishing pressure continues to shift towards lower trophic levels as apex predators decline; this has been termed 'fishing down food webs' (Pauly et al, 1998). Furthermore, the global fishing fleet is far larger than what is required although its effectiveness at catching fish continues to increase through technological innovations. This surplus fishing capacity is underwritten by $30–40 billion in annual subsidies by most fishing nations, thus providing no incentive to reduce fishing effort. Since the mid-1990s the world's fish capture production has peaked and has now levelled off at 70–80 million tonnes annually. However, this is only part of the equation as evidence suggests that fishing effort from port-based fisheries has increased over the past three decades at upwards of 3 per cent per decade. As such, more fishers are chasing fewer fish, and if the global fishing fleet were a country, it would be the 18th largest oil consumer on earth.

An estimated 20 per cent of the global fisheries catch constitutes unwanted by-catch that is discarded. Shrimp fisheries exhibit the largest amount of by-catch and small pelagic fisheries the least amount. By-catch also consists of the incidental take of endangered marine mammals, turtles and seabirds although recent advances in gear technology are reducing these impacts. Lastly, approximately 35 per cent of the global annual catch is captured on non-tropical continental shelves, which occupy only 5 per cent of the total ocean area. It is estimated that these levels of harvest on non-tropical continental shelves require 36 per cent of the total primary productivity in these areas to sustain these fisheries (Frid et al, 2005).

There are many books dedicated entirely to the plight of marine fisheries (e.g. Roberts, 2007), and a detailed description on how humanity found itself in its present predicament will be left to other texts. While many authors and governments have conducted post-mortem examinations on how humans either consciously or unconsciously ignored the warning signs of fisheries collapse, the following reasons are generally accepted as foundational to the current situation:

- Fisheries have traditionally focused on biomass and yield without proper consideration or understanding of the impacts of fisheries on marine ecosystems.
- Most fisheries operate as common property resources managed primarily under international agreements with minimal compliance and enforcement.
- Most fisheries in poorer countries are subsidence fisheries necessary to sustain life and therefore conservation provisions are secondary to sustenance needs.
- Fishing impacts are not readily apparent to the public in the same way that terrestrial losses of species or habitats are.
- Fishery management decisions by scientific bodies are often overturned through political interference either by ignoring or invoking uncertainty in scientific estimates.
- Traditional approaches to fishery management lag far behind those in terrestrial environments.

The above explanations for the global collapse of fisheries underpin the remainder of this chapter.

Impact of marine capture fisheries on biodiversity

Abundant evidence exists demonstrating the unintended impacts of fishing on marine biodiversity. Well-known impacts include bottom trawling, by-catch and reductions in target species, all of which are widely reported on in the popular media and have been incorporated into the school curriculums of many countries. In the past two decades, however, a more thorough understanding of the true impacts of fishing has begun to emerge that outlines the evolutionary shifts in population dynamics and changes to the structure and function of marine ecosystems (Pikitch et al, 2004). In particular, efforts have been directed at identifying and describing the causal links between fishing and the consequences to the broader marine ecosystem. This is not a trivial exercise given that fisheries impacts can either be direct (e.g. by-catch) or indirect (e.g. reducing prey items) and also exhibit different spatial and temporal dimensions that may or may not be detectable (Rosenberg, 2002). The following sections briefly summarize the current knowledge of the effects of marine capture fisheries on biodiversity at the genetic, community and ecosystem levels, following the framework and nomenclature that underpins the structure of this book.

Genetic effects of fishing

While there is no direct evidence that fishing modifies the genotypes of the target stocks or populations, there is no doubt that conventional single-species fishery management has resulted in the evolution of phenotypic traits, especially changes in life history features, in a way that favours traits such as early maturity, slow growth and smaller absolute maximum size (Law, 2000; Jennings and Revill, 2007). Thus the concern is that beneficial traits, such as fast growth to avoid predation and diversity of alleles (e.g. to adapt to changes in environmental conditions), are being selected out of the stock such that diversity acquired over long periods of evolution may be quickly lost with significant fitness consequences (Frid et al, 2006). Other genetic consequences of fishing that are assumed to be occurring, but lack observational or anecdotal evidence at this time, include: range reductions, population loss and fragmentation leading to inbreeding and genetic isolation; hybridization between species as a result of difficulties in finding mates; and changes in sex ratios (Law and Stokes, 2005). Furthermore, populations that have recovered from heavy exploitation may never recover their previous genotypic and phenotypic diversity (see Chapter 10).

Community-level effects of fishing

Fishing inevitably alters the food web in which a target species resides but the effects of fishing on marine community structure and food webs has only been quantified in the last decade. While a detailed discussion on the effects of fishing on marine trophic structure is beyond the scope of this chapter, a brief discussion on the community-level impacts of fishing is warranted. Fishing has the potential to affect marine biodiversity through altering energy flows and species interactions (including the strength of these interactions) through predation and competition. These changes may be triggered by direct (e.g. removing predators, by-catch) or indirect (e.g. indirectly benefiting a species either through a reduction in predation or reduced competition) effects leading to a number of biodiversity impacts, including: reductions in apex predators; increases in lower and middle trophic level biomass; changes in species interactions; reductions in the number and length of pathways in food webs; declines in species richness and density; changes in patterns of predation mortality leading to effects on recruitment and productivity; and reductions in food web resiliency (Pauly et al, 2002).

Continuous overexploitation may also trigger even larger, more serious events including trophic cascades and regime shifts. Trophic cascades can be created when fishing has unintended consequences for species that are one or more trophic levels removed from the target species. In a cascade situation, management actions may have unpredictable and surprising results thus introducing additional uncertainty into fishery management decisions. Fishing in Caribbean reefs has been found, through a trophic cascade, to increase primary producers (seagrass, macroalgae) and reduce populations of alligators, sharks, Caribbean monk seals (*Monachus tropicalis*), predatory fishes, invertebrates, corals, manatees (*Trichechus manatus*) and sea turtles (Jackson et al, 2001).

There is less agreement on whether fishing, in the absence of other natural or human-induced oceanographic events (e.g. climate change, decadal oscillations), can cause regime shifts, which are rapid reorganizations of food webs from one stable state to another. Probably the strongest case for fishing-induced regime shifts is in the northwest Atlantic where overfishing and the subsequent collapse of groundfish (primarily cod) stocks has resulted in what appears to be a permanent or semi-permanent shift from a groundfish-dominated to an invertebrate-dominated food web (Collie et al, 2004; Frank et al, 2005; Mangel and Levin, 2005).

While it is well known that fishing impacts food webs through removal of biomass, and that up to 40–50 per cent of the observed ecological impacts are then transmitted through indirect linkages (Schoener, 1983), a comprehensive understanding of the effects of fishing on ecosystems is not likely in the near future. For example, whale populations at present levels of abundance are thought to consume 3–5 times the global fisheries landings and, as whale populations continue to recover from past harvesting, fisheries catches will need to be adjusted downward to compensate for their additional consumption (Tamura, 2003). In addition, while fishing has only been directly responsible for a single extinction (the Steller's sea cow), many non-target species have been fished to commercial and perhaps ecological extinction from which populations may never recover. In recognition that marine ecosystems have a theoretical limit to the productivity that can be removed without serious consequences to marine biodiversity, an understanding of marine trophic interactions will be fundamental to effecting fisheries management.

Habitat and ecosystem effects of fishing

While the direct effects of fishing on habitats through bottom trawling, dynamite and cyanide fishing are well documented, fishing can also affect the abiotic components of ecosystem structure and function in other ways. Bottom trawling is by far the greatest threat to habitats and ecosystems through the scraping, scouring and resuspension of bottom sediments, and has the ability to change the bottom structure, microhabitats and associated benthic and epi-benthic fauna (Frid et al, 2000). There is considerable debate around the impacts of different types of bottom trawling; flat, mud or sand areas that are frequently trawled are generally of less concern than trawling on areas with more complex habitats or those areas (e.g. continental slopes, deep water) where trawling has not yet occurred but there is interest in doing so (Watling and Norse, 1998).

Other direct impacts to pelagic and benthic habitats may occur as a result of vessel disturbance (e.g. prop scarring, anchoring), nutrient cycling, pollution (from at-sea processing, greenhouse gas emission) and direct dumping of debris (gear, twine, food containers, plastic bands etc.). The indirect effects of fishing on the structure and function of marine habitats and ecosystems are difficult to evaluate, but are expected to have some impact on marine biodiversity (Gislason, 2002) and may include: changes in the rates of accretion and bioerosion in reef habitats (Pearson, 1981); contributing to hypoxia in benthic areas as a result of the release of hydrogen sulphide by trawls (Caddy, 2004); and elimination of macrophytes, benthos and demersal fish as a result of resuspension of sediments enriched in organic matter (Ball et al, 2000).

Current approaches to and difficulties with fishery management

The previous two sections explored the current state of the world's fisheries as well as the impact of fishing on marine ecosystems and biodiversity. This section will explore how fisheries are currently operated and the consequences and limitations of this approach. The human history of fishing is relatively straightforward and tends to repeat itself. As humans begin to overexploit a species, they: (a) exploit the same species somewhere else; (b) exploit less preferred species locally; and (c) increase local resource production through aquaculture (Lotze, 2004). Thus, if fishery resources are limited and effort is uncontrolled, mortality will increase until the fishery becomes economically non-viable or the population collapses. Early pre-Western cultures dependent on the sea for protein had the capacity to overexploit nearshore fishery resources and instituted rights-based systems of ownership and tenure that conferred title to marine resources at a local and tribal level. Perhaps the last example of this governance system at a true ecosystem scale was traditional Hawaiian marine management prior to switching to a commons model of marine stewardship in 1866 (Juvik et al, 1998).

Many definitions of 'fishery management' exist but the Food and Agriculture Organization (FAO) defines fishery management as:

> *The integrated process of information gathering, analysis, planning, consultation, decision-making, allocation of resources and formulation and implementation, with enforcement as necessary, of regulations or rules which govern fisheries activities in order to ensure the continued productivity of the resources and the accomplishment of other fisheries objectives.* (FAO, 1997)

Most fisheries throughout human history have been operated under the principle of single-species management, where each target stock or population is managed independently of other targeted stocks that may overlap in space and time (Larkin, 1977). Thus, fishery management advice under this model is provided on a stock-by-stock basis – a model under which most species continue to be managed today. Under moderate fishing pressure the single-species approach is an acceptable approach to management; it has been demonstrated to work well in terrestrial environments and, if properly applied, for the management of certain marine species (e.g. whales). However, as discussed in the previous section, global fishing pressures are now at a level where species-by-species management has the potential to seriously undermine the structure and function of marine ecosystems.

The purpose of this section is to review the single-species approach, its strengths and weaknesses, and show how this approach has been improved over the past two decades to include community and ecosystem considerations.

The single-species approach to fisheries

The purpose of single-species fishery management is to control fishing effort in order to avoid overexploitation of a stock and to ensure that fishers are provided with a suitable return on their investment. The alternative to single-species management is a situation where fishers make minimal or no profits while the fishery becomes biologically overexploited through overfishing, resulting in fish being caught at a size before they have realized their full growth and spawning potential. Therefore the goal of single-species management is to invoke technical conservation measures to prevent overexploitation by protecting young and/or spawning fish and/or making the fishery sufficiently inefficient that the zero profit level is reached before the stock is overexploited (Larkin, 1977; Cochrane, 2002).

The fundamental objective of traditional single-species management is to manage a fishery to MSY, which assumes that every fish stock generates 'surplus production' and that this production can be fished down to a biomass necessary to maintain the stock at some sustainable level. As such, traditional applications of MSY usually result in biomass reductions of the target stock of 30–50 per cent of unfished levels to maximize production (Mace, 2001, 2004). Given that managing multiple stocks to MSY in the same geographic (food web) area will inevitably result in consequences to the food web structure, fisheries managers have begun to realize that MSY is an upper threshold that should not be exceeded and thus have begun managing to a fishing mortality less than MSY, often re-expressed as F_{MSY}. A further improvement on the single-species approach is multi-species fishery management, where fishery management decisions are made in consideration of other target species under harvest in the same area, and may include ecotrophic considerations in management decisions. Regardless of the improvements to the MSY concept since its introduction in the 1930s, the following assumptions underpin single-species management (summarized by Babcock and Pikitch, 2004):

- The objective of management is to maximize the long-term average yields of the fishery.
- A population biomass level exists that will maximize the long-term average yield.
- Fish growth, natural mortality and fecundity are constant and do not change over time, irrespective of the abundance of other species, environmental changes or the effects of fishing.
- The total fishing mortality can be controlled by regulation of the fishery.

Current fishery management practices in most countries and international agencies are based on the single-species approach and generally follow the model below (adapted from Hilborn, 2004):

- Single-species stock assessments are undertaken for each stock or stock complex to set maximum sustained yield (MSY) or some other maximum threshold such as F_{MSY}.
- Regulations that determine, for various users, allowable time, area, gear and catch limits are recommended by fishery managers but are ultimately established through a political process.
- A centralized management agency responsible for science, decision-making and enforcement with costs paid by government is responsible for operating the fishery.
- Stakeholders are involved in decision-making either through policy, legal or political means.

However, in reality the traditional assumptions that underpin the single-species management model have resulted in the following characteristics (adapted from Pavlikakis and Tsihrintzis, 2000; Marasco et al, 2007):

- A disproportionate focus on short-term economic objectives and the maintenance of a single species.
- Removal of humans as ecosystem components from fishery models and decision-making.
- Political, economic and social values are either discounted or ignored.
- Needs of commercial fishing stakeholders take precedence over the public interest.
- Science and management occurs at local scales, where scaling up to regional, national or international interests is difficult.
- A focus on population ecology rather than community ecology in combination with oceanography, fisheries economics, ecology and fisheries biology.
- Modern tools, such as geographic information systems, oceanographic models and economic techniques are not applied.

Single-species fishery management is primarily based on developing stock assessment models to establish reference points that define overfishing limits based on (spawning) stock biomass and fishing mortality (Link, 2005). When these reference points are exceeded, control rules are triggered to reduce fishing either through input (e.g. number and size of vessels, time allotted to fish, gear restrictions) or output (e.g. amount of fish that can be caught) controls.

Properly applied, the single-species approach is a viable model for the management of certain fisheries and can provide significant economic and ecological value to the sustainable management of a fishery (Marasco et al, 2007). Single-species approaches can work where stock status can be determined, life histories are well known, the effects of environment changes on stocks is understood and compliance with regulations can be enforced. Notable examples of species that can be managed under this approach include certain marine mammals, such as those covered by the International Whaling Commission's Revised Management Procedure, and nearshore shellfish stocks where stock status and environmental conditions can be easily determined.

Most fisheries, however, do not meet the conditions for being managed under the single-species approach set out above and thus decisions based on this approach must face the reality that, in certain circumstances, single-species approaches will lead to overexploitation of a stock or population or economic failure of the fishery (Mace, 2004). Failures of the approach have generally not been the fault of the scientific and management aspects of the approach, but rather a failure of a lack of political will and data limitations (Marasco et al, 2007). For example, a review of the scientific advice provided to European decision-makers on 18 fish stocks in 2002 found that scientists provided the correct advice 53 per cent of the time, provided the wrong advice resulting in a detriment to the fishery 23 per cent of the time and 24 per cent of the advice consisted of 'false alarms' where catch reductions were recommended but later found to be unnecessary (Frid et al, 2005). Thus, scientific advice on single-stocks was either correct or neutral to the health of the stock 77 per cent of the time.

Perhaps the most contentious aspect of the single-species approach is the reliance on MSY or MSY variants in order to set fishery harvests. Many jurisdictions continue to use stock assessment models to attempt to determine MSY or the rate of fishing mortality F_{MSY} that a population or stock can sustain (Symes, 2007). As discussed earlier in this chapter nearly 30 per cent of the world's fisheries are overfished, suggesting that attempting to manage to MSY is failing either as a result of how MSY is calculated, uncertainty in the knowledge of how marine ecosystems function, or political interference resulting in allocations above MSY. Many stock assessment models used to calculate MSY result in typical biomass reductions of 50–70 per cent below unfished levels, which has been demonstrated to be biologically unfeasible for slow growing, late maturing species such as sharks and rays (Hirshfield, 2005). As such, setting catches below MSY resulting in larger stock sizes should result in fewer stock collapses and more positive medium and longer-term outcomes (Hillborn, 2004). In addition, traditional application of the MSY concept has led to presumptions that all mature female fish are of equal importance while in fact older females have a disproportionate spawning potential and thus may be critical to maintain population viability. In addition, the MSY concept ignores ecosystem effects of high harvest rates on fast growing, high biomass species, where populations appear to be resilient to high levels of fishing (Hirshfield, 2005).

Much of the difficulty in managing to MSY is related to uncertainty in single-species stock assessment models. These models have: been ignored by governments due to perceived uncertainty; failed to provide sound advice in a few important cases resulting in rapid fishery declines thus further contributing to the perception that they are inaccurate; failed to consider ecosystem effects and food web consequences that may negatively affect stocks; and failed to assist governments in designing regulatory systems to achieve fishery targets (Pauley et al, 2002). In addition, funding limitations often result in models that lack the necessary input data, scientific analysis, and applications across spatial and temporal scales to meet the needs of properly managing a stock. Further issues with the current stock assessment approach include a lack of involvement by stakeholders in the collection, analysis and interpretation of data.

There are a number of recent and proposed improvements to the single-species model, some of which are now enshrined in national and international law. Most suggested improvements to the model fall under the guise of establishing MPAs and improving fishery management regulations, which, while useful, are generally used in specific circumstances and have limited scope to address the broader challenges of ocean conservation and management (Day et al, 2008). A broader consideration of fundamental improvements to the approach fall under the following types (adapted from Hilborn, 2004; Symes, 2007):

- Elimination of the approximately $30bn in annual subsidies for fishing fleets.
- Reduction of target fishing mortalities through adopting a more precautionary approach to setting harvest levels.
- Establishing no-fishing areas over a significant (20 per cent or more) area of the world's oceans either through MPAs or other means.
- Elimination of destructive fishing practices, including prohibiting bottom trawling of previously untrawled areas and reducing by-catch.
- Weakening the current command and control fisheries management model to a system of co-management where responsibilities are shared between stakeholders.
- Establishing new forms of marine tenure where individuals or groups of fishers are guaranteed a specific share of a future catch thus removing incentives to overcapitalize and permit alignment of economic interests with long-term conservation goals.
- Establishing incentive-based versus regulatory-based management approaches.

Additional improvements to the single-species approach may include consideration of the species of interest in the broader ecosystem context. Examples of recent efforts towards these ends include: modifying stock assessment models to account for density dependence for the target species that arises from predation from another species (community effects); considering time-dependent disease and predation effects when modelling natural mortality; incorporating perceived regime shifts in defining biological reference points; incorporating considerations for low productivity or endangered species that may be affected by harvest of the target species in stock models (Marasco et al, 2007).

In conclusion, the single-species approach, however well intentioned and applied, is subject to the political whims of governments, which are rarely willing to fundamentally reform fisheries policies (Symes, 2007). Improvements on the single-species approach, including re-expressions of MSY to include a more precautionary approach to fisheries management and account for uncertainty (e.g. recruitment failure) in the dynamics of stocks and populations, is a step in the right direction but has not yet been proven to result in more sustainable fisheries management outcomes, most likely due to political interference and illegal fishing. Similarly, the multi-species approach has also shown little promise that it is a sustainable tool for fisheries management. The US Northeast multispecies groundfish management plan consists of 24 target species managed collectively since 1986. A 20-year review of the plan concluded that indicators of population and community health (e.g. spawning stock biomass) were not improving as projected and that illegal harvest was responsible for 12–24 per cent of total harvest (King and Sutinen, 2010). There is no shortage of national and international fishery conservation laws, conventions, agreements and policies that should be sufficient to control overfishing and ensure that community and ecosystem considerations are incorporated into decision-making but, for whatever reasons, fisheries continue to be overexploited thus requiring a new approach to managing both fisheries and oceans.

Guerry (2005) eloquently summarized the reasons for the failure of traditional species-based approaches as follows: fragmented ocean governance where the fishery as a commons resource has led to jurisdictions competing for the resources; inability to maintain ecosystem elements, such as water quality or spawning habitat necessary to sustain successful fisheries; inability to manage diverse, non-fisheries-related impacts, including pollution, habitat loss, overharvesting, climate change and introduced species; and lack of recognition of connections between ecosystem structure, functioning, and services. These connections include those between marine systems and land, marine habitats, species, other stressors, and knowledge and uncertainty (Guerry, 2005).

The ecosystem approach to fisheries

In light of the limitations of the single-species approach to fisheries management discussed above and, as discussed in Chapter 4, the recognition that a global and holistic approach to marine management that simultaneously considers multiple ecological and socio-economic objectives in the management of either a geographic area (e.g. protected area) or ecosystem (however defined) is necessary, the remainder of this chapter will explore the ecosystem approach to management as it applies to fisheries. The EAF will be discussed as fisheries are the most challenging aspect of implementing ecosystem approaches in marine environments.

The EAF stems from the realization that single- (and multi-) species approaches to fisheries management have, regardless of the reasons, resulted in a situation where most of the world's fish stocks are fully exploited or overfished (FAO, 2009). The term 'EAF' is used in this book because, after much debate, the FAO concluded that 'EAF' is a more appropriate term than ecosystem-based fisheries management as EAF explicitly considers ecosystem processes in the formulation of management actions.

Defining the ecosystem approach to fisheries

While the term 'EAF' is relatively new, the EAF is fundamentally a re-expression of the principles of sustainable development, which have been enshrined in international conventions since the 1972 United Nations Conference on the Human Environment (Scandol et al, 2005).

General concepts and definitions of the EAF tend to converge around the need either to alter existing practices or to develop a new fishery management paradigm that explicitly recognizes the interrelationships and dependences within food webs and that humans and human activities are a significant component in these systems and that they have the ability to quickly affect the sustainability of these ecosystems and ecosystem processes (Pitcher et al, 2009). The EAF is predicated on the assumption that an improved understanding and management of stock interactions, stock–prey relationships and stock-habitat requirements will result in more accurate fishery assessment models (Christie et al, 2007). Activities and concepts that roll up into the EAF include: establishing a sustainable fishery resource for future generations; inclusion of humans in ecosystems; an emphasis on ecosystem sustainability over ecosystem products; understanding the dynamic nature of marine systems; establishing clear management goals and objectives; and precautionary and adaptive management (Pilling and Payne, 2008).

Thus, the overall goal of the EAF is to achieve ecologically sound resource conservation that is responsive to the reality of ecosystem processes (Marasco et al, 2007). The EAF objectives therefore include (modified from Pikitch et al, 2004):

- Avoiding degradation of, and potentially restoring, ecosystems as measured by indicators of environmental quality and system status.
- Maintaining ecosystem structure, process and function at the community and ecosystem level.
- Obtaining and maintaining long-term socioeconomic benefits without compromising the sustainability of the ecosystem.
- Generating knowledge of ecosystem processes sufficient to understand the likely consequences of human actions.

Elements of the EAF are summarized as follows (adapted from Marasco et al, 2007 and Sissenwine and Murawski, 2004):

- Ensuring that broader societal goals are taken into account and balance social objectives.
- Employing geographic (spatial) representation.
- Recognizing the importance of climatic-oceanic conditions.
- Emphasizing food-web interactions and pursuing ecosystem modelling and research.

- Incorporating improved habitat information (for target and non-target species).
- Expanding monitoring and ecosystem assessments.
- Acknowledging and responding to higher levels of uncertainty.
- Employing adaptive management.
- Taking into account ecosystem knowledge and uncertainties.
- Considering multiple external influences.

Perspectives on the EAF range from the EAF being simply an incremental extension of current fisheries management approaches to a complete redesign of marine management (Pitcher et al, 2009). Regardless of which perspective is applied, the EAF necessitates that short-term socio-economic benefits will be reduced due to decreased harvests as a result of the application of the precautionary principle and consideration of the ecosystem effects of fishing (Marasco et al, 2007). Furthermore, uncertainty cannot be used as an excuse to maintain the status quo.

There have been many criticisms of the EAF. Perhaps the most significant issue with the EAF is that, given humanity's dismal record with single-species management, the likelihood of successfully implementing the EAF is low, especially given that the same fishery scientists, politicians and stakeholders responsible for single-species declines are now in charge of implementing the EAF (Mace, 2004; Murawski, 2007). Compounding this problem is the lack of a general theory of the functioning of marine ecosystems, which limits the ability to explain and predict even simplified single-species impacts on marine systems. As such, the expectation that the EAF will be able to cope with the increased uncertainty, due to the additional need to model community and ecosystem considerations, will severely limit the usefulness of the tool (Curry et al, 2005; Valdermarsen and Suuronen, 2003). Others argue that improvements in single-species approaches, such as the application of F_{MSY}, negate the need to implement EAF. While these criticisms are valid, many of the primary issues to be addressed under EAF, including by-catch, indirect effects of harvesting, and interactions between biological and physical components of ecosystems are already being addressed (Sissenwine and Murawski, 2004).

International instruments that enable the EAF

While a thorough treatment and discussion of international 'soft-law' agreements that enable the EAF is beyond the scope of this text, a short description of the relevant agreements is provided below with specific attention to those parts of the agreement that mandate the use of the EAF.

The 1982 United Nations Convention on the Law of the Sea (UNCLOS), which came into force in 1994, replaced four 1958 treaties, and establishes rights and responsibilities for use of the world's oceans. While UNCLOS is primarily concerned with the exclusive economic zone (EEZ) and mostly ignores the issues of high seas fishing, the convention does include some specific references to species that are associated with, or dependent upon, the harvested target species; therefore the convention is the basis for a number of significant international fisheries agreements discussed below (Caron-Lorimier et al, 2009). UNCLOS has been complemented by the Agreement to Promote Compliance with International Conservation and Management Measure by Fishing Vessels on the High Seas (Compliance Agreement) which is intended to improve the regulation of fishing vessels on the high seas by strengthening 'flag-state responsibility'. Parties to the agreement must ensure that they maintain an authorization and recording system for high seas fishing vessels and that these vessels do not undermine international conservation and management measures. The agreement aims to deter the practice of 're-flagging' vessels with the flags of states that are unable or unwilling to enforce such measures.

The purpose of the 1992 United Nations Conference on Environment and Development was to reconcile environment and development concerns in the context of disparities between nations' abilities to implement sustainable development. The conference launched Agenda 21 aimed at preparing the world for the challenges of the following century. Agenda 21 contained a chapter titled 'Protection of the Oceans', which addresses: integrated management and sustainable development of coastal areas; marine environmental protection; sustainable use and conservation of marine living resources of the high seas; sustainable use and conservation of marine living resources under national jurisdiction; critical uncertainties for the management of the marine environment and climate change; strengthening international, including regional, co-operation and co-ordination; and sustainable development of small islands (FAO, 2005).

The 1992 United Nations Convention on Biodiversity (UNCBD) came into force in 1993 and is the global umbrella convention for the protection of living resources both terrestrial and marine. The convention addresses biodiversity at the genetic, species and ecosystem or seascape levels, and directs signatories to establish systems of protected areas and to protect endangered species. A 1995 meeting in Jakarta, Indonesia led to the Jakarta Ministerial Statement on the implementation of the CBD, which includes protection of marine biodiversity within sustainable fisheries practices (Sinclair et al, 2002).

Broadly, the principles and objectives of the EAF from UNCLOS, UNCED and UNCBD can be summarized as follows (adapted from Sainsbury et al, 2000):

- Manage marine living resources sustainably for human nutritional, economic and social goals.
- Protect and conserve the coastal and marine environment.
- Protect rare or fragile ecosystems, habitats and species.
- Use preventative, precautionary and anticipatory planning and management implementation.
- Protect and maintain the relationships and dependencies among species.
- Conserve genetic, species and ecosystem biodiversity.
- Strengthen cooperation and coordination.

While UNCED, UNCLOS and UNCBD laid the foundation for the ecosystem approach to management is wasn't until the 1995 FAO Code of Conduct for Responsible Fisheries and the UN Fish Stocks Agreement that soft law or voluntary agreements began to outline the principles and operational procedures for what would become marine EBM.

The FAO Code of Conduct contains Article 7 (Fisheries Management) that commits signatories to adopt measures for the long-term conservation and sustainable use of fisheries resources including: effective compliance and enforcement; transparent science and decision-making; reductions in excess fishing capacity; protection of aquatic habitats and endangered species; minimizing by-catch; and implementation of the precautionary approach. Specifically, the agreement states that there is a 'conscious need to avoid adverse impacts on the marine environment, preserve biodiversity, and maintain the integrity of marine ecosystems.' (Valdimarrson and Metzner, 2005)

Following the development of the FAO Code of Conduct, the FAO prepared a background paper in 2001 titled 'Towards Ecosystem-Based Fisheries Management' for the Reykjavik Conference on Responsible Fisheries in the Marine Ecosystem (FAO, 2003). The key messages of this document are captured as the 2001 Reykjavik Declaration on Responsible Fisheries in the Marine Ecosystem, which commits signatories to:

- Advance the scientific basis for developing and implementing management strategies that incorporate ecosystem considerations, and which will ensure sustainable yields while conserving stocks and maintaining the integrity of ecosystems and habitats on which they depend.
- Identify and describe the structure, components and functioning of relevant marine ecosystems, diet composition and food webs, species interactions and predator–prey relationships, the role of habitat and the biological, physical and oceanographic factors affecting ecosystem stability and resilience.
- Build or enhance systematic monitoring of natural variability and its relations to ecosystem productivity.
- Improve the monitoring of by-catch and discards in all fisheries to obtain better knowledge of the amount of fish actually taken.
- Support research and technology developments of fishing gear and practices to improve

gear selectivity and reduce adverse impacts of fishing practices on habitat and biological diversity.
- Assess adverse human impacts of non-fisheries activities on the marine environment as well as the consequences of these impacts for sustainable use.

Following the Reykjavik Declaration, the FAO published the FAO Operational Guidelines on the ecosystem approach to fisheries (FAO, 2003) that defines the EAF as:

> '*An ecosystem approach to fisheries strives to balance diverse societal objectives, by taking into account the knowledge and uncertainties about biotic, abiotic and human components of ecosystems and their interactions and applying an integrated approach to fisheries within ecologically meaningful boundaries.*' (FAO, 2003)

Murawski (2007) summarizes the FAO process for operational implementation into the following steps:

- Set high level policy goals.
- Identify broad objectives.
- Prioritize issues to be addressed in management.
- Set operational objectives.
- Develop indicators and reference points.
- Develop decision rules for application of measures.
- Monitor and evaluate performance.

The Agreement for the Implementation of the Provisions of the United Nations Convention on the Law of the Sea of 10 December 1982 relating to the Conservation and Management of Straddling Fish Stocks and Highly Migratory Fish Stocks (The 1995 UN Fish Stocks Agreement), which came into force in 2001, primarily addresses the management of straddling and highly migratory stocks but also provides specific language related to protecting the marine environment. The agreement contains the same general measures as the FAO Code of Conduct discussed above and, as such, will not be reiterated here (FAO, 2003).

Implementing the EAF

With few exceptions that will be discussed below, the EAF to date has consisted of a sporadic application of marine protection areas (MPAs), limits on destructive fishing practices and efforts to limit by-catch (Shelton, 2009). The EAF, being relatively new, has not yet realized its potential primarily because the EAF sustainability objectives have not yet been clearly articulated for most fisheries, the MSY (or F_{MSY}) concept has not been quantitatively put into the context of ecosystem objectives, and therefore setting EAF-informed MSYs has not been undertaken for most fisheries. In addition, the realization that the EAF, if applied properly and with a precautionary focus, will almost always lead to reduced fishery targets has led to bureaucratic inertia by governments and stakeholder gfoups unwilling to accept reduced catch allocations (Shelton, 2009).

A number of perspectives exist on how the EAF should be implemented. Implementation closest to existing single- and multi-species management views the EAF as simply a recalculation of MSYs for target fisheries that considers direct and indirect impacts of the fishery on the local and regional ecosystem. Adjusting MSYs to limit known fishery impacts (e.g. by-catch, benthic disturbance) is straightforward and would likely result in catch limits at or below F_{MSY} (Link, 2005; Shelton, 2009). This approach is similar to the 'weak stock management' approach used in the singe-species approach where all fisheries are regulated at levels to prevent any one stock from overexploitation and to produce single-species MSYs for target fisheries (Hillborn, 2004). This approach also, compared to more ambitious applications of the EAF, requires minimal new data collection and avoids the need for additional scientific studies to fully understand the relationships between reducing a stock and impacts on the broader ecosystem. The disadvantages of simply recalculating MSYs to reduce known impacts are that significant community and ecosystem impacts may still occur under this management scenario, and that the EAF was advanced primarily because these types of limited approaches have not resulted in sustainable fisheries or stable ecosystems.

A

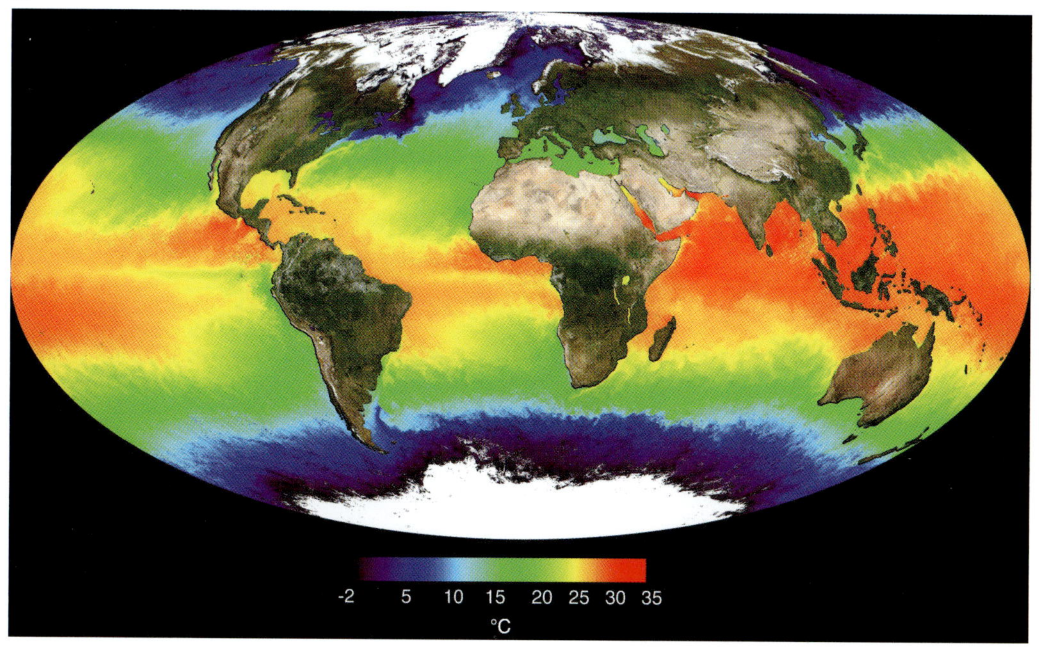

B

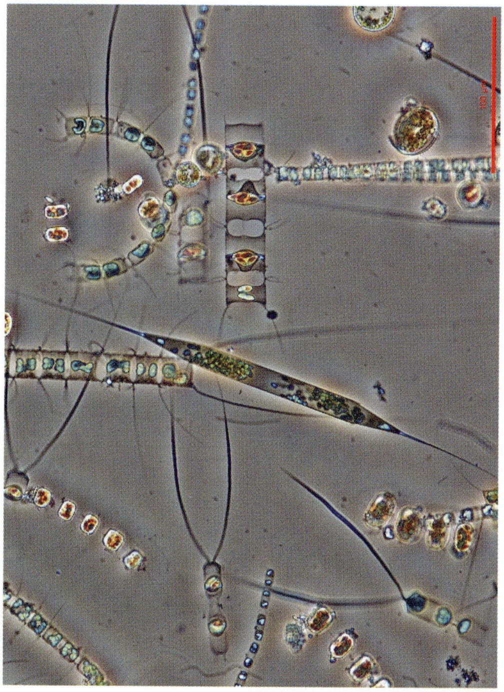

C

Plate 1a. Composite view (Mollweide homolographic projection) of Planet 'Water' showing the range of surface temperatures. The oceans cover 70.6 per cent of the planet's surface. *Source*: NASA. 1b. The most abundant plants in the world. A sample of marine net phytoplankton showing the diversity of cell forms. Cells are about 5–30 µm in size. *Source:* Karl Bruun. 1c. The most abundant animals in the world. The marine copepod *Calanus finmarchicus* (about 3mm in length). *Source*: Michael Bok.

A

B

C

Plate 2. Some of the macrophytic primary producers of the coastal fringing communities. Most of their production enters detrital food webs. 2a. Salt marsh in New Brunswick, Canada. *Source*: M. Buzeta-Innes and M. Strong. 2b. Macrophytic algae of the inter-tidal zone, New Brunswick, Canada. *Source*: M. Buzeta-Innes and M. Strong. 2c. Mangrove forests in Jamaica. *Source*: authors.

Plate 3. Underwater images of a diversity of epibenthic invertebrate species from Atlantic Canada. Source: M. Buzeta-Innes and M. Strong.

A

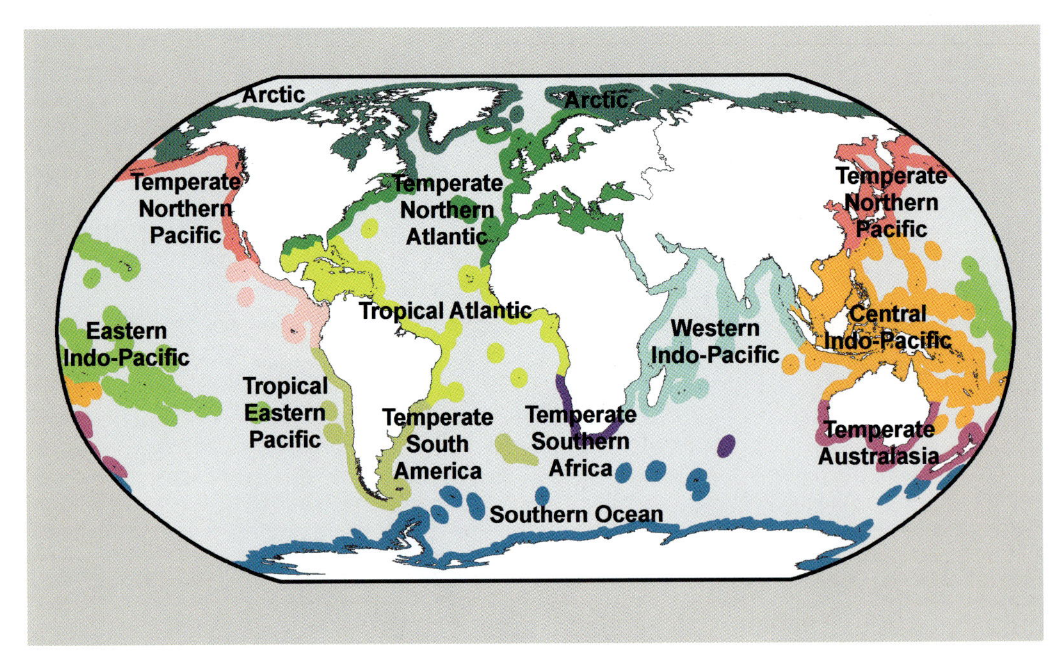

B

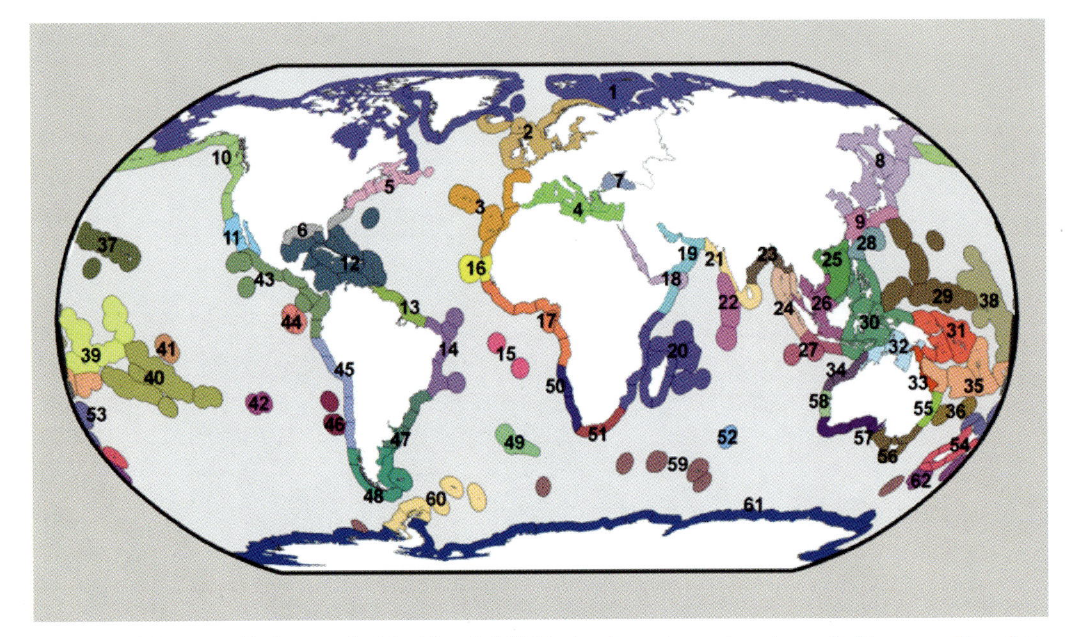

Plate 4. Final biogeographic framework of Spalding et al (2007), showing realms and provinces. 4a. Biogeographic realms with ecoregion boundaries outlined. 4b. Provinces with eco-regions outlined. For further description see source. *Source:* Spalding *et al* (2007).

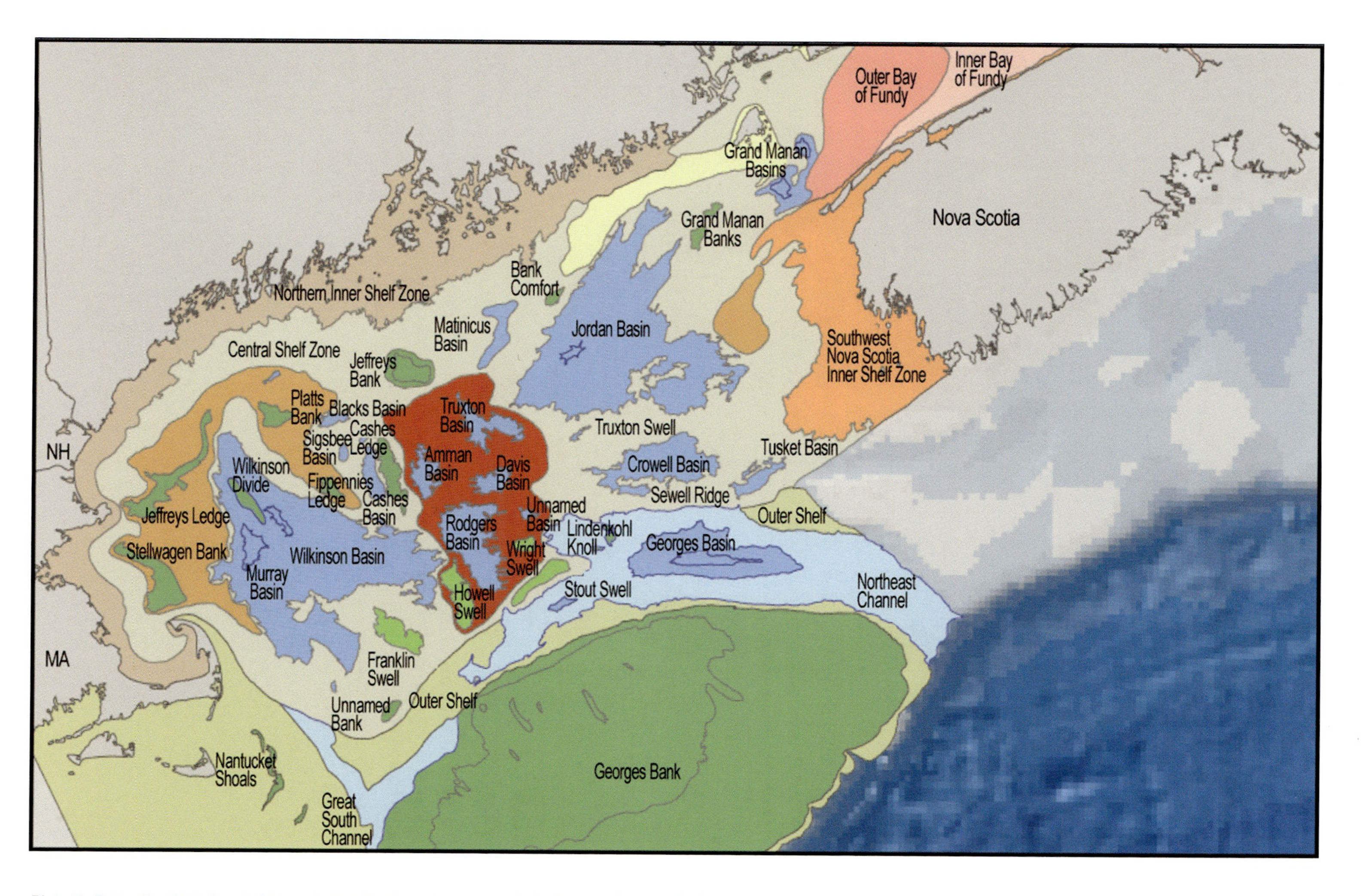

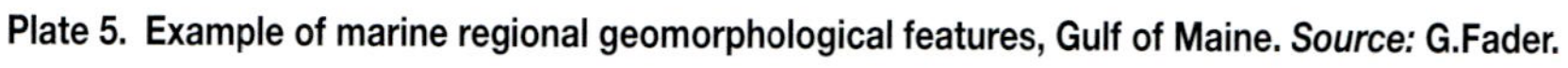
Plate 5. Example of marine regional geomorphological features, Gulf of Maine. *Source:* G.Fader.

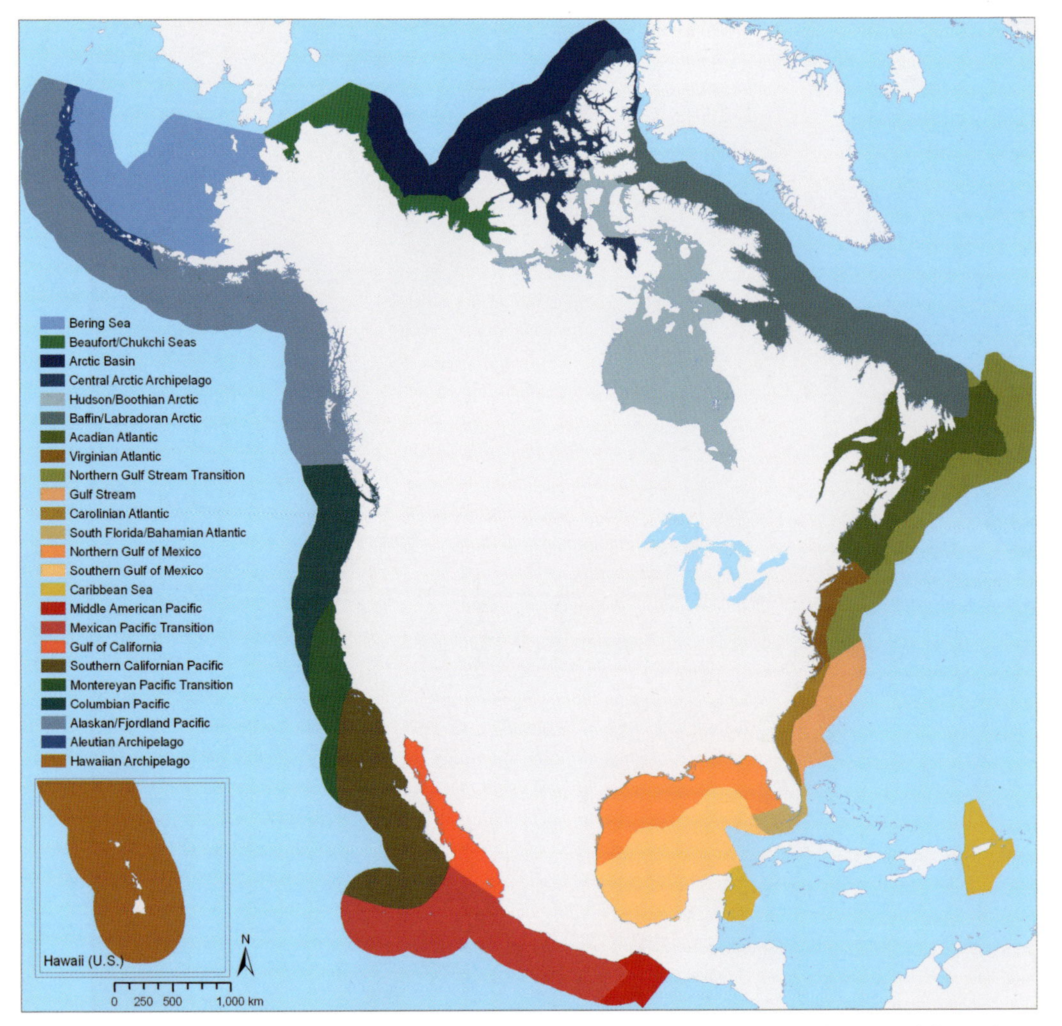

Plate 6. North American Environmental Atlas – Marine Ecoregions Level I. The 24 Level 1 eco-regions were delineated based on similarities in biological, physiographic and oceanographic characteristics (Wilkinson et al, 2009). *Source*: Commission for Environmental Cooperation.

Plate 7. North American Environmental Atlas – Marine Ecoregions Level II. The 86 Level 2 eco-regions capture the break between near-shore and oceanic areas, with the boundaries determined by large-scale features such as the continental shelf, continental slope, major trenches, and other features. Level II reflects the importance of depth as well as the importance of major physiographic features in determining current flows and upwelling, delineated based on similarities in biological, physiographic, and oceanographic characteristics (Wilkinson et al, 2009). *Source*: Commission for Environmental Cooperation.

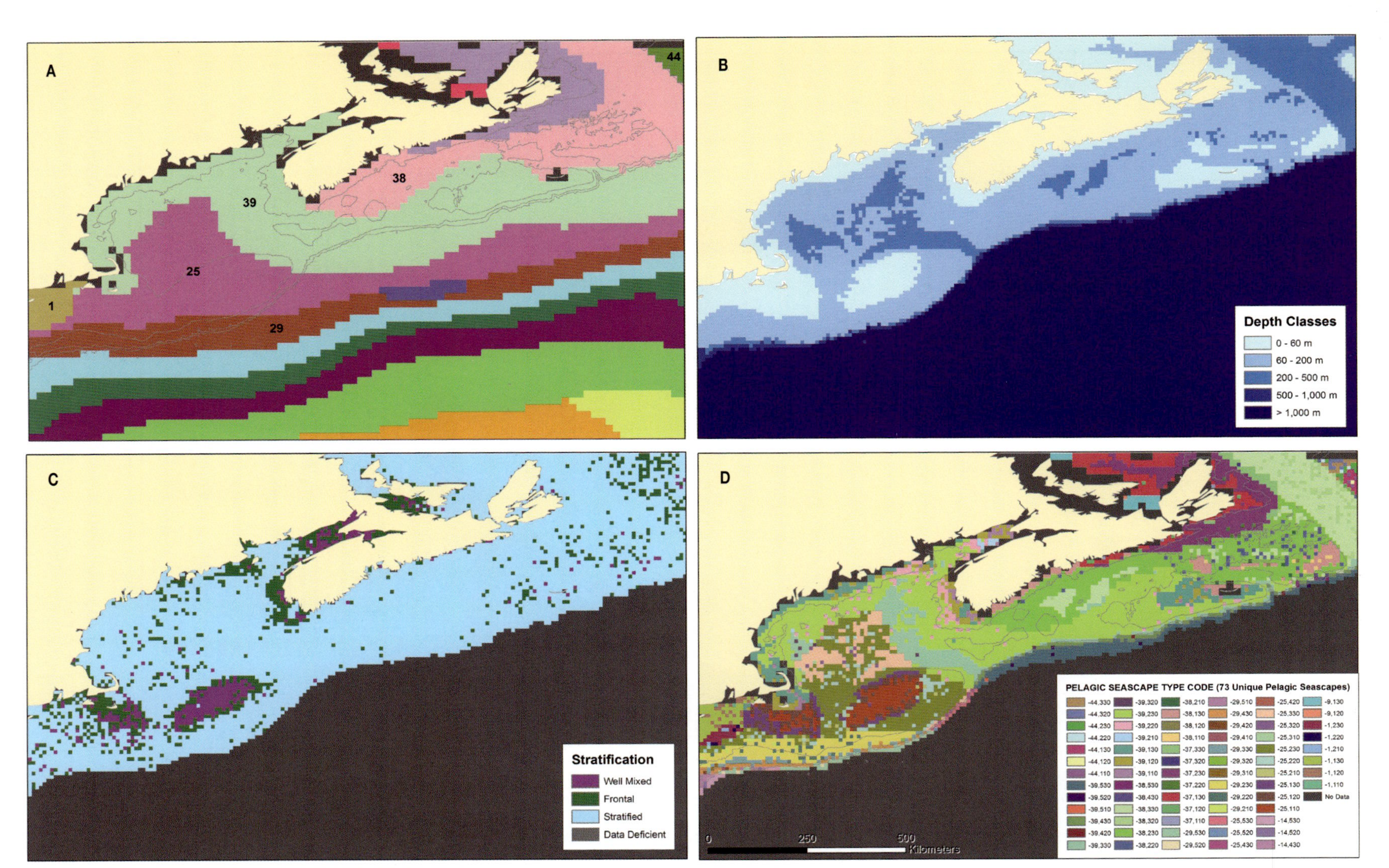

Plate 8. Pelagic seascape classification of Scotian Shelf and Gulf of Maine. *Source*: H. Alidina and J. Roff. See CLF/WWF (2006) for further details. 8a. Pelagic temperature-salinity zones. Each color corresponds to one cluster, or a zone of similar temperature and salinity characteristics (water masses) as identified through cluster analysis. 8b. Water depth zones used for defining pelagic seascape classes. 8c. Distribution of stratification classes (as calculated from a vertical density anomaly of $\Delta\sigma t/\Delta z$) used for defining pelagic seascapes. 8d. Distribution of pelagic seascape classes as defined by pelagic temperature-salinity zones, depth and stratification classes.

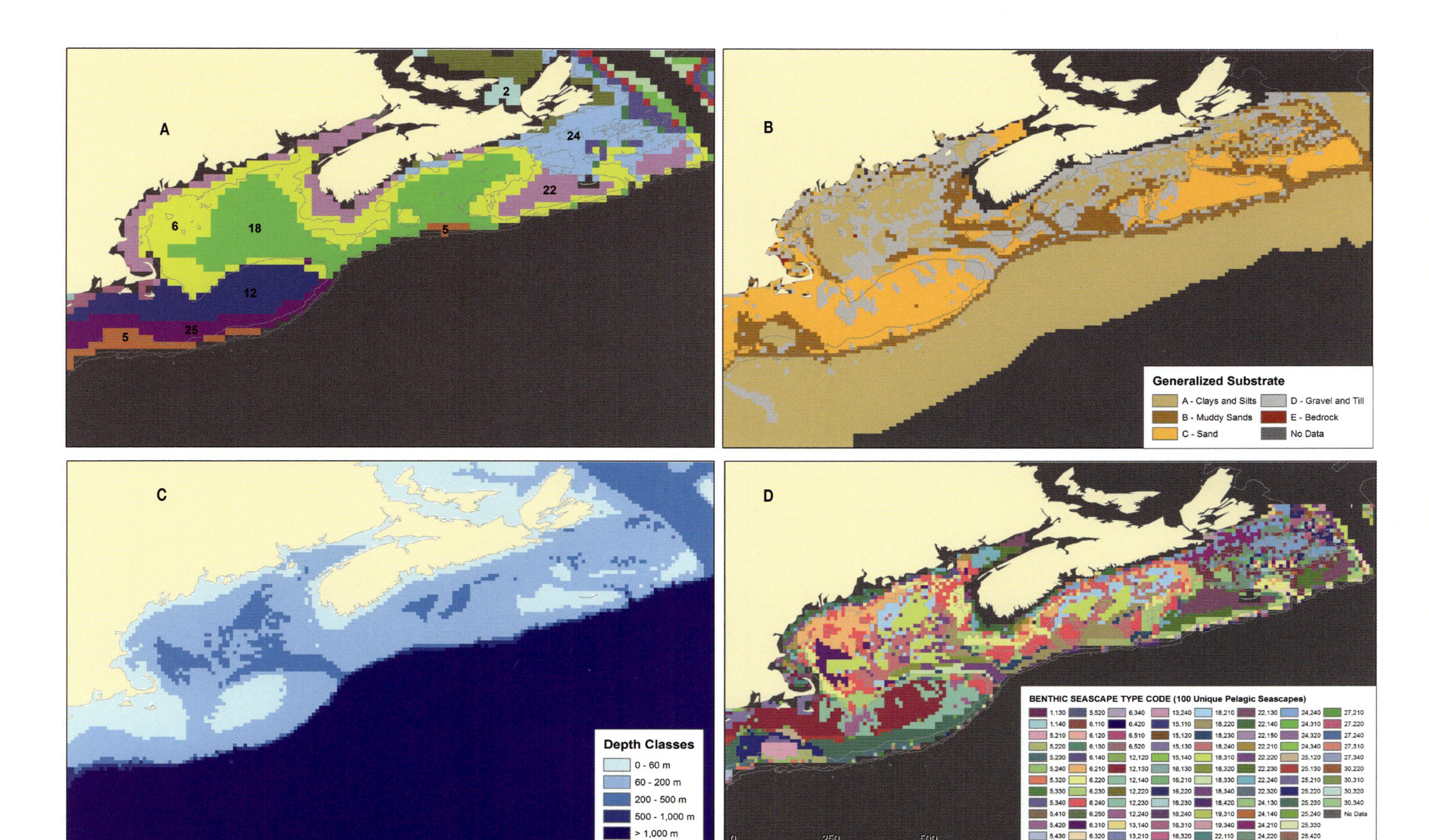

Plate 9. Benthic seascape classification of Scotian Shelf and Gulf of Maine. *Source*: H. Alidina and J. Roff. See CLF/WWF (2006) for further details. 9a. Benthic temperature-salinity zones. Each color corresponds to one cluster, or a zone of similar temperature and salinity characteristics (water masses) as identified through cluster analysis. 9b. Substrate classification used for defining seascapes, illustrated in a grid of 5-minute squares. 9c. Water depth zones used for defining pelagic seascape classes (repeat of plate 8b). 9d. Distribution of benthic seascape classes as defined by benthic temperature-salinity zones, substrate classes and water depth.

Plate 10. Examples of biogenic (or foundation or engineering) species that form spatially complex, heterogeneous habitats for a diversity of other species. 10a. From the Great Barrier Reef, Australia. *Source*: Great Barrier Reef Marine Park Authority. 10b. From Andros, Bahamas. *Source*: M. Buzeta-Innes and M. Strong.

A B

C

Plate 11. Example of community succession within a habitat type (rocky temperate sub-tidal). *Source*: R. Scheibling. 11a. Substrate dominated by macrophytic laminarians. 11b. Substrate co-dominated by laminarians and green sea-urchins. 11c. Substrate dominated by green sea-urchins.

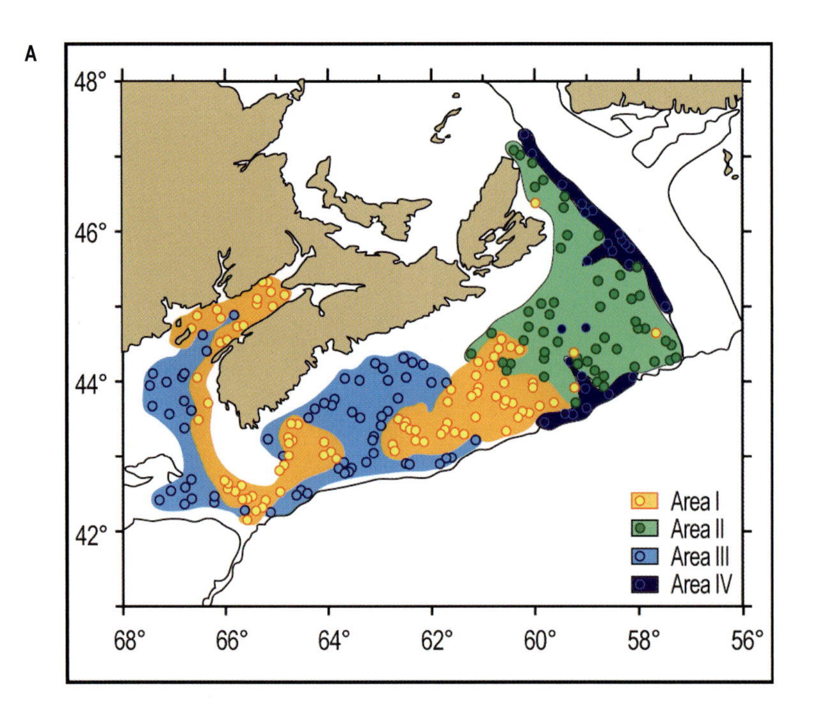

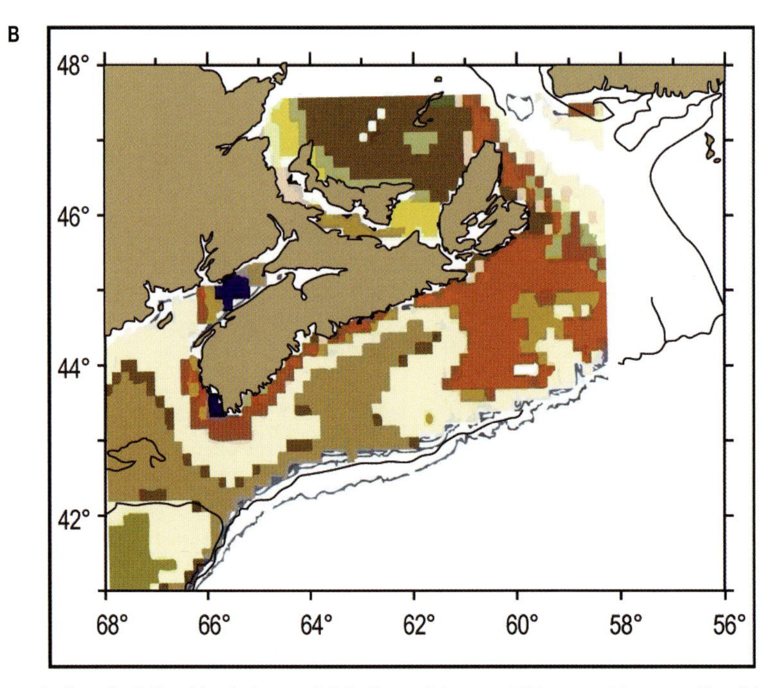

Plate 12. Demonstration of relationships between distributions of demersal fish assemblages and benthic water masses on the Scotian Shelf, Canada (identified from Bray-Curtis similarity matrices and detrended canonical correspondence analysis). 12a. Distribution of four demersal fish assemblages. *Source*: K. Zwanenburg and A. Jaureguizar. 12b. Distribution of bottom water masses (note correspondence to distributions of fish assemblages). *Source*: H. Alidina and J. Roff.

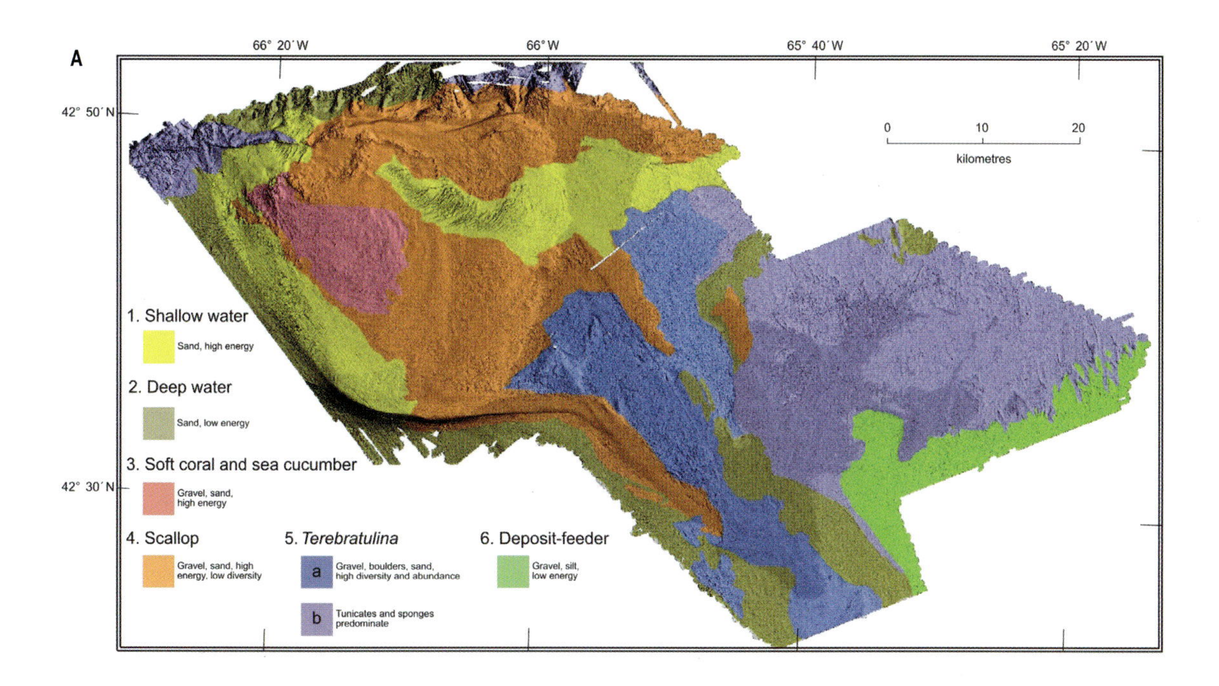

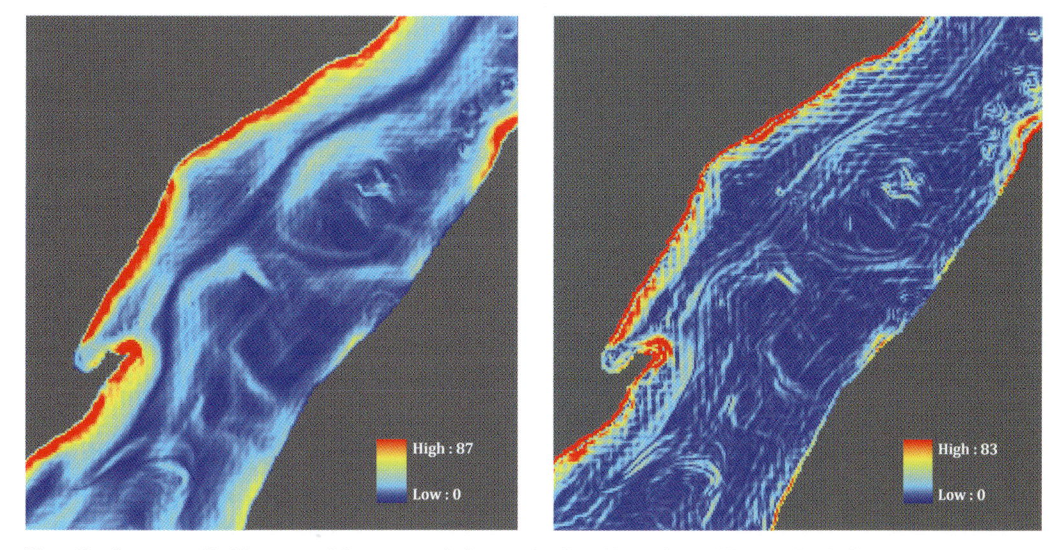

Plate 13a. Interpreted habitat map of the geomorphology and epibenthic ecology of Browns Bank. Six colour-coded benthic habitats were defined, distinguished on the basis of substrate type, benthic assemblage, habitat complexity, relative current strength and depth. *Source*: V. Kostylev. 13b, 13c. Comparison of benthic slope surface (b) and complexity surface (c). The complexity surface picks up small complex areas in the sea floor. *Source*: Adapted from Ardron and Sointula (2002) and M. Greenlaw and J. Roff.

Plate 14. Examples of Focal Species. 14a. Grey Seal. *Source*: authors. 14b. Harbour Porpoise. *Source*: M. Buzeta-Innes and M. Strong. 14c. Ridley's Turtle. *Source*: M. Buzeta-Innes and M. Strong. 14d. Humpback Whale. *Source*: A. Crisp.

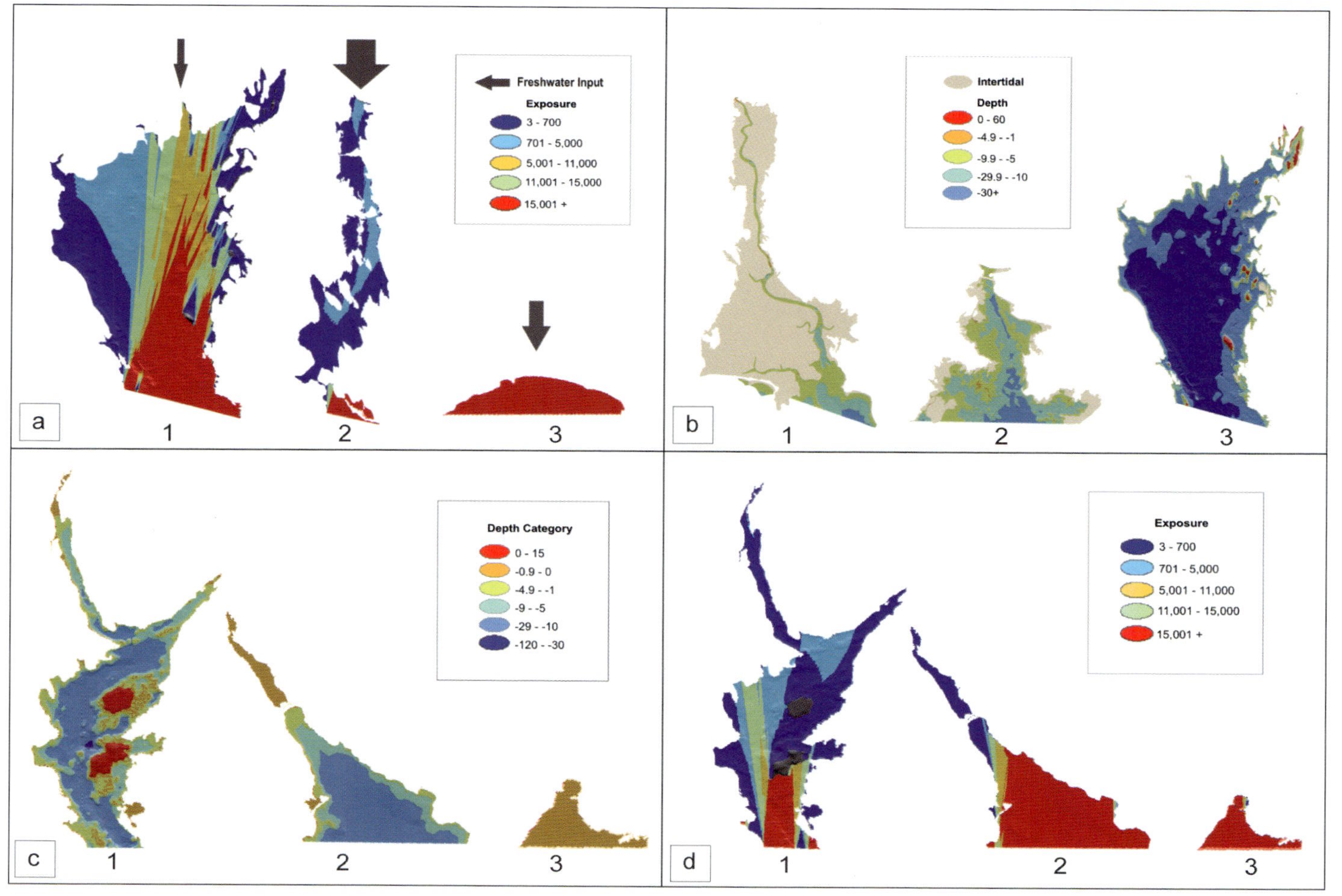

Plate 15. Classification of coastal inlets. *Source*: M. Greenlaw and J. Roff. 15a. Visualization of the three hydrographic inlet archetypes: bays (1), estuaries (2) and coves (3). The width of arrows represents level of freshwater input; exposure is represented from low to high by blue to red colors, respectively (in m). Bays are characterized by low freshwater input compared to their tidal volume and mid levels of exposure, estuaries by their low levels of exposure and high freshwater input, while coves are characterized by mid levels of freshwater input and high levels of exposure. 15b. Visualization of the three productivity regime archetypes: benthic (1), mixed (2) and pelagic (3). The benthic dominant productivity regime type has high proportions of intertial zone and low average depths (in m); the pelagic dominant productivity regime type has low proportions of intertidal zone and high average depths; the mixed productivity regime type has intermediate levels of both factors. Brown = intertidal zone, red = supra-tidal, yellow = 1-5m, light blue = 5-10m, med blue = 10-30m, dark blue = 30^{+} m. 15c. Visualization of the variation in depth categories of the three inlet archetypes (1,2,3). Depth categories are in meters. 15d. Visualization of variation in exposure categories of the three inlet archetypes (1,2,3). Exposure categories are in meters. Six variables were used to classify overall complexity, including depth and exposure diversity and evenness, topographic complexity and shoreline complexity.

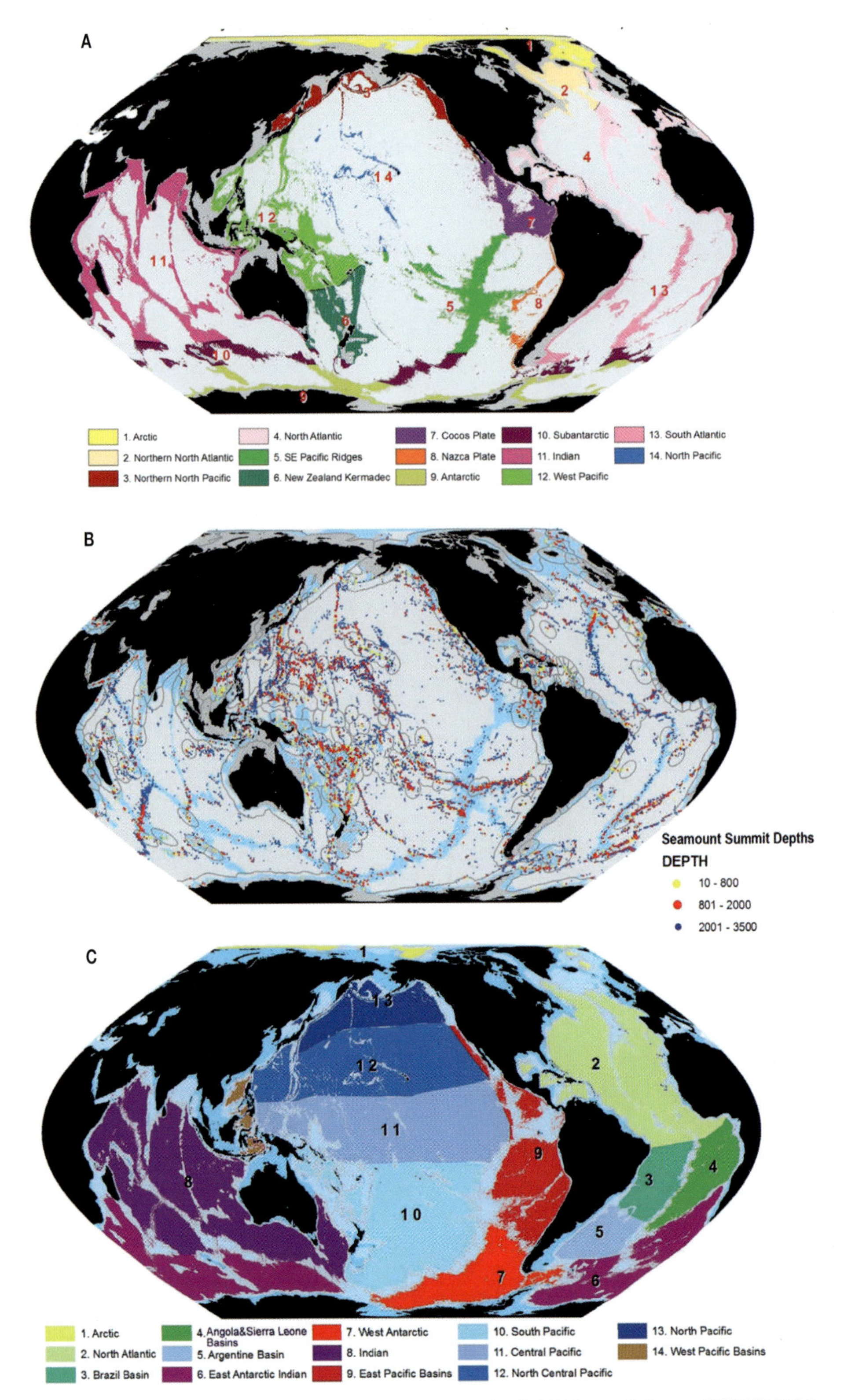

Plate 16. Global Open Oceans and Deep Seabed biogeographic classification (GOODS report). *Source*: UNESCO (2009). 16a. Map of lower bathyal provinces at a depth range from 800 to 3000m. 16b. Map of seamounts with summits shallower than 3500 m. Bottom depths 2000 to 3500m are shown in light blue. Seamount locations are according to Kitchingman and Lai (2004), and bathymetry is according to ETOPO2. 16c. Map of abyssal provinces at a depth range from 3500 to 6500 m.

A more robust application of the EAF that expressly captures ecosystem interactions involves developing ecological objectives for target and non-target stocks and populations that consider diversity, productivity and resilience at the genetic, population and community levels (Rogers et al, 2007). However, developing objectives requires sufficient understanding of fish population dynamics and the structural and functional aspects of life history, which are required to define ecosystem overfishing in a way that relates to setting MSYs in the single-species context. Thus, setting objectives will likely be one of the more difficult aspects of implementation of the EAF (Link, 2005; Rogers et al, 2007).

The most fulsome application of the EAF involves integrating the target stock with community dynamics and environmental variables into a stock assessment before determining catch limits and management measures. While this approach requires a level of inventory and scientific understanding that may not currently exist, it does attempt to identify the surplus production required to satisfy ecosystem needs (Goodman et al, 2002). Additional challenges of this application of the EAF are addressing the high levels of uncertainty as a result of integrating biological and oceanographic processes into stock assessment models, which are likely to have limited predictive power given most currently available data and understanding of marine ecosystem function (Jennings and Revill, 2007). For example, Georges Bank, off the east coast of North America, is one of the world's most studied ecosystems yet the proper application of the EAF has been hampered by a lack of sufficient data (Froese et al, 2008).

Regardless of what 'flavour' of the EAF is selected for a particular application, the EAF faces a number of implementation challenges, including: developing long-term, ecosystem-related objectives; developing meaningful indicators that clearly demonstrate when a threshold has been exceeded; and developing a stronger scientific foundation to inform EAF decision (Cury et al, 2005).

The science gaps to fully operationalize the EAF can be summarized as follows (modified from Frid et al, 2006):

- Understanding the relationships between hydrographic regimes and fish stock dynamics.
- Understanding and inventorying habitat distributions.
- Establishing 'design rules' for MPAs.
- Understanding ecological dependencies within marine communities.
- Developing predictive capabilities in complex systems.
- Incorporating uncertainty into management advice and direction.
- Understanding the genetics of target and non-target stocks, populations, and species.
- Understanding the response of fishers to management measures.

Regardless of how the EAF is used, a number of implementation tools have been identified, many of which are currently in practice and discussed elsewhere in this book (modified from Pikitch et al, 2004):

- Marine habitat delineation at multiple spatial and temporal scales.
- Identification and protection of essential fish habitat.
- Marine protected areas.
- Marine spatial planning and zoning.
- Impacts of fisheries on endangered species.
- Reductions in by-catch.
- Managing target species.
- Precautionary approach.
- Adaptive management.
- New analytical models and management tools.
- Integrated management plans.

As discussed previously, while dozens of publications and conferences have explored the principles and implementation of the EAF, there have been few examples of the EAF actually being applied to real-world management decisions. Pitcher et al (2009) evaluated the performance of the EAF using a number of principles, criteria and quantifications. Only two countries were rated as 'good' (Norway, USA), four countries (Iceland, South Africa, Canada, Australia) were rated as 'acceptable' and over half of the countries received failing grades.

However, there are a number of EAF successes, including Alaskan groundfish management and the Convention for the Conservation of Antarctic Marine Living Resources (CCAMLR). CCAMLR was established primarily to ensure that krill harvests in the southern ocean did not significantly affect other marine life. The treaty is signed by 31 nations with interests in the Antarctic regions and came into force in 1982, which predates the EAF concept. The treaty, which is part of the Antarctic Treaty system, applies to all living marine resources (except seals and whales, which were covered in existing treaties) south of 50 degrees south latitude, and has a mandate to conserve and manage mainly the high seas based on the ecosystem and precautionary approaches. The treaty has a separate ecosystem monitoring programme that aims to detect changes in stock status and determine whether these changes arise from natural or anthropogenic sources (Constable et al, 2000).

CCAMLR has demonstrated that the EAF and precautionary approach can be applied in high-seas environments. Under the treaty, the scientific committee has developed a number of innovative methods to manage prey species to protect dependant predators, limit by-catch and develop precautionary-based protocols prior to the exploitation of new fisheries. Advice from the CCAMLR scientific committee is almost always followed and the quality of science is high (Agnew, 1997; Constable, 2004). Continuing challenges to implement the EAF under the treaty include data limitations, controlling fishing effort and incorporating climate change into ecosystem and stock assessments.

Another example of a successful implementation of the EAF is the management of groundfish in the northeast Pacific, where there are currently no overfished groundfish stocks; however, four species of crabs are considered to be overexploited (Dew and Austring, 2007). As part of the EAF approach in this region a full ecosystem assessment was developed (PICES, 2004) and significant research has been undertaken to address particular challenges, including explaining how threatened Steller sealions have been impacted by the combined effects of fishing, predation, competition and ocean productivity in the Aleutian Islands and the eastern Gulf of Alaska (Christensen et al, 2007). Much of the success in applying the EAF techniques in this region has been ascribed to the fact that the northeast Pacific is primarily a bottom-up-driven ecosystem, thus food chains are shorter and inferences on ecosystem health and setting catch limits can be at least partially predicted using primary production estimates (Ware and Thompson, 2005).

In conclusion, the EAF concept has been part of the suite of fishery management tools for nearly two decades and has been successfully applied in certain ecosystems with the characteristics that they are generally bottom-up structured, their EEZs fall within developed nations (CCAMLR excepted), and there is genuine stakeholder and political will to move away from traditional, single-species management and towards a more holistic approach to ocean conservation and management. The degree and speed of uptake of the EAF will depend on the willingness of fishing industries and governments to bear the high short-term costs associated with reducing fishing efforts and moving towards sustainability.

References

Agnew, D. J. (1997) 'The CCAMLR ecosystem monitoring programme', *Antarctic Science*, vol 9, no 3, pp235–242

Babcock, E. A. and Pikitch, E. K. (2004) 'Can we reach agreement on a standardized approach to ecosystem-based fishery management?' *Bulletin of Marine Science*, vol 74, no 3, pp685–692

Ball, B. J., Fox, G. and Munday, B. W. (2000) 'Long- and short-term consequences of a Nephrops trawl fishery on the benthos and environment of the Irish Sea', *ICES Journal of Marine Science*, vol 57, pp1315–1320

Caddy, J. F. (2004) 'Current usage of fisheries indicators and reference points, and their potential application to management of fisheries for marine invertebrates', *Canadian Journal of Fisheries and Aquatic Sciences*, vol 61, no 8, pp1307–1324

Caron-Lormier, G., Bohan, D. A., Hawes, C., Raybould, A., Haughton, A. J. and Humphry, R. W. (2009) 'How might we model an ecosystem?' *Ecological Modelling*, vol 220, no 17, pp1935–1949

Christensen, V., Aiken, K. A. and Villanueva, M. C. (2007) 'Threats to the ocean: On the role of ecosystem approaches to fisheries', *Social Science*

Information sur les Sciences Sociales, vol 46, no 1, pp67–86

Christie, P., Fluharty, D. L., White, A. T., Eisma-Osorio, L. and Jatulan, W. (2007) 'Assessing the feasibility of ecosystem-based fisheries management in tropical contexts', *Marine Policy*, vol 31, no 3, pp239–250

Cochrane, K. L. (2002) 'A fishery manager's guidebook. Management measures and their application', *FAO Fisheries Technical Paper*, vol 424, FAO, Rome

Collie, J. S., Richardson, K. and Steele, J. H. (2004) 'Regime shifts: Can ecological theory illuminate the mechanisms?', *Progress in Oceanography*, vol 60, nos 2–4, pp281–302

Constable, A. J. (2004) 'Managing fisheries effects on marine food webs in Antarctica: Trade-offs among harvest strategies, monitoring, and assessment in achieving conservation objectives', *Bulletin of Marine Science*, vol 74, no 3, pp583–605

Constable, A. J., de la Mare, W. K., Agnew, D. J., Everson, I. and Miller, D. (2000) 'Managing fisheries to conserve the Antarctic marine ecosystem: Practical implementation of the Convention on the Conservation of Antarctic Marine Living Resources (CCAMLR)', *ICES Journal of Marine Science*, vol 57, pp778–791

Commission of the European Communities (2008) *Council Facts and Figures on the CFP: Basic data on the Common Fisheries Policy. Edition 2008*, Office for Official Publications of the European Communities, Luxemburg

Cury, P. M., Mullon, C., Garcia, S. M. and Shannon, L. J. (2005) 'Viability theory for an ecosystem approach to fisheries', *ICES Journal of Marine Science*, vol 62, no 3, pp577–584

Davies, R. W. D. and Rangeley, R. (2010) 'Banking on cod: Exploring economic incentives for recovering Grand Banks and North Sea cod fisheries', *Marine Policy*, vol 34, pp92–98

Day, V., Paxinos, R., Emmett, J., Wright, A. and Goecker, M. (2008) 'The Marine Planning Framework for South Australia: A new ecosystem-based zoning policy for marine management', *Marine Policy*, vol 32, no 4, pp535–543

Dew, C. B. and Austring, R. G. (2007) 'Alaska red king crab: A relatively intractable target in a multispecies trawl survey of the eastern Bering Sea', *Fisheries Research*, vol 85, pp265–173

FAO (1997) *FAO Technical Guidelines for Responsible Fisheries Management No 4, Fisheries Management*, FAO, Rome

FAO (2003) 'The ecosystem approach to fisheries', *FAO Technical Guidelines for Responsible Fisheries*, vol 4, no 2, FAO, Rome

FAO (2005) *Progress in the Implementation of the Code of Conduct for Responsible Fisheries and Related Plans of Action*, FAO, Rome

FAO (2009) *The State of World Fisheries and Aquaculture 2008*, FAO, Rome

Frank, K. T., Petrie, B., Choi, J. S. and Leggett, W. C. (2005) 'Trophic cascades in a formerly cod-dominated ecosystem', *Science*, vol 308, no 5728, pp1621–1623

Frid, C. L. J., Harwood, K. G., Hall S. J. and Hall, J. A. (2000) 'Long-term changes in the benthic communities on North Sea fishing grounds', *ICES Journal of Marine Science*, vol 57, pp1303–1309

Frid, C. L. J., Paramor, O. A. L. and Scott, C. L. (2005) 'Ecosystem-based fisheries management: Progress in the NE Atlantic', *Marine Policy*, vol 29, no 5, pp461–469

Frid, C. L. J., Paramor, O. A. L. and Scott, C. L. (2006) 'Ecosystem-based management of fisheries: Is science limiting?', *ICES Journal of Marine Science*, vol 63, no 91, pp567–572

Froese, R., Stern-Pirlot, A., Winker, H. and Gascuel, D. (2008) 'Size matters: How single-species management can contribute to ecosystem-based fisheries management', *Fisheries Research*, vol 92, nos 2–3, pp231–241

Gislason, H. (2002) 'The Effects of Fishing on Non-target Species and Ecosystem Structure and Function', in M. Sinclair and G. Valdimarsson (eds) *Responsible Fisheries in the Marine Ecosystem*, CAB International, Wallingford

Goodman, D., Mangel, M., Parkes, G., Quinn, T., Restrepo, V., Smitch, T. and Stokes, K (2002) *Scientific Review of the Harvest Strategy Currently Used in the BSAI and GIA Groundfish Fishery Management Plans*, North Pacific Fishery Management Council, Anchorage, AL

Guerry, A. D. (2005) 'Icarus and Daedalus: Conceptual and tactical lessons for marine ecosystem-based management', *Frontiers in Ecology and the Environment*, vol 3, pp202–211

Hilborn, R. (2004) 'Ecosystem-based fisheries management: The carrot or the stick?', *Marine Ecology Progress Series*, vol 274, pp275–278

Hirshfield, M. F. (2005) 'Implementing the ecosystem approach: Making ecosystems matter', *Marine Ecology Progress Series*, vol 300, pp253–257

Jackson, J. B. C., Kirby, M. X., Berger, W. H., Bjorndal, K. A., Botsford, L. W., Bourque, B. J., Bradbury, R.

H., Cooke, R., Erlandson, J., Estes, J. A., Hughes, T. P., Kidwell, S., Lange, C. B., Lenihan, H. S., Pandolfi, J. M., Peterson, C. H., Steneck, R. S., Tegner, M. J. and Warner, R. R. (2001) 'Historical overfishing and the recent collapse of coastal ecosystems', *Science*, vol 293, no 5530, pp629–638

Jennings S. and Revill A. S. (2007) 'The role of gear technologists in supporting an ecosystem approach to fisheries', *ICES Journal of Marine Science*, vol 64, pp1525–1534

Juvik, S., Juvik, J. and Paradise, T. (1998) *Atlas of Hawaii. Third Edition*, University of Hawaii Press, Honolulu, HI

King, D. M. and Sutinen, J. G. (2010) 'Rational noncompliance and the liquidation of Northeast groundfish resources', *Marine Policy*, vol 34, no 1, pp7–21

Larkin, P. A. (1977) 'An epitaph for the concept of maximum sustainable yield', *Transactions of the American Fisheries Society*, vol 106, pp1–11

Law, R. (2000) 'Fishing, selection, and phenotypic evolution', *ICES Journal of Marine Science*, vol 57, pp659–668

Law, R. and Stokes, K. (2005) 'Evolutionary Impacts of Fishing on Target Populations', in E. A. Norse and L. B. Crowder (eds) *Marine Conservation Biology: The Science of Maintaining the Sea's Biodiversity*, Island Press, Washington, DC

Link, J. S. (2005) 'Translating ecosystem indicators into decision criteria', *ICES Journal of Marine Science*, vol 62, no 3, pp569–576

Lotze, H. K. (2004) 'Repetitive history of resource depletion and mismanagement: The need for a shift in perspective', *Marine Ecology Progress Series*, vol 274, pp282–285

Mace P. M. (2001) 'A new role for MSY in single-species and ecosystem approaches to fisheries stock assessment and management', *Fish and Fisheries*, vol 2, pp2–32

Mace, P. M. (2004) 'In defence of fisheries scientists, single-species models and other scapegoats: Confronting the real problems', *Marine Ecology Progress Series*, vol 274, pp285–291

Mangel, M. and Levin, P. S. (2005) 'Regime, phase and paradigm shifts: Making community ecology the basic science for fisheries', *Philosophical Transactions of the Royal Society B*, vol 360, pp95–105

Marasco, R. J., Goodman, D., Grimes, C. B., Lawson, P. W., Punt, A. E. and Quinn, T. J. (2007) 'Ecosystem-based fisheries management: Some practical suggestions', *Canadian Journal of Fisheries and Aquatic Sciences*, vol 64, no 6, pp928–939

Mueller-Dombois, D. and Wirawan, N. (2005) 'The Kahana Valley Ahupua`a, a PABITRA study site on O`ahu, Hawaiian Islands', *Pacific Science*, vol 59, no 2, pp293–314

Murawski, S. A. (2007) 'Ten myths concerning ecosystem approaches to marine resource management', *Marine Policy*, vol 31, no 6, pp681–690

Myers, R. A. and Worm, B. (2003) 'Meta-analysis of cod–shrimp interactions reveals top–down control in oceanic food webs', *Ecology*, vol 84, pp162–173

NAFO (2008) *Report of the Fisheries Commission Intersessional Meeting, 30 April – 07 May 2008 Montreal, Quebec, Canada*, North Atlantic Fisheries Organization, Dartmouth, Canada

National Marine Fisheries Service (2009) 'Fisheries of the United States, 2009', www.st.nmfs.noaa.gov/st1/fus/fus09/fus_2009.pdf, accessed 10 September 2010

New Zealand Ministry of Fisheries (2009) http://fs.fish.govt.nz/Page.aspx?pk=16, accessed February 2010

Pauly, D., Christensen, V., Guenette, S., Pitcher, T. J., Sumaila, U. R., Walters, C. J., Watson, R. and Zeller, D. (2002) 'Towards sustainability in world fisheries', *Nature*, vol 418, no 6898, pp689–695

Pauly, D. V., Christensen, V., Dalsgaard, J., Froese, R. and Torres, F. Jr. (1998) 'Fishing down marine food webs', *Science*, vol 279, pp860–863

Pavlikakis, G. E. and Tsihrintzis, V. A. (2000) 'Ecosystem management: A review of a new concept and methodology', *Water Resources Management*, vol 14, no 4, pp257–283

Pearson, R. G. (1981) 'Recovery and recolonisation of coral reefs', *Marine Ecology Progress Series*, vol 4, pp105–122

PICES (2004) *Marine Ecosystems of the North Pacific* (edited S. M. McKinnell), PICES Special Publication 1, North Pacific Marine Science Organization, Sidney, British Columbia, Canada

Pikitch, E. K., Santora, C., Babcock, E. A., Bakun, A., Bonfil, R., Conover, D. O., Dayton, P., Doukakis, P., Fluharty, D., Heneman, B., Houde, E. D., Link, J., Livingston, P. A., Mangel, M., McAllister, M. K., Pope, J. and Sainsbury, K. J. (2004) 'Ecosystem-based fishery management', *Science*, vol 305, no 5682, pp346–347

Pilling, G. M. and Payne, A. I. L. (2008) 'Sustainability and present-day approaches to fisheries management: Are the two concepts irreconcilable?', *African Journal of Marine Science*, vol 30, no 1, pp1–10

Pitcher, T. J., Kalikoski, D., Short, K., Varkey, D. and Pramod, G. (2009) 'An evaluation of progress in implementing ecosystem-based management of fisheries in 33 Countries', *Marine Policy*, vol 33, no 2, pp223–232

Roberts, C. (2007) *An Unnatural History of the Sea*, Island Press, Washington, DC

Rogers, S. I., Tasker, M. L., Earll, R. and Gubbay, S. (2007) 'Ecosystem objectives to support the UK vision for the marine environment', *Marine Pollution Bulletin*, vol 54, no 2, pp128–144

Rosenberg, A. A. (2002) 'The precautionary approach from a manager's perspective', *Bulletin of Marine Science*, vol 70, pp577–588

Sainsbury, K. J., Punt A. E. and Smith, A. D. M. (2000) 'Design of operational management strategies for achieving fishery ecosystem objectives', *ICES Journal of Marine Science*, vol 57, pp731–741

Scandol, J. P., Holloway, M. G., Gibbs, P. J. and Astles, K. L. (2005) 'Ecosystem-based fisheries management: An Australian perspective', *Aquatic Living Resources*, vol 18, no 3, pp261–273

Schoener, T. W. (1983) 'Field experiments on interspecific competition', *American Naturalist*, vol 122, pp240–285

Shelton, P. A. (2009) 'Eco-certification of sustainably managed fisheries: Redundancy or synergy?', *Fisheries Research*, vol 100, no 3, pp185–190

Sinclair, M., Arnason, R., Csirke, J., Karnicki, Z., Sigurjonsson, J., Skjoldal, H. R. and Valdimarsson, G. (2002) 'Responsible fisheries in the marine ecosystem', *Fisheries Research*, vol 58, no 3, pp255–265

Sissenwine, M. P. and Murawski. S. (2004) 'Moving beyond "intelligent tinkering": Advancing an ecosystem approach to fisheries', *Marine Ecology Progress Series*, vol 274, pp291–295

Symes, D. (2007) 'Fisheries management and institutional reform: A European perspective', *ICES Journal of Marine Science*, vol 64, no 4, pp779–785

Tamura, T. (2003) 'Regional Assessments of Prey Consumption and Competition by Marine Cetaceans in the World', in M. Sinclair and J. W. Valdermarson (eds) *Responsible Fisheries in the Marine Ecosystem*, FAO, Rome

Valdermarsen, J. W. and Suuronen, P. (2003) 'Modifying Fishing Gear to Achieve Ecosystem Objectives', in M. Sinclair and G. Valdimarsson (eds) *Responsible Fisheries in the Marine Ecosystem*, FAO, Rome

Valdimarrson, G. and Metzner, R. (2005) 'Aligning incentives for a successful ecosystem approach to fisheries management', *Marine Ecology Progress Series*, vol 300, pp286–291

Ware, D. M. and Thomson, R. E. (2005) 'Bottom–up ecosystem trophic dynamics determine fish production in the Northeast Pacific', *Science*, vol 308, pp1280–1284

Watling, L. and Norse, E. A. (1998) 'Disturbance of the seabed by mobile fishing gear: A comparison with forest clear-cutting', *Conservation Biology*, vol 12, no 6, pp1189–1197

14

Size and Boundaries of Protected Areas

Rationale for function, location, dimensions

If size did matter, the dinosaurs would still be alive.

Wendelin Wiedeking (1952–)

Introduction

The main questions to which we shall seek answers in this chapter are: how large should a marine protected area (MPA) actually be; and how can its boundaries be defined? More specifically we shall ask: what is the relationship between the size and intended purpose of a given MPA? Unfortunately, all too often, although planning authorities describe the locations, characteristics and 'importance' of existing conservation areas, there is simply no rationale for purpose, regional context, importance, boundaries or size of MPAs. Thus we have merely description, inventory and assertion. This is a beginning, but is not sufficient scientific rationale for effective coastal zone management and marine conservation. Even where we have some regional context or declared purpose for MPAs, the size and boundaries of areas may not be rationalized. This is a major omission, because purpose and function of an MPA cannot be properly evaluated without explicit consideration of its size and delineation. In this respect it is interesting to contrast Roberts and Hawkins (1997) – 'How small can a marine reserve be...?', with Walters (2000) '...how large should protected areas be?'

Statements such as: 'An MPA should be large enough to capture local representative biodiversity'; 'An MPA should be large enough to capture local significant features'; 'An MPA should be large enough to capture local ecosystem processes and ecological integrity' and so on abound in the literature. However, almost invariably the authors offer no guidance as to how such objectives might be achieved based on any ecological or environmental analysis. In addition, we typically encounter faulty logic in such statements. For example: preservation of ecological integrity of a region is generally acknowledged as a prime function for an MPA. Ecological integrity is known to be (at least in large part) a function of size. Nevertheless, in practice, the size of almost all MPAs is determined as a pragmatic compromise between socio-economic and environmental constraints, and ecological integrity is never assessed.

Although we can reasonably well identify the factors that should lead to decisions on the size of MPAs of some identified types, it is important to note that the actual size of an individual MPA will need to be established by study of the particular environmental characteristics of a region. Here we attempt to draw out the fundamental principles whereby the size of MPAs can be established, and matched to their designated purpose. This chapter is therefore intended not as a *review* of the justifications of sizes of existing MPAs, but rather as a *perspective* on the ecological and environmental principles that should be applied to determine the appropriate and effective size of an MPA of any type and intended purpose.

There can be no unique answer to the question: how large should an MPA actually be, and how should its boundaries be defined? because there are many 'kinds' of MPAs – that is, they are designated and designed for varied purposes (see Table 14.1). It is vital to define the function (purpose) and specify the type of MPA before making decisions as to size. All MPAs should not be created equal. The preference is that location, scale, size and boundaries for MPAs should be determined as far as possible by regional 'natural' conditions, according to defensible ecological and environmental principles, and taking into account the purpose for which each was designated. The concept of ecological scale (see e.g. Angel, 1994; May, 1994), which will not be explicitly considered here, therefore underlies decisions on MPA size and boundaries.

We are therefore asking: what are the principles from which size is determined, and how can these be put into practice? We need a statement of 'principles' to guide thinking and decision-making. Such principles need to be rational and defensible – so that they can be used to justify decisions to policy makers and politicians; and they need to be transparent – to the public and affected 'stakeholders' who might challenge the size of an existing or proposed MPA, either because it is deemed as too large or too small.

The various types of MPAs and their sizes

Many authors have recognized that all MPAs should not be created equal. Palumbi (2001) recognised three major types of MPAs: fisheries management tools; biodiversity protection; special feature. The list in Table 14.1 describes more precisely all these types of MPAs, and summarizes their primary purposes and suggested methods to determine appropriate size. Additional purposes for MPAs that feature in the IUCN (1994) list are not considered here.

Size of MPAs in distinctive areas (those with focal species, anomalies etc.)

This category includes protection of larger mobile, entirely marine, generally seasonally migrant species within distinctive habitats, primarily larger marine mammals such as whales and porpoises, although larger migratory fish such as tuna may also be included.

It has been recognized for some time that some focal species may be seasonally associated with marine areas exhibiting unique properties and processes. However, no general synthesis of relationships between focal species (Chapter 9) and distinctive areas (Chapter 7) had been carried out before Roff and Evans (2002). Distinctive habitats are often distinguished by their oceanographic *processes* which exhibit various anomalies, such as temperature, chlorophyll a, topography and isolation. These anomalies may define either permanent locations or temporary developments.

The size and boundaries of such distinctive areas, and the corresponding minimum size of any MPA designed to capture such an area, can therefore be readily defined from aerial or satellite images of the sea surface, from local topography or some combination of these features. Note, however, that marine mammals may seasonally use areas larger than those defined by such anomalies. In such cases, the area actually used and its boundaries can generally be defined by simple observation of the distribution of the focal species in question (see Brown et al, 1995). Some distinctive areas may not exhibit surface anomalies. Accordingly, it is important to define and categorize the various kinds of distinctive habitats (see Chapter 7).

Although it is relatively easy to define the size and boundaries of the distinctive area itself – from one or more of the regional anomalies, the area may be variable in development in both time and space. Temporal variation may be interannual, seasonal, tidal, daily, aperiodic and so on,

Table 14.1 Summary of some MPA types and strategies for the determination of their size

MPA type	Basic purpose	Strategies for size determination	Effective size	Proportion of habitat	Effectiveness assessed?
1. Distinctive areas, entirely marine	Protection of larger mobile, entirely marine, generally seasonally migrant species – mainly marine mammals	Measurement of anomaly: temperature, topography, sea colour, sea height	Variable, determined by extent of anomaly	Not specified	No, effectiveness may be confused with other means of protection for focal species
2. Distinctive areas, land-referenced	Protection of mobile marine species referenced to the land environment – mainly seals and birds	Land area from topography or observation of utilization. Marine area from anomaly or observation of foraging area	Variable, determined by extent of usage	Not specified	No, effectiveness may be confused with other means of protection for focal species
3. Areas for protection of rare/ endangered/ isolated populations and communities of benthic species/ areas of high local biodiversity	Protection of rare and endangered invertebrate species	N/A	N/A	N/A	N/A
	Protection of isolated populations of benthic species and communities	Distribution of colony itself. Topographic anomaly and high currents	Not determined	Not determined	No
	Protection of areas of high biodiversity (species diversity)	Not determined	Not determined	Not determined	No
4. Representative areas	Protection of specific habitats and their associated communities of the wider marine environment	1. Protect an entire ecosystem	Prime example: the GBRMP with a size of 344,000km^2	Not determined	No
		2. Select the largest observed unit of a habitat	Not determined	Not determined	No
		3. Use community composition indicator species	Not determined	Not determined	No
		4. Select areas equal to or greater than S–A curve asymptote	See Section 5 below	Not determined	No
		5. Select according to disturbance regimes	Not determined	Not determined	No
5. Fisheries habitat/ stocks/spawning/ recruitment areas	Protection of fisheries habitat and stocks	1. Habitat suitability indices	Not determined	Not determined	No
		2. Traditional scientific studies	N/A	20–50% of fishing grounds	Yes, modelling exercises
		3. Traditional ecological knowledge	Small areas, variable size, a few kms	N/A	Yes, well documented effects
		4. Spawning areas	From 2km^2 to 10^{4}km^2	N/A	No

Table 14.1 Summary of some MPA types and strategies for the determination of their size (*continued*)

MPA type	Basic purpose	Strategies for size determination	Effective size	Proportion of habitat	Effectiveness assessed?
		5. Minimum viable population/home range	5–7km^2, and even as small as 0.72km^2	N/A	Yes, well documented effects
		6. Models to prevent overexploitation	N/A	> 20% of fishing grounds	Yes, modelling exercises
		7. Stock- recruitment relationships	400km^2	N/A	Yes, modelling exercises
		8. Species-area curves for fish	Highly variable, from 5km^2 to 10,000km^2	N/A	No
		9. Pelagic MPA's	Parts of entire ocean basin, on seasonal basis	N/A	No
		10. Size of non-MPA areas	N/A	20–50% of fishing grounds	Yes, modelling exercises
6. Combination areas	Multi-purpose	Modelling exercises	Depends on regional characteristics	Depends on regional characteristics	No
7. Areas of anthropological/ archaeological/ sociological interest	Protection of shipwrecks, sunken cities etc.	Arbitrary	1nm diameter	N/A	No
8. Scenic areas, land-referenced	Protection of local 'beauty spots' and recreation areas	Land area of arbitrary size. Marine region from natural 'coastal cells'	Variable but easily determined	Arbitrary	No
9. High production/ upwelling/retention areas	Not specified	Measurement of anomaly: temperature (+\-), topography, sea colour, sea height	Variable, determined by extent of anomaly	Not specified	No
10. Areas for other purposes	Not specified	N/A	N/A	N/A	No

and these variations should be identified, for example from monitoring and/or satellite data.

Because these areas may be susceptible to environmental disturbance, some sort of buffer zone may also be needed. The extent of this zone could be determined from a risk assessment process, based on the nature of the threat, and some prediction of distance of effect based on current speed and direction (i.e. the rate of advection and diffusion of impact source).

Unfortunately, even where areas have been established for the express purpose of protecting flagship species, their size and boundaries are generally not explained. This is surprising because the association between populations of marine mammals and oceanographic features has been recognized for some years (e.g. Brown and Winn, 1989). Observer surveys of marine mammals can readily lead to the delineation of distinctive areas (e.g. Beckmann, 1995) even in remote and extensive areas of the Arctic. In the Canadian Arctic, it has been well established that seasonal aggregations of marine mammals are nearly always associated with upwelling regions (low temperature anomaly), or interfaces between marine and freshwaters or polynyas (leads of open water), characterized by high currents, mixing and year-round open water. Even areas of international significance – such as the islands of the Gulf of California, with high breeding concentration of sea-lions, turtles, birds, porpoises and whales – have not been well documented with respect to the physical or biological properties underlying their appeal to these groups of animals (Anaya et al, 1998). Until this is done, designation of areas to be protected will remain somewhat arbitrary.

A prime example of a proposed MPA, set as a combined function area (distinctive because of use by several whale species but also comprising representative features) is the Gully area on the Scotian Shelf, Canada. The limits of the core and buffer zones of the proposed MPA are proposed from a combination of natural physical, geographic and biological characteristics (the slope of the Gully walls and the 200m isobath; Harrison and Fenton, 1998). Subsequently, the use of the Gully region by cetaceans has been more fully documented by Hooker et al (1999). They have proposed an MPA also based on the geographic extent of cetacean sightings, and on the geophysical characteristics of the area including gully wall slopes and the 200m isobath.

Good examples of distinctive areas that have been properly designated in terms of size and boundaries are given by Brown et al (1995). Two areas, one in the outer Bay of Fundy and the other on the southern Scotian Shelf, were established by DFO Canada in 1993, to protect marine mammals – right whales – with an estimated current population of 350. The size and boundaries of these two areas was determined from sighting data collected during annual photographic surveys of the whale pods, and boxes were drawn around the 95 per cent limits of these observations. The corresponding geophysical properties associated with these areas have not been adequately defined, although the area within the Bay of Fundy is clearly centred on the gyre with high concentrations of the copepod *Calanus finmarchicus* (Roff, 1983).

Larger species such as marine mammals cannot be effectively protected throughout the year in distinctive MPAs because the majority are migratory. Other legislation is required to protect them outside distinctive areas; for example, specified shipping channels to avoid ship collisions (see e.g. Brown et al, 1995) and restrictions on types of fishing gear in selected locations. In Canada, there has now been agreement reached with shipping companies that they will seasonally adjust their routes to avoid the two designated MPAs for right whales in the Bay of Fundy and off southwest Nova Scotia.

These types of distinctive areas can also be seasonally designated to permit other activities. In addition, some distinctive areas may be used in alternate years as focal species change their patterns of resource use. Designation of an MPA may therefore be on an annual basis only.

In conclusion, the size and appropriate boundaries of such MPAs can be determined by the natural geophysical anomalies they exhibit. In addition, there is likely to be considerable local traditional knowledge of the actual seasonal distribution patterns of any focal species.

The level of protection, and human uses permitted within a region, should be consistent with the biological requirements of the focal species. Because most focal species will be seasonal migrants, establishment of local MPAs will not by itself be an adequate or complete conservation strategy.

Size of MPAs in distinctive areas – land referenced

This category includes protection of mobile marine species referenced to the land environment – primarily animals such as seals and birds. Although referenced to the land for some part of the year, they obtain their food resources mainly from the sea. There are two components to MPAs of this type – a land component and a marine component. Species using such areas typically do so either during intense feeding bouts in preparation for migration and/or for breeding.

The size of the land component can be established simply by observation of the area seasonally occupied by the focal species that use it. A buffer zone of some type should be established around this core area to prevent the effects of human disturbance on behaviour and reproduction. Many seabirds are extremely sensitive to disturbance by humans, especially at breeding times (e.g. Anderson and Keith, 1980).

These sensitivities can be respected by the establishment of buffer zones or exclusion zones surrounding the colony itself. Nesting sites for seabirds are generally well documented at the national and international level. Because nesting tends to be in remote areas of difficult access (e.g. cliffs, small islands), delineation of the size and boundaries of the area used is straightforward.

The size of the marine component of an MPA can be determined from the foraging area used by birds or seals. Many species of birds will routinely travel substantial distances (50 to 100km) over the sea to their preferred foraging areas to collect food for their young (Zurbrigg, 1996). The properties of these marine areas are generally poorly documented (apart from the nature of the food resource itself), but are likely to be areas of high productivity or high biomass (see Roff and Evans, 2002, for examples) exhibiting some kind of geophysical anomaly. Thus Haney et al (1995) showed that seabird abundance and biomass was 2.4- and 8-fold higher respectively, within a 30km radius of a seamount (a sub-surface topographic anomaly) in the North Pacific Ocean. These increases were attributed to the topographic effects of the seamount in increasing zooplankton resources for the birds.

In such areas of important resources for marine birds, the size and boundaries of an MPA can be set to conform to some function of resource concentrations and foraging distances. Brown and Gaskin (1988) showed that phalaropes made extensive use of regions of the outer Bay of Fundy, where tidally mixed and upwelling water brings copepods and euphausiids close to the surface. Several other species of migratory birds also take advantage of these advected resources in the same area. In these areas, a temperature anomaly or sea height anomaly (in combination with in situ monitoring and verification of increased zooplankton abundance and biomass) should define the boundaries and extent of the area in which resources are elevated.

Protection of larger mobile, seasonally migrant pelagic species while beyond distinctive areas

Most MPAs have focused on nearshore habitats, or at least those within a 200nmi exclusive economic zone (EEZ). Hyrenbach et al (2000) propose the establishment of pelagic MPAs that might be seasonally adjusted to accommodate the migration routes of larger marine vertebrates including birds, mammals and fish. Such an MPA could potentially be very large, though only part of it would be protected at any time, and then only against specified activities. The essential purpose of such vast MPAs would be to protect target focal species during their migrations, while they are not associated with distinctive areas in specific locations generally closer to the coastal environment.

There is currently no experience with this type of MPA which would need to be under international jurisdiction. However, the technology to make it practicable does exist. The migration routes of many of the larger marine species are becoming known, and can be tracked in real time (see Ocean Tracking Network http://oceantrackingnetwork.org). The locations to be protected can therefore potentially be specified in great detail. As well, vessel identification systems reveal the position, course and speed of ships, and analysis of its wake pattern can potentially reveal its activity. Ships can therefore be contacted – again in realtime to avoid interference with migrant species.

On the east coast of Canada, the position of shipping lanes has been adjusted, and advice issued to shipping in order to reduce the probability of collision with endangered populations of right whales. There is no fundamental reason why this kind of initiative should not be extended to the high seas to create a series of temporary mobile pelagic MPAs.

Size of areas for protection of rare/endangered or isolated populations and communities of benthic species

The major groups of interest here are the deep-sea corals and sponges, but where areas have been established for the express purpose of protecting such unique or unusual communities of marine organisms, the size of the areas set aside is generally not well rationalized. Reasons for the distribution patterns of these communities are becoming known, and involve some combination of depth, temperature, slope and topography, and currents, although we may find that these can be explained on the basis of some aspects of topography and flow fields (Bryan and Metaxas, 2006, 2007). These communities can have a high species diversity associated with them (Roff and Evans, 2002). The appropriate minimum size for such regions can be defined from geophysical predictions of their occurrences, and verified by surveys of the extent of the colonies themselves. Beyond this, borders can only be set by analysis of threats from human activities to establish buffer zones.

There is also some information on isolated communities of benthic invertebrates, such as deep sea vents (see Tunnicliffe, 1988), submarine caves (see Vacelet et al, 1994) and cold seep areas. In the first two cases, the extent of the area occupied can be described in terms of topography, and in the last case in terms of anomalously high concentrations of hydrocarbons. Their size and natural boundaries are readily measured by surveys.

The first report of anomalous abundances of deep-sea fauna from non-hydrothermal vent (or other chemosynthetically enriched) sites was from depths in excess of 3.5km on rocky cliffs of the Atlantic continental margin (Genin et al, 1992). Here, on the Blake Spur, high abundances of sponges and gorgonian corals were found. These rich communities, unexpected at such depths, were attributed to the anomalously high current speeds (that may exceed 100cm sec^{-1}) that eliminate sediments and enhance flux of food to the dominant suspension feeders.

Such anomalous communities of suspension feeders in the deep ocean are likely to be discovered in several places, probably in locations that exhibit some combination of anomalous local substrate (e.g. rock rather than sediment) and high currents. Most of these areas will not lie within the coastal or EEZ zones of any country. They will only be detected by in situ observation although, like deep-sea vents, coral and sponges, their distributions could probably be predicted. Their conservation is unlikely to be of high priority, but the size of any MPA designated to protect them can be determined by survey work to establish their size and extent.

The subject of genetic variation among populations of marine organisms and ESUs (evolutionary significant units) is a complex emerging field that has not yet been thoroughly reviewed with respect to conservation practices (see Chapter 10). However, this is the discipline that is most likely to define degrees and times of isolation of marine species. The extent of several kinds of distinctive areas (e.g. vents, deep sea corals, seamounts) is already set by topography. However, when considering designation of MPAs in such regions, their natural 'life-span', as well as their spatial extent and probable human impacts upon them, need to be assessed. For example, seamounts are of the order of 10^6 years old, deep-sea corals about 10^2–10^3 years of age, but vents communities have a life of only between 10^1 and 10^2 years.

Areas of high local species diversity and relations to production and biomass: Representative or distinctive areas?

The scientific literature on biodiversity and its relationships to characteristics of the marine environment is still extremely confused. Despite considerable interest in the distribution of species diversity, we still have a limited understanding of why it varies among habitats and why some habitats support a greater diversity of species than others. Species diversity may be either higher or lower in distinctive habitats than in representative ones (see Chapter 8). Here we consider several types of habitats with respect to their diversity and production or biomass.

Coastal upwelling areas and oceanic divergence zones

Coastal upwelling areas and oceanic divergence zones represent two major classes of distinctive habitats. Here production is increased due to nutrient additions, but species diversity is generally reduced (Margalef, 1978; Sakko, 1998). This lower diversity may extend through higher trophic levels to fish communities. However, at the highest trophic levels exploiting such areas we may again see higher diversity, for example in migrant birds of several species or mixed populations of marine mammals (e.g. Brown and Gaskin, 1988; and references therein). Thus the 'exploiters tend to become more diversified among themselves than the exploited' (Margalef, 1997). If diversity is reduced in these distinctive habitats, as it substantially is in the benthos below some upwelling areas (e.g. Sanders, 1968), then protecting them may in fact be less effective in conserving species diversity than in protecting a representative habitat of similar area. In general, these are among the most productive areas of the world's oceans and the most heavily exploited as fisheries, because they provide the best economic return for effort. However, if such habitats are targets for conservation, they are readily recognized because of their anomalous temperature and chlorophyll signatures. The size and boundaries of the area to be protected can therefore be readily determined, although it may vary greatly in time and space (see Sakko, 1998).

Distinctive habitats that accumulate biomass

In distinctive habitats that accumulate biomass without an increase in the rate of production, species diversity should be unchanged from that of surrounding representative habitats (see Roff, 1983). However, diversity at higher trophic levels (e.g. birds) may again be higher because of the rich resources available (e.g. Brown and Gaskin, 1988). The size of such gyral systems is poorly reported, but larger ones (> 10s of km) can be measured from sea height anomalies (e.g. by RADARSAT), and the extent and boundaries of smaller ones can be determined by surveys of current speeds and direction over a complete tidal cycle. Setting the size of an MPA to contain such gyral systems is therefore relatively straightforward.

Frontal zones

Frontal zones in coastal waters can be generated by a variety of mechanisms; their relation to the distribution of species diversity has not been systematically investigated. Complex frontal zones of mixing between water masses, for example the area off southwest Nova Scotia, may represent areas of high species diversity. This particular area contains many expatriate larval forms advected here by the Gulf Stream and Nova Scotia Currents (Roff et al, 1986). Where frontal zones are generated by tidal currents dominated by the M2 component, their position and geographic extent can be modelled from the stratification parameter (H/U^3) following Hunter and Simpson (see Pingree, 1978). The east coast of Scotland for example shows several meso-scale tidally induced frontal zones. The spatial extent of such zones can also be identified from satellite images as temperature anomalies between warmer stratified and colder unstratified waters. Determination of the size and boundaries of an MPA to protect such a frontal system is therefore straightforward.

Coral reefs

Coral reefs are a clearly recognizable category of habitats, where diversity is high but production is high as well (Muscatine and Weis, 1992). However, nutrient addition to coral reefs, although first increasing production (as carbon fixation rate, Muscatine and Weis, 1992), also causes a replacement of corals by seaweeds and decline in coral diversity. The extent of coral reefs is generally well known and mapped. Because coral reefs are so extensive in tropical and sub-tropical waters of the world, they can be considered as representative areas.

Underwater caves

Underwater caves are habitats where biomass and production of resources are low, and where species diversity and endemicity may be high (e.g. Vacelet et al, 1994). Because these habitats are directly a consequence of the physical topography of a region, the size and boundaries of an MPA to protect them is automatically determined by submarine topography and physiography.

Areas where production and biomass are not increased

Areas where production and biomass are not increased and where community types can actually or potentially be related to geophysical features constitute the representative habitats.

Other issues

A remaining issue is whether we can account for all distinctive habitats in terms of recognizable anomalies, or whether there are other areas (where production and biomass of resources is not increased) of high species diversity or 'hotspots' (e.g. Norse, 1993) without recognizable anomalies, that are unaccounted for.

Species diversity of the deep-sea benthos – which is considerably higher than earlier thought – presents particular problems. Here the separation between representative and distinctive communities becomes blurred. This is because each area sampled contains many new species, and species-accumulation (or species-area, or species-richness) curves do not reach an asymptote over many samples and broad sampling areas (e.g. Gage and Tyler, 1991). Therefore, each sampling area (or set of samples) constitutes a distinctive habitat (according to our definition of being different from its surroundings at some scale) and is a distinctive community. Thus the apparently 'homogeneous' environment of the deep sea, without identifiable anomalies, contains many contiguous species assemblages – each of high diversity – rather than the more clearly defined community types related to substrate type in shallower waters (see Chapter 8). Such local endemicity might be expected (and is observed) on individual widely spaced seamounts (e.g. de Forges et al, 2000), but was not expected in the flat and featureless deep-sea bottom. Such areas would require explanation of their origins, persistence and resource supply, or revelation of their anomalies or separate surveys to document their extent.

Finally, some sort of buffer zones may be important for all such areas; these will need to be defined locally depending upon an analysis of the nature of threats. There may be additional types of distinctive areas, containing communities of high biodiversity that cannot be described on the basis of their physical or oceanographic anomalies. The extent of any MPA to protect them can only be established by local in situ survey work.

Size of representative areas

The category of representative habitats is perhaps the most difficult, but largest and potentially the most important, for which to define MPA size. The prime function of such areas is to protect the representative components of biodiversity. This means that each representative MPA should contain all (or a majority of) the regional species and communities typically found in its designated habitat type or types.

At least five potential strategies for determination of the appropriate size of such areas are possible:

1. Protect an entire 'ecosystem'
2. Select the largest observed unit of a given habitat within a region
3. Select the largest habitat unit based on community composition indicator species
4. Select areas that are equal to or greater than the asymptote of species-area (S-A) curves
5. Select areas according to the extent and frequency of disturbance regimes.

Protect an entire 'ecosystem'

The concept of an ecosystem is firmly lodged in the ecological literature. Unfortunately, in marine systems which are open and extensively interconnected, the physical dimensions of an ecosystem are not easy to define, although marine ecologists often use the term. It is important to note: first that marine ecosystems are rather arbitrarily defined; and second that even if defined they cannot be considered as hierarchically organized – that is, they are discontinuous, not contiguous and not hierarchically nested.

At present, a prime example of designation of an entire marine ecosystem as a marine park (with some level of environmental protection) is the Great Barrier Reef Marine Park (GBRMP) in Australia (Ottesen and Kenchington, 1995). The GBRMP, with a size of 344,000km^2, is probably as close as we can come to a recognizable and acceptably defined marine ecosystem. A variety of practices are permitted within its borders that are defined by natural physiographic and biological characteristics of the habitats. Smaller MPAs in Australia cover a span of sizes from < $2km^2$ to over $5000km^2$, generally of a multiple-use type with integrated management permitting a variety of activities.

It is not likely that many entire marine 'ecosystems' will become completely protected (although the issue of what we pragmatically term an 'ecosystem' makes this a moot point). Rather, we can expect to see some sort of practical trade-off between smaller areas with high degrees of protection, and larger areas with lower degrees of protection. In addition, we can expect that some units of a representative habitat type will receive some degree of protection, rather than that the entire ecosystem will be protected. This is basically the foundation for the compromise practice of designation of 'core' MPA areas surrounded by buffer zones. In contrast to representative areas which may be partially protected, the appropriate practice for distinctive areas is that the entire area delineated should be protected, because it is a functional whole.

Largest observed habitat unit

A relatively simple strategy would be to conduct mapping of the habitats of a region, at the finest scale of available data, based on their geophysical properties (e.g. Roff and Taylor, 2000; Roff et al, 2003). The largest units of each representative habitat would then become candidates for establishment of MPAs, based on the theory of species–area relationships that predicts more species in larger units of a given habitat type (see Neigel, 2003, and below). Major advantages of this simplistic approach are that this exercise can be carried out wherever suitable data exist on fundamental geophysical properties, and that it maps the natural heterogeneity of marine environments and recognizes natural boundaries. Such an approach could be useful for example in coastal regions, because it inherently recognizes the natural habitat alternation between rocky headlands, sandy beaches and muddy bays (e.g. Carter, 1988).

A major problem with this approach is that, within any region, such geophysical data is only available at a relatively coarse scale. All marine habitats are in fact highly heterogeneous, and much of this heterogeneity will go 'unseen' in coarse-scale maps. The approach may therefore be somewhat arbitrary as a determinant of the size of habitat units. However, in combination with species–area relationships (or species-accumulation curves) it could be a useful method to determine both the boundaries and size of representative habitats for MPAs. In addition, coarser scale mapping can direct our efforts for more intensive in situ sampling and mapping, using techniques such as multibeam sonar (e.g. Foster-Smith et al, 2000).

Community composition indicator species

Indicator species (of community composition, Zacharias and Roff, 2001a; see also Chapter 9) may be used to define biogeographic boundaries. Where biogeographic boundaries coincide with geophysical boundaries, community composition indicators can be of value. However, the ecology of indicator species is not nearly as well developed in marine waters as it is in freshwaters. Biogeographic boundaries as indicated by single species ranges should be interpreted with caution, and indicator species are probably best used as adjuncts to habitat recognition (see Chapter 6).

An alternative approach, based on zoogeographic information over broad marine regions, has been followed by Haedrich et al (1995), Mahon et al (1998), Perry and Smith (1994) and Horn and Allen (1978). These authors mapped the distributions of fish assemblages and their associated habitat characteristics. Within such areas, the establishment of MPAs of defined size, proportion and purpose could serve important functions, especially for fish and biodiversity conservation. It is most likely that the size of MPAs within such biogeographic regions would be set by scientific survey or traditional ecological knowledge (TEK) of spawning areas or from knowledge of species–area relationships.

Species–area (and species–accumulation) relationships

A prime consideration for determination of the size of MPAs within representative habitats is to know the relationships between species richness (S = the number of species observed) and area (A), that is the shape of species-area, or species-accumulation (S-A) curves, and, most importantly, the size at which these curves reach an asymptote. A species–area curve relates species richness to the area of progressively larger natural habitat units, while a species–accumulation curve represents the addition to species richness of successive independent samples taken from within a defined habitat type.

The area at which an asymptote is reached represents the area of a habitat that should contain all the species characteristic of that habitat. The species–area relationship is derived from island biogeography theory of McArthur and Wilson (1967), and is considered one of the cornerstones of modern ecological science and conservation biology. The shapes of S-A curves will vary among taxa and among habitat types, but take the general form:

$$S = cA^z$$

where z = the slope of the relationship

It is expected that, within taxonomic groups, the asymptote of an S-A curve will increase as a function of depth (Sanders, 1968) although this view has been challenged (Gray, 1994). If Sanders is correct, then the size of MPAs, based on the S-A curves for any taxonomic group, would increase with depth. If this is not so, then size of MPAs (based on S-A curves) would be invariant with respect to depth. Clearly this important issue needs to be resolved; for further review in deep sea communities see Etter and Mullineaux (2001) and Chapter 8.

Each taxonomic group will have its own characteristic S-A curve for a given habitat. Among taxa, there is a strong relationship between the size of organisms and the area asymptote of the S-A curve. Bacteria will exhibit the lowest, and whales the largest area asymptote. From the point of view of the smaller burrowing zoobenthos, an MPA of only a few tens of square metres might be adequate to represent all species of a given habitat type, and even for the larger epifaunal macrobenthos an MPA of several hundred square metres would likely capture all species in a given habitat type. However, such a small size would not capture the demersal fish, nor would it allow for the effects of periodic disturbances.

Any representative MPA should be large enough to accommodate the entire array of species normally resident within the type of habitat it represents, and the boundaries of a habitat type can be set by its geophysical characteristics (e.g. Roff and Taylor, 2000; Roff et al, 2003; and see Chapter 6). Size will be determined by the area requirements of the largest resident species; this generally means fish. Because the largest marine species of all – marine mammals – are generally migrants, their requirements can be accommodated, at least seasonally, under distinctive areas, and at other times within pelagic MPAs (see above).

Each species of fish has slightly different habitat requirements and natural range of population densities, and each species may use several types of related habitats. This means that the appropriate analyses of size and boundaries of MPAs need to be based on the entire fish community or the various component species assemblages.

The relevant species then are fish, primarily those resident within a region. There are several reasons why attention will focus on them at some point, including the following:

- These are commercially by far the most important marine species.
- They are among the largest marine organisms, and correspondingly, their S-A curves can be expected to show an asymptote corresponding to a large area. Therefore, for conservation purposes, the fish community can be considered as an 'umbrella taxon' (Zacharias and Roff, 2001a, and Chapter 9).
- We know a great deal about the abundances, biogeography, distributions, life cycles, larval ecology and recruitment patterns, migrations, habitat requirements, taxonomy, and genetics of stocks and meta-populations of many

fish species. This means that we can actually or potentially map 'representative fish communities or assemblages', and correlate these distributions with benthic habitat types based on geophysical characteristics (see Chapter 6).

- We also usually know enough about the fish community that we can construct S-A curves over broad areas within a region, for several types of habitats, based on commercial and scientific survey data.
- Information is also available on the effects of both natural and human-induced disturbance regimes, for example North Atlantic oscillation indices, decadal shifts in oceanographic regimes, upwelling regimes and seasonal stratification cycles, and fishing, dumping and pollution effects respectively.

Such an analysis of S-A curves for marine fish communities in temperate waters has been carried out by Frank and Shackell (2001). Assessing the diversity of fish species on the banks along the Scotian Shelf (considered as 'islands' according to island biogeography theory), they showed that the number of fish species increased as a function of the bank area (r^2 = 82 per cent). This increase was due primarily to the increase in the uncommon species in the larger banks, and may be related to the higher habitat diversity of the larger banks. Frank and Shackell (2001) attribute this relationship between bank size and species diversity to the tendency of larger banks to support greater abundances of individual fish species that are less prone to population extinctions. It is sobering to note that the largest bank on the Scotian Shelf with the greatest fish diversity is Sable, at approximately 10,000km^2. This is also the bank with the highest species diversity of epibenthic invertebrates (Figure 14.1). Thus areas for MPAs derived in this manner from such S–A relationships could potentially be very large in temperate waters.

However, this does not mean that entire banks (i.e. an entire geomorphological feature) need to be protected. Lewin and Roff (unpublished) re-examined the fish data of the Scotian Shelf (essentially the same data set as Frank and Shackell, 2001) and the companion data for epibenthic invertebrates in an analysis

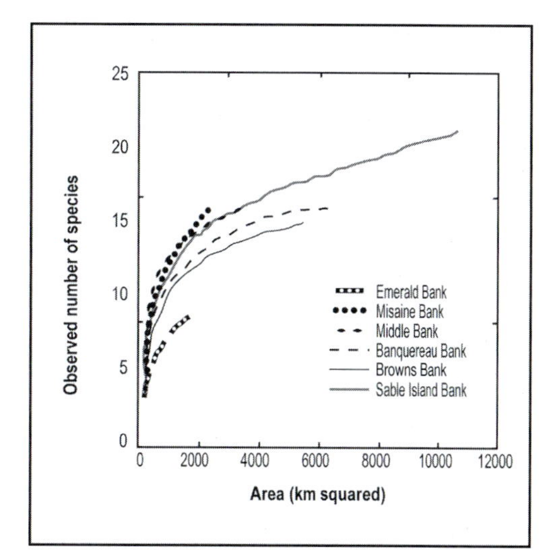

Note: Taxonomic diversity shows a significant relationship (R2 = 0.66, p<0.05) to bank size

Source: Lewin and Roff, unpublished

Figure 14.1 Species–area curves for epibenthic macroinvertebrates for six banks on the Scotian Shelf and the Bay of Fundy

similar to that of Zwanenburg and Jaureguizar (unpublished). Based on extrapolations of the species-accumulation curves from the Chao I equation (Colwell and Coddington, 1994), the area asymptotes at which all species in the region should be represented (based on the benthic area actually estimated to have been sampled) were 56.3 and 34.4km^2 for fish (129 species) and epibenthic invertebrates (34 species) respectively (Figure 14.2).

Further analysis of the same data by Lewin and Roff, using a multi-dimensional scaling ordination, suggested four separate species assemblages clearly geographically demarcated. The fish assemblages were similar to those of Zwanenburg and Jaureguizar (unpublished) presented in Colour plate 12a. Again, using the Chao I estimator the area asymptotes at which all species of an assemblage should be represented ranged from 5.7 to 49.4km^2 for fish, and from 2.4 to 26.1km^2 for epibenthic invertebrates respectively. This clearly shows that the estimate for size of an area that would 'capture' all regional species depends on the taxon being sampled.

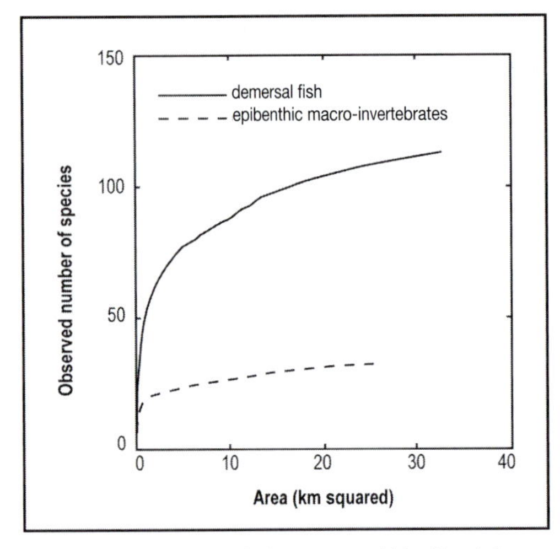

Note: Neither relationship reached an asymptote. Using Chao 1, the predicted number of taxa for demersal fish was 129 and the area at a predicted asymptote was $56.29km^2$; the predicted number of taxa for epibenthic macro-invertebrates was 40 and the area at a predicted asymptote was $34.41km^2$

Source: Lewin and Roff, unpublished

Figure 14.2 A comparison of the species–area relationships for all demersal fish and all epibenthic macro-invertebrates along the Scotian Shelf and Bay of Fundy from 1999 to 2002

However, these estimates of size of area and species numbers observed, based only on species-accumulation curves, are composites and should be regarded cautiously. They represent absolutely minimal estimates for species representation, and should only be used for comparative purposes, in conjunction with other estimates such as minimum viable populations (MVPs) of appropriate species (see Conclusions). Simple presence of a species without knowledge of its population structure does not ensure viability.

For tropical seas, estimates of areas that capture the entire species diversity of fish may be even smaller, probably because a higher proportion of these species are non-migratory and have much smaller home ranges. For example, in a study of demersal fishes off Costa Rica, Wolff (1996) calculated a maximum species diversity of 306 species for the whole region between depths of 20 and 200m. From that data, an area of only $5km^2$ should contain all the species found.

Disturbance regimes

The effects of disturbance and an analysis of implications for the size of MPAs is an extremely complex subject. A framework adapted from Sousa (2001) is presented in Box 14.1. All marine biological communities are subject to a multitude of disturbance types on various time and space scales (Mann and Lazier, 1996; Sousa, 2001). These disturbances may be: physically induced by storms, tides, upwelling events, seasonal stratification cycles, atmosphere–ocean regime shifts (e.g. Steele, 1996) and so on; biologically induced by competition, predation, keystone effects, species invasions and so on; or induced by human actions from fishing, dumping, pollution effects and so on. The effects of any disturbance regime are a function of magnitude, type, periodicity and organism size. Sessile communities are more severely affected than motile ones. Modelling exercises (e.g. Caswell and Cohen, 1991) make clear that the diversity of metapopulations reflects the interactions among organisms with respect to the scales of disturbance and dispersal.

The intermediate disturbance hypothesis (Connell, 1978) has significant explanatory power for the diversity of marine communities (see e.g. Zacharias and Roff, 2001b), with both low and high frequencies and severities of disturbance leading to lower species diversity. Accordingly, the disturbance regime (whether physical, biological or human) and the scale of the effects experienced by a particular kind of habitat have important consequences for the determination of the size of an MPA within that habitat.

It is not possible to draw any general conclusions as to the appropriate size of an MPA based on disturbance regimes. However, we can establish some pragmatic guidelines. In most cases physical disturbances will overwhelm biological ones in scale and effects, though not in frequency. Within any region where MPAs are to be established, the types, frequencies and probable extents of disturbance regimes should be documented. Any single MPA should be larger in size than the maximum extent of the maximum disturbance regime expected within a region. When this is not possible, several MPAs of a given type should be established within a region such that not all

Box 14.1 A proposed framework for marine disturbance regimes

The impact of any disturbance on a community is a function of:

- the physical scale of the disturbing process;
- the spatial variation in disturbance;
- the severity of the disturbance;
- its periodicity, whether predictable, seasonal or not;
- the susceptibility of the affected community in terms of the following characteristics of the biota and its habitat:
  - morphological
  - physiological
  - physical
  - aggregation or fragmentation of species populations
  - species trophic interactions
  - species diversity and successional stage of the community
  - substrate type.
- correlations among disturbance characteristics. Generally large, severe disturbances occur less frequently than smaller ones.

The rate of recovery of any community following disturbance is a function of:

- severity of the damage;
- environmental alterations caused;
- changes in species composition and trophic interactions produced;
- existence of refuges for species in heterogeneous environments;
- resulting patch shape, size and type following disturbance;
- life-history characteristics including:
  - mode of dispersal and re-establishment of biota by propagules
  - the source and quantity of recolonizing individuals.

Source: Largely following Sousa (2001)

of them can be impacted simultaneously by any single disturbance event, and so that recolonization of fauna and flora in any member of such a set of MPAs is ensured from another source. In this way we can ensure that it is extremely unlikely that the entire community type which the MPAs are designed to protect can be simultaneously disturbed – that is, there will always be a habitat unit of a representative community type that remains unaffected by a disturbance within a given region. This is basically planning for 'an insurance factor' (see Allison et al, 2003).

The coastal zone

Typically, temperate coastlines alternate between exposed rocky headlands, exposed sandy beaches, and protected muddy bays and estuaries. These form natural repeating units of the coastal zone of varied size and character (see Chapter 11). It should therefore be relatively straightforward to devise a conservation strategy and decide on the size of coastal MPAs – indeed it should be simpler than for offshore regions. However, surprisingly little effort has been directed to this issue, despite the fact that the coastal zone is where humans interact most strongly with the oceans. An integrated plan for coastal zone protection

Box 14.2 A method to estimate the size and number of conservation units required in the coastal zone. Based on a study of juvenile fish of the Atlantic coast of Nova Scotia (O'Connor and Roff, unpublished).

The Atlantic coast of Nova Scotia is probably as complex as any in the world, with a highly convoluted coastline of estuaries, bays and coves (see chapter 11). Within the selected study area, there were 53 bays and 62 coves. Relationships between the juvenile fish and physiographic features of this nearshore environment were explored, in order and to show that it may not be necessary to survey all types of habitat along an entire coastline in order to devise an overall conservation strategy for a coastal region. Juvenile fish were selected as the target community for several reasons. Nearshore areas provide multiple benefits to the early life stages of many fish species; their habitats provide cover to hide from predators, warmer waters for faster potential growth, and a larger quantity of food resources. Importantly, the general public identifies immediately with these organisms.

Twenty bays were repeatedly sampled, chosen to represent the variability in coastal bay size, shape and habitat in the province. Quantitative seine hauls (of standard 50m length) were made in sub-tidal waters, at low tide, over a series of substrates ranging from cobble/ pebble to sand/ mud combinations. Nearly 20,000 juvenile fish of 35 species were caught.

The survey results suggested that the majority of juvenile fish species are distributed across the entire study area, and the community as a whole showed no clear biogeographic tendencies. There was as much variability in juvenile fish species composition within a bay as there was among bays, the fish assemblages were not related to a set of physiographic features of the bays examined,nor was there any apparent effect of substrate type on species composition. More information on the characteristics of these bays is presented in chapter 11. The bays sampled can be treated as a set of representative types, at least as far as the fish are concerned, however they differ considerably in size, from small (<40 km^2), to medium (40-200 km^2) and large (>200 km^2).

According to theories of relationships between species and area, we should expect that as the size of a bay increases, more species will be caught. In fact, when equivalent lengths of shoreline in each bay type were examined, more species were caught in the set of smaller bays than in the set of larger ones (see Neigel 2003 for suggested reasons). It would seem that along the Atlantic coast of Nova Scotia small bays are providing important habitat, where many juvenile fish species are assembled. This suggests that in order to maximize protection of coastal fish species, locating MPAs in a set of small bays would be more advantageous than protecting a smaller number of large bays. This is certainly an appropriate strategy because larger bays are also likely to be more heavily impacted by human activities (e.g. wharfs, shipping, aquaculture operations, recreational activities, etc.). But how many small bays should be protected? We proceeded as follows, taking certain liberties with species-accumulation curves.

Species-accumulation curves for the samples of juvenile fish were constructed, and using the Chao I equation (Chao 1984), an estimate was made of the likely total number of juvenile fish species in the region. Chao 1was chosen as the most appropriate estimator for the data, as it works well on data with a large number of rare species (Colwell and Coddington 1994).

$$\text{Chao } 1 = S_{obs} + (a^2/2b)$$

Where: S_{obs} is the number of species actually observed during sampling;
a is the number of species represented by a single individual (i.e. singletons); and
b is the number of species represented by two individuals (i.e. doubletons).

From the estimated total number of species in the bays of the region, the length of shoreline (not the bay area) that would need to be sampled to catch 'all' species of juvenile fish can now be calculated from a modification of the Preston (1962) equation:

$$S = cL^z$$

Where: the equation is solved for L, the length of shoreline;
S is the number of species; and
c and z are the y-intercept and slope, respectively, of the linear regression on the observed number of species caught.

For the large bays, an average shoreline length of 1.5 to 1.7 km would be required to theoretically encompass all of the species found within the bays of this size group, while small bays needed between 3.4 to 6 km of protected shoreline. Although a greater length of shoreline is needed for small bays, a greater total number of fish species would be contained in their protected coastlines (26 to 35 species in small bays versus only 14 to 17 in large bays).

Proceeding further, and making certain assumptions, it is now possible to calculate the number of entire bays required in order to afford protection to all fish species (and all habitats) whether included in the sampled habitat type or not.

From the Preston (1962) equation, the total length of shoreline needed to protect the overall species richness of juvenile fish within the study area was determined (as above). However, this value was calculated based only on data from those substrates which could be sampled with a beach seine (i.e. cobble/ pebble to sand / mud). The habitats found at either end of the substrate spectrum (e.g. rocky headlands and mudflats) were not examined because of the difficulties in sampling these areas. With estimates of the total shoreline length which could have been sampled with a beach seine, and the proportion of each bay that this type of shoreline encompasses, the predicted shoreline length needed to protect all juvenile fish species can be extrapolated to the entire shoreline of an average bay, including its rocky and muddy habitats. By encompassing all habitats in the total shoreline protected, the species assemblages of these additional habitats would also be protected.

The appropriate number of bays to conserve was determined from the number of small bays along the Atlantic coast of Nova Scotia and the length of shoreline (within these small bays) which could be sampled with a beach seine. When this proportion was compared to the predicted shoreline length required to include all estimated species, the average number of bays required for conservation was determined.

This was done separately for each group of small bays, for the Eastern Shore and the Yarmouth area, and provided the average number of bays needed as follows:

$$\text{No. Small Bays}_{TC} = \frac{\text{No. Small Bays}_{NS} \times \text{Shoreline length}_{TC}}{\text{Shoreline length}_{SS}}$$

Where: No. Small Bays$_{NS}$ is the number of small bays found within the study area of mainland Nova Scotia;
Shoreline length$_{SS}$ is the total shoreline length of substrates which could have been sampled by beach seine;
No. Small Bays$_{TC}$ is the average number of small bays that should be conserved;
Shoreline length$_{TC}$: the shoreline length which should be conserved, as calculated by the Preston (1962) equation.

The average number of small Nova Scotia bays needed to attain a shoreline length of 3.4 to 6 km was then determined. The calculation includes data on the number of small bays found along the Atlantic coast of Nova Scotia, the total length of shoreline within these small bays which could have been sampled with a beach seine, and the total predicted shoreline length required (i.e. 3.4 to 6 km). Data was available from studies described in chapter 11. According to these calculations, for two sets of small bays (<40 km^2) around the Atlantic coast of Nova Scotia, in order to attain these shoreline lengths, between 0.3 and 0.5 of a bay in one set, and between 0.98 and 1.7 bays in the second set would be needed.

A study designed in this way can therefore estimate both the overall length of coastline needed for conservation, and the proportion or number of bays to conserve. These proportions of small bay are based on estimates of shoreline length which only include those substrate types which can be sampled with a beach seine. There are, obviously, many other substrates found within these bays which will likely have their own set of coastal fishes associated with them. However, if we assume that all substrates are randomly distributed within bays (a reasonable assumption for a set of representative bays), then protecting bays according to this strategy should afford protection to both the sampled and un-sampled substrates and their fish species.

This should cover not only the juvenile fish community, but would extend protection to other representative communities of the shoreline subtidal (e.g. macrophytes, seagrasses, saltmarshes etc.).

References

Chao, A. (1984) 'Nonparametric estimation of the number of classes in a population', *Scandinavian Journal of Statistics*, vol 11, pp265–270

Colwell, R. K. and J. A. Coddington (199 4) 'Estimating terrestrial biodiversity through extrapolations', *Philosophical Transactions of the Royal Society B: Biological Sciences*, vol 345, pp101–118

Neigel, J. E. (2003) 'Species-area relationships and marine conservation', *Ecological Applications*, vol 13, S138–S145

Preston, F. W. (1962) 'The canonical distribution of commoness and rarity: Part 1', *Ecology*, vol 43, pp185–215

would carry out an analysis of the size, boundaries and degree of replication of such units, their ecological representativity or distinctiveness, and combine them into a network of MPAs. Unfortunately, the majority of MPAs in the immediate coastal zone seem to have been designated as much for their scenic appeal as for their ecological value.

We present three strategies for deciding on the size or number of MPAs in the coastal zone, but others should be devised.

Coastal retention cells

The size of individual habitat units is set by the natural folding and indentations of the coastline itself. The processes that lead to natural 'coastal cells' (see e.g. Carter 1988), involving local sediment transport, are well known and can form the foundation for coastal zone management. Cell boundaries may be either 'fixed' – set by geomorphological and topographic features – or 'free' – more difficult to locate and determined by local wave fields. The movement of inorganic material within and between cells is determined by the drift pulse (see Carter, 1988), but these geophysical concepts have rarely been directly applied to biological materials. The size and interaction among these cells will change depending upon the local wind and current patterns, but Sotka et al (2004) showed remarkably strong correspondence between drifter trajectories (defining coastal circulation patterns) and genetic structure in coastal populations of *Balanus glandula*. Further integrated study of coastal circulation and the genetics of selected species should be revealing at more local scales.

Genetic clines

Genetic tools provide a way to define the spatial spread of larvae (see Chapters 10 and 17). Genetic clines (geographic zones in which genetically differentiated populations interbreed) provide opportunities to quantify the relative roles of selection and dispersal (see Sotka and Palumbi, 2006). The geographic width of a stable genetic cline is determined by a balance between the homogenizing effects of dispersal and the diversifying effects of selection. The theory of genetic clines indicates that the average dispersal distance of larvae is some fraction (generally ~ 35 per cent) of the clinal width. The width of a cline in gene frequency is approximately proportional to gene flow (σ) divided by the square root of per-locus selection (s). Measures of σ and s are especially useful because they can be made from collections (Mallet et al, 1990).

Although cline theory is based on several underlying assumptions, the dispersal distances inferred from empirical data should be of the correct order though not precise. Even so, such estimates of larval dispersal are valuable, as they can be utilized to design appropriate scales for future investigations and provide some guidance to conservation efforts including the size and spacing of marine reserves.

Species–coast-length relationships

Other methods to estimate the appropriate size, boundaries or number of MPAs in the coastal zone will probably involve geophysical surveys, as described in Chapter 11, direct biological surveys of communities or a combination of both. However, although direct biological surveys may be necessary, this does not necessarily mean that all coastal communities need to be surveyed. The survey of a single community type, along an entire length of coastline, may be sufficient. An example of how this could be accomplished to yield estimates of requirements for conservation is given in Box 14.2, from an unpublished study by O'Connor and Roff.

Size of fisheries habitat/stocks/ spawning/recruitment areas

This includes the sustainable management of natural marine resources especially fisheries and fishing/spawning areas. Nine potential strategies for determination of the appropriate size of such areas seem possible:

1 Habitat suitability indices (HSI) within a region
2 Traditional scientific studies on stock sustainability
3 Traditional ecological knowledge (TEK)
4 From knowledge of spawning areas
5 Minimum viable population and home range of target species
6 Models to prevent overexploitation
7 Stock–recruitment relationships of target species
8 Species–area (S–A) curves
9 Size of non-MPA areas.

Habitat suitability indices within a region

Habitat suitability indices (HSI) can be mapped for one or more species in a region, from some combination of appropriate environmental and habitat characteristics. The 'best' habitat for a given species can then be identified, including its small-scale topographic heterogeneity. In addition, the best combinations of habitat characteristics for several species can be 'overlaid' to produce composite HSI maps. Such maps (e.g. Brown et al, 1997) automatically yield the boundaries and size of areas that could be designated as conservation areas, spawning areas, recruitment areas, fishing grounds and so on.

Analysis of habitat suitability, and matching of habitat to the various phases of a specific life history, can be critical to the effective planning and management of MPAs, especially for those designed to enhance populations of commercially exploited species (e.g. Fernandez and Castilla, 2000). Even small areas of the right kind of substrate may enhance survivorship and recruitment of juveniles of commercially important species, for example a combination of sheltered habitats with a substrate of sand and boulders for the stone crab *Homalaspis plana* (e.g. Fernandez and Castilla, 2000), and rock 'bowls' that act as shelter for juvenile red sea urchins *Strongylocentrotus franciscanus* (Rogers-Bennet et al, 1995).

Traditional scientific studies on stock sustainability

The science of management and protection of individual fish stocks has had a long, detailed but ultimately largely unsuccessful history. Successful resource management involves knowledge of all phases of the life history of a species, including: location and size of spawning areas; patterns of larval drift and recruitment; adult migration/counter-migration routes; spawning biomass estimates; estimates of minimum viable population size and so on. Protection of spawning and recruitment areas has again become an important management tool.

Recent expert opinion on the size of fisheries reserves has emphasized the proportion of habitat that should be set aside, rather than the absolute size of individual marine reserves (e.g.

NRC, 2001; and references therein). The consensus view is that, if reserves are designed for fisheries enhancement and sustainability, then setting aside 20–50 per cent of existing fishing grounds will minimize the risk of fisheries collapse and maximize long-term sustainable catches. This view is based on a combination of modelling studies and empirical observations, but there is considerable uncertainty around the estimates, and the proportion of areas set aside may need to be a function of species requirements. Working 'backwards' from the total size of a region under consideration, from estimates of the proportion of habitat to be protected and from a rather arbitrarily decided number of replicates of each habitat type to be protected, the National Research Council (NRC) panel suggested that marine reserves designated for fisheries conservation would vary in size from 25 to 225nmi^2.

Traditional ecological knowledge

The significance of traditional ecological knowledge (TEK) for management of natural resources is now widely appreciated by scientists, sociologists and economists. TEK has been defined as 'the sum of the data and ideas acquired by a human group on its environment as a result of the group's use and occupation of a region over many generations' (Neis, 1995). It is apparent that local fishers always have a significant level of knowledge of both fish and their environment, and the causal interactions between the two (e.g. Neis et al, 1996).

There are now several examples from tropical and subtropical areas where reef fisheries have been effectively managed on the basis of TEK (e.g. Warner, 1997). In the Philippines, there has been considerable success in establishing coral reef sanctuaries based on TEK, in collaboration with local fishing communities (Walters and Butler, 1995). In each case, increases in fish abundance followed establishment of reserves, and led to new incentives. Such management may be simply on the basis of the 'law of the commons', or may be mediated by the broader local community and/or regulatory authorities. Effective management and authority, originally in the hands of indigenous peoples, then supplanted by regulatory bodies, is now often devolved back to the local community of fishers. Regulation of location, size and boundaries of protected areas and fishing quotas can vary widely depending upon the biological characteristics of the resource fishery and the habitat. It may involve protection of areas, restrictions on gear type and size, seasons and so on.

From knowledge of spawning areas

This is generally covered under either scientific ecological knowledge (SEK) or TEK or both. However, it has been explicitly considered in relation to size of MPAs by Haedrich et al (1995). They propose a scheme similar to that used in the Great Barrier Reef of Australia (Ottesen and Kenchington, 1995) for use on the Grand Banks of Newfoundland. In type B zones within a region, fishing would be allowed on a seasonal basis. Within these zones, fishing would not be permitted at all in smaller type A. These type A zones would correspond to spawning areas. Protection of them would afford protection not only to the spawning stocks themselves, but would also eliminate habitat destruction due to fishing gear. The size and boundaries of such areas can only be established by scientific survey, in combination with TEK. Using these principles for the major fisheries of temperate waters, the sizes of areas to be protected range over seven orders of magnitude (Haedrich et al, 1995). Type A zones might be as small as 2km^2, and type B areas as large as 10^{4}km^2, depending upon habitat characteristics, heterogeneity and the fish assemblage under consideration.

Minimum viable population and home range of target species

Analysis of MVPs has been carried out extensively for larger terrestrial animals. Although there is also considerable experience with such analyses for pelagic, demersal and reef fish populations, the concept does not seem to have been widely applied to MPA planning. A major reason is that population size is not the only (or perhaps even the most important) factor determining viability (see Chapter 10). The question of how large an area of habitat should be protected depends partly on how the estimate of MVP was made; if it was made using a model, then that model can be used in a spatial context to determine habitat needs.

The more general question of habitat required as a function of population size depends on the quality and spatial distribution of the habitat, and the expected density of the species (at its carrying capacity) per unit area of optimal habitat. Both of these are species- and seascape-specific, so a generalization is difficult, and this approach would lead us back to analysis of 'essential fish habitat' or HSI. Such an approach could, however, be very valuable in specific locations designed for the protection of individual species, where habitats are fragmented and a metapopulation modelling approach to risk analysis is required.

The possibility that the home range of a fish could be used as a guide to determination of MPA size has been examined by Kramer and Chapman (1999). Many fish species, especially reef fish, are known to have well-defined home ranges, the size of which increases as a function of body size. For reef fishes in temperate and tropical waters, reserves as small as 5–7km^2, and even as small as 0.72km^2, can have dramatic effects in increasing the average size and population abundances (Dugan and Davis, 1993). Unfortunately, although even small reserves can be effective, there does not seem to be any clear relationship between the size of the reserve and its incremental effectiveness in terms of promoting fish size or biomass (Dugan and Davis, 1993).

The concept of MVP must be linked with knowledge of larval dispersal characteristics and metapopulation theory. Although the parameters needed to assess an MVP and the size of an MPA to protect a given species seem clear, theory is not yet sufficiently well developed that this can be deemed a general approach to species conservation. It seems likely that for the immediate future, this approach to marine conservation will be largely empirical and experimental, and will require continued evaluation of the performance of existing MPAs (see e.g. Carr and Reed, 1992).

Models to prevent overexploitation

A potentially more robust approach, better in accord with the precautionary principle, attempts to define the size of an area that should be set aside to prevent a stock collapse due to over-exploitation. The concept of fisheries reserves to prevent over-exploitation is an ancient one practised in artisanal fisheries around the world. In recent years it was first proposed by Beverton and Holt (1957), but the practice of regulation of maximum sustainable yield gained favour instead.

Fisheries reserves are now increasingly being examined as alternate management options, but because of resistance from the fishing community itself, most studies are carried out as mathematical simulations (NRC, 2001). An example of such a model is by Guenette et al (2000) for populations of northern cod off the east coast of Newfoundland. This showed that reserve areas equivalent to 20 per cent of the cod fishing grounds, combined with a reduction in fishing capacity, could have succeeded not only in preventing the collapse, but would also have aided in rebuilding the stock.

Such modelling exercises will clearly be required for each target species and type of fishery but could, in combination with planning for other mixed types of MPAs, be powerful tools for fishery conservation and biodiversity conservation combined. Although the size of such reserves can be calculated, and their boundaries can be established from known patterns of fish distributions and/or habitat characteristics, locating them and ensuring that they are adequately connected as a network of sites that will enhance survivorship of all phases of a life history is more complex (e.g. Dayton et al, 2000). Even for highly mobile fish, models show that fish stocks protected within a marine reserve are more resilient to exploitation than when managed without reserves (Guenette and Pitcher, 1999).

Stock–recruitment relationships of target species

One interesting application of the concept of MVP (although not called that by the authors) and subsequent analysis of the size of MPA required has been carried out by McGarvey and Willison (1995) for scallops in the Gulf of Maine. Two functions were proposed for an MPA along the 1977 Hague line (separating the Canadian and US sectors of Georges Bank). First, such an MPA would separate fishing disputes, and second it would provide a protected area for scallop recruitment. Using a simple stock–recruitment model, involving age structure of natural scallop populations and age-specific fecundities and mortalities, the authors calculated annual recruitment of scallops as a function of the size of the protected area. From these calculations, they concluded that an MPA about 10km wide and about 40km long (i.e. about 400km^2), located along the Hague line, could significantly enhance the scallop fisheries of Georges Bank (currently worth about $100 million annually). It would be instructive to carry out such calculations for combinations of commercially fished stocks in various areas in terms of cost-benefit ratios for MPAs of various sizes.

The size of MPAs designated from stock–recruitment calculations will naturally vary, but for sets of ecologically similar species, areas might converge in size to the same order of magnitude. The general form of survivorship in relation to NER size (non-extractive reserve) and their potential efficacy to enhance harvested populations has been considered by Auster and Malatesta (1995). Which of the patterns of survivorship are seen (types I to III in Figure 14.3) will depend upon the mobility and recruitment characteristics of the species. A type I pattern would be shown by highly mobile species, while a type III pattern would be shown by a relatively sessile species. Unfortunately the numerical models and field experiments necessary to quantify the actual sizes of NERs required to effectively enhance exploited populations have not yet been carried out.

The general question of how large an MPA should be in order to significantly enhance recruitment processes has been addressed by Planes et al (2000). They showed that the issue is not just one of MPA size, but also involves knowledge of location, habitat type, physical oceanography regime, life history processes affecting recruitment and so on (see Figure 17.3). As Planes et al (2000) note, there is virtually no empirical evidence to assess the relative importance of the elements of their general model at present, despite the centrality of this subject to marine conservation. The same subject has been tackled by Lindholm et al (2001), who show that fish movement rates and post-settlement density are critical for predicting the effects of MPA size on survivorship.

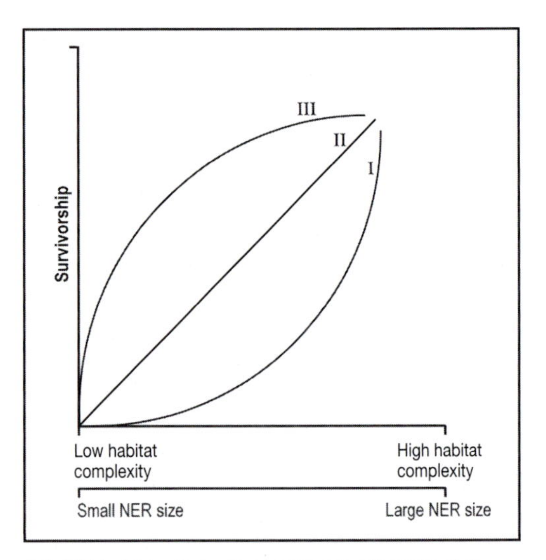

Note: Which of the patterns of survivorship occurs (types I to III) depends upon the mobility and recruitment characteristics of the species. A type I pattern would be shown by highly mobile species, while a type III pattern would be shown by a relatively sessile species

NER = non-extractive reserve

Source: Modified from Auster and Malatesta (1995)

Figure 14.3 The size of MPAs required will vary, but for sets of ecologically similar species may converge in size to the same order of magnitude, depending on the general form of survivorship in relation to NER size (see Auster and Malatesta, 1995)

Species–area curves

Species–area and species–accumulation curves (with a major emphasis on fish) have been considered above, under representative areas.

Summary

There are obviously several ways in which we could estimate the size of an MPA designed for fisheries conservation, but the size of effective MPAs may vary greatly depending upon objectives. Ideally, several independent estimates of MPA size for fisheries conservation would lead to similar estimates of effective size of MPAs designed for the same objectives. Pauly et al (2000) and co-workers have developed various ecosystem simulation models that will predict the effects of fisheries reserves of various sizes and help determine how large MPAs should be for effective fisheries conservation. We suggest the following as likely viable combinations of approaches that warrant further study:

To protect non-migrant fish (e.g. reef fish), areas based on estimates of home ranges and MVPs might be sufficient to establish a minimum size for a local or regional MPA. These estimates could be compared with habitat suitability analyses and the proportion of habitat that should be protected – from modelling studies, and with estimates of the size of habitat that should be protected based on S-A curves of the non-migrant fish community. Even small areas may prove useful for local conservation, although benefits of small MPAs are unlikely to extend to a stock or meta-population as a whole. Where MPAs are isolated from one another by non-protected areas, there may be consequences for the genetic identity of stocks (see Chapter 10). Some analysis of meta-population structure and gene flow may be required unless a set of MPAs is organized as an effective network of MPAs, or unless 20–50 per cent of total habitat can be protected.

For regionally migrant populations (e.g. the major temperate waters of commercial fisheries), TEK and scientific knowledge of spawning and recruitment areas and their size is required. These studies can again be compared with modelling exercises designed to estimate the proportion of total habitat that should be protected. In addition, for this category of fish, estimates of the size of habitat to be protected should be obtained from knowledge of S-A curves of the species assemblages of the corresponding fish community. Present evidence suggests that some combination of sets of large MPAs (covering > 20 per cent of the entire range of a species) that exclude fishing, together with restrictions on fishing quotas, is the emerging fisheries conservation practice of choice. The fisheries reserves (MPAs) designated would be located primarily so as to protect spawning and recruitment areas. Much of the present evidence in favour of this combined tactic comes from recent modelling studies, and empirical evidence (e.g. Fisher and Frank, 2002) may not yet strongly support this approach in terms of rebound of fish communities in recently protected areas in temperate seas.

For broadly migratory species (i.e. those exploiting a significant portion of an entire ocean basin, such as tuna and marine mammals), very large areas of the pelagic habitat may need to be established and regulated in terms of seasonal activities.

It should be realized that many of the issues that are critical to the prediction of appropriate size of MPAs for fisheries conservation have not yet been adequately studied. Even though this section has dealt with MPAs as a means of fish conservation, in practice most MPAs will be expected to perform a combination of functions, including protection of fisheries and biodiversity conservation.

Size of areas of anthropological/archaeological/sociological interest

This type of protected areas would include shipwrecks and sunken towns or cities (e.g. Port Royal, Jamaica, West Indies). The size of the area to be directly preserved is set primarily by the nature of distribution of artefacts and physical topographic limits. In the case of towns, although the whole area may not be visible, historical or archaeological records should allow an approximate extent to be described. Beyond this, the major question is of the extent of the buffer zone which is required for several reasons, including imprecise navigation skills, pollution from surrounding areas, site protection from unauthorised visitors and so on.

An example of a very small protected area (1nmi in diameter!) is the Monitor National Marine Sanctuary off Cape Hatteras, North Carolina, designed to protect the Civil War ironclad, the USS Monitor (http://monitor.noaa.gov/).

Size of scenic areas (land-referenced)

This category includes local 'beauty spots' and recreation areas (generally land-referenced), which may or may not be distinctive for some reason. If it can be shown that they are distinctive (in other words they exhibit geophysical anomalies/possess focal species such as marine mammals or bird colonies), then they belong in those categories (see sections above). If this is not the case, then the extent of the area to be protected may depend primarily upon aesthetic features and local patterns of land-uses. Many such areas are designated or desired by local natural history groups. However, the rationale for their designation as MPAs may not ecologically or environmentally be evident; it may rather be founded in protection of property rights. The extent of such MPAs cannot be set objectively; it will simply be as large as the local interest group thinks it can 'defend'.

Unfortunately, the overall ecological effects in such areas are just as likely to be negative as they are positive, because of increased human interest and usage. The literature on terrestrial conservation now has many studies on the effects of human impact on park areas. This is a very recent and emerging area of study for marine conservation, but experience with the effects of human diving on benthic communities (primarily on coral reefs and kelp beds) is now accumulating (see e.g. Schaeffer et al, 1999; Rouphael and Inglis, 2001). Evidence shows that the effects of diver activity can be controlled within a well managed marine park.

Some combination of one or more of these types of areas

The process of integrating MPAs into sets will be examined in Chapter 16, and networks of MPAs are considered in Chapter 17.

Conclusions and management implications

Individual MPAs may vary widely in their purpose, size and how their boundaries are determined. The primary purpose of an MPA is generally the driving motivation for its establishment, however, even this is frequently not well explained in ecological terms. This chapter covers the most important types of MPAs and suggests how their size and boundaries can be estimated and defended on ecological and environmental grounds. Marine reserves are clearly beneficial and benefits increase directly with reserve size (Halpern, 2003). However, some words of caution are warranted.

Not all the methods suggested here may be directly applicable to all ecoregions. For example, there will be differences between 'ecosystems' and implications for MPA design from the tropics to polar regions. Laurel and Bradbury (2006) suggest that dispersal and gene flow in marine fish populations increase with latitude. For example, north temperate fish species at latitudes between 40° and 45° had about three times greater dispersal potential (as planktonic larval duration) and genetic homogeneity than fish species in equatorial regions. Dispersal was estimated to increases at a rate of ~ 8 per cent per degree of latitude north or south of the equator. Therefore they suggest that size of tropical MPAs should not serve as direct scalar templates in other regions, but rather at higher latitudes MPAs should be implemented at significantly larger scales. Consistent with this suggestion, on the basis of single-cohort modelling exercises, Gerber et al (2003) conclude that marine reserves will provide fewer benefits for species with greater adult rates of movement.

The size of conservation reserves, the significance of species–area relationships and especially the relative merits of 'single large versus many small' (the SLOSS debate) has had a chequered history in the ecological literature. This debate seems to have now come full circle to a point where the value of species–area relationships is again appreciated as a planning tool (Neigel, 2003), but where the SLOSS issue has become essentially passé for the marine environment because of requirement for multiple sites and connectivity among them (see Chapters 16 and 17).

Each of the methods suggested above as potential estimates of the size of MPAs has its attendant biases. Analysis of regional structure of the target community or communities is essential, in order to identify component species assemblages. Estimates of areas required to represent all species differ between taxa, and are larger for entire communities than for the individual species assemblages, as indicated above by species-accumulation estimates for the entire fish and epibenthic communities versus their component assemblages.

The species–accumulation curve is by itself a rather artificial construct as an estimator of size for a potential protected area. The discrepancy between species–area estimates for the greatest number of fish (some 10,000km^2) and the species–accumulation estimate for 'all' species of fish (56.3km^2) clearly point to this. Species–accumulation curves are composites of data from independent samples within a habitat, whereas species–area curves represent an entire area of a given habitat type. To extrapolate from species–accumulation curves as if these composite samples were contiguous is not appropriate. Species–accumulation curves do, however, yield valuable comparative estimates of species richness among areas of the same habitat, among taxa and among species assemblages, and can therefore point to areas and sizes of habitats for conservation.

For representative areas, it is necessary to define the number and the set of MPAs and to define the spatial relationships among them (connectivity) while defining the size of each. Although the size of each MPA may be independent of the others in the set, its position is not. The intention of this chapter is therefore to attempt to define a 'minimum viable size' for MPAs rather than the preferred size, which may be larger and should be determined in the context of a regional set of MPAs, including the number of replicates of each type.

For distinctive areas, the number and set of such MPAs should emerge directly from ecological and environmental analyses. The connectivity among them may or may not be defined. For management purposes, many MPAs in a region may be of some composite type, encompassing characteristics of both (or several) distinctive and representative areas.

In summary then, marine conservation does not end with decisions on the size or locations of individual MPAs; the marine environment is continuous and connected. Any individual MPA, whatever its size, can only have limited and local value. There can be no such thing as 'ecological integrity' of an individual MPA. It is now widely recognized that, for marine conservation to be effective, marine reserves must be considered as sets, or more functionally as networks of MPAs. Our overall strategy needs to be to define networks of MPAs and the total size of the marine environment that should be protected (in other words the proportion of a region that should be occupied by MPAs) rather than to simply define the size of individual MPAs. These subjects will be covered in the following chapters.

References

Allison, G. W., Gaines, S. D., Lubchenco, J., Possingham, H. P. (2003) 'Ensuring persistence of marine reserves: Catastrophes require adopting an insurance factor', *Ecological Applications*, vol 13, pp8–24

Anaya, G., Arizpe, O., Figueroa, A. L., Niembro, E., Robles, A. and Zavala, A. (1998) 'Working Towards the Conservation and Sustainable Use of the Islands of the Gulf of California, Mexico: The importance of managing insular environments to marine and coastal biodiversity conservation', in W. P. Munro and J. H. Willison (eds) *Linking Protected Areas with Working Landscapes Conserving Biodiversity*, Science and Management of Protected Areas Association, Acadia University, Wolfville, Nova Scotia

Anderson, D. W. and Keith, J. O. (1980) 'The human influence on seabird (*Pelicanus occidentalis californicus* and *Larus heermanni*) nesting success: Conservation implications', *Biological Conservation*, vol 18, pp65–80

Angel, M. V. (1994) 'Spatial Distribution of Marine Organisms: Patterns and processes', in P. J. Edwards, R. M. May and N. R. Webb (eds) *Large-Scale Ecology and Conservation Biology*, Blackwell Science, London

Auster, P. J. and Malatesta, R. J. (1995) 'Assessing the Role of Non-extractive Reserves for Enhancing Harvested Populations in Temperate and Boreal Systems', in N. L. Shackell and J. H. M. Willison (eds) *Marine Protected Areas and Sustainable Fisheries*, Science and Management of Protected Areas Association, Acadia University, Wolfville, Nova Scotia

Beckmann, L. (1995) 'Marine Conservation in the Canadian Arctic', in N. L. Shackell and J. H. M. Willison (eds) *Marine Protected Areas and Sustainable Fisheries*, Science and Management of Protected Areas Association, Acadia University, Wolfville, Nova Scotia

Beverton, R. J. H. and Holt, S. J. (1957) 'On the dynamics of exploited fish populations', *Fishery Investigations (London)*, vol 2, no 19

Brown, C. W. and Winn, H. E. (1989) 'Relationships between the distribution patterns of right whales (*Eubalena gracilis*) and satellite-derived sea surface thermal structure in the great south channel', *Continental Shelf Research*, vol 9, pp247–260

Brown, R. G. B. and Gaskin, D. E. G. (1988) 'The pelagic ecology of the gray and red-necked phalaropes *Phalaropus fulicarius* and *Phalaropus lobatus* in the Bay of Fundy, eastern Canada', *Ibis*, vol 130, pp234–250

Brown, M. W., Allen, J. M. and Kraus, S. D. (1995), The Designation of Seasonal Right Whale Conservation Areas in the Waters of Atlantic Canada', in N. L. Shackell and J. H. M. Willison (eds) *Marine Protected Areas and Sustainable Fisheries*, Science and Management of Protected Areas Association, Acadia University, Wolfville, Nova Scotia

Brown, S. K., Buja, K. J., Jury, S. H, Monaco, M. E. and Banner, A. (1997) 'Habitat suitability index models for Casco and Sheepscot bays, Maine', National Oceanic and Atmospheric Administration, Falmouth, ME

Bryan, T. L. and Metaxas, A. (2006) 'Distribution of deep-water corals along the North American continental margins: Relationships with environmental factors', *Deep-Sea Research*, vol 53, pp1865–1879

Bryan, T. L. and Metaxas, A. (2007) 'Predicting suitable habitat for deep-water gorgonian corals on the Atlantic and Pacific Continental Margins of North America', *Marine Ecology Progress Series*, vol 330, pp113–126

Carr, M. H. and Reed, D. C. (1992) 'Conceptual issues relevant to marine harvest refuges: Examples from temperate reef fishes', *Canadian Journal of Fisheries and Aquatic Sciences*, vol 50, pp2019–2028

Carter, R. W. G. (1988) *Coastal Environments. An Introduction to the Physical, Ecological and Cultural Systems of Coastlines*, Academic Press, London

Caswell, H. and Cohen, J. E. (1991) 'Disturbance, interspecific interaction and diversity in metapopulations', *Biological Journal of the Linnean Society*, vol 42, pp193–218

Chao, A. (1984) 'Nonparametric estimation of the number of classes in a population', *Scandinavian Journal of Statistics*, vol 11, pp265–270

Colwell, R. K. and Coddington, J. A. (1994) 'Estimating terrestrial biodiversity through extrapolation', *Philosophical Transactions of the Royal Society (Series B)*, vol 345, pp101–118

Connell, J. H. (1978) 'Diversity in tropical rainforests and coral reefs', *Science*, vol 199, pp1302–1310

Dayton, P. K., Sala, E., Tegner, M. J. and Thrush, S. (2000) 'Marine reserves: Parks, baselines and fishery enhancement', *Bulletin of Marine Science*, vol 66, pp617–634

de Forges, R., Koslov, B., and Poore, J. A. (2000) 'Diversity and endemism of the benthic fauna in the southwest Pacific', *Nature*, vol 405, pp944–947

Dugan, J. E. and Davis, G. E. (1993) 'Application of marine refugia to coastal fisheries management', *Canadian Journal of Fisheries and Aquatic Sciences*, vol 50, pp2029–2042

Etter, R. J. and Mullineaux, L. S. (2001) 'Deep Sea Communities', in M. D. Bertness, S. D. Gaines and M. E. Hay (eds) *Marine Community Ecology*, Sinauer Associates, Sunderland, MA

Fernandez, M. and Castilla, J. C. (2000) 'Recruitment of *Homalaspsi plana* in intertidal habitats of central Chile and implications for the current use of management and marine protected areas', *Marine Ecology Progress Series*, vol 208, pp157–170

Fisher, J. A. D. and Frank, K. T. (2002) 'Changes in finfish community structure associated with the implementation of a large offshore fishery closed area on the Scotian Shelf', *Marine Ecology Progress Series*, vol 240, pp249–265

Foster-Smith, R. L., Davies, J. and Sotheran, I. (2000) 'Broad scale remote survey and mapping of the sublittoral habitats and biota: Technical report of the Broadscale Mapping Project', *Scottish Natural Heritage Research, Survey and Monitoring Report No. 167*, Edinburgh

Frank, K. T. and Shackell, N. L. (2001) 'Area-dependant patterns of finfish diversity in a large marine ecosystem', *Canadian Journal of Fisheries and Aquatic Sciences*, vol 58, pp1703–1707

Gage, J. D. and Tyler, P. A. (1991) *Deep Sea Biology. A Natural History of Organisms at the Deep Sea Floor*, Cambridge University Press, Cambridge

Genin, A., Paull, C. K. and Dillon, W. P. (1992) 'Anomalous abundances of deep-sea fauna on a rocky bottom exposed to strong currents', *Deep Sea Research*, vol 39, pp293–302

Gerber, L. R., Botsford, L. W., Hastings, A., Possingham, H. P., Gaines, S. D., Palumbi, S. R. and Andelman, S. (2003) 'Population models for marine reserve design: A retrospective and prospective', *Ecological Applications*, vol 13, pp47–64

Gray, J. S. (1994) 'Is deep sea species diversity really so high? Species diversity of the Norwegian continental shelf', *Marine Ecology Progress Series*, vol 112, pp205–209

Guenette, S. and Pitcher, T. J. (1999) 'An age-structured model showing the benefits of marine reserves in controlling overexploitation', *Fisheries Research*, vol 39, pp295–303

Guenette, S., Pitcher, T. J. and Walters, C. J. (2000) 'The potential of marine reserves for the management of northern cod in Newfoundland', *Bulletin of Marine Science*, vol 66, pp831–852

Haedrich, R. L., Villagarcia, M. G. and Gomes, M. C. (1995) 'Scale of Marine Protected Areas on Newfoundland's Continental Shelf', in N. L. Shackell and J. H. M Willison (eds) *Marine Protected Areas and Sustainable Fisheries*, Science and Management of Protected Areas Association, Acadia University, Wolfville, Nova Scotia

Halpern, B. S. (2003) 'The impact of marine reserves: Do marine reserves work and does reserve size matter?', *Ecological Applications*, vol 13, pp117–137

Haney, J. C., Haury, L. R. Mullineaux, L. S. and Fey, C. L. (1995) 'Sea-bird aggregation at a deep North Pacific seamount', *Marine Biology*, vol 123, pp1–9

Harrison, G. and Fenton, D. (1998) *The Gully Science Review*, Fisheries and Oceans Canada Maritimes Region, Halifax, Canada

Hooker, S. K., Whitehead, H. and Gowans, S. (1999) 'Marine protected area design and the spatial and temporal distribution of cetaceans in a submarine canyon', *Conservation Biology*, vol 13, pp592–602

Horn, M. H. and Allen, L. G. (1978) 'A distributional analysis of California coastal marine fishes', *Journal of Biogeography*, vol 5, pp23–42

Hyrenbach, K. D., Forney, K. A. and Dayton, P. K. (2000) 'Marine protected areas and ocean basin management', *Aquatic Conservation: Marine and Freshwater Ecosystems*, vol 10, pp437–458

IUCN (1994) 'Guidelines for protected area management categories', World Conservation Union (IUCN), Gland, Switzerland

Kramer, D. L. and Chapman, M. R. (1999) 'Implications of fish home range size and relocation for marine reserve function', *Environmental Biology of Fishes*, vol 55, pp65–79

Laurel, B. J., and Bradbury, I. R. (2006) '"Big" concerns with high latitude marine protected areas (MPAs): Trends in connectivity and MPA size', *Canadian Journal of Fisheries and Aquatic Sciences*, vol 63, no 12, pp2603–2607

Lindholm, J. B., Auster, P. J., Ruth, M. and Kaufman, L. (2001) 'Modelling the effects of fishing and implications for the design of marine protected areas: Juvenile fish responses to variations in seafloor habitat', *Conservation Biology*, vol 15, pp424–437

MacArthur, R. H. and Wilson, E. O. (1967) *The Theory of Island Biogeography*, Princeton University Press, Princeton, NJ

Mahon, R., Brown, S. K., Zwanenburg, K. C. T., Atkinson, D. B., Burj, K. R., Caflin, L., Howell, G. D., Monaco, M. E., O'Boyle, R. N. and Sinclair, M. (1998) 'Assemblages and biogeography of demersal fishes of the east coast of North America', *Canadian Journal of Fisheries and Aquatic Sciences*, vol 55, pp1704–1738

Mallet, J., Barton, N., Gerardo, L. M., Jose, S. C., Manuel, M. M. and Eeley, H. (1990) 'Estimates of selection and gene flow from measures of cline width and linkage disequilibrium in heliconius hybrid zones', *Genetics*, vol 124, pp921–936

Mann, K. H. and Lazier, J. R. N. (1996) *Dynamics of Marine Ecosystems. Biological–Physical Interactions in the Oceans*, Blackwell Science, Cambridge, MA

Margalef, R. (1978) 'Phytoplankton communities in upwelling areas: The example of NW Africa', *Oecologia Aquatica*, vol 3, pp97–132

Margalef, R. (1997) *Our Biosphere, Excellence in Ecology 10*, Ecology Institute, Oldendorf, Germany

May, R. M. (1994) 'The Effects of Spatial Scale on Ecological Questions and Answers', in P. J. Edwards, R. M. May and N. R. Webb (eds) *Large-Scale Ecology and Conservation Biology*, Blackwell, Oxford

McGarvey, R. and Willison, J. H. M. (1995) 'Rationale for Marine Protected Area along the International Boundary between U.S. and Canadian Waters in the Gulf of Maine', in N. L. Shackell and J. H. M Willison (eds) *Marine Protected Areas and Sustainable Fisheries*, Science and Management of Protected Areas Association, Acadia University, Wolfville, Nova Scotia

Muscatine, L. and Weis, V. M. (1992) 'Productivity of Zooxanthellae and Biochemical Cycles', in P. G. Falkowski and A. D. Woodhead (eds) *Primary Productivity and Biogeochemical Cycles in the Sea*, Plenum Press, New York

Neigel, J. E. (2003) 'Species–area relationships and marine conservation', *Ecological Applications*, vol 13, Supplement, S138–145

Neis, B. (1995) 'Fishers' Ecological Knowledge and Marine Protected Areas', in N. L. Shackell and J. H. M. Willison (eds) *Marine Protected Areas and Sustainable Fisheries*, Science and Management of Protected Areas Association, Acadia University, Wolfville, Nova Scotia

Neis, B., Felt, L., Schneider, D. C., Haedrich, R., Hutchings, J. and Fischer, J. (1996) 'Northern cod stock assessment: What can be learned from interviewing resource users?', DFO Atlantic Fisheries, Research Document 96/45, Bedford, Nova Scotia, Canada

Norse, E. A. (1993) *Global Marine Biological Diversity. A Strategy for Building Conservation into Decision Making*, Island Press, Washington, DC

NRC (2001) *Marine Protected Areas: Tools for Sustaining Ocean Ecosystems*, National Academy Press, Washington, DC

Ottesen, P. and Kenchington, R. (1995) 'Marine Protected Areas in Australia: What is the future?', in N. L. Shackell and J. H. M. Willison (eds) *Marine Protected Areas and Sustainable Fisheries*, Science and Management of Protected Areas Association, Acadia University, Wolfville, Nova Scotia

Palumbi, S. R. (2001) 'The Ecology of Marine Protected Areas', in M. D. Bertness, S. D. Gaines and M. E. Hay (eds) *Marine Community Ecology*, Sinauer Associates, Sunderland, MA

Pauly, D., Christensen, V. and Walters, C. (2000) 'Ecopath, Ecosim, and Ecospace as tools for evaluating ecosystem impact of fisheries', *ICES Journal of Marine Science*, vol 57, pp697–706

Perry, R. and Smith, S. J. (1994) 'Identifying habitat associations of marine fishes using survey data: An application to the Northwest Atlantic', *Canadian Journal of Fisheries and Aquatic Sciences*, vol 51, pp589–602

Pingree, R. D. (1978) 'Mixing and Stabilization of Phytoplankton Distributions on the Northwest European Continental Shelf', in J. H. Steele (ed) *Spatial Patterns in Plankton Communities*, Plenum Press, New York

Planes, S., Galzin, R., Rubies, A. G., Goni, R., Harmelin, J. G., Le Direach, L., Lenfant, P. and Quetglas, A. (2000) 'Effects of marine protected areas on recruitment processes with special reference to Mediterranean littoral ecosystems', *Environnemental Conservation*, vol 27, pp126–143

Preston, F. W. (1962) 'The canonical distribution of commoness and rarity: Part 1', *Ecology*, vol 43, pp185–215

Roberts, C. M. and Hawkins J. P. (1997) 'How small can a marine reserve be and still be effective?', *Coral Reefs*, vol 16, p150

Roff, J. C. (1983) 'The Microzooplankton of the Quoddy Region', in M. Thomas (ed) *Marine Biology of the Quoddy Region*, special publication, Natural Sciences and Engineering Research Council of Canada, Ottawa, Canada

Roff, J. C. and Evans, S. (2002) 'Frameworks for marine conservation: Non-hierarchical approaches and distinctive habitats', *Aquatic Conservation: Marine and Freshwater Ecosystems*, vol 12, no 6, pp635–648

Roff, J. C. and Taylor, M. (2000) 'A geophysical classification system for marine conservation', *Journal of Aquatic Conservation: Marine and Freshwater Ecosystems*, vol 10, pp209–223

Roff, J. C., Fanning, L. P. and Stasko, A. B. (1986) 'Distribution and association of larval crabs (*Decapoda:Brachyura*) on the Scotian Shelf', *Canadian Journal of Fisheries and Aquatic Sciences*, vol 43, pp587–599

Roff, J. C., Taylor, M. E. and Laughren, J. (2003) 'Geophysical approaches to the classification, delineation and monitoring of marine habitats and their communities', *Aquatic Conservation, Marine and Freshwater Ecosystems*, vol 13, pp77–90

Rogers-Bennett, L., Bennett, W. A., Fastenau, H. C. and Dewees, C. M. (1995) 'Spatial variation in red sea urchin reproduction and morphology: Implications for harvest refugia', *Ecological Applications*, vol 5, pp1171–1180

Rouphael, A. B. and Inglis, G. J. (2001) 'Take only photographs and leave only footprints? An experimental study of the impacts of underwater photographers on coral reef dive sites', *Biological Conservation*, vol 100, pp281–287

Sakko, A. L. (1998) 'The influence of the Benguela upwelling system on Namibia's marine biodiversity', *Biodiversity and Conservation*, vol 7, pp419–433

Sanders, H. L. (1968) 'Marine benthic diversity: A comparative study', *American Naturalist*, vol 102, pp243–281

Schaeffer, T. N., Foster, M. S., Landau, M. E. and Walder, R. K. (1999) 'Diver disturbance in kelp forests', *California Fish and Game*, vol 85, pp170–176

Sotka, E. E. and Palumbi, S. R. (2006) 'The use of genetic clines to estimate dispersal distances of marine larvae', *Ecology*, vol 87, pp1094–1103

Sotka, E. E., Wares, J. P., Barth, J. A., Grosberg, R. K. and Palumbi, S. R. (2004) 'Strong genetic clines and geographical variation in gene flow in the rocky intertidal barnacle *Balanus glandula*', *Molecular Ecology*, vol 13, pp2143–2156

Sousa, W. P. (2001) 'Natural Disturbance and the Dynamics of Marine Benthic Communities', in M. D. Bertness, S. D. Gaines and M. E. Hay (eds) *Marine Community Ecology*, Sinauer Associates, Sunderland, MA

Steele, J. (1996) 'Regime shifts in fisheries management', *Fisheries Research*, vol 25, pp19–23

Tunnicliffe, V. (1988) 'Biogeography and evolution of hydrothermal-vent fauna in the eastern Pacific Ocean', *Proceedings of the Royal Society: London B*, vol 233, pp347–366

Vacelet, J., Boury-Esnault, N. and Harmelin, J. G. (1994) 'Hexactinellid cave, a unique deep-sea habitat in the scuba zone', *Deep-Sea Research*, vol 41, pp965–973

Walters, C. J. (2000) 'Impacts of dispersal, ecological interactions, and fishing effort dynamics on efficacy of marine protected areas: How large should protected areas be?', *Bulletin of Marine Science*, vol 66, pp745–757

Walters, B. B. and Butler, M. (1995) 'Should We See Lobster Buoys in a Marine Park?', in N. L. Shackell and J. H. M. Willison (eds) *Marine Protected Areas and Sustainable Fisheries*, Science and Management of Protected Areas Association, Acadia University, Wolfville, Nova Scotia

Warner, G. (1997) 'Participatory management, popular knowledge, and community empowerment: The case of sea urchin harvesting in the Vieux-Fort area of St. Lucia', *Human Ecology*, vol 25, pp29–46

Wolff, M. (1996) 'Demersal fish assemblages along the Pacific Coast of Costa Rica: A quantitative and multivariate assessment based on the Victor Hensen Costa Rica expedition', *Revista de Biologica Tropical*, vol 44, pp187–214

Zacharias, M. A. and Roff, J. C. (2001a) 'Use of focal species in marine conservation and management: A review and critique', *Aquatic Conservation: Marine and Freshwater Ecosystems*, vol 11, pp59–76

Zacharias, M. A. and Roff, J. C. (2001b) 'Explanations of patterns of intertidal diversity at regional scales', *Journal of Biogeography*, vol 28, pp471–483

Zurbrigg, E. (1996) *Towards an Environment Canada Strategy for Coastal and Marine Protected Areas*, Canadian Wildlife Service, Environment Canada, Hull, Quebec, Canada

15

Evaluation of Protected Areas

The concept of 'value' as applied to marine biodiversity

Your true value depends entirely on what you are compared with.

Bob Wells (1966–)

Introduction

In order to develop management strategies for sustainable use and conservation in the marine environment, meaningful, reliable and integrated ecological information is required. Environmental maps that compile and summarize all available biological and ecological information for a study area, and that allocate an overall biological value to subzones, can be used as baseline maps for future spatial planning. Such spatial planning is of prime importance for two kinds of operations. The first is to evaluate the properties of subzones for their significance in terms of both fisheries conservation and biodiversity conservation. The second is to establish the priorities – including the sequence, timing of establishment and regulations – for protected areas. As the process of developing national strategies for marine conservation unfolds, it is unlikely that all areas that might warrant some form of protection would be established simultaneously. Thus immediate questions become: what are the priorities for conservation and what should be done first? This involves the concept of evaluation of areas of the marine environment according to defined criteria.

The concept of 'value' for regions of the marine environment has received uneven attention and different interpretations around the world. As usual with other terminology in marine conservation, different words are used in different ways. For example 'priority conservation areas' (PCAs or PACs) can mean areas of the marine environment with valued characteristics that warrant protection (e.g. as marine protection areas (MPAs)) or special management (e.g. under some scenario of ecosystem-based management). Such PACs would recognize some component or combination of components of marine biodiversity, and acknowledge that they require special attention. Such PACs may be recognized on the basis of both 'species' and/or 'spaces', because of their importance to fisheries, to species diversity, to focal species and so on. PACs, whether in existence or planned, may therefore be either rather small and specific in nature (e.g. a turtle nesting beach) or large and only seasonally regulated (e.g. a fisheries spawning/recruitment area). At present, however, there is really no systematic way of acknowledging that 'All areas of the marine environment are created equal, but some are more equal than others'. There is a need to evaluate their value.

At a series of international meetings held at the University of Gent, continued by correspondence and expert review, the subject of how to rank the 'value' of marine environments was discussed. Following these deliberations, the consensus of the group was published by Derous et al (2007a) and by Derous et al (2007b). This chapter relies heavily on the outcome of those meetings, correspondence and publications.

The value of value

There is widespread recognition of the benefits of spatial planning for sustainable use and conservation of the sea (e.g. Tunesi and Diviacco, 1993; Vallega, 1995; Ray, 1999) – for example as embodied in the concept of 'ecosystem-based management'. Meaningful biological and ecological information is required to inform and underpin sustainable management practices. Biological valuation maps (BVMs) – that is, maps showing the intrinsic value of the components of biodiversity within the subzones of a study area – would provide indispensible information for managers and decision-makers. Such maps would need to make best use of available data, by compiling and summarizing relevant biological and ecological information for a study area, and by allocating an overall biological/ecological value to different subzones. Rather than being a general strategy for protecting areas that have some ecological significance, biological valuation is a tool for calling attention to areas that have particularly high ecological or biological significance. This facilitates provision of a greater-than-usual degree of risk aversion in management of activities in such areas, and aids in the establishment of priorities.

BVMs have been developed primarily for terrestrial systems and species (e.g. De Blust et al, 1985, 1994), but criteria developed for identifying terrestrial species and habitats for conservation cannot easily be applied to the marine environment. Any relevance of terrestrial approaches for marine systems requires an understanding of both the nature and degree of differences between marine and terrestrial systems (see especially Chapters 2 and 3). The significance of these environmental differences has become clear in difficulties encountered when applying terrestrial-based assessments to marine areas. For example, the EC Habitats Directive (92/43/EEC), written from a terrestrial viewpoint, has proved problematic in application to the more dynamic marine systems (Hiscock et al, 2003). Lacking an integrated policy focused on the protection of the European marine environment, the European Commission is currently developing a Marine Strategy Directive.

Coastal planners and marine resource managers have utilized various tools for assessing the biological value of subzones in the past. These approaches vary in information content, scientific rigour and the level of technology used. The most simple approach is low-tech participatory planning, which often occurs in community-based MPA design (e.g. the Mafia Island Marine Park Plan, described in Agardy, 1997), but the selection of such priority areas is very ad hoc, opportunistic or even arbitrary, resulting in decisions which are often difficult to defend to the public. The chance of selecting the areas with the highest intrinsic biological and ecological value through these methods is small (Fairweather and McNeill, 1993; Ray, 1999; Roberts et al, 2003a). An alternative Delphic-judgemental approach has been advocated. In this approach, an expert panel is consulted to select areas for protection, based on expert knowledge. The method is relatively straightforward and easily explained, which may indicate why it is still common (Roberts et al, 2003a). However, owing to the urgency for site selection, the consultation process is usually too short, the uncertainty surrounding decisions is too high and the information input is too generalized to permit defensible, long-term recommendations (Ray, 1999).

The disadvantages of these aforementioned methods for assessing the value of marine areas have led to an increasing awareness that a more objective valuation procedure is needed. Other existing methodologies utilize a variety of tools to optimize site selection through spatial analysis, such as geographic information system (GIS)-based multicriteria evaluation (e.g. Villa et al, 2002). The most sophisticated methods are those where planning is driven in part by high-tech decision-support tools. One such tool is MARXAN (Possingham et al, 2000) – a systematic conservation planning software program used to locate and design reserves that maximize the number of species or communities contained within a designated level of representation. This technique can be used to select those subzones which contribute most to specified conservation goals, while minimizing the costs for conservation (see Chapter 16).

Without denying the merits of MARXAN and similar mathematical tools, such techniques cannot be applied for the purpose of biological valuation of areas themselves. Biological valuation is not a process to select areas for conservation according to quantitative objectives; rather it should yield an integrated view of the biological value of the individual subzones within a study area – relative to each other. Conversely, the decision to include one or more subzones in a marine reserve cannot be made on the basis of the outcome of a biological valuation, because the latter process does not take into account management criteria and quantitative conservation targets. These subjects will be considered in the next chapter.

The element common to all the above approaches is the identification of criteria to discriminate between marine areas and to guide the selection process. While the vast majority of these efforts are relevant to MPA design, there is no reason why such criteria cannot be equally helpful in coastal zone and ocean management (ecosystem-based management) more generally. It is therefore necessary that the definition of the value of marine areas should be based on the assessment of areas against a set of objectively chosen ecological criteria, making best use of scientific monitoring and survey data (Mitchell, 1987; Hockey and Branch, 1997; Ray, 1999; Connor et al, 2002; Hiscock et al, 2003).

This chapter presents concepts for marine biological valuation, used to determine which marine areas have special ecological values in terms of high biodiversity. It draws on existing valuation criteria and methods, and attempts to rationalize them into a single model that should be ecologically sound and widely applicable. We trust that, along with its 'parent papers' (Derous et al, 2007a, 2007b), it will also be regarded as an incentive to further discussion on marine biological valuation.

Definition of marine biological value

Various definitions of 'marine biological value' are currently found in the literature. These are listed in Table 15.1, along with their final use in the present valuation process. What different authors mean by 'value' is directly linked to the objectives behind the process of valuation (e.g. fisheries conservation, sustainable use, preservation of biodiversity). Discussions on the value of marine biodiversity almost always refer to the socio-economic value of biodiversity – that is, the so-called goods and services provided by marine ecosystems, or the value of an area in terms of importance for human use, and subsequent attempts to attach a monetary value to the biodiversity in an area (Bockstael et al, 1995; King, 1995; Edwards and Abivardi, 1998; Borgese, 2000; Nunes and van den Bergh, 2001; De Groot et al, 2002; Turpie et al, 2003). Many approaches try to highlight only the most important sites in a region in order to designate priority sites for conservation. These priority sites are often chosen on the basis of the 'hotspot' approach, which is used to select sites with high numbers of rare/endemic species or high species richness (e.g. Myers et al, 2000; Beger et al, 2003; Breeze, 2004).

For the purpose of this chapter, 'marine biological value' is defined as follows: 'the intrinsic value of marine biodiversity, without reference to anthropogenic use'. This definition is similar to the definition of value of natural areas of Smith and Theberge (1986): 'the assessment of ecosystem qualities *per se*, (i.e. their intrinsic value) regardless of their social interests'. By 'ecosystem qualities' the authors of the latter paper implied all levels of biodiversity, from genetic diversity to ecosystem processes.

The purpose of marine biological valuation is to describe subzones within the target study area in terms of their intrinsic biological value (on a continuous or discrete value scale, e.g. high, medium and low value). Subzones or subregions within the study area can then be scored relative to each other, against a set of biological valuation criteria. The size of these subzones depends on the size of the study area, on the biodiversity components under consideration and on the scale and density of available data, and should therefore be decided on a case-by-case basis. The advantage of this approach, as compared to the hotspot approach (identification of priority areas for conservation) is that it describes all subzones and does not highlight solely the most valuable ones. The product of the valuation process – that

Table 15.1 Summary of existing ecological criteria for selection of valuable marine areas, or those deserving of protection

Criterion	Included in final set of original criteria?
Rarity	Yes, 1st order criterion
(Bio)diversity	Not as a criterion itself, but all hierarchical levels of biodiversity are implicitly included in the valuation strategy (see text and Figure 15.1)
Naturalness	Yes, modifying criterion
Proportional importance	Yes, modifying criterion, or under 'fitness consequences and aggregation' as 1st order criterion
Ecosystem functioning	Yes, under 'fitness consequences' as 1st order criterion
Reproductive/bottleneck areas	Yes, under 'fitness consequences' as 1st order criterion
Density	Yes, under 'aggregation' as 1st order criterion
Dependency	Yes, under 'fitness consequences and/or aggregation' as 1st order criterion
Productivity	Yes, under 'fitness consequences and/or aggregation' as 1st order criterion
Special features present	Yes, under 'rarity' 1st order criterion
Uniqueness	Yes, under 'rarity' 1st order criterion
Irreplaceability	Yes, under 'rarity' 1st order criterion
Isolation	Yes, under 'rarity' 1st order criterion
Extent of habitat type	Yes, under 'proportional importance', modifying criterion
Biogeography	No, this is an MPA selection criterion – see Chapter 16
Representativity	No, this is an MPA selection criterion – see Chapter 16
Integrity	EITHER: Yes, as 'naturalness', modifying criterion OR: No, this is an MPA selection criterion – see Chapter 16
Vulnerability	No: related to 'resilience' criterion which is excluded from valuation criteria – see text
Decline	No: related to 'resilience' criterion which is excluded from valuation criteria – see text
Recovery potential	No: related to 'resilience' criterion which is excluded from valuation criteria – see text
Degree of threat	No, management criterion
Protection level	No, management criterion
International significance	EITHER: No, management criterion OR: Yes, modifying criterion, under proportional importance
Economic interest	No, socio-economic criterion

is, the intrinsic values of each of the subzones – can then be presented on marine BVMs. The BVMs can serve as baseline maps showing the distribution of complex biological and ecological information.

Selection of valuation criteria

Several suggestions for selecting biological criteria and for developing valuation methods already exist in the literature. These were reviewed by Derous et al (2007a) and the criteria deemed most appropriate (see Table 15.1) were selected for incorporation into a BVM system. Appropriateness was decided on the basis of description of intrinsic biological/ecological properties, redundancy and synonymy of terms and how well criteria could be interpreted with respect to the whole spectrum of biodiversity components.

Some of the selected criteria have already been included in international legislation in Europe (e.g. EC Habitat – 92/43/EEC and Bird – 79/409/EEC Directives) (Brody, 1998). This latter point is very important, because any workable valuation assessment for marine areas should ideally mesh with relevant international protection or management initiatives (such as OSPAR, 1992). This should aid in consistency of approaches to conservation and regulation throughout the territorial waters of continental shelf and adjacent waters, especially where initiatives overlap (Laffoley et al, 2000).

Three distinct types of literature were included in the review of Derous et al (2007a): articles on the assessment of valuable ecological marine areas; literature on selection criteria for MPAs and international legislative documents that include selection criteria (EC Bird/Habitat

Directives; RAMSAR 1971 Convention; OSPAR, 1992 and 2003 guidelines; UNEP, 1990; 2000 Convention on Biological Conservation; etc.).

Only ecological criteria were considered for the purposes of this study; others – for example socio-economic or practical considerations – were not included. It is extremely important to distinguish carefully between the process of evaluating the inherent or intrinsic biological and ecological components of regions and their properties, and the separate process of evaluating possible attendant human influences and their consequences. The 'natural' properties of a region and human influences should not be confused – though they may be superimposed. For example, Sealey and Bustamante (1999) described a set of indicators that are indirect or direct measures of biological and ecological value, and whose assessment allows a ranking of the marine study area into subzones with different values. Following this first step, they applied a subsequent set of prioritizing criteria to the list of high-ranked areas to identify the priority areas for conservation. The criteria used to determine the conservation need of the area (essentially socio-economic) were based on changes induced by human activities, an evaluation of the potential threats to the area, the political and public concern to protect the area and the feasibility of designation.

The objective of Derous et al (2007a) was the same as for the first step of Sealey and Bustamante's (1999) work – that is, the ranking of areas according to their inherent biological and ecological value – but the issues of determination of conservation status, or the socio-economic criteria, were not addressed since these also involve social and management decisions. An additional difference between Sealey and Bustamante (1999) and Derous et al (2007a) is that the former authors scored the different valuation criteria through expert judgement (a Delphic process), whereas the latter authors tried to establish a valuation concept which is as objective as possible.

The valuation concept was developed, based in part on a framework developed for the identification of ecologically and biologically significant areas (EBSAs) (DFO (Department of Fisheries and Oceans), 2004), using five criteria: uniqueness, aggregation, fitness consequences, resilience and naturalness. The first three criteria were considered the first-order (main) criteria to select EBSAs, while the other two were used as modifying criteria to upgrade the value of certain areas when they scored highly for these criteria.

Derous et al (2007a) decided that, for the purposes of marine biological valuation, the criterion of 'resilience' (the degree to which an ecosystem or a part/component of it is able to recover from disturbance without major persistent change, as defined by Orians (1974)) should not be included. The rationale is that resilience is closely related to the assessment of (future) human impacts; this is not an appropriate criterion for determining the current and inherent biological value of an area – although it is an important (perhaps vital) consideration in formulating practical management strategies.

It can of course be argued that resilience is also an intrinsic quality of a certain biological entity to be able to resist or to recover from natural stresses (e.g. resilience of mangrove communities to climate change stress), but since the term 'resilience' is used to define resistance to both natural and anthropogenic stresses, it is excluded as an ecological valuation criterion. In contrast, the criterion 'naturalness' was retained, because it is an index of the degree to which an area is currently (though not inherently) in a pristine condition. In this way, unaltered areas with a high degree of resilience against natural stresses will still be covered by the valuation concept. The criterion 'uniqueness' was renamed 'rarity' as this term is more frequently used in literature and encompasses unique features.

The criteria listed in Table 15.1 were then cross-referenced with the selected valuation criteria – rarity, aggregation, fitness consequences and naturalness – to see if additional criteria needed to be included in order to produce a comprehensive valuation concept for the marine environment. There is considerable redundancy in the valuation criteria; however, most of the terminology and criteria mentioned in the literature are accounted for – as synonyms – by the selected valuation criteria. One additional criterion was added to the framework to make it fully comprehensive: 'proportional importance' – included as a modifying criterion.

The concept of 'biodiversity' – including all organizational levels of biodiversity, from the genetic to the ecosystem level, separated into biodiversity structures and processes – must also be included in the valuation framework, though not as a criterion (see below). Biodiversity is the framework against which selected criteria must be interpreted. Table 15.2 gives an overview of the chosen set of valuation criteria together with a brief definition of each, and Figure 15.1 shows an overview of the biological valuation concept proposed. Each criterion is defined and discussed in further detail in the text below. In summary, the valuation criteria selected for the development of marine BVMs are: rarity, aggregation, fitness consequences (as main criteria), naturalness and proportional importance (as modifying criteria).

Rarity

Rarity can be assessed in a variety of contexts. However, from an ecological viewpoint the goal is to be as objective as possible so as to remove or contain biases from aesthetics or socio-economics. These can be added later as they may become relevant to planning considerations. Rarity can be considered on different scales, for example national, regional, global. In order to be able to assess the rarity of marine species or communities on a regional or global scale, it is clear that international lists of rare species, habitats or communities are required. Unfortunately, unlike the terrestrial environment, rather few marine species are included in Red Data Books, like the IUCN Red Lists or the appendices of CITES, CMS (RAMSAR COP 7, 1999) and the Bern Convention (1979). This is primarily due to the lack of systematic assessment and study of marine species at a regional scale (Sanderson, 1996a, b; Ardron et al, 2002).

It should be noted that most species or communities that are mentioned in the above lists are 'rare' because their numbers have been depressed by human actions, while other species or communities are just not numerous (in other words they may be naturally rare). Thus rarity and naturalness may inevitably become confounded. If such lists of rare species or habitats are not available on a local or regional scale, data on species rarity within subzones may still be available.

Reliable and comprehensive survey data on species populations and communities are frequently lacking, which only leaves the concept of 'area of occupancy' as a proxy to assess the number and location of rare species within a study area (Sanderson, 1996a, b; Connor et al, 2002). The application of this concept is shown in Table 15.3. This approach has been adopted for the UK's Review of Marine Nature Conservation (Golding et al, 2004; Vincent et al, 2004; Lieberknecht et al, 2004a) and the UK Biodiversity Action Plan for marine species and habitats (www.ukbap.org.uk/), both in combination with other criteria.

A species described by the method of Sanderson (1996a, b) as nationally rare or scarce is not necessarily regionally or globally rare or scarce; it may simply have been reported at the edge of its range, or else this designation may indicate subtle adversity such as stress caused by human activities in the study area. However, it could also be important to give a high value to subzones containing species at the margins of their range, because these sites could host important stocks of a species that are genetically distinct. Also, populations of sessile southern or northern species may have a poor capacity for recovery, and may recruit slowly at the margins of their distributions. They are therefore particularly vulnerable to even the most minor, infrequent impacts (Sanderson 1996a, b). Nationally rare or scarce species may also be restricted to specific habitat types that themselves may be rare in the study area and need to be given a high value.

A disadvantage of rarity assessment is that it may overlook local densities. Locally abundant species (in one or several subzones of a study area) which are restricted in their range might be considered to conflict with assertions made about national rarity, should population-based methods of assessment ever be used (Sanderson, 1996a, b).

Uniqueness and distinctiveness (Roff and Evans, 2002) are also considered under this criterion which should be used to assess the number and location of unique or distinct features at any

Table 15.2 Final set of valuation criteria and their definitions

Valuation criterion	Definition	Source
First order criteria		
Rarity	Degree to which an area is characterized by unique, rare or distinct features (landscapes/habitats/communities/species/ecological functions/geomorphological and/or hydrological characteristics) for which no alternatives exist.	DFO (2004); Rachor and Günther (2001); modified and complemented after Salm and Clark (1984); Salm and Price (1995); Kelleher (1999); UNESCO (1972)
Aggregation	Degree to which an area is a site where most individuals of a species are aggregated for some part of the year or a site which most individuals use for some important function in their life history or a site where some structural property or ecological process occurs with exceptionally high density.	DFO (2004)
Fitness consequences	Degree to which an area is a site where the activity(ies) undertaken make a vital contribution to the fitness (= increased survival or reproduction) of the population or species present.	DFO (2004)
Modifying criteria		
Naturalness	The degree to which an area is pristine and characterized by native species (in other words absence of perturbation by human activities and absence of introduced or cultured species).	DFO (2004); DEFRA (2002); Connor et al (2002); JNCC (2004); Laffoley et al (2000)
Proportional importance	Global importance: proportion of the global extent of a feature (habitat/seascape) or proportion of the global population of a species occurring in a certain subarea within the study area. Regional importance: proportion of the regional (e.g. NE Atlantic region) extent of a feature habitat/seascape) or proportion of the regional population of a species occurring in a certain sub-area within the study area. National importance: proportion of the national extent of a feature (habitat/seascape) or proportion of the national population of a species occurring in a certain sub-area within territorial waters.	Connor et al (2002); Lieberknecht et al (2004a, b); BWZee workshop definition (2004)

Source: Derous et al (2007a)

level of the ecological hierarchy, including genetic, stocks, species, communities and ecosystems within the study area.

Aggregation

The 'aggregation' and 'fitness consequences' criteria will mainly identify subzones that have high ecological importance for the wider environment. In fact these criteria span the ecological hierarchy from genes to ecosystems, and should identify and include all distinctive areas (e.g. Roff and Evans, 2002) that are important for fisheries (including stocks) and focal species. Note again that there will be some overlap between areas identified on the basis of these criteria and areas identified on the basis of 'rarity'. Whatever criteria are used in an evaluation process, there will inevitably be some redundancy and confounding in results.

Evaluation of the criteria of 'aggregation' and 'fitness consequences' lies at the heart of an ecosystem-based approach to management and will be especially valuable for fisheries management. Indeed these criteria, which originated

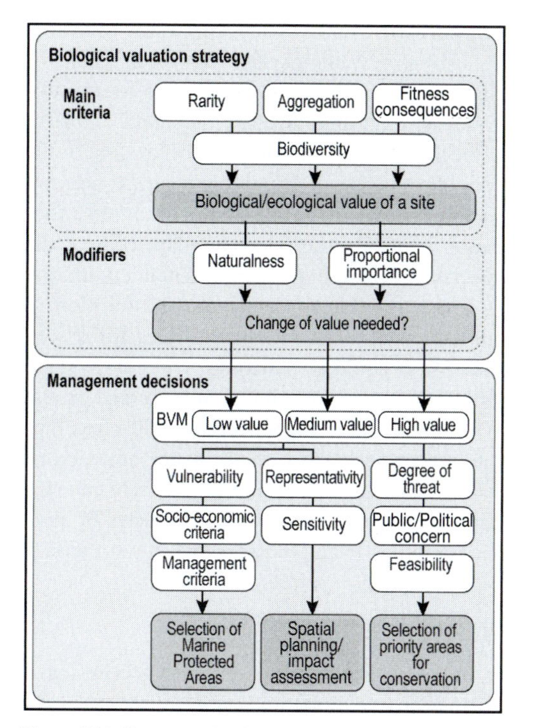

Figure 15.1 Conceptual scheme of the biological valuation method and possible future steps in developing decision support tools for managers

from DFO in Canada, were largely designed for the purposes of ecosystem-based management of fisheries. These criteria can be considered as descriptors of factors that 'drive' ecological processes. Assigning value to subzones containing them is therefore one way to achieve preservation of the larger marine ecosystem (Brody, 1998). Ecosystem-based management forces us to adopt a holistic view of the components of biodiversity as parts of the larger system, rather than the reductionist view of single-species management, which ignores the fact that species exist only as part of the ecosystem (Simberloff, 1998). This is in conformity with the present objective of including as many components of biodiversity (both structural and processes) in the criteria assessment as possible.

If data on the population size of a species are available at the scale of the study area, it is then possible to determine whether a high percentage of the population of a given species is located within subzones of the study area. If these data are lacking but qualitative information exists on certain areas where the members of a species may aggregate (e.g. wintering, resting, feeding, spawning, breeding, nursery, rearing area or migration

Table 15.3 Approaches to apply the rarity criterion

Rare species	**Regionally rare** (sessile or of restricted mobility) **species** = species occurring in less than 2% of the 50×50 km UTM grid squares of the following bathymetric zones in the region (e.g. North East Atlantic): littoral/sublittoral/bathyal, abyssal	Connor et al (2002) (only applicable to sessile species; no guidelines available for mobile species); Lieberknecht et al (2004a, b)
	Nationally rare species = species occurring in less than 0.5% of the 10 km×10 km squares within the study area	Sanderson (1996a,1996b); Lieberknecht et al (2004a,b)
	Nationally scarce species = species occurring in less than 3.5% of the 10 km x 10 km squares within the study area	Hiscock et al (2003); DEFRA (2002)
Rare habitats	**Regionally rare habitat** = habitat type occurring in less than2% of the 50 x 50 km UTM grid squares of the following bathymetric zones in the region(e.g. North East Atlantic): littoral/sublittoral/bathyal, abyssal	Connor et al (2002)
	Nationally rare habitat = habitat type restricted to a limited number of locations in territorial waters	DEFRA (2002)

routes), then this information can be used as an alternative or addition to broad-scale quantitative abundance data. When the location of such areas is not documented, their existence and location may be predicted by examination of physical processes (including modelling or remote sensing data), for example as indicated by Roff and Evans (2002) in their survey of distinctive marine areas.

Alternatively, traditional ecological knowledge (TEK) may assist in the definition of aggregation areas. It needs to be emphasized that any data, modelled or otherwise, should be assessed for its reliability and degree of confidence. The inclusion of aggregation as a criterion for biological valuation introduces a certain degree of connectivity into the valuation concept, because this criterion is used to determine the aggregation value of subzones relative to adjacent subzones, thus allowing the clustering of subzones with equal value.

The aggregation criterion is especially important for highly mobile species such as birds, mammals or fish. For the preservation of such wide-ranging species, information on their full distribution is less useful than the localization of areas which are critical for foraging, nursing, haul-out, breeding or spawning; it is these areas that should be included when a biological valuation is done (Connor et al, 2002; Roff and Evans, 2002).

Owing to the continuous nature of the marine environment, it is often difficult to identify the boundaries of such aggregation areas, especially for widely dispersed, highly mobile species (Johnston et al, 2002; Airamé et al, 2003). This can be seen in the difficulties encountered by many countries to implement the EC Bird Directive (1979) and RAMSAR Convention (1971), which both select important bird areas based on high densities of bird species (Johnston et al, 2002). However, see Chapter 14 for suggestions as to how the size and boundaries of population aggregations can be determined.

Fitness consequences

This criterion distinguishes subzones where natural activities take place that contribute significantly to the survival or reproduction of a species or population (DFO, 2004). These are not necessarily areas where species or individuals aggregate. When genetic data are available for the study area, these can be used to locate subzones where a high diversity of genetic stocks of a species occurs. The occurrence of genetically variable individuals could significantly improve the survival of a species in the study area, because it enables the selective adaptation of the species to changing environmental conditions. It is also possible to determine the location of subzones with fitness consequences for a species. These could be subzones where individuals stop for a certain amount of time to feed or rest, which will lead to higher reproduction (e.g. bigger/more young). Also, the presence of structural habitat features or keystone species may enhance the survival or reproduction of species by providing refuge from predators or key resources.

Naturalness

A major problem with assessing this criterion is the fact that it is often unknown what the natural state of an area actually is or should be. In addition, definitions of 'natural' may vary considerably (e.g. Bergman et al, 1991; Hiscock et al, 2003). There are also few if any completely natural areas left any more (Ray, 1984), and it is difficult to assess the degree of naturalness in areas at great depth or in areas of poor accessibility (Breeze, 2004). In order to assess the naturalness of a subzone, there is a need for comparison to appropriate pristine areas or reference sites. However, there are ways in which the degree of naturalness of environments may be assessed.

If areas within a region cannot be defined as natural (in other words control sites), then an alternative way to assess naturalness may be to use information on native/introduced or cultured species in the study area, which can be seen as proxies for the degree of naturalness. Another approach to assess the naturalness of a subzone is to examine the 'health' or composition of the inhabiting communities and species. For instance, healthy, natural benthic communities are in many cases characterized by a high biomass (dominated by long-lived species) and a high species richness (Dauer, 1993). Deviations from this pattern, resulting in a reduced macrobenthic biomass and a species richness dominated by opportunistic species, could be assigned to a certain level of stress and could be used to index the naturalness of a subzone. Such health indices, however, still require some reference to a baseline level of naturalness.

Lacking even this information, it is possible to use data on the location and intensity of human activities. The environmental and ecological state of subzones which are characterized by the absence of human disturbance can be used as a rough index of the degree of naturalness (e.g. Ban and Alder, 2008). Naturalness should be considered not only in terms of the degree of disturbance to attributes of species, but also to functional processes of marine ecosystems.

Proportional importance

Proportional importance measures the proportion of the national, regional and/or global resource of a species or feature which occurs within a subzone of the study area. This should be distinguished from the aggregation criterion. While the 'aggregation' criterion defines whether a high percentage of the population of a species (at the scale of the study area) is clustered within certain subzones of that area, the 'proportional importance' criterion defines whether a high percentage of the population of a species at a national (provided that the national scale is greater than the scale of the study area), regional and/or global scale can be found within the entire study area, regardless of whether this proportion is clustered within some subzones rather than others (in other words regionally aggregated).

To assess this criterion, data on the extent of marine features or population data of individual species are needed. When population data are lacking, it may be possible to use available abundance data for species within the study area, and determine the national importance of subzones for these species. This criterion was first defined by Connor et al (2002) and adapted by Lieberknecht et al (2004a, b), who also defined thresholds for the term 'high proportion'. These thresholds are similar to those in the criteria guidance of OSPAR (2003). However, thresholds can be very scale-dependent and should therefore be set for each case study separately, taking into account the spatial extent of the study area.

A map depicting biological valuation attempts to represent the biological values of the different subzones considered, relative to each other, but incorporation of the proportional importance criterion aims at comparing certain features or properties with the wider environment of the study area, attaching extra value to subzones where a high proportion of the population of a species occurs. It would also be possible to include the genetic (e.g. restricted distribution of a certain genetic stock) or community (e.g. restricted distribution of a defined community type) levels of the ecological hierarchy under this criterion.

Biodiversity: A valid valuation criterion?

When evaluating marine areas, it is important to 'capture' as many attributes of biodiversity as possible, since biological structures and processes exist on different organizational levels (namely genes, species, population, community and ecosystem) across the ecological hierarchy (Zacharias and Roff, 2000, 2001a). According to Roberts et al (2003), valuable marine areas should be characterized by high biodiversity and properly functioning ecological processes which support that diversity (in other words a higher proportion of the structural and process components of the biodiversity hierarchy). This is precisely the concept that is advocated here. According to many authors the biodiversity of an area is simply a function of the species diversity; however, a valuation framework that incorporates as many organizational levels of biodiversity as possible is far preferable.

Although the concept of biodiversity as a valuation criterion is useful for management purposes, the practice of distilling biodiversity to a single index or a few dimensions is unjustified (Margules and Pressey, 2000; Purvis and Hector, 2000; Price, 2002). This is the reason why biodiversity itself was not used directly as a criterion in the valuation concept. However, biodiversity still is integrated in the concept, but in a different way (see below). Yet, because of its frequent use (e.g. HELCOM, 1992; IUCN, 1994; Brody, 1998; UNEP, 2000; GTZ GmbH, 2002) or misuse, some further argument for not including biodiversity directly as a valuation criterion in the concept is in order.

In most research studies only the species richness of a subzone is assessed (Humphries et al, 1995; Woodhouse et al, 2000; Price, 2002), but biodiversity manifests itself on many more levels of organization – from the genetic to the ecosystem. Simply counting the number of species in a subzone (as a measure of species diversity and biodiversity as a whole) can be misleading because subzones with a high species richness do not necessarily exhibit a high diversity at other levels of the ecological hierarchy (Attrill et al, 1996; Hockey and Branch, 1997; Vanderklift et al, 1998; Purvis and Hector, 2000; Price, 2002).

Several authors have tried to find surrogate measures for biodiversity in general, in order to decrease the sampling effort or data requirements (Purvis and Hector, 2000). For example, Ray (1999) used species richness of birds as a surrogate for overall biodiversity, an approach which is based on the fact that birds have dispersed to and diversified in all regions of the world. Yet analyses revealed that species richness hotspots of birds coincided poorly with those of other biota. Hotspots of species richness, endemism or rarity are often less discernible in continuous marine ecosystems than in terrestrial environments. Turpie et al (2000) used the hotspot approach for species richness (and weighting all species equally) and did not achieve good representation for coastal fish species. Thus the hotspot approach based on species richness alone is not a useful starting point for the selection of biologically valuable marine areas. This was also noted by Breeze (2004), who found the traditional hotspot approach to be narrowly defined and species focused, while the criteria used for identification of highly valuable marine areas should be much broader.

The use of focal species (e.g. indicators, umbrellas, flagship species), which has been developed mainly from a terrestrial viewpoint, is not straightforward to apply in the marine environment (but see Zacharias and Roff, 2001b). Since connectivity is very different in the marine environment, the concept of a particular species indicating a certain size of intact habitat is not readily applicable (Ardron et al, 2002), in part because of likely aliasing of data. Ward et al (1999) also investigated the use of surrogates for overall biodiversity, and found that habitat types suited this function best. However, no surrogate was able to cover all species, from which it can be concluded that the hotspot paradigm, based on individual surrogates of biodiversity, is problematic to apply.

The concept of 'benthic complexity' was introduced by Ardron et al (2002) as a proxy for benthic species diversity. The authors assume that the bathymetric (topological) complexity of an area is a measure of benthic habitat complexity, which in turn would represent benthic species diversity. However, the data needed to perform the spatial variance analyses needed to quantify 'benthic complexity' are usually lacking. Because detailed data on the diversity of species or communities are often scarce or nonexistent, Airamé et al (2003) proposed to assess the habitat diversity as a proxy for overall biodiversity, because data on habitat distributions are generally available or can be constructed.

A more general framework for the assessment of biodiversity is clearly needed (see e.g. Humphries et al, 1995), and such a framework should encompass all available information from a range of organizational levels (genes, species, communities, ecosystems); relationships among these levels also need to be examined. In addition to biodiversity structures, there is also a need to include biodiversity processes such as aspects of the functioning of ecosystems, which could even be more important than high species richness or diversity indices in certain low biodiversity sites such as estuaries (Attril et al, 1996; Bengtsson, 1998). Bengtsson (1998) also stated that biodiversity is an abstract aggregated property of species in the context of communities or ecosystems, and that there is no mechanistic relationship between single measures of biodiversity and the functioning of the entire ecosystem. Ecosystem functioning can, however, be included indirectly in an assessment of biodiversity value, through the identification of functional species or groups and critical areas. Zacharias and Roff (2000) visualized the various components of biodiversity in their 'marine ecological framework' – going from the genetic to the ecosystem level, and including both biodiversity structures and processes (see Tables 1.1 and 1.2). Each of these components can be linked to one or more of the selected valuation criteria, which makes it unnecessary to include biodiversity as a separate valuation criterion. Rather, it is the sum total of the components of biodiversity which are being evaluated, according to the selected criteria, for each subzone (Figure 15.1). By using this framework it is therefore possible to apply the valuation criteria while integrating the various components of biodiversity.

Potential application of the biological valuation concept

The criteria of the biological valuation process can be applied to a marine study area, and the result of this process can be visualized on marine BVMs. Marine BVMs can act as a kind of baseline describing the intrinsic biological and ecological value of subzones within a study area. They can be used for various purposes including: planning for marine conservation and evaluation of locations for MPAs; as a foundation for ecosystem-based management, especially for fisheries; as warning systems for marine managers who are planning new resource explorations and in general helping to indicate potential conflicts between human uses and a subzone's high biological value during any form of spatial planning.

It should be explicitly stated that these BVMs give no information on the potential impacts that any activity could have on a certain subzone; this is because criteria like vulnerability or resilience are deliberately not included in this valuation scheme – because the determination of the 'vulnerability' of a system is mainly a human value judgement (McLaughlin et al, 2002).

These human-value and human-impact criteria should therefore be considered in a later phase of site-specific management (e.g. selection of protected areas) rather than in the assessment of the value of marine subzones themselves (Gilman, 1997, 2002). The BVMs could be used as a framework to evaluate the effects of certain management decisions (e.g. implementation of MPAs or a new quota for resource use), when BVMs are revisited and revised after a period of time to see if value changes have occurred in subzones where certain management actions were implemented. However, these value changes may not be directly attributable to specific impact sources, but may only give an integrated view of the effect of all impact sources in a subzone.

The development of decision support tools for marine management could build on these BVMs by adding other criteria to the assessment process. For example, when developing a framework suitable for the selection of MPAs, representation, ecological integrity, and socio-economic and management criteria should also be taken into account (Rachor and Günther, 2001), especially when considering the need for management for sustainable use (Hockey and Branch, 1997). Managers may also want to know which areas should get the highest priority for specific purposes. Therefore, the sites that attained the highest biological and ecological value could be screened, with the application of additional criteria such as 'degree of threat', 'political/public concern' and 'feasibility of conservation measures'. Thus, although the ultimate selection of the priority areas may be a political decision (Agardy, 1999), selection can still have a solid scientific base through the use of BVMs. An overview of the possible steps beyond the development of a marine BVM is given in the lower part of Figure 15.1, which shows that, although these following steps should be founded on scientific principles of biological valuation, they cannot be based solely on such criteria.

A refinement to the biological valuation process

Following submission of the original paper by Derous et al (2007a) further thought was given to how a methodology for marine biological valuation would be realistically applied. Some modifications to the process were deemed desirable, largely for practical reasons, and these are summarized in Box 15.1.

Box 15.1 Modifications to the process of marine biological valuation (see Derous et al (2007b)

- The two criteria 'aggregation' and 'fitness consequences' are strongly linked, and to avoid double counting they are merged into a single criterion. Although the two criteria attempt to define ecologically different factors, in reality it may be almost impossible to distinguish between them and they are necessarily frequently allied. Note that the criteria aggregation and fitness, originally applied predominantly to the level of processes that are important for fisheries. However this notion should be expanded to consider ALL processes that are significant for the ecological support of ALL species in a region.
- In many cases it is very difficult to define the natural state of a marine area. The assessment of (un) naturalness – in relation to various types of impacts – should therefore be seen as a second step, after biological valuation itself, to produce an overlying layer on the biological valuation map.
- All other criteria assess the value of subzones relative to each other; the inclusion of a wider scale (global, national, regional etc.) under 'proportional importance' can therefore be misleading and introduce bias. Consequently, it is recommended that valuation should be done first at the level of the (local) whole study area, and only afterwards should it be done on the broader (e.g. ecoregional) scale to allow a wider perspective.

The final revised process of marine biological evaluation appears in Derous et al (2007b).

Conclusions and management implications

The concept of marine biological valuation provides a comprehensive means for assessing the intrinsic value of the subzones within a study area. Marine biological valuation is not a strategy for protecting all habitats and marine communities that have some ecological significance, but rather it is a tool for calling attention to subzones that have particularly high ecological or biological significance. It is a tool that can be used to ask and answer questions such as: what should be done first in marine conservation? Are some components of marine biodiversity more important than others? What components of marine biodiversity are most important to humanity? This in turn can provide the rationale for provision of a greater-than-usual degree of risk aversion in spatial planning activities in these subzones.

The concept is based on a thorough review and clarification of existing criteria and definitions of value (Derous et al, 2007a, b), and the selection of criteria has been rationalized to develop a widely applicable valuation process. However, because this biological valuation concept is based on the consensus reached by a group of experts (i.e. a Delphic process), some refinement of the methodology is inevitable once it has been evaluated on the basis of case studies.

The concept of value can make significant contributions to marine spatial planning and ecosystem-based management. It allows explicit and simultaneous evaluation of all components of marine biodiversity in the ecological hierarchy (Tables 1.1 and 1.2). It therefore presents a potential mechanism to evaluate spatially defined areas, either in terms of fisheries conservation or the broader conservation of biodiversity as a whole. Perhaps most importantly, it allows concurrent evaluation of the relationships between fisheries conservation and biodiversity conservation as a whole, and the potential conflicts or synergies between them. In addition, it can also be used as a tool or process to evaluate existing protected areas and conservation policies, in order to discover whether appropriate or efficient conservation decisions have been made or are being made in a planning process. In other words, an evaluation process can provide a form of GAP analysis between existing and desirable conservation strategies.

Other research and management groups may use processes of evaluation of regions that differ from that of Derous et al (2007a, b). However, whatever the actual process, if it is applied to the entire array of components of marine biodiversity, the end product of designation and selection of zones is likely to be very comparable with similar membership.

It should be clearly noted that 'value' alone cannot be the only criterion for selection of areas to be protected. Although most, perhaps all, distinctive areas would likely be selected based on an assessment of value (e.g. for fisheries or focal species), other components of biodiversity – including representative areas – may not. We should therefore consider that the evaluation of value is only one part of the process of ensuring that all components of biodiversity are captured or managed within adequately protected areas. Further processes to integrate the selection of protected areas (Chapter 16) and to design networks of MPAs (Chapter 17) must ensue.

References

Agardy, T. S. (1997) *Marine Protected Areas and Ocean Conservation*, Academic Press, Austin, TX

Agardy, T. S., (1999) 'Global Trends in Marine Protected Areas', in B. Cicin-Sain, R. W. Knecht and N. Foster (eds) *Trends and Future Challenges for U.S. National and Coastal Policy: Trends in Managing the Environment*, National Oceanic and Atmospheric Administration, Silver Spring, MD

Airamé S., Dugan J. E., Lafferty, K. D., Leslie, H., McArdle, D. A. and Warner, R. R. (2003) 'Applying ecological criteria to marine reserve design: A case study from the California Channel Islands', *Ecological Applications*, vol 13, pp170–184

Ardron, J. A., Lash, J. and Haggarty, D. (2002) *Modelling a Network of Marine Protected Areas for the Central Coast of British Columbia*, Living Oceans Society, Sointula, BC, Canada

Attrill, M. J., Ramsay, P. M., Myles Thomas, R. and Trett, M. W. (1996) 'An estuarine biodiversity hotspot', *Journal of the Marine Biological Association of the UK*, vol 76, pp161–175

Ban, N. and Alder, J. (2008) 'How wild is the ocean? Assessing the intensity of anthropogenic marine activities in British Columbia, Canada', *Aquatic Conservation: Marine and Freshwater Ecosystems*, vol 18, no 1, pp55–85

Beger, M., Jones, G. P. and Munday, P. L. (2003) 'Conservation of coral reef biodiversity: A comparison of reserve selection procedures for corals and fishes', *Biological Conservation*, vol 111, pp53–62

Bengtsson, J. (1998) 'Which species? What kind of diversity? Which ecosystem functioning? Some problems in studies of relations between biodiversity and ecosystem function', *Applied Soil Ecology*, vol 10, no 3, pp191–199

Bergman, M. N., Lindeboom, H. J., Peet, G., Nelissen, P. H. M., Nijkamp, H. and Leopold, M. F. (1991) 'Beschermde gebieden Noordzee: Noodzaak en mogelijkheden', *NIOZ Rep*, vol 91, no 3

Bern Convention (1979) 'Convention on the conservation of European wildlife and natural habitats, 19 IX1979', *European Treaty Series – No. 104*, Council of Europe, Strasbourg, France

Bockstael, N., Costanza, R., Strand, I., Boynton, W., Bell, K. and Wainger, L. (1995) 'Ecological economic modeling and valuation of ecosystems', *Ecological Economics*, vol 14, pp143–159

Borgese, E. M. (2000) 'The economics of the common heritage', *Ocean & Coastal Management*, vol 43, pp763–779

Breeze, H. (2004) 'Review of criteria for selecting ecologically significant areas of the Scotian Shelf and Slope: A discussion paper', Bedford Institute to Oceanography, Bedford, Nova Scotia, Canada

Brody, S. D. (1998) 'Evaluating the role of site selection criteria for marine protected areas in the Gulf of Maine', *Gulf of Marine Protected Areas Project Report*, vol 2

Connor, D. W., Breen, J., Champion, A., Gilliland, P. M., Huggett, D., Johnston, C., Laffoley, D. d'A., Lieberknecht, L., Lumb, C., Ramsay, K. and Shardlow, M. (2002) 'Rationale and criteria for the identification of nationally important marine nature conservation features and areas in the UK. Version 02.11', Joint Nature Conservation Committee, Peterborough

Dauer, D. M. (1993) 'Biological criteria, environmental health and estuarine macrobenthic community structure', *Marine Pollution Bulletin*, vol 26, pp249–257

De Blust, G., Froment, A. and Kuijken, E. (1985) 'Biologische waarderingskaart van Belgi¨e. Algemene verklarende tekst', Min. Volksgezondheid et Gezin, Inst. Hyg. Epidemiol., Cöordinatiecentrum van de Biologische Waarderingskaart, Brussels

De Blust, G., Paelinckx, D. and Kuijken, E. (1994) 'Up-to-date Information on Nature Quality for Environmental Management in Flanders', in F. Klijn (ed) *Ecosystem Classification for Environmental Management*, Kluwer, Boston, MA

DEFRA (2002) 'Safeguarding our seas. A strategy for the conservation and sustainable development of our marine environment', Department for Environment, Food and Rural Affairs, London

De Groot, R. S., Wilson, M. A. and Boumans, R. M. J. (2002) 'A typology for the classification, description and valuation of ecosystem functions, goods and services', *Ecological Economics*, vol 41, pp393–408

Derous, S., Agardy, T., Hillewaert, H., Hostens, K., Jamieson, G., Lieberknecht, L., Mees, J., Moulaert, I., Olenin, S., Paelinckx, D., Rabaut, M., Rachor, E., Roff, J., Willem, E., Stienen, M., Tjalling, J., van der Wal, J. T., van Lancker, V., Verfaillie, E., Vincx, M., Węsławski, J. M. and Degraer, S. (2007a) 'A concept for biological valuation in the marine environment', *Oceanologia*, vol 49, pp99–128

Derous, S., Austen, M., Claus, S., Daan, N., Dauvin, J. C., Deneudt, K., Sepestele, J., Desroy, N., Heessen, H., Hostens, K., Husum Marboe, A., Lescrauwaet, A.-K., Moreno, M. P., Moulaert, I., Paelinckx, D., Rabaut, M., Rees, H., Ressurreição, A., Roff, J., Talhadas Santos, P., Speybroeck, J., Willem, E., Stienen, M., Tatarek, A., Ter Hofstede, R., Vincx, M., Zarzycki, T. and Degraer, S. (2007b) 'Building on the concept of marine biological valuation with respect to translating it to a practical protocol: Viewpoints derived from a joint ENCORA–MARBEF initiative', *Oceanologia*, vol 49, pp579–586

DFO (Department of Fisheries and Oceans Canada) (2004) 'Identification of ecologically and biologically significant areas', *Canadian Science Advisory Secretariat Ecosystem Status Report*, 2004/06

EC Bird Directive (1979) 'Council Directive 79/409/EEC of 2 April 1979 on the conservation of wild birds', OJ L 103, 25.04.1979, p1

EC Habitat Directive (1992) 'Council Directive 92/43/EEC of 21 May 1992 on the conservation of natural habitats and of wild fauna and flora', OJ L 206, 22.07.1992, p7

Edwards P. J. and Abivardi, C. (1998) 'The value of biodiversity: Where ecology and economy blend', *Biological Conservation*, vol 83, no 3, pp239–246

Fairweather, P. G. and McNeill, S. E. (1993) 'Ecological and Other Scientific Imperatives for Marine and Estuarine Conservation', in A. M. Ivanovici, D. Tarte and M. Olsen (eds) *Protection of Marine and Estuarine Environments – A Challenge for Australians*, Department Environment, Sport and Territories, Canberra, Australia

Gilman, E. (1997) 'Community-based and multiple purpose protected areas: A model to select and manage protected areas with lessons from the Pacific islands', *Coastal Management*, vol 25, no 1, pp59–91

Gilman, E. (2002) 'Guidelines for coastal and marine site-planning and examples of planning and management intervention tools', *Ocean & Coastal Management*, vol 45, pp377–404

Golding, N., Vincent, M. A. and Connor, D. W. (2004) 'The Irish Sea Pilot: A marine landscape classification for the Irish Sea', *Joint Nature Conservation Committee Report*, vol 346, Peterborough

GTZ GmbH (2002) *Marine Protected Areas: A Compact Introduction*, Deutsche Gesellschaft fürTechnische Zusammenarbeit (GTZ) GmbH, Eschborn, Germany

HELCOM (1992) *Convention on the Protection of the Marine Environment of the Baltic Sea Area*, Helsinki Commission, Helsinki, Finland

Hiscock, K., Elliott, M., Laffoley, D. and Rogers, S. (2003) 'Data use and information creation: Challenges for marine scientists and for managers', *Marine Pollution Bulletin*, vol 46, no 5, pp534–541

Hockey, P. A. R. and Branch, G. M. (1997) 'Criteria, objectives and methodology for evaluating marine protected areas in South Africa', *South African Journal of Marine Science*, vol 18, pp369–383

Humphries, C. J., Williams, P. H. and Vane-Wright, R. I. (1995) 'Measuring biodiversity value for conservation', *Annual Review of Ecology, Evolution and Systematics*, vol 26, pp93–111

IUCN (1994) *Guidelines for Protected Area Management Categories*, International Union for the Conservation of Nature, Gland, Switzerland

JNCC (2004) 'Developing the concept of an ecologically coherent network of OSPAR marine protected areas', Paper 04 N08, Joint Nature Conservation Committee, Peterborough

Johnston, C. M., Turnbull, C. G. and Tasker, M. I. (2002) 'Natura 2000 in UK offshore waters: Advice to support the implementation of the EC Habitats and Birds Directives in UK offshore waters', Report No 235, Joint Nature Conservation Committee, Peterborough

Kelleher G. (1999) *Guidelines for Marine Protected Areas*, IUCN, Gland, Switzerland

King, O. H. (1995) 'Estimating the value of marine resources: A marine recreation case', *Ocean & Coastal Management*, vol 27, nos 1–2, pp129–141

Laffoley, D., Connor, D. W., Tasker, M. L. and Bines, T. (2000) 'An implementation framework for the conservation, protection and management of nationally important marine wildlife in the UK', *English Nature Research Report No 394*, Joint Nature Conservation Committee, Peterborough

Lieberknecht, L. M., Vincent, M. A. and Connor, D.W. (2004a) 'The Irish Sea pilot. Report on the identification of nationally important marine features in the Irish Sea', Report No 348, Joint Nature Conservation Committee, Peterborough, UK

Lieberknecht, L.M., Carwardine, J., Connor, D.W., Vincent, M.A., Atkins, S.M. and Lumb, C.M. (2004b) 'The Irish Sea pilot. Report on the identification of nationally important marine areas in the Irish Sea', Report No 347, Joint Nature Conservation Committee, Peterborough, UK

Margules, C. R. and Pressey, R. L. (2000) 'Systematic conservation planning', *Nature*, vol 405, pp243–253

McLaughlin, S., McKenna, J. and Cooper, J. A. C. (2002) 'Socio-economic data in coastal vulnerability indices: Constraints and opportunities', *Journal of Coastal Research*, vol 36, pp487–497

Mitchell, R. (1987) *Conservation of Marine Benthic Biocenoses in the North Sea and the Baltic: A framework for the establishment of a European network of marine protected areas in the North Sea and the Baltic*, Council of Europe, Strasbourg, France

Myers, N., Mittermeier, R. A., Mittermeier, C. G., da Fonseca, G. A. B. and Kent, J. (2000) 'Biodiversity hotspots for conservation priorities', *Nature*, vol 403, pp853–858

Nunes, P. A. L. D. and van den Bergh, J. C. J. M. (2001) 'Economic valuation of biodiversity: Sense or nonsense?', *Ecological Economics*, vol 39, pp203–222

Orians, G. H. (1974) 'Diversity, Stability and Maturity in Natural Ecosystems', in W. H. van Dobben and R. H. Lowe-McConnell (eds) *Unifying Concepts in Ecology*, Junk, The Hague

OSPAR (1992) 'Convention for the protection of the marine environment of the North-East Atlantic, 22 September 1992, Paris', OSPAR, London

OSPAR (2003) 'Criteria for the identification of species and habitats in need of protection and their method of application (The Texel-Faial criteria)', OSPAR, London

Possingham, H. P., Ball, I. R. and Andelman, S. (2000) 'Mathematical Methods for Identifying Representative Reserve Networks', in S. Ferson and M. Burgman (eds) *Quantitative Methods for Conservation Biology*, Springer-Verlag, New York

Price, A. R. G. (2002) 'Simultaneous 'hotspots' and 'coldspots' of marine biodiversity and implications for global conservation', *Marine Ecology Progress Series*, vol 241, pp23–27

Purvis, A. and Hector, A. (2000) 'Getting the measure of biodiversity', *Nature*, vol 405, pp212–219

Rachor, E. and Günther, C. P. (2001) 'Concepts for offshore nature reserves in the southeastern North Sea', *Senckenb. Marit.*, vol 31, no 2, pp353–361

RAMSAR (1971) 'Convention on wetlands of international importance especially as waterfowl habitat', Ramsar, Gland, Switzerland

RAMSAR (1999) 'Strategic framework and guidelines for the future development of the list of wetlands of international importance of the Convention on Wetlands', Ramsar, Gland, Switzerland

Ray, G. C. (1984) 'Conservation of Marine Habitats and Their Biota', in A. V. Hall (ed) *Conservation of Threatened Natural Habitats*, CSIR, Pretoria, South Africa

Ray, G. C. (1999) 'Coastal-marine protected areas: Agonies of choice', *Aquatic Conservation*, vol 9, pp607–614

Roberts, C. M., Andelman, S., Branch, G., Bustamante, R. H., Castilla, J. C., Dugan, J., Halpern, B. S., Lafferty, K. D., Leslie, H., Lubchenco, J., McArdle, D., Possingham, H. P., Ruckelshaus, M. and Warner, R. R. (2003) 'Ecological criteria for evaluating candidate sites for marine reserves', *Ecological Applications*, vol 13, pp199–214

Roff, J. C. and Evans, S. M. J. (2002) 'Frameworks for marine conservation: Nonhierarchical approaches and distinctive habitats', *Aquatic Conservation: Marine and Freshwater Ecosystems*, vol 12, pp635–648

Salm, R.V. and Clarke, J. R. (1984) *Marine and Coastal Protected Areas: A Guide for Planners and Managers*, IUCN, Gland, Switzerland

Salm, R.V. and Price, A. (1995) 'Selection of Marine Protected Areas', in S. Gubbay (ed) *Marine Protected Areas, Principles and Techniques for Management*, Chapman & Hall, London

Sanderson, W. G. (1996a) 'Rarity of marine benthic species in Great Britain: Development and application of assessment criteria', *Aquatic Conservation*, vol 6, pp245–256

Sanderson, W. G. (1996b) 'Rare marine benthic flora and fauna in Great Britain: The development of criteria for assessment', Report No 240, Joint Nature Conservation Committee, Peterborough

Sealey, S. K. and Bustamante, G. (1999) *Setting Geographic Priorities for Marine Conservation in Latin America and the Caribbean*, The Nature Conservancy, Arlington, VA

Simberloff, D. (1998) 'Flagships, umbrellas, and keystones: Is single-species management passé in the landscape era?', *Biological Conservation*, vol 83, no 3, pp247–257

Smith, P. G. R. and Theberge, J. B. (1986) 'A review of criteria for evaluating natural areas', *Environmental Management*, vol 10, pp715–734

Tunesi, L. and Diviacco, G. (1993) 'Environmental and socio-economic criteria for the establishment of marine coastal parks', *International Journal of Environmental Studies*, vol 43, pp253–259

Turpie, J. K., Beckley, L. E. and Katua, S. M. (2000) 'Biogeography and the selection of priority areas for conservation of South African coastal fishes', *Biological Conservation*, vol 92, pp59–72

Turpie, J. K., Heydenrych, B. J. and Lamberth, S. J. (2003) 'Economic value of terrestrial and marine biodiversity in the Cape Floristic Region: Implications for defining effective and socially optimal conservation strategies', *Biological Conservation*, vol 112, pp233–251

UNEP (1990) 'Protocol concerning specially protected areas and wildlife to the Convention for the Protection and Development of the Marine Environment of the Wider Caribbean Region', United Nations Environment Programme, Nairobi, Kenya

UNEP (2000) 'Progress report on the implementation of the programmes of work on the biological diversity of inland water ecosystems, marine and coastal biological diversity, and forest biological diversity (Decisions IV/4, IV/5, IV/7)', Conference of the Parties to the Convention on Biological Diversity 5 (INF/8), UNEP, Nairobi, Kenya

UNESCO (1972) 'Convention concerning the protection of the world cultural and natural heritage', United Nations Educational, Scientific and Cultural Organization, Paris, France

Vallega, A. (1995) 'Towards the sustainable management of the Mediterranean Sea', *Marine Policy*, vol 19, no 1, pp47–64

Vanderklift, M. A., Ward, T. J. and Phillips, J. C. (1998) 'Use of assemblages derived from different taxonomic levels to select areas for conserving marine biodiversity', *Biological Conservation*, vol 86, pp307–315

Villa, F., Tunesi, L. and Agardy, T. (2002) 'Zoning marine protected areas through spatial multiple-criteria analysis: The case of the Asinara Island National Marine Reserve of Italy', *Conservation Biology*, vol 16, no 2, pp515–526

Vincent, M. A., Atkins, S., Lumb, C., Golding, N., Lieberknecht, L. M. and Webster, M. (2004) *Marine Nature Conservation and Sustainable Development: The Irish Sea Pilot*, Joint Nature Conservation Committee, Peterborough

Ward, T. J., Vanderklift, M. A., Nicholls, A. O. and Kenchington, R. A. (1999) 'Selecting marine reserves using habitats and species assemblages as surrogates for biological diversity', *Ecological Applications*, vol 9, no 2, pp691–698

Woodhouse, S., Lovett, A., Dolman, P. and Fuller, R. (2000) 'Using a GIS to select priority areas for conservation', *Computers, Environment, and Urban Systems*, vol 24, pp79–93

Zacharias, M. A. and Roff, J. C. (2000) 'A hierarchical ecological approach to conserving marine biodiversity', *Conservation Biology*, vol 13, no 5, pp1327–1334

Zacharias, M. A. and Roff, J. C. (2001a) 'Zacharias and Roff vs. Salomon et al: Who adds more value to marine conservation efforts?', *Conservation Biology*, vol 15, no 5, pp1456–1458

Zacharias, M. A. and Roff, J. C. (2001b) 'Use of focal species in marine conservation and management: A review and critique', *Aquatic Conservation: Marine and Freshwater Ecosystems*, vol 11, pp59–76

16

Sets of Protected Areas

Integrating distinctive and representative protected areas

To manage a system effectively, you might focus on the interactions of the parts rather than their behaviour taken separately.

Russell L. Ackoff (1919–2009)

Introduction

This chapter and the following chapter concern the process of selecting the members of networks of protected areas that will collectively address the requirements for preservation of biodiversity and conservation of fisheries. It is first necessary to carefully define a series of terms so that concepts in these two chapters follow a logical path.

The term 'network of marine protected areas' has several definitions (see Box 16.1), but the most important element in any definition is the concept of connectivity. Thus a series of MPAs in a network should be oceanographically linked to ensure recruitment of propagules from one protected area to another. This is very different from the concept of a 'set of protected areas' which are not designed or known to be connected. The ideas of network and connectivity have become so fundamental to marine conservation that they will rate their own chapter (Chapter 17 following).

In contrast to a network, a 'set' of MPAs is simply a group of MPAs within a defined ecoregion, each designed for a specific purpose, to capture some specified component(s) of biodiversity, and each of some planned size based on ecological principles. More importantly a 'coherent set' of MPAs is a set that captures all valued and identified components of marine biodiversity in a region (Box 16.1).

There has been much confusion on this issue of definitions, and many proponents of networks of MPAs are in fact really referring to sets of MPAs. Currently there are several examples described as networks of MPAs that are in fact simply sets of MPAs because patterns of connectivity among them remain undefined. A set of MPAs cannot be considered as a network of MPAs unless we have defined and demonstrated the patterns of connectivity among the component members.

The term 'ecological integrity' is also frequently used in conjunction with networks of MPAs. This term is defined and explained in Box 16.1, but in essence it means that a protected region should be established and managed in such a way as to ensure the continued functioning of its natural ecosystem processes. The natural functioning of any region of the marine environment is, however, completely dependent upon processes of reproduction and recruitment of its populations, by propagules from both within and beyond its own borders. Any portion of the marine environment must, at least in some part, be reliant upon recruitment of its populations from outside its own spatial limits. The idea of ecological integrity of an isolated portion of the

Box 16.1 Some definitions concerning sets and networks of marine protected areas

Ecological integrity

Ecological or biological integrity originated as an ethical concept. The generic concept of integrity connotes a valuable whole, the state of being whole or undiminished, unimpaired, or in perfect condition. Among the most important aspects of integrity are the autopoietic (self-creative) capacities of life to organize, regenerate, reproduce, sustain, adapt, develop and evolve over time at a specific location. Thus integrity defines the evolutionary and biogeophysical processes of a system as well as its parts or elements at a specific location (Westra, 2005).

According to the Canada National Parks Act (www.canlii.org/en/ca/laws/stat/sc-2000-c-32/latest/sc-2000-c-32.html, the law governing national parks in Canada), 'ecological integrity' means, with respect to a park, '...a condition that is determined to be characteristic of its natural region and likely to persist, including abiotic components and the composition and abundance of native species and biological communities, rates of change and supporting processes.'

Following from these (and similar) definitions, there can in fact be no such thing as ecological integrity of an isolated MPA – because it is not self-sufficient as a location, and will depend on its connectivity to other parts of the ocean for recruitment of its component species.

Sets of MPAs

A set of MPAs is any group of protected areas within a geographic region or regions that collectively represent the components of marine biodiversity of that region.

Coherent sets of MPAs

A coherent set of MPAs is a set of MPAs within a defined region that collectively achieve a defined goal. For example, they may achieve a level of protection for 20 per cent of the region, and collectively represent all the identifiable biotic and abiotic components of marine biodiversity within that region. Such a set will be defined according to stated geographic, geological, oceanographic and ecological principles, and will include both representative and distinctive areas (modified from J. Ardron, personal communication).

Networks of MPAs

'A collection of individual marine protected areas that operates cooperatively and synergistically, at various spatial scales, and with a range of protection levels, in order to fulfill ecological aims more effectively and comprehensively than individual sites could alone' (IUCN, 2007).

An 'ecologically representative network of MPAs' (or a 'representative network of MPAs' for short) comprises a coherent set of MPAs whose members should allow them to support each other by taking advantage of ocean currents, migration routes and other natural ecological connections. This would help provide much-needed resilience against a range of threats. For example, if one MPA is damaged by a storm, oil spill, coral bleaching event or other disaster, it could be recolonized by fish and other species from an up-current MPA in the network. And by protecting multiple sites within the ecosystem, the overall damage caused by a disaster in one MPA is reduced.

More extensively: A proper national or regional network of MPAs must consist of: multiple sites with replicates of all habitat types, that are oceanographically connected; individually or in aggregate they are of sufficient size to sustain minimum viable populations of the largest species in a region (including those of seasonal migrants to the region) and their resident species can sustain their populations by recruitment from one MPA to another. Again: the member sites of a network of MPAs should comprise replicates of all representative habitats, and various distinctive habitats, that are mutually supporting in recruitment processes through connectivity. In short, a network of MPAs should capture and be able to sustain the regional elements of marine biodiversity (Roff, 2005).

Thus the members of a representative network of MPAs should be known to be oceanographically connected by recruitment processes.

Efficient networks of MPAs

An efficient network of MPAs is a network of MPAs whose members collectively can be shown (by studies of current patterns, models and genetics of selected species) to recruit efficiently from one member MPA to another. A measure of efficiency is that the proportion of propagules recruiting among the protected areas is at least equal to or greater than the proportion of the region protected within the member MPAs (see Chapter 17).

Priority areas for conservation (PAC)

This term is now widely used but infrequently defined where used. It can mean several things but is generally applied to areas of the (marine) environment with valued (combinations of) characteristics that warrant protection (e.g. as MPAs), or as special management areas, for example under some scenario of ecosystem-based management. Such PACs would recognize some component or combination of components of marine biodiversity, and acknowledge that they require special attention. Such PACs may be recognized on the basis of both 'species' and/or 'spaces', because of their importance to fisheries, to species diversity, to focal species and so on. PACs, whether in existence or planned, may range from rather small and specific in nature (e.g. a turtle nesting beach) to large and only seasonally regulated (e.g. a fisheries spawning/recruitment area). The fundamental concept of a PAC is therefore that it has 'value' greater than other surrounding areas (see Chapter 15). In addition to inherent value, other attributes of a PAC may include threats (actual or expected), and opportunity to afford protection (because of public interest, socio-economic concerns etc.)

marine environment is therefore a chimera (see Roff, 2009).

All areas of the oceans are ultimately connected, but on various time scales. A critical point is that the protected areas within a network should be oceanographically connected on a time frame consistent with the life cycles and dispersal abilities of the flora and fauna (see Chapter 17). In this chapter we shall examine the concept and planning for a coherent set of representative MPAs, and in the following chapter define the requirements for an efficient network of MPAs.

The Great Barrier Reef: A set of representative protected areas

A prime, world-leading example of a set of representative MPAs in tropical and subtropical waters is the Great Barrier Reef. Details of the planning process that lead to the creation and protection of the Great Barrier Reef can be found in the websites noted in Box 16.2.

In 1994, The 25-Year Strategic Plan for the Great Barrier Reef World Heritage Area was produced to outline strategies and objectives for managing and preserving the Great Barrier Reef World Heritage Area. It provides the basis to ensure wise use and protection of the Great Barrier Reef World Heritage Area for the future. The strategic plan involved significant public consultation, allowing everyone who has a stake in the reef's long-term future to say how the Great Barrier Reef World Heritage Area should be managed over the following 25 years. This approach was taken to ensure the reef remains in a healthy state and can be enjoyed by future generations.

From the beginning, emphasis was placed on the concerns and opinions of all stakeholders. These included governments, Aboriginal and Torres Strait Islander communities, conservationists, scientists, recreational users and established reef industries such as fishing, shipping and tourism. Overall, the strategic plan was endorsed by almost 70 organizations representing all levels of government, recreational and commercial users, conservation and scientific groups and Aboriginal and Torres Strait Islander communities.

In the mid-1990s concerns were raised that the levels of protection provided by the zoning at the time were inadequate to protect the range of biodiversity components that existed in the marine park. This was recognized as important in order to ensure that the Great Barrier Reef remained a healthy, productive and resilient ecosystem that would continue to support a range of activities.

Between 1999 and 2004, the Great Barrier Reef Marine Park Authority undertook a systematic planning and consultative programme to develop new zoning for the marine park. The primary aim of the programme was to better protect the biodiversity components of the Great Barrier Reef, by increasing the extent of no-take areas (or highly protected areas, locally known as 'green zones'), ensuring that they included 'representative' examples of all different habitat types – hence the name, the Representative Areas Programme or RAP (see Box 16.2).

While increasing the protection of biodiversity, a further aim of the RAP was to maximize the benefits and minimize the negative impacts of rezoning on the existing multiple users of the marine park. Both these aims were achieved by a comprehensive programme of scientific input, community involvement and innovation. The planning process therefore involved a combination of both ecological and public/socio-economic research.

Example of planning for a coherent set of MPAs based on ecological principles

Perhaps the best way to present how a set of MPAs can be integrated into a coherent set of MPAs is to follow a specific example; here the emphasis is on the ecological aspects. One of the most thorough such planning studies in temperate waters, undertaken over several years, has been carried out on parts of the east coast of Canada and the USA by the Conservation Law Foundation USA and World Wildlife Fund Canada. A full report of this planning process (referred to as a network in that report, but actually comprising a set – see Box 16.1) is contained in CLF/WWF (2006). The objective was to define priority areas for conservation in a broad area of the New England region of the USA and Maritime Canada.

Overview of the planning process

Developing plans for biodiversity conservation is challenging because knowledge of the components of the diversity is almost always incomplete, ecosystems are dynamic and complex and the requirements for targets of conservation are poorly defined. In order to guide the process, a set of operating principles was adopted by CLF/WWF (2006) – see Box 16.3.

Representation is a widely accepted strategy for countering this uncertainty in planning for biodiversity conservation (e.g. Noss, 1983; Groves et al, 2000; Roff et al, 2003). However, by itself representation is not a sufficient strategy to capture all aspects of biodiversity. A systematic effort to conserve enduring examples of the full range of communities, habitats and ecological processes in a region is a precautionary approach. An ecosystem that includes a representative network of protected areas should be better able to withstand shock without fundamental change and without sacrificing the provision of ecosystem services.

Physical habitat mapping – which is based upon various types of information, including vegetation and substrate types, altitude, grade, rainfall, temperature and, in the marine environment, salinity and water depth, temperature and stratification – allows planners to design networks that include some minimum amount of each habitat type when data on the distribution of biological communities is lacking (Leslie et al, 2003; Roff et al, 2003; Soule and Terborgh, 1999). In principle, good habitat mapping should lead to representative networks that include all the types of areas that support the various ecological communities. This approach, which is sometimes called a 'coarse-filter' approach, has been used for both terrestrial and marine conservation.

Box 16.2 The overall plan and vision for The Great Barrier Reef World Heritage Area

The plan states that in 25 years there will be:

- a healthy environment: an area which maintains its diversity of species and habitats, and its ecological integrity and resilience, parts of which are in pristine condition;
- sustainable multiple use;
- maintenance and enhancement of values;
- integrated management;
- knowledge-based but cautious decision-making in the absence of information;
- an informed, involved, committed community.

To realise this vision, the plan identifies eight broad strategy areas:

- Conservation
- Resource management
- Education, communication, consultation and commitment
- Research and monitoring
- Integrated planning
- Recognition of Aboriginal and Torres Strait Islander interests
- Management processes
- Legislation.

For each of these broad areas, the plan provides the rationale, 25-year objective, five-year objectives and strategies to fulfill these objectives.

Great Barrier Reef Marine Park Authority, Representative Areas Programme: Key phases

(See: www.gbrmpa.gov.au/corp_site/management/representative_areas_program.)

Step 1 – The need for rezoning
Realization the existing plan was inadequate
Original zoning plans 1983–1988
Commencement of Representative Areas Programme (1998)

Step 2 – Research and planning
Collation of datasets (1998–1999)
Bioregion mapping (1999–2000)
New coastal zones added (1998–2001)
Development of operational principles by independent committee

Step 3 – First community participation phase
Public Notice issued (May 2002)
Over 10,000 public submissions
Zoning options identified

Step 4 – Developing the draft zoning plan
Draft zoning plan (late 2002–mid 2003)

Step 5 – Second community participation phase
21,000 public submissions (June–August 2003)

Step 6 – Further development of the plan
Revised zoning plan (November 2003)
Regulatory impact statement (November–December 2003)
Submission to Parliament (December 2003)

Step 7 – Implementing the plan
New zoning plan took effect on 1 July 2004
Operation of the zoning plan
Monitoring the zoning plan

Given the great diversity of species and their habitat requirements, the variables chosen to define habitats will be more suitable for some species than for others. Consequently, using habitat representation as a strategy for biodiversity conservation is more effective when integrated with information about the distributions of key species or biological communities (Hunter, 1991; Day and Roff, 2000; O'Connor, 2002; Meir et al, 2004; Stevens and Connolly, 2004). For example, a particular fish species might occupy only some portion of the 'habitat' that was otherwise suitable; a representative approach based only on habitat could therefore miss these areas. Additionally, even in a case where the set of variables used for habitat definition is perfectly matched to a given species, that species may fail to fully saturate all of the available habitat (O'Connor, 2002). Conservation planning that is based upon both representation of habitat and upon the distributions of life forms is less susceptible to these pitfalls, and thus more powerful than if based on only one or the other.

The value of a well-designed representative network for conservation planning at large scales has been discussed by a number of authors, and the representative approach has become fundamental in conservation theory and practice (Noss, 1987; Franklin, 1993; Pressey et al, 1993; Noss and Cooperrider, 1994; Maybury, 1999). Day and Roff (2000) set out a framework for designing networks of MPAs that included a case study for the Canadian portion of the northwest Atlantic Shelf region. They outlined the use of what they called 'enduring and recurring' environmental features for habitat classification as a basis for achieving representation. These features were non-living – or abiotic – and described sea water conditions and the seafloor.

Day and Roff (2000) also highlighted the need to consider distinctive areas in the development of marine conservation plans (see also Roff and Evans, 2002, and Chapter 7). A distinctive area is distinguished by the presence of one or more unique biological or physical attributes, such as a known spawning area for a fish, a known feeding ground for an endangered whale, a location where cold-water corals survive, or a rare habitat such as a seamount or a particular submarine canyon.

The integrated approach of CLF/WWF to planning for biodiversity conservation involved combining coarse-scale habitat representation with data depicting distinctive areas to define an ecologically diverse set of conservation features (Table 16. 1). Habitat representation was achieved on the basis of a suite of abiotic characteristics that were known to be fundamental components of habitats for a wide diversity of marine life, and for which adequate spatial data were available. Maps of marine habitat, referred to as seascapes (Day and Roff, 2000; Roff et al, 2003), were derived from physical parameters that describe the seawater (temperature, salinity, stratification, depth) and the seafloor (substrate type) – see Chapter 5.

Sets of MPAs were selected that met goals for both the representative and distinctive conservation features simultaneously (see Chapters 5 and 7 respectively). However, some areas were selected primarily because they were essential to meeting goals for particular biological conservation features, or because they were only required for achieving goals for representation of the seascapes. In defining seascapes, CLF/WWF recognized that the seafloor (benthic realm) and water column (pelagic realm) each display distinct seascape mosaics.

Biologically distinctive areas were identified as those high in relative abundance and/or species richness for a number of cetacean and fish species, and phytoplankton. For example, the distinctive areas for particular fish and whale species were defined as those areas where abundance was at or above the average for the species. Areas of high species richness for fishes were similarly selected, and locations where chlorophyll concentrations were persistently in the top 10 per cent were taken as distinctive for primary production (Table 16.1).

These features are expected to function as ecological indicators, serving as 'umbrella species' for other components of the ecosystem (see Primack, 2002, and Chapter 9). Geologically distinctive areas were not included, nor were localities known to be biologically distinctive but for which a systematic survey covering most of the region was not available (e.g. isolated deep sea coral areas). It was judged that such localized

Box 16.3 Operating principles for the CLF/WWF (2006) study

The following operating principles guided WWF/CLF in developing a method for identifying a 'network' of priority areas for conservation:

- Engage in conservation planning at an ecoregional scale. The ecologically unified analysis region covers the shelf waters ranging from Cape Cod, Massachusetts to Cape North, Nova Scotia, an area of some 277,388km^2 (=80,886nmi^2; =107,100mi^2).
- Recognize biogeographic areas. The analysis region includes three biogeographic areas – the Gulf of Maine (including Bay of Fundy), Georges Bank and the Scotian Shelf – which were distinguished based upon studies of biological communities and ecologically significant habitat features (e.g. water temperature, currents). These three areas were explicitly recognized as distinct biogeographic areas by setting area-specific conservation goals.
- Use the best-available spatial data for biological conservation features. Take into account an ecologically diverse set of biological conservation features while maintaining standards for region-wide sampling. Data were excluded that did not have sufficient spatial extent or resolution for assessing large-scale distributional patterns. Isolated areas known for their ecological significance (e.g. localities for hard corals) were not included as explicit conservation features, but it was recognized that it may be desirable to add such localities to future analyses.
- Use the best-available abiotic data for classification of marine habitat or seascapes. Data were selected for defining benthic and pelagic seascapes that are based on: (1) variables of demonstrated ecological significance; and (2) datasets that have adequate spatial resolution and extent for the analysis region.
- Use biological and abiotic data simultaneously to design a network that is representative of habitat and includes biologically distinctive areas.
- Design a network of a size that is sufficient for meeting objectives for biodiversity conservation and for playing a role in sustaining the region's ecosystems.

areas could be added in the future, during the further development of conservation plans.

Scale is a critical aspect of habitat, population viability and conservation planning. The issues around defining habitat on scales that are appropriate for particular species are important and complex (e.g. Warman et al, 2004), and they relate to the complicated question of how much habitat is needed to attain the broad goals of ecosystem conservation associated with a network of MPAs. The analyses undertaken by CLF/WWF were at a coarse scale, aimed at capturing a broad spectrum of habitat types and the associated biodiversity components, to the extent possible with currently available data. In the classification of habitats, the grain of analysis had a minimum resolution determined by the size of the seascape grid (5 geographic minutes on a side, or about 58km^2) and planning units (10 geographic minutes on a side or about 234km^2). These limitations on grain were determined by the data available at an ecoregional scale. The identification of habitat then refers to relatively large areas that are dominated by habitat conditions of the specified type, but these areas were not assumed to be exclusively of one habitat type throughout. At this coarse grain of analysis, individual squares will often contain a variety of conditions (in other words they are heterogeneous) that would be revealed if a finer-grained analysis were possible.

Habitat is most commonly defined by the characteristics of those places where a particular species lives, both abiotic and biotic. However, at the coarse scale of mapping for conservation purposes, we are generally left with correlations between species abundance and those attributes of the environment that scientists have the ability to measure (e.g. seafloor type, depth, salinity). This complex issue is examined in Chapter 6. Some of these attributes may not directly determine where a given species is found, and some of them that do directly influence distribution may

Box 16.4 How much is enough? The process to determine the total area to be protected within a region

There are at present perhaps three ways in which to decide how much of a region should be (i.e. needs to be) protected in order to achieve sustainable marine ecosystem preservation.

1. At the Fifth World Parks Congress in 2003 the recommendation was made to 'greatly increase the marine and coastal areas managed in marine protected areas by 2012; these networks should include strictly protected areas that amount to at least 20–30% of each habitat'. Most recently, at the 8th Ordinary Conference of the Parties to the Convention on Biological Diversity (CBD) in 2006, a target that 'at least 10% of each of the world's ecological regions [including marine and coastal] be effectively conserved [by 2010]' was adopted (CBD, 2006). These decisions recommending very different and variable proportions of total areas to be protected are quite arbitrary. They have not been subjected to any rigorous form of analysis. In some regions 10 per cent of a region may be sufficient to preserve all distinctive and selected representative areas; in other regions even 30 per cent may not be sufficient. Ultimately the number and spacing of MPAs will be determined from connectivity studies (see 3 below).
2. Based on models of fisheries conservation, the area within a region that needs to be protected in order to ensure sustainability of resources can be estimated. Such models (e.g. Guenette et al, 1998; Beattie et al, 2002) suggest that 10–50 per cent of a region, or more precisely 25–40 per cent of a region, should be protected from fishing activity (along with catch regulations) in order to ensure sustainability of fisheries. It has not yet been determined how such spatial estimates could be combined with conservation requirements for all other components of marine biodiversity.
3. Based on ecological analysis of patterns and efficiency of connectivity among candidate MPAs. This has only rarely been done and requires information on regional oceanographic current patterns, together with iterative modelling of connectivity among candidate MPAs and/or data on genetic relationships (see Chapter 17).

be missing. In order to include as many components of biodiversity as possible within their networks, CLF/WWF (2006) combined maps based on several key marine habitat parameters with maps of a diversity of life forms.

There is no simple answer yet to the question of how much is enough (Roff, 2009), in terms of the sizes of individual areas making up a network, the total number of areas or the overall spatial extent of a network (but see Box 16.4 and Chapters 14 and 17) . However, based on available data, CLF/WWF (2006) focused on a 'network' design that included approximately one-fifth of each of the three biogeographic areas (Table 16.1).

A set of MPAs of this scale is expected to represent much of the region's biodiversity, but may possibly contribute more to the less mobile (usually smaller) species than to more mobile species. However, even highly mobile species will benefit from the protection of areas where they spend part of their life cycle, for example through increased availability of smaller prey species or of areas where human disturbance is minimal. CLF/WWF (2006) sought to balance the benefits that emerge from achieving conservation objectives through a 'network' of distributed areas (Roberts et al, 2003) against the potential costs of including some areas that may be too small to support some of the mobile species.

The task of identifying a network of priority areas for conservation based upon many types of information is complex. A summary of the steps involved is presented in Box 16.5. The ecological planning process entails meeting a large number of goals efficiently yet keeping the overall area to a minimum. This was accomplished using MARXAN, a powerful computer-based site-selection program (Possingham et al, 2000).

Box 16.5 Summary of the ecological steps required for planning sets and networks of MPAs

There are three essential phases to defining a regional network of MPAs:

1 Data assembly and mapping
2 Define coherent sets of candidate MPAs
3 Select an efficient network of MPAs

Phase one: Data assembly and mapping of candidate areas

1.1 Assemble data on oceanography, physiography, biology
1.2 Define biogeographic regions (within ecoregions)
1.3 Define regional geomorphological units
1.4 Map distinctive areas (from geophysical anomalies, TEK and SEK); these will include areas for cetaceans, fish, birds etc.
1.5 Map existing protected areas (e.g. fishing closed areas)
1.6 Map representative areas from geophysical data and biological surveys.

Phase two: Define coherent sets of MPAs

2.1 Decide on overall targets of representation of distinctive and representative areas – in other words the proportion of the total region and of each category of area to be protected (see Box 16.4). Alternatively or in addition go to step 3.5.
2.2 Decide which distinctive and fisheries areas could become MPAs (MARXAN or other similar analyses of each type of area) and define possible alternatives.
2.3 Decide which representative areas could become MPAs (MARXAN or other similar analyses) and define possible alternatives and sizes and numbers required.
2.4 Conduct combined analyses of distinctive and representative areas to determine the candidate/alternative sets of MPAs that would meet overall targets (under 2.1). This should define possible coherent sets of MPAs.

Phase three: Select an efficient network of MPAs

3.1 Define the appropriate size and boundaries for each MPA (e.g. by applying criteria from Chapter 14) and check that candidates selected can meet criteria of size for stated purpose.
3.2 Obtain data on oceanographic flow patterns among MPAs and regional meteorological data.
3.3 Determine meroplanktonic/larval development times (at regional water temperatures) for macrobenthos and demersal fish species.
3.4 Define the patterns of recruitment and connectivity among candidate MPAs of a set, from in situ studies, regional oceanographic models, genetic studies etc. (see Chapter 17).
3.5 Iterate the models/studies until an efficient network can be defined (see Box 16.1 and Chapter 17).
3.6 Note that: smaller species will auto-recruit within each MPA; larger species will allo-recruit among MPAs; no species should lose all its recruits to areas beyond the network unless they recruit to another network.

Establishing biogeographic regions

Biogeographic regions were established according to recognized ecological boundaries. The work focused on the continental shelf waters defined by a boundary from Cape Cod through the Great South Channel to the seaward edge of Georges Bank, and extending northeast along the edge of the Scotian Shelf to the Laurentian Channel (Figure 16.1). The total area of the region is approximately 277,388km^2

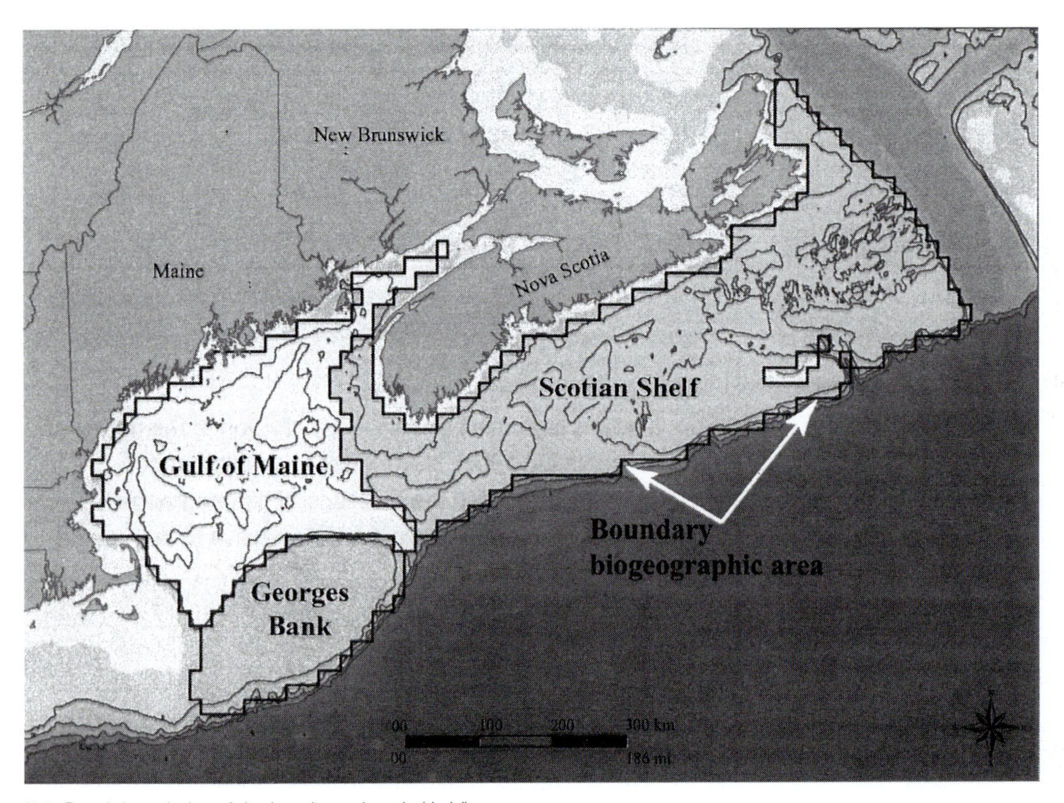

Note: Boundaries and edges of planning units are shown by black lines.

Source: Reproduced with permission, CLF/WWF (2006)

Figure 16.1a Showing the region of analysis of sets of MPAs and biogeographic regions

(= 80,886nmi^2 = 107,100mi^2). Separation of biogeographic areas was based on a synthesis of the literature on the region's biogeography, and the faunal regions recognized conform closely to those adopted by Cook and Auster (2005).

The recognition of biogeographic areas is essential to the goal of achieving the best possible representation of habitat and marine biodiversity because, by definition, biogeographic areas are characterized by distinct floral and faunal assemblages. Achieving conservation goals within each biogeographic area is expected to provide some insurance against possible failure to capture important variation in population structure and species distributions. Distributing protection amongst biogeographic areas, and among multiple areas within biogeographic areas, may also contribute to overall resilience and connectivity of the network, and provide additional insurance against localized disasters (Roberts et al, 2003).

The Great South Channel to the southwest of Nova Scotia provided a natural faunal break, as did the Laurentian Channel to the northeast between Cape Breton Island and Newfoundland (Figure 16.1). The faunal composition of this region can be distinguished from that of the warmer waters to the south and of the colder boreal waters to the north, though many species from within the region range beyond these boundaries. On the seaward side, the ecological region was demarcated by the shelf edge (200m isobath), beyond which deepwater communities characteristic of the slope and abyss predominate. The coastal and estuarine zones of the inshore were excluded by means of a shoreline buffer of 15km approximating the 30–50m

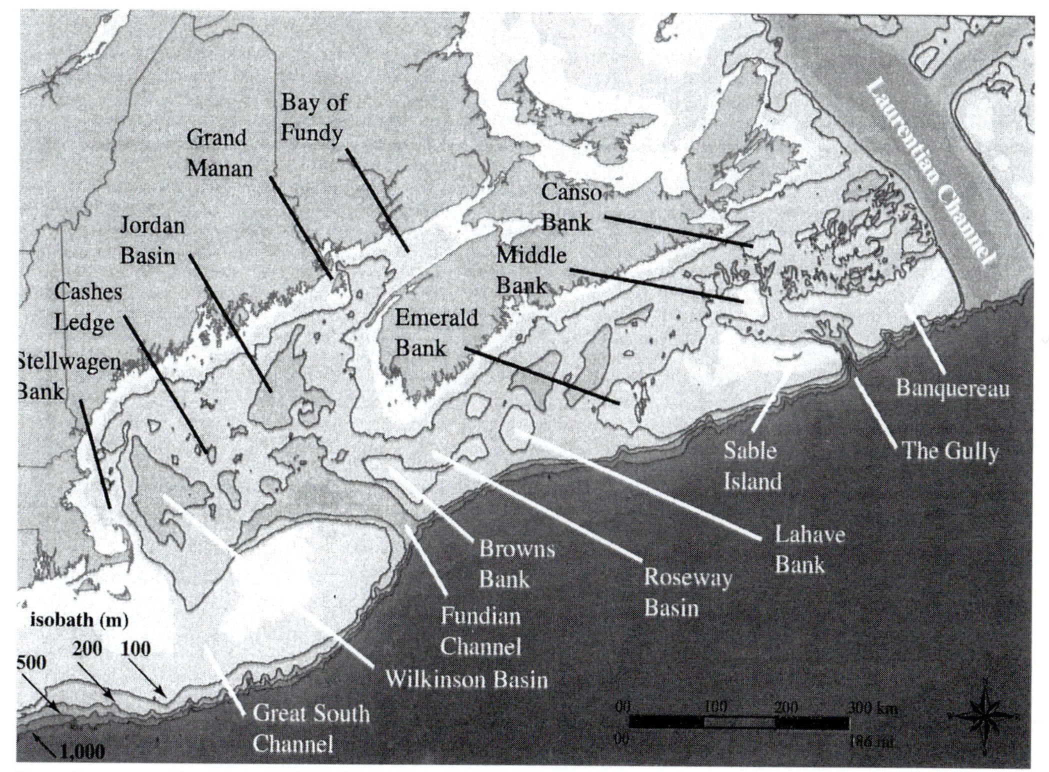

Source: Reproduced with permission, CLF/WWF (2006)

Figure 16.1b Showing the region of analysis for sets of MPAs with some of the prominent marine geomorphological features

isobath. The coastal zone is considered separately in Chapter 11.

In deriving a coherent set of priority areas for conservation, CLF/WWF recognized three biogeographic areas (or regions): Georges Bank, the Gulf of Maine (including the Bay of Fundy) and the Scotian Shelf (Figure 16.1).

- Georges Bank (42,343km^2). The Georges Bank region is ovoid in shape, one of the most productive fishing areas of the northwest Atlantic and characterized by currents that retain biological material including primary producers and larval fishes (Backus and Bourne, 1987).
- Gulf of Maine (87,156km^2). The Gulf of Maine region is an irregularly shaped area that includes the Bay of Fundy and the Northeast Channel, and which is bounded by the Scotian Shelf and Georges Bank regions.
- Scotian Shelf (147,889km^2). The Scotian Shelf region is a large elongate area extending from the waters off southwestern Nova Scotia northeast to the Laurentian Channel.

The differences in biota among these biogeographic areas are probably substantial, but elucidation of biological diversity of the region has only begun. Theroux and Wigley (1998) have examined the region's biogeography based on oceanography and a survey of the macrobenthic invertebrate fauna. Much of the current knowledge is based on inferences from studies of oceanographic features that are known to influence the distributions of biological communities, such as ocean currents, water temperature, salinity, stratification, depth and substrate types.

Research on demersal fishes indicates that several species have distinct subpopulations within these regions, and fisheries managers treat some species as consisting of distinct stocks within these areas (e.g. Collette and Klein-MacPhee, 2002). Analyses of demersal fish distributions (see below) also indicate differences between areas in species composition.

A decision support tool – MARXAN

CLF/WWF (2006) set a goal of including a representative proportion of each of the seascape classes (i.e. 20 per cent of each habitat type) and for distinctive areas. To ensure that goals would be met for conservation features within each of the biogeographic areas, a unique code for each feature within each area was assigned, and area-specific goals were set. The decision support tool used for overall planning was MARXAN. Examples and details of MARXAN procedures are provided in the manual (Ball and Possingham, 2000) and on several websites, e.g. www.ecology.uq.edu.au/marxan.htm and www.mosaic-conservation.org/cluz/marxan1.html.

Computer-based site selection is essential because large-scale, systematic marine conservation planning is a difficult task that demands many goals be met simultaneously and efficiently.

With computer-based site selection, data are used objectively by a program that follows fully specified rules, making the method of network design transparent.

MARXAN performs site selection based on a set of conservation features, given a quantitative goal specified for each. MARXAN is by no means the only quantitative modelling tool for conservation. However, it has the virtues of being ecologically based, able to accept a wide variety of types of data (both abiotic and biotic), understandable, relatively easy to learn and use – and freely available. It was first developed for use in planning the Great Barrier Reef Marine Park, and has become the 'industry standard'. Note that MARXAN is described as a 'decision support tool', not as a decision-making process. It also produces 'efficient' planning solutions not optimal ones (because of the enormous number of potential solutions).

MARXAN repeatedly searches through all of the information provided to it, seeking combinations of areas that attained the specified conservation goals in a spatially efficient manner. The program could achieve the goals in a variety of ways because most habitat types and marine life are found in a number of locations within any biogeographic region, and because each network of areas only needs to capture some portion of each. MARXAN allows the performance of each network to be evaluated with respect to the specified conservation goals and degree of spatial efficiency and thereby identify the 'best' performers. Because the method can generate several networks that all perform reasonably well, it has the added benefit of providing planners with a choice of viable networks. This can be essential in a public planning process.

A conservation planning problem often begins by partitioning the region into manageable geographic planning units. These may take on any size or shape, but squares of 10 geographic minutes (i.e. 10-minute squares, as used in this plan, are approximately 16km per side) have often been used for coarse-scale planning. Each planning unit is characterized in terms of a list of its conservation features. In the CLF/WWF work, these included habitat features such as depth and seafloor type, and biological features such as abundance of fish and whale species, as proxies for ecological community types. Conservation networks are developed by evaluating different combinations of planning units in terms of these conservation features and goals.

MARXAN simply takes the set of geographically referenced data it is given (maps), follows the instructions provided and uses explicit mathematical functions to find solutions (in other words networks) to a given planning problem. Using MARXAN, CLF/WWF produced several networks of priority areas, each consisting of multiple areas, and each area being composed of one or more planning units. These priority areas, taken together, meet all the planning goals whereas individual priority areas do not. The networks identified by MARXAN are evaluated based on a combination of how well

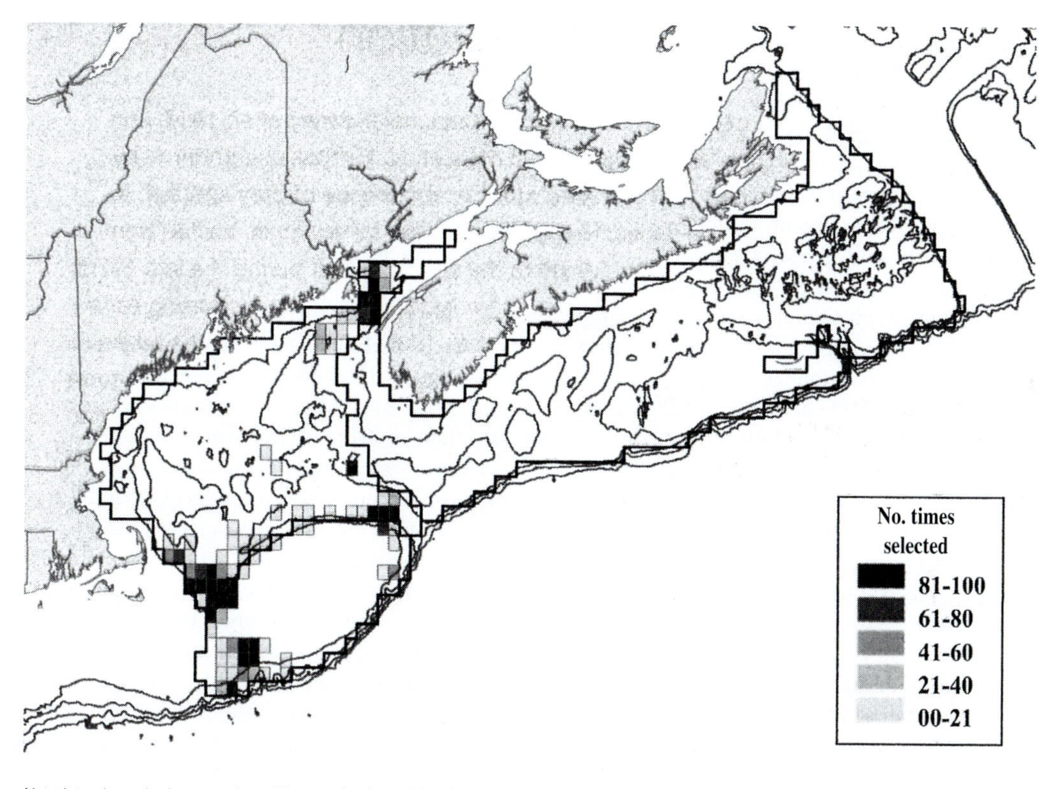

Note: Intensity scale shows number of times a planning unit is selected

Source: Reproduced with permission, CLF/WWF (2006)

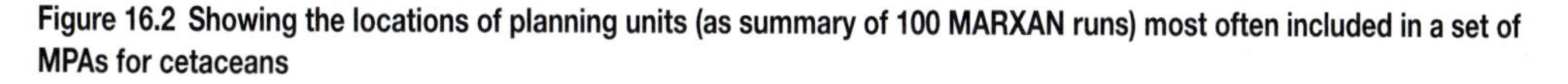

Figure 16.2 Showing the locations of planning units (as summary of 100 MARXAN runs) most often included in a set of MPAs for cetaceans

conservation goals are met and on other factors related to spatial efficiency. Other factors that MARXAN evaluates include: total area required to meet goals; boundary length (the total perimeter length of all areas); the number of priority areas within a region; irreplaceability (those areas that are unique and must be included). The 'best' result is often the one that meets conservation targets with the least number of planning units in the smallest total area and the lowest number of priority areas.

A single 'best' network emerges from the evaluation process, as well as other networks that do not perform quite as well. However, it should be realized that the 'best' network produced by MARXAN is not necessarily the optimum one. Often, a number of networks are very nearly as good as the best one and these serve as viable alternatives.

The network of priority areas for conservation identified by CLF/WWF was based on all of the different biological conservation features and seascapes. In addition, MARXAN was used to identify networks of areas based only on the individual component data layers (in other words whales, fishes and seascapes) in order to provide insight into these layers and how site selection proceeds based on these layers in isolation. These exploratory MARXAN analyses also provided an opportunity to compare the resulting networks with the network of priority areas for conservation based on all of the different types of data.

MARXAN is not expected to produce a single, true, optimum network in any given run.

Thus it performs many iterations of the site-selection process (100 times for each of the CLF/WWF analyses, in combination running into millions of runs), and uses the scores from the output to evaluate each network generated.

The examination of multiple MARXAN networks permits the identification of planning units that are repeatedly included, and others that may be selected for only a minority of the networks. This is most readily observed on a map where each planning unit is colour-intensity coded to show the number of times it was included in a set of 100 networks. Such maps are referred to as 'summed solution maps' (Stewart and Possingham, 2002) – see example in Figure 16.2.

Typically, a core subgroup of planning units is included in nearly all the networks. These are planning units whose characteristics are such that one or more goals cannot be attained without them. Because they are essential to arriving at a network solution they are sometimes referred to as 'irreplaceable' (Stewart and Possingham, 2002). Planning units with the highest irreplaceability are most likely to be needed as part of a network; conversely, the ability to achieve conservation goals will be most substantially decreased if these planning units are not available for protection. A planning unit may be irreplaceable because it contains a rare feature that is not found in other units. Planning units may also be (relatively) irreplaceable because they contain an unusually rich combination of features. In a complex MARXAN analysis, the degree to which goals are met across the many conservation features varies, with some goals being exceeded (or overshot) and others being attained just under the specified goal. The areal extent of a final MARXAN network, based on both seascapes and biotic data (CLF/WWF, 2006), was actually very close to the area-based seascape goal of 20 per cent.

The types of areas to be included and overall 'target' of proportions

Overall suggested targets for marine conservation networks have ranged from 10 per cent to over 30 per cent. In fact most such numbers are arbitrary, do not respect biogeographical differences and have rarely been defended on ecological grounds (see Roff, 2009, and Box 16.4). However, studies on models of sustainable fisheries suggest that from about 10 per cent to 50 per cent of a region, PLUS fisheries quotas, should lead to sustainability for some species (see e.g. Guenette et al, 1998). As we shall see in the next chapter there are alternative ways of achieving more defensible, less arbitrary and more precise overall spatial targets.

In the CLF/WWF study, goals were specified for each of the individual conservation features (see Table 16.1). For the seascapes, goals were specified as a simple proportion of each class of seascape within each of the biogeographic areas. Because the entirety of each biogeographic area was classified by seascapes, setting goals in this fashion in turn determined the minimum area required for a network. For example, with a goal of 20 per cent for seascapes, the network will include at least 20 per cent of the analysis region as a whole. For the biological conservation features, goals were set as a proportion of some descriptive metric, such as relative abundance, and the selection of planning units was restricted to a subset of those determined to be of high quality, as discussed further below.

Each of the contributing data sets is now briefly considered in turn. For further details refer to CLF/WWF (2006).

Areas of persistently high chlorophyll a

Primary production forms the base of the food chain in marine ecosystems. The areas of highest production support the highest overall biomass at higher trophic levels, although relations between production and species diversity are complex (see Chapter 8). Shallow areas associated with submarine banks and upwelling currents have high primary production and are widely recognized for their unusually abundant marine life (Thurman and Trujillo, 2002). The survival of larval fishes also depends on the timing and seasonal abundance of phytoplankton (Platt et al, 2003). Moreover, recent work suggests that there can be a direct, bottom-up control of fisheries production by phytoplankton (Ware and Thompson, 2005). Incorporation of some of the areas of highest

Table 16.1 Conservation feature classes and goals for the CLF/WWF (2006) study of priority areas for conservation

Conservation feature class	Description and data source	Goals
Primary production	Areas of persistently high chlorophyll concentration – SeaWIFS satellite images	20% of planning units exhibiting high concentrations
Demersal fishes		
Species richness	Number of species per trawl, average by planning unit – NMFS and DFO research surveys	20% of the richness contained in those planning units at or above the mean for the biogeographic area
Juvenile abundance	Number of individuals per trawl, average of log-normalized counts by planning unit – NMFS and DFO research surveys	20% of the relative abundance contained in those planning units at or above the mean for the biogeographic area, goals set by species
Adult abundance	Number of individuals per trawl, average of log-normalized counts by planning unit – NMFS and DFO research surveys	20% of the relative abundance contained in those planning units at or above the mean for the biogeographic area, goals set by species
Cetacean abundance	Number of sightings per 1000km of survey transect, average of log-normalized counts by planning unit – NARWC database	20% of the relative abundance contained in those planning units at or above the mean for the biogeographic area, goals set by species
Seascapes	Habitat types classified from abiotic data, benthic and water characteristics	20% of each seascape, goals set by seascape type

Key: NFMS = National Marine Fisheries Service, USA
DFO = Department of Fisheries and Oceans, Canada
NARWC = North Atlantic Right Whale Consortium

primary production was thus deemed essential by CLF/WWF (2006) to the design of an effective network of priority areas for conservation.

Beyond the immediate coastal zone, primary production depends upon the unicellular phytoplankton. Their light-capturing molecule chlorophyll can be detected from the colour of the seawater by remote optical sensing (Platt et al, 1995; Sathyendranath et al, 2001). Thus, satellite imagery has been used to estimate the *potential* primary production in different regions of the ocean.

The goal for CLF/WWF (2006) was to identify planning units that persistently exhibited unusually high chlorophyll concentrations, which were taken to be areas of anomalously high production. The data were derived from satellite images of the sea surface obtained from the Sea-Viewing Wide Field-of-View Sensor Project (NASA, 2006) and consisted of two-week composite images for a period of five-and-a-half years from September 1997 through to March 2003 (133 images). Measurement resolution was approximately 1.1km. Allowing for various correction procedures, all those pixels with chlorophyll concentrations in the top 10 per cent were identified – that is, for a given two-week period and a given biogeographic area for three or more of the five years – and were flagged as corresponding to persistently high locations (see Figure 16.3).

Nearshore and other relatively shallow areas such as Georges, Browns and St Ann's banks were commonly associated with high chlorophyll concentrations, although spatial patterns were seasonally dynamic. Chlorophyll concentrations were at a maximum during the spring and early autumn, and reached their minima during the late autumn–winter period and during midsummer.

Distinctive areas for demersal fish

The well-known groundfish (bottom-living fishes) or demersal fishes represent a major component of the marine ecosystems of the northwest Atlantic Shelf and have supported human populations along the region's coasts for thousands of years (Jackson et al, 2001). The demersal fishes are an important yet compromised component of regional biodiversity, and occupy a range of habitats and ecological niches. The diversity of this group

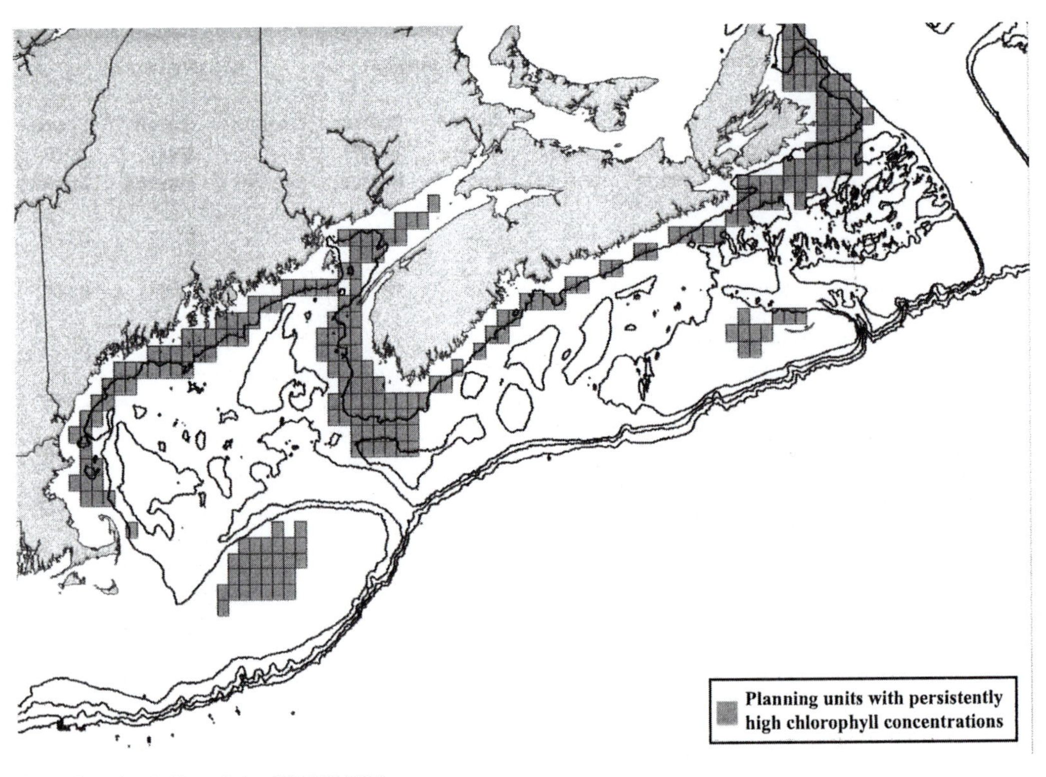

Source: Reproduced with permission, CLF/WWF (2006)

Figure 16.3 Areas of persistently high chlorophyll concentration in the region of analysis

of species makes them indicators for biodiversity and benthic communities as a whole.

During the past two centuries a substantial industrial economy has developed around demersal and other fishes. Unfortunately, several historically important fisheries have vanished or become compromised. The composition of present-day catches has changed markedly, reflected in significant changes in the relative abundances of species, shifts in population size structure and the trophic organization of ecosystems (e.g. Collette and Klein-MacPhee, 2002; Rosenberg et al, 2005). The removal of cod and other top predators has also produced ecosystem-altering trophic cascades, the consequences of which we are only beginning to understand (Frank et al, 2005).

The high commercial value of these fishes motivated the Canadian and United States fisheries services to carry out systematic trawl surveys over much of the past century. These surveys have yielded excellent quantitative data on the distributions of a large number of the region's fish species, including both commercially valuable and a great many others (see CLF/WWF, 2006 for species lists). CLF/WWF used these research trawl data to map two types of biologically distinctive areas for demersal fishes: *species richness* (average number of species per trawl, see Figure 16.4) and *relative abundance* (average number of individuals per trawl).

The site-selection process favoured places that were both areas of high relative abundance and high species richness because they contributed to both goals. Areas of high species richness have high conservation value because a relatively large number of species can be protected in a given location.

Relative abundance was mapped for juveniles and adults of each species separately because fishes have well-known differences in habitat utiliza-

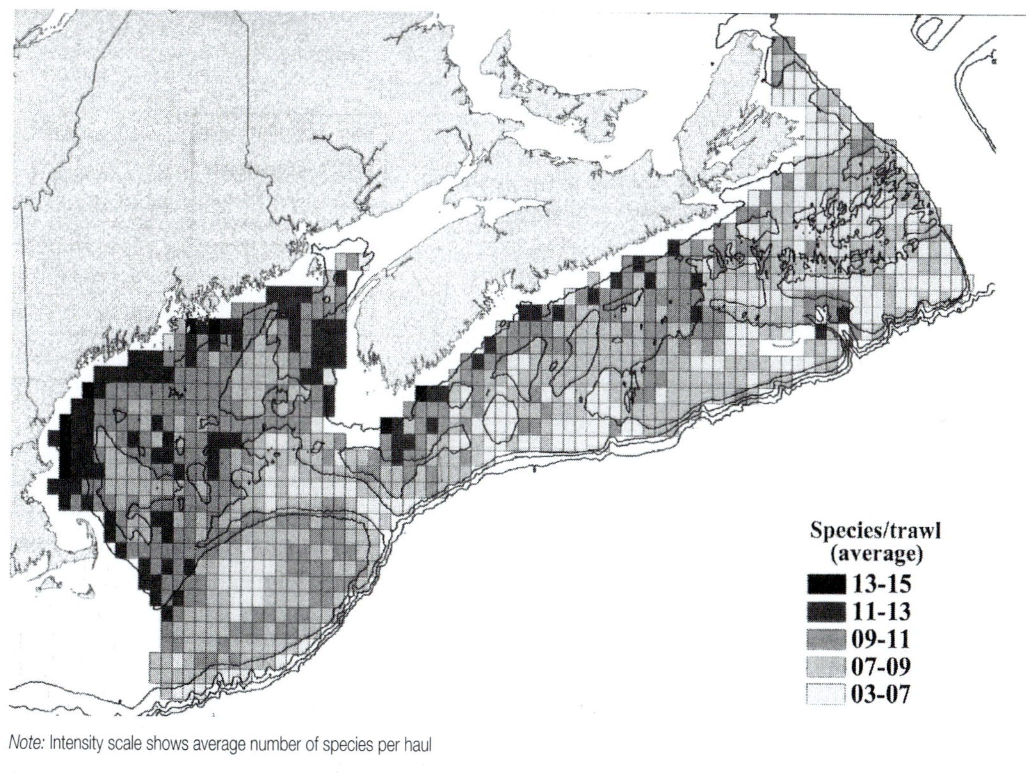

Note: Intensity scale shows average number of species per haul

Source: Reproduced with permission, CLF/WWF (2006)

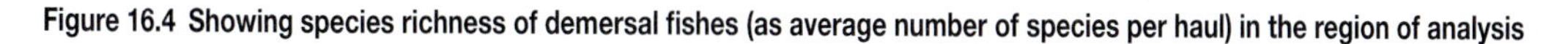

Figure 16.4 Showing species richness of demersal fishes (as average number of species per haul) in the region of analysis

tion between stages (e.g. Cook and Auster, 2005). For example, bottom features provided by gravel, rock and biogenic structures such as sponges, corals or plants define habitat areas that are particularly important for juveniles because they provide protection from predators and currents (Lindholm et al, 2001). Adults are influenced by other constraints such as food availability and availability of areas suited to spawning. By incorporating abundance layers for individual species CLF/WWF (2006) ensured that important areas for all species were included in site selection, including species that may not have been associated with areas of high species richness.

CLF/WWF used data collected during surveys conducted by the Canadian and United States governments between 1970 and 2002; Scotian Shelf data were collected by the Department of Fisheries and Oceans Canada (DFO), and the Gulf of Maine and Georges Bank data were collected by the US National Marine Fisheries Service (NMFS). Survey methods and references to methods are summarized in the CLF/WWF (2006) report. Briefly, surveys employed a random sampling design with stratification by depth and location. Standard, bottom-trawl gear was towed at 10.5km/h for a 30-minute trawl of 5.25km (3.0nmi). For each biogeographic area, only demersal fishes that were classified as residents according to published accounts (e.g. Scott and Scott, 1988; Auster, 2000; Collette and Klein-MacPhee, 2002) were included, thus avoiding selection of areas that might be marginal for non-resident species.

Areas with the highest average species richness formed a crescent-shaped band starting near the Great South Channel, roughly following the 100-mi isobath within the Gulf of Maine, and ending near the Bay of Fundy, off southeastern Nova Scotia (Figure 16.4). This band of richness continued in a less-pronounced fashion along the nearshore boundary of the Scotian Shelf biogeographic area. A second prominent peak in richness occurred in the centre of the Gulf of Maine near certain topographic ridges and banks. The deeper basins within the Gulf of Maine tended to display lower average species richness. The areas of highest average species richness appeared to be concentrated near ecological transition zones (ecotones) – for example, transitions between coastal and shelf regions or shelf and slope regions, or around distinctive features such as Stellwagen Bank and Cashes Ledge.

However, including only the areas with highest species richness does not ensure that all species are represented, nor does it ensure that those areas that support high abundances are included.

Thus, the use of abundance distributions for individual species is also important for achieving conservation objectives. Abundances of the fish species showed a variety of patterns. Although the overlap between areas of high abundance and high overall species richness was substantial, the use of these separate layers in analyses ensured that as many species as possible were captured in site selection. MARXAN was again used to identify combinations of areas that efficiently met the goals for all of the resident dermersal fish, both juveniles and adults (see CLF/WWF for figures).

Distinctive areas for whales and dolphins

As a group, whales and dolphins (see CLF/WWF, 2006 for species lists) have played a prominent role in the ecology of the northwest Atlantic. The total cetacean biomass for the Gulf of Maine and Georges Bank was estimated to be on the order of 200,000 tons during the period 1979 to 1982 (Kenney et al, 1997). These predators consume over one million tons of prey annually – as much as one-fifth of the total net primary production in the Gulf of Maine region – feeding on zooplankton, larger invertebrates such as squid, and a number of fishes. They have few predators and hold an apical position within marine ecosystems (Kenney et al, 1997).

These cetaceans are an important part of regional biodiversity. Their spatial distribution in the oceans correlates with components of the marine ecosystems for which we currently lack good data, including invertebrates and some of the smaller fishes. As such, cetaceans are valuable as habitat and biodiversity indicators, or umbrella species, and are clearly worthy targets of conservation efforts in their own right. Current populations of some species in the analysis region are dangerously low (Kraus et al, 2005), including the North Atlantic right whale *(Eubalaena glacialis glacialis)* and blue whale *(Balaenoptera musculus)*. A number of species are considered to be at risk by United States (United States Fish and Wildlife Service, 2006) and Canadian (Committee on the Status of Endangered Wildlife, 2006) agencies.

CLF/WWF (2006) mapped important habitats for cetacean populations based on data from the North Atlantic Right Whale Consortium (NARWC) database and the Cetacean and Turtle Assessment Programme (CETAP, 1982) which include data from numerous smaller scale intensive surveys. The CETAP dataset includes more than 10,000 sightings distributed throughout the shelf waters from Cape Hatteras to Nova Scotia.

The spatial pattern of observations is uneven but was corrected for sightings per unit of effort (SPUE) and for seasonal changes. However, no specific attempt was made to capture the well-known seasonal (Kenney et al, 1997) and longer-term cycles in the habitat-use patterns of cetaceans. The sightings were judged to be the best available data for estimating relative abundance and have been the basis for a number of important published studies (Kenney and Winn, 1986; Kenney et al, 1997). It was assumed that sightings correlated with actual abundance, but the correlation is not expected to be perfect. For example, differences in behaviour of a given species, such as feeding and migration, may influence sighting rates, which may also vary among species.

An overall picture of the usage of the region by cetacean species was gained through maps of species richness and a relative abundance summary for all the cetaceans combined. For the abundance summary, the relative abundance values for each species were first divided by the maximum value for a species (in other words corrected), thereby setting the maximum to unity. Next, the sum of these corrected values was determined for each planning unit across all species and mapped, which provided a summary view of relative abundance. This view was quite similar to that provided by cetaceans species richness. Both of these statistics revealed distinctive areas around the Great South Channel, Georges and Stellwagen banks, Jeffreys Ledge, the outer Bay of Fundy and Roseway Basin.

MARXAN was again used to identify sets of areas that efficiently met goals based on the abundance distributions for all cetacean species. For each species, those planning units for which the SPUE was equal to or greater than the mean for the biogeographic area were identified as important habitat areas, and MARXAN selected from among these planning units. Goals for each species were set as a proportion (20 per cent) of the sum of the relative abundance values among those planning units that were at or above the mean for each biogeographic area.

The best network based on conservation features for the cetacea (Figure 16.2) included areas in the outer Bay of Fundy, around Georges Bank and other smaller areas. The Great South Channel and Outer Bay of Fundy areas correspond directly to areas with high species counts and corrected abundance and areas of high cetacean biomass (Kenney and Winn, 1986).

Seascapes and geomorphological features: Representation and replicates

The classification system and mapping for representation in CLF/WWF (2006) grew out of an approach advanced by Day and Roff (2000) that is based on the axiom that physical habitat types can be used to partially predict distributions of marine life; that is, they can act as surrogates for the biota (see Chapter 6). This is necessary because the data available for spatial distributions of marine biota are both sparse and variable. There are at least two ways of classifying geophysical characteristics to 'capture' habitat representation within biogeographic areas – by mapping geomorphological units or by constructing seascapes (see Chapter 5 and Table 5.1). The two approaches can be complementary, and combined in MARXAN analyses.

In the seascape approach used by CLF/WWF, physical habitat types were characterized based on a suite of relatively enduring and recurrent characteristics (see examples in Table 5.5) that are known to influence the distribution of species and biological communities (see Chapter 6). These included oceanographic and physiographic factors, including composition of the seafloor and depth.

Using an approach based on physical habitat types defined by enduring and recurrent abiotic characteristics is advantageous in the region for two reasons. First, the use of these characteristics makes the classification relatively stable (or naturally adaptable) through time. Second, the approach can be implemented using physical datasets for which there was relatively good coverage throughout the region. The seascapes developed by CLF/WWF represent the first effort to provide region-wide habitat maps for the shelf waters of the greater Gulf of Maine and Scotian Shelf. Examples of the recurrent and enduring factors used to produce seascape maps for the Scotian Shelf are given in Chapter 5.

The seascape classification system characterizes physical habitats at each geographic location within the region, and it distinguishes between the pelagic and benthic realms. The distributions of demersal and benthic communities are most strongly shaped by the characteristics of the seafloor, while the distribution of pelagic communities is more heavily influenced by the physical parameters of the water column (Cox and Moore, 2000). Nevertheless, the interactions between these realms are important (e.g. Wahle et al, 2006).

In this classification system, each pelagic and benthic seascape is defined by a unique combination of characteristics: surface water temperature–salinity zone, depth class and degree of stratification within the pelagic realm; and bottom temperature–salinity zone, depth class and substrate type in the benthic realm. Zones of similar water temperature and salinity are in some respects analogues of major climatic regions in terrestrial environments, and at broad scales they correlate well with the differences in biological community types (McGowan, 1985; Day and Roff, 2000; Breeze et al, 2002). Pronounced differences in the temperature and salinity of ocean water can occur at a single geographic location where the water is vertically stratified; this is another reason why the benthic and pelagic realms are considered separately.

A number of studies have demonstrated the association between seawater temperature and salinity characteristics, or water masses, on biogeography (see Chapter 6). However, it is challenging to produce a static classification scheme for seascape mapping because water masses are dynamic in space and time. Thus in classifying seascapes, CLF/WWF defined zones that experienced similar ranges of temperature and salinity conditions over the course of the full year.

The range of values of each characteristic was split into ecologically meaningful classes appropriate for the analysis region, as defined through a review of the literature and an analysis of the data. These values were mapped, which created a separate layer for each characteristic. Finally, these layers were combined to create seascape maps for the benthic and pelagic realms. The seascapes were created by overlaying the maps of each characteristic (Day and Roff, 2000) on a grid of five-minute squares. Note that this five-minute square grid is finer than the ten-minute square planning unit grid used for mapping the biological conservation features and for identifying the network of priority areas for conservation.

Zones of similar temperature and salinity were defined based on data provided in the *Hydrographic Atlas for the Eastern Continental Shelf of North America* (NOAA, 2005). This atlas consists of monthly average values in ten-minute squares for the surface and bottom, and for various intermediate depths, and spans more than 30 years of data from several Canadian and United States sources. A multivariate cluster analysis (see e.g. Hargrove and Hoffman, 2004) was employed to identify geographic zones in both the benthic and pelagic realms that followed similar seasonal regimes of temperature and salinity.

The stratification parameter – based on depth, tidal current velocity and drag coefficient – is valuable for predicting the locations of fronts and for delineating spawning areas of fish (Iles and Sinclair, 1982) and pelagic communities (Pingree, 1978; Day and Roff, 2000). It was calculated here based on the difference between seawater density at the surface and seawater density at a depth of 100m as a proxy for the stratification parameter itself.

A bathymetric dataset was compiled from a number of sources from the Bedford Institute of Oceanography, Canada. The substrate characteristics of the seafloor were combined from several disparate data sources, including those of Fader et al (1977), Poppe and Polloni (2000) and Poppe et al (2003). Reconciliation of differences in descriptions and spatial resolution was undertaken in consultation with expert marine geologists. This resulted in a generalized classification scheme from combined datasets. The scheme involved an amalgamation of different substrate classes into five broad categories: (1) clays and silt; (2) muddy sands; (3) sand; (4) gravel and till; and (5) bedrock.

MARXAN was again used to identify sets of areas that were representative of each of the seascape classes resulting from overlay of the individual geophysical characteristics. Conservation planning was done at the level of the biogeographic areas, thus a seascape class that occurred for example in Georges Bank was treated as distinct from identical classes that occurred in one of the other areas. The goal for each seascape class was again set to 20 per cent.

CLF/WWF (2006) first developed a representative 'network' for the benthic seascapes alone, then one for the pelagic, and finally a network that is representative of all of the classes for both benthic and pelagic (Figure 16.5).

The best representative 'network' for benthic seascapes consisted of 29 areas distributed throughout the analysis region. The selected areas covered a total of approximately 56,091km^2 (= 16,356nmi^2; = 21,657mi^2), or about 20 per cent of each of the biogeographic areas. In contrast, the representative 'network' for pelagic seascapes consisted of only 13 areas, spanning a combined area of 53,744km^2 (15,672nmi^2; 20,751mi^2), or again about 20 per cent of each of the biogeographic areas.

The analysis based on both benthic and pelagic seascapes resulted in a final 'network' that was fully representative of marine habitats as defined by the initial CLF/WWF (2006) goals. The 'network' consisted of 31 areas, covering an area of 57,414km^2 (= 16,742nmi^2; = 22,167mi^2), or about 20 per cent of each of the biogeographic areas (Figure 16.5).

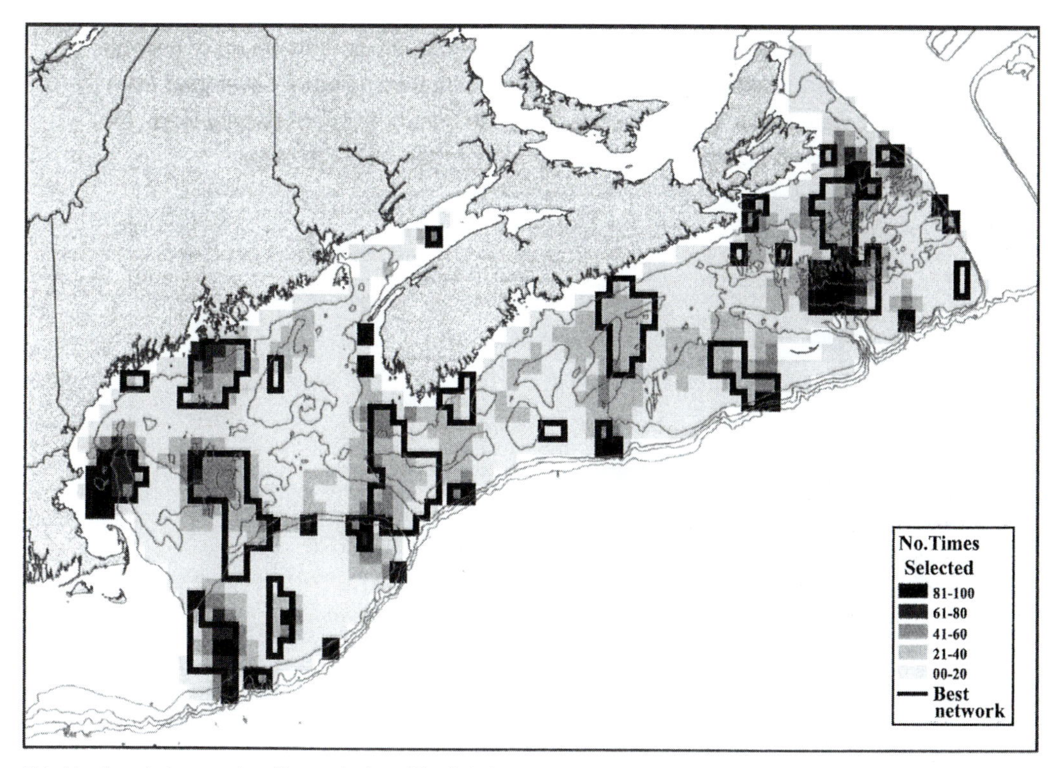

Note: Intensity scale shows number of times a planning unit is selected.

Source: Reproduced with permission, CLF/WWF (2006)

Figure 16.5 A representative set of MPAs for benthic and pelagic seascapes combined, showing best 'network' with summed solution

A coherent set of MPAs – priority areas

Following analysis of each of the data sets individually, the combined data sets were next analysed collectively using MARXAN. The 'best' result of these combined analyses (see Figure 16.6) was selected as the set (referred to in the CLF/WWF report as a 'network') of priority conservation areas that in combination would well protect the selected components of marine biodiversity in the region by achieving goals for all conservation features simultaneously. Thus, the network of priority areas for conservation developed by CLF/WWF (based on 100 MARXAN runs) includes distinctive areas for marine life and is representative of the range of physical habitat types.

The network consisted of 30 priority areas based on 237 of the 1057 planning units, representing approximately 22 per cent of the whole area, or 62,449km^2 (= 18,210nmi^2; = 24,112mi^2). The constituent priority areas ranged from small areas of one or two planning units to larger multi-unit areas – with the largest consisting of 46 planning units – extending over 12,279km^2 (= 3581 nmi^2; = 4741mi^2) and straddling all three biogeographic areas. The network included a wealth of ecologically diverse and productive areas which are fully described in the CLF/WWF (2006) report. Georges Bank is pre-eminent among the areas, with unusual primary production, a history of important fisheries resources and a diversity of whales and other marine life.

In general, larger priority areas for conservation contributed proportionately more to meeting goals than did smaller ones, resulting in a clear

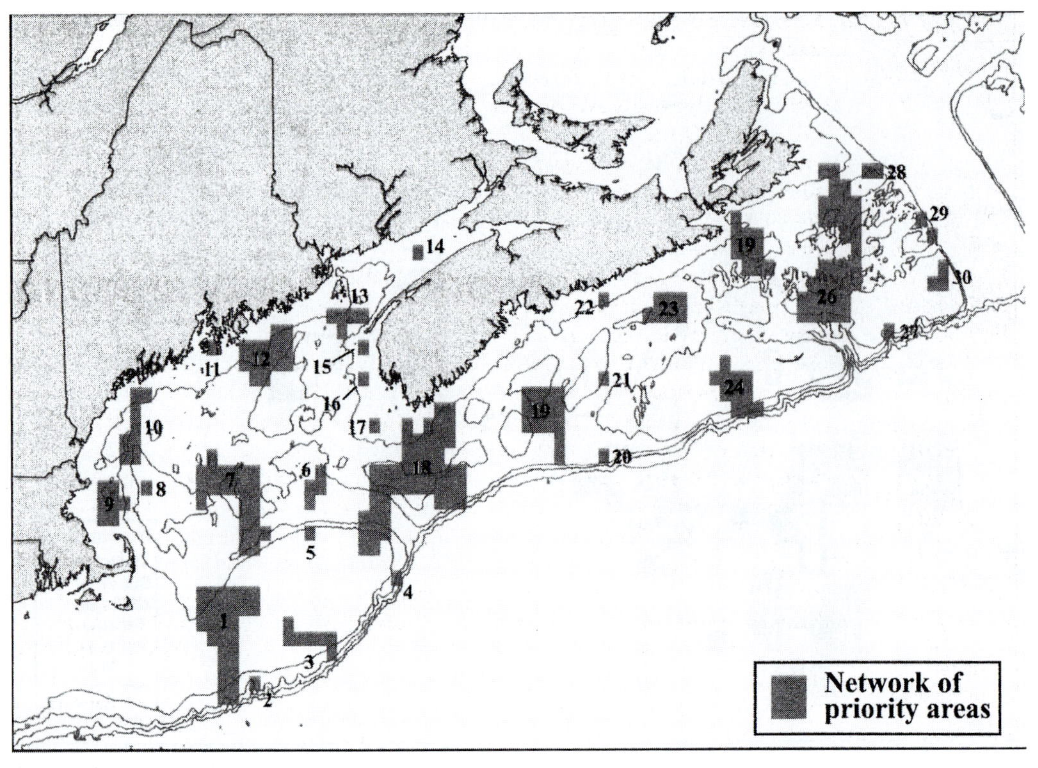

Note: The 30 priority areas are described in detail in CLF/WWF (2006)

Source: Reproduced with permission, CLF/WWF (2006)

Figure 16.6 A set of priority areas for conservation in the greater Gulf of Maine and Scotian Shelf region

linear correlation between the size of a priority area and the number of individual conservation features to which a priority area contributed. This was true for both the seascapes and for the cetaceans and demersal fishes. Additionally, large priority areas tended to contribute more towards the goals for particular conservation features than did smaller priority areas.

Goal attainment within the network

The CLF/WWF network of priority areas for conservation performed well in the sense that the inclusion of the various conservation features was generally close to the goal – that is, it was most commonly between the actual goal and 1.5 times the desired goal. Of the very few under-represented biological features, all were within 10 per cent of the goal. For juvenile and adult fish relative abundance, only 1 per cent of the conservation features were slightly under the goal in terms of representation; and for all fish species, richness, cetacean and primary production goals were met or exceeded. Other conservation features were sometimes over-represented as a consequence of the need to attain goals for other features. Over-representation also occurred in cases where a large proportion of a goal fell in a restricted area (e.g. within one or a few planning units). This tendency to over-represent uncommon conservation features was an unintended consequence of the method. To the extent that these correspond to ecologically distinct but uncommon areas, this over-representation may be beneficial to the performance of the network in terms of connectivity (see Chapter 17) and redundancy.

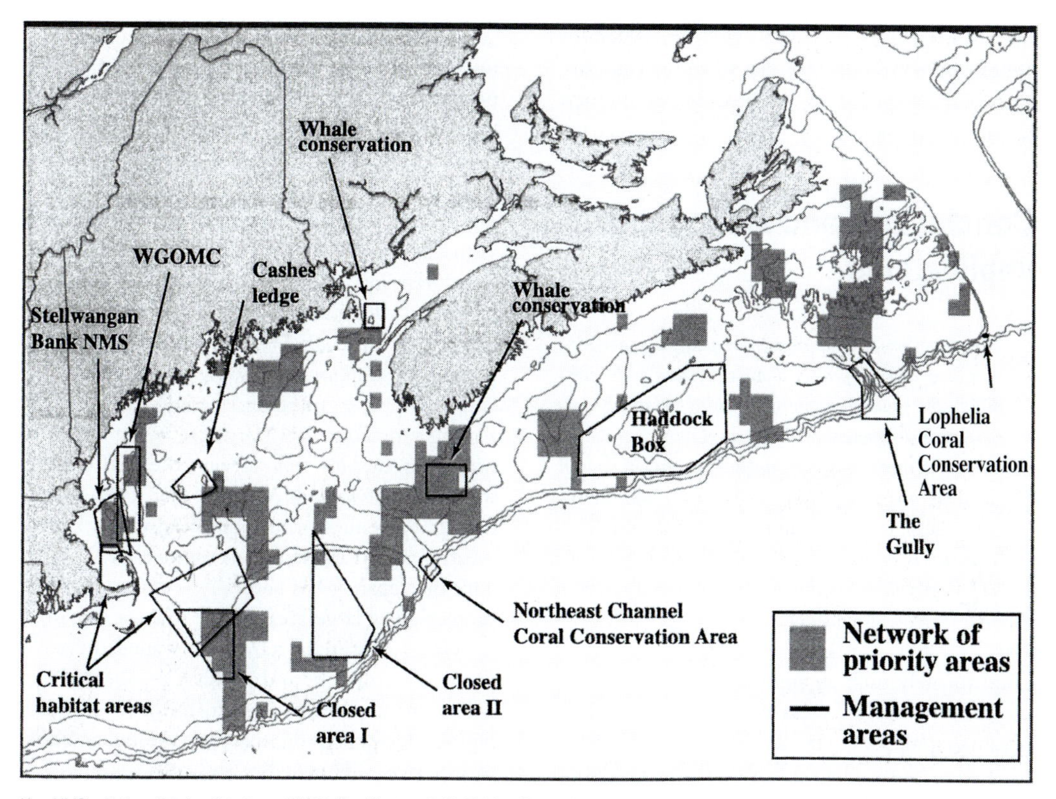

Key: NMS = National Marine Sanctuary; WGOMC = Western Gulf of Maine Closed Area

Source: Reproduced with permission, CLF/WWF (2006)

Figure 16.7 Showing locations of some existing management areas relative to the set of priority areas for conservation from Figure 16.6

Comparison of network of priority areas with known significant areas

The network of priority areas for conservation included areas that are well known for reasons ranging from historical importance for fisheries to their current importance for whale watching. The network includes areas overlapping with, or adjacent to, some existing marine management areas, and also includes areas in and around designated whale conservation areas at Grand Manan and Roseway Basin, and critical habitat areas at Cape Cod and the Great South Channel. The congruence of some of the priority areas for conservation with areas of previously recognized biological significance provides some added confidence in the approach to selection since the methods used by CLF/WWF were quite different from those contributing to local historical knowledge or those used by resource managers to date (Figure 16.7). However, the differences between the CLF/WWF (2006) proposed network of MPAs and the widely dispersed existing management areas point to the latter as individually established and not according to any integrated regional plan.

Potential connectivity among priority areas for conservation

The movement of organisms among suitable habitat areas is a critical consideration in the design of conservation networks in both terrestrial and marine settings (Roberts et al, 2003).

Although the report did not explicitly include information about connectivity in the design of the priority areas, the network is expected to benefit from connectivity due to the ocean currents of this region. This will be considered much further in Chapter 17.

Conclusions and management implications

> *The time-honored approach for planning a representative marine protected areas system is to subdivide the marine environment into 'relatively homogeneous' geographic units displaying similarity among a number of oceanographic and biological elements, and to represent each unit or 'marine region' by at least one marine protected area. This approach is derived from stratified sampling theory and biogeography, and is in use by several jurisdictions...* Mondor (1997)

Unfortunately, such an approach leads to a series of isolated protected areas that are not linked or oceanographically connected – the current state in most ecoregions. The CLF/WWF report (2006) is an example of the kind of comprehensive regional planning that is required for effective and integrated marine conservation. Biogeographic classification forms a basis for the application of the representative areas approach. This, combined with analysis of distinctive areas, can lead to the identification of coherent sets of MPAs.

The authors of the report acknowledge that the data for large-scale marine conservation planning are not perfect; much remains unknown, but this will always be the case. Nevertheless, the data used by CLF/WWF have proven sufficient because the resulting network identified a number of priority areas that coincided with places that were already well known for their ecological significance, in some cases dating back to the earliest historical accounts of the region. The site-selection process was also strengthened through the integration of a relatively large number of data layers.

Limited knowledge leads to the need to deal with uncertainty in management of ocean resources. The management and protection of a wide, representative range of biodiversity and ecosystem processes is one way to deal with this uncertainty. This approach will ensure that important but poorly understood ecological processes, or poorly studied areas, are protected.

In addition to ecological planning there are political and socio-economic realities in defining not only member MPAs, but also in dealing with the timing of their implementation. The Great Barrier Reef Marine Park (GBRMP) is presently a foremost example of such planning. Success of the GBRMP was based on extensive public consultation and analyses of overlapping and competing interests, leading to complex zoning regulations. However, it is important to note that despite this success the GBRMP is still not a network of MPAs, rather it is a coherent set. Revenues from tourism in the GBRMP greatly exceed those from commercial fisheries. This discrepancy has naturally lent favour to the protection of large areas of the reef for tourism.

Unfortunately, in temperate waters, the prospects for tourism and public engagement in MPAs are much less, even in immediate coastal regions. However, the financial benefits of MPAs to fisheries, as opposed to the collapse of fisheries in unprotected regions, are becoming evident.

To this point we have dealt only with the ecological process in defining sets of candidate MPAs that could contribute to an efficient network of MPAs. One important reason for defining alternate sets of MPAs from MARXAN runs is so that each can be tested for the best network in terms of connectivity and recruitment processes. The following chapter deals explicitly with the process of defining and planning networks.

References

Auster, P. (2000) 'Representation of Biological Diversity of the Gulf of Maine Region at Stellwagen Bank National Marine Sanctuary (northwest Atlantic): Patterns of fish diversity and assemblage composition', in S. Bondrup-Nielsen, N. W. P. Munro, G. Nelson, J. H. M. Willison, T. B. Herman and P. Eagles (eds) *Fourth International Conference on Science and Management of Protected Areas*, Science and Management of Protected Areas Association, Acadia University, Wolfville, Nova Scotia, Canada

Backus, R. H. and Bourne, D.A. (eds) (1987) *Georges Bank*, MIT Press, Cambridge, MA

Ball, I. and Possingham, H. (2000) 'Marine reserve design using spatially explicit annealing', www.uq.edu.au/marxan/docs/marxan_manual_1_8_2.pdf, accessed 9 November 2009

Beattie, A., Sumaila, U. R., Christensen, V. and Pauly, D. (2002) 'A model for the bioeconomic evaluation of marine protected area size and placement in the North Sea', *Natural Resources Modelling*, vol 15, pp413–437

Breeze, H., Fenton, D. G., Rutherford, R. J. and Silva, M. A. (2002) 'The Scotian Shelf: An ecological overview for ocean planning', *Canadian Technical Report on Fisheries and Aquatic Sciences*, vol 2393, pp1–259

CBD (2006) 'Decisions adopted by the conference of the parties to the Convention on Biological Diversity at its eighth meeting (Decision VIII/15, Annex IV)', IUCN, Gland, Switzerland

CETAP (1982) 'A characterization of marine mammals and turtles in the mid- and north-Atlantic areas of the U.S. outer continental shelf', Cetacean and Turtle Assessment Program, Bureau of Land Management, Washington, DC

CLF/WWF (2006) 'Marine ecosystem conservation for New England and maritime Canada: A science-based approach to identifying priority areas for conservation', Conservation Law Foundation and World Wildlife Fund Canada, Toronto, Canada

Collette, B. B. and Klein-MacPhee, G. (2002) *Bigelow and Schroeder's Fishes of the Gulf of Maine*, Smithsonian Press, Washington, DC

Cook, R. R. and Auster, P. J. (2005) 'Use of simulated annealing for identifying essential fish habitat in a multispecies context', *Conservation Biology*, vol 19, pp876–886

Committee on the Status of Endangered Wildlife Canada (2006) 'About COSEWIC', www.cosewic.gc.ca/eng/sct6/index_e.cfm, accessed 10 December 2009

Cox, C. B. and Moore, P. D. (2000) *Biogeography: An Ecological and Evolutionary Approach*, Oxford, Blackwell Science

Day, J. and Roff, J. C. (2000) *Planning for Representative Marine Protected Areas: A Framework for Canada's Oceans*, World Wildlife Fund Canada, Toronto, Canada

Fader, G. B., King, L. H. and MacLean, B. (1977) *Surficial Geology of the Eastern Gulf of Maine and Bay of Fundy*, Fisheries and Oceans Canada, Ottawa, Canada

Frank, K. T., Petrie, B., Choi, J. S. and Leggett, W. C. (2005) 'Trophic cascades in a formerly cod-dominated ecosystem', *Science*, vol 308, pp1621–1623

Franklin, J. F. (1993) 'Preserving biodiversity: Species, ecosystems or landscapes?', *Ecological Applications*, vol 3, pp202–205

Groves, C., Valutis, L., Vosick, D., Neely, B., Wheaton, K., Touval, J. and Runnels, B. (2000) *Designing a Geography of Hope: A Practitioner's Handbook for Ecoregional Conservation Planning*, The Nature Conservancy, Arlington, VA

Guenette, S., Lauck, T. and Clark, C. (1998) 'Marine reserves: From Beverton and Holt to the present', *Reviews in Fish Biology and Fisheries*, vol 8, pp251–272

Hargrove, W.W. and Hoffman, F.M. (2004) 'Potential of multivariate quantitative methods for delineation and visualization of ecoregions', *Environmental Management*, vol 34, pp39-60

Hunter, M. L. (1991) 'Coping with Ignorance: The coarse-filter strategy for maintaining biodiversity', in L. A. Kohn (ed) *Balancing on the Brink of Extinction*, Island Press, Washington, DC

Iles, T. D. and Sinclair, M. (1982) 'Atlantic herring: Stock discreteness and abundance', *Science*, vol 215, pp627–633

IUCN (2007) 'Establishing networks of marine protected areas: Making it happen – a guide for developing national and regional capacity for building MPA networks', cmsdata.iucn.org/downloads/nsmail.pdf, accessed 15 January 2010

Jackson, J. B. C., Kirby, M. X., Berger, W. H., Bjorndal, K. A., Botsford, L. W., Bourque, B. J., Bradbury, R. H., Cooke, R., Erlandson, J. and Estes, J. A. (2001) 'Historical overfishing and the recent collapse of coastal ecosystems', *Science*, vol 293, pp629–638

Kenney, R. D. and Winn, H. E. (1986) 'Cetacean high-use habitats of the northeast United States continental shelf', *Fisheries Bulletin*, vol 84, pp345–357

Kenney, R. D., Scott, G. P., Thompson, T. J. and Winn, H. E. (1997) 'Estimates of prey consumption and trophic impacts of cetaceans in the USA northeast continental shelf ecosystem', *Journal of Northwest Atlantic Fishery Science*, vol 22, pp155–171

Kraus, S. D., Brown, M. W., Caswell, H., Clark, C. W., Fujiwara, M., Hamilton, P. K., Kenney, R. D., Knowlton, A. R., Landry, S., Mayo, C. A. (2005) 'North Atlantic right whales in crisis', *Science*, vol 309, pp561–562

Leslie, H., Ruckelshaus, M., Ball, I. R., Andelman, S. and Possingham, H. P. (2003) 'Using sitting algorithms in the design of marine reserve networks', *Ecological Applications*, vol 13, pp185–198

Lindholm, J. B., Auster, P. J., Ruth, M. and Kaufman, L. (2001) 'Modeling the effects of fishing and implications for the design of marine protected areas: Juvenile fish responses to variations in seafloor habitat', *Conservation Biology*, vol 15, pp424–437

Maybury, K. P. (1999) *Seeing the Forest and the Trees: Ecological Classification for Conservation*, The Nature Conservancy, Arlington, VA

McGowan, J. (1985) 'The Biogeography of Pelagic Ecosystems', in S. van der Spoel and A. Pierrot-Bults (eds) *Pelagic Biogeography*, UNESCO Technical Papers in Marine Science 49

Meir, E., Andelman, S. and Possingham, H. P. (2004) 'Does conservation planning matter in a dynamic and uncertain world?', *Ecology Letters*, vol 7, pp615–622

Mondor, C. A. (1997) 'Alternative Reserve Designs for Marine Protected Area Systems', in M. P. Crosby, K. Greenen, C. Mondor and G. O'Sullivan (eds) *Proceedings of the Second International Symposium and Workshop on Marine and Coastal Protected Areas: Integrating Science and Management*, National Oceanic and Atmospheric Administration, Silver Spring, MD

NASA (2006) 'SeaWiFS Project Information', http://oceancolor.gsfc.nasa.gov/SeaWiFS/BACKGROUND/, accessed 9 November 2009

NOAA (2005) 'Hydrographic Atlas for the eastern continental shelf of North America', www.dynalysis.com/Projects/projects.html, accessed 9 November 2009

Noss, R. F. (1983) 'A regional landscape approach to maintain diversity', *Bioscience*, vol 33, pp700–706

Noss, R. F. (1987) 'From plant communities to landscapes in conservation inventories: A look at the Nature Conservancy (USA)', *Biological Conservation*, vol 41, pp11–37

Noss, R. F. and Cooperrider, A. Y. (1994) *Saving Nature's Legacy: Protecting and Restoring Biodiversity*, Island Press, Washington, DC

O'Connor, R. J. (2002) 'GAP conservation and science goals: Rethinking the underlying biology', *GAP Analysis Bulletin*, vol 11, pp2–6

Pingree, R. D. (1978) 'Mixing and Stabilization of Phytoplankton Distributions on the Northwest European Shelf', in J. H. Steele (ed) *Spatial Patterns in Plankton Communities*, Plenum Press, New York

Platt, T. C., Fuentes-Yaco, C. and Frank, K. T. (2003) 'Spring algal bloom and larval fish survival', *Nature*, vol 423, pp398–399

Platt, T., Sathyendranath, S. and Longhurst, A. (1995) 'Remote-sensing of primary production in the ocean – promise and fulfillment', *Philosophical Transactions of the Royal Society B*, vol 348, pp191–201

Poppe, L. J. and Polloni, C. F. (2000) 'USGS east-coast sediment analysis: Procedures, database, and georeferenced displays', http://pubs.usgs.gov/of/2000/of00-358/, accessed 10 November 2009

Poppe, L. J., Paskevich, V. F., Williams, S. J., Hastings, M., Kelly, J. T., Belknap, D. F., Ward, L. G., FitzGerald, D. M. and Larsen, P. F. (2003) *Surficial Sediment Data from the Gulf of Maine, Georges Bank, and Vicinity: A GIS Compilation*, Woods Hole Field Center, Woods Hole, MA

Possingham, H., Ball, I. and Andelman, S. (2000) 'Mathematical Methods for Identifying Representative Reserve Networks', in S. Ferson and M. Burgman (eds) *Quantitative Methods for Conservation Biology*, Springer-Verlag, New York

Pressey, R. L., Humphries, C. J., Margules, C. R., Vane-Wright, R. I. and Williams, P. H. (1993) 'Beyond opportunism: Key principles for systematic reserve selection', *Trends in Ecology and Evolution*, vol 8, pp124–128

Primack, R. B. (2002) *Essentials of Conservation Biology*, Sinauer Associates, Sunderland, MA

Roberts, C. M., Branch, G., Bustamante, R. H., Castilla, J. C., Dugan, J., Halpern, B. S., Lafferty, K. D., Leslie, H., Lubchenco, J., McArdle, D., Ruckelshaus, M. and Warner, R. R. (2003) 'Application of ecological criteria in selecting marine reserves and developing reserve Networks', *Ecological Applications*, vol 13, no 1, pp215–228

Roff, J. C. (2005) 'Conservation of marine biodiversity: Too much diversity, too little co-operation', *Aquatic Conservation: Marine and Freshwater Ecosystems*, vol 15, pp1–5

Roff, J. C. 2009 'Conservation of marine biodiversity: How much is enough?', *Aquatic Conservation: Marine and Freshwater Ecosystems*, vol 19, pp249–251

Roff, J. C. and Evans, S. M. J. (2002) 'Frameworks for marine conservation: Nonhierarchical approaches and distinctive habitats', *Aquatic Conservation: Marine and Freshwater Ecosystems*, vol 12, pp635–648

Roff, J. C., Taylor, M. E. and Laughren, J. (2003) 'Geophysical approaches to the classification, delineation and monitoring of marine habitats and their communities', *Aquatic Conservation: Marine and Freshwater Ecosystems*, vol 13, pp77–90

Rosenberg, A. A., Bolster, W. J., Alexander, K. E., Leavenworth, W. B., Cooper, A. B. and

McKenzie, M. G. (2005) 'The history of ocean resources: Modeling cod biomass using historical records', *Frontiers in Ecology and the Environment*, vol 3, pp84–90

Sathyendranath, S., Cota, G., Stuart, V., Maass, H. and Platt, T. (2001) 'Remote sensing of phytoplankton pigments: A comparison of empirical and theoretical approaches', *International Journal of Remote Sensing*, vol 22, pp249–273

Scott, W. B. and Scott, M. G. (1988) *Atlantic Fishes of Canada*, University of Toronto Press, Toronto, Canada

Soule, M. E. and Terborgh, J. (1999) 'Conserving nature at regional and continental scales: A scientific program for North America', *Bioscience*, vol 49, pp809–817

Stevens, T. and Connolly, R. M. (2004) 'Testing the utility of abiotic surrogates for marine habitat mapping at scales relevant to management', *Biological Conservation*, vol 119, pp351–362

Stewart, R. R. and Possingham, H. P. (2002) 'A Framework for Systematic Marine Reserve Design in South Australia: A case study', in J. P. Beumer, A. Grant and D. C. Smith (eds) *Proceedings of the World Congress on Aquatic Protected Areas*, North Beach, WA

Theroux, R. B. and Wigley, R. L. (1998) 'Quantitative composition and distribution of the macrobenthic invertebrate fauna of the continental shelf ecosystems of the northeastern United States', *NOAA Technical Report NMFS 140*, National Marine Fisheries Service, Seattle, WA

Thurman, H. V. and Trujillo, A. P. (2002) *Essentials of Oceanography*, Prentice-Hall, Upper Saddle River, NJ

United States Fish and Wildlife Service (2006) 'The endangered species program', www.fws.gov/endangered/, accessed 10 November 2009

Wahle, C., Grober-Dunsmore, R. and Wooninck, L. (2006) 'MPA perspective: Managing recreational fishing in MPAs through vertical zoning: The importance of understanding benthic-pelagic linkages', *MPA News*, vol 7, no 5

Ware, D. M. and Thompson, R. E. (2005) 'Bottom-up ecosystem trophic dynamics in the northeast Pacific', *Science*, vol 308, pp1280–1284

Warman, L. D., Sinclair, A. R. E., Scudder, G. G. E., Klinkenberg, B. and Pressey, R. L. (2004) 'Sensitivity of systematic reserve selection to decisions about scale, biological data, and targets: Case study from southern British Columbia', *Conservation Biology*, vol 18, pp655–666

Westra, L. (2005) 'Ecological integrity', *Encyclopaedia of Science, Technology, and Ethics*, Macmillan Reference, Detroit, MI

17

Networks of Protected Areas

Patterns of connectivity in the oceans

Call it a clan, call it a network, call it a tribe, call it a family. Whatever you call it, whoever you are, you need one.

Jane Howard (1935–1996)

Introduction

There exists widespread recognition for the need and a commitment to developing new approaches for the conservation of marine biodiversity. But despite commitments by participating governments to implement networks of marine protected areas (MPAs), there has been little progress towards their implementation. In order to adequately justify or defend proposals for networks of MPAs, answers to several critical questions are required. For example: what is the purpose of an MPA network? Where should MPAs be located? How many MPAs will be enough? In response to these kinds of challenges, a network of MPAs should be ecologically based and scientifically defensible, making use of the best existing information. In this chapter we are dealing with the specification and design of networks of MPAs within an ecoregion.

The term 'network of MPAs' is now widely used, but still often misunderstood, poorly defined or not defined at all. Even where several MPAs may be designated within a region, they only ever constitute a 'set' unless their connectivity as a true 'network' has been evaluated (see Box 16.1).

In essence, a proper national or regional network of MPAs must consist of multiple sites with replicates of all habitat types, that are oceanographically connected; individually or in aggregate they are of sufficient size (see Chapter 14) to sustain minimum viable populations of the largest species in a region – including those of seasonal migrants to the region – and their resident species can sustain their populations by recruitment from one MPA to another. Again: the member sites of a network of MPAs should comprise replicates of all representative habitats (see Chapter 5), and various distinctive habitats (see Chapter 7), that are mutually supporting in recruitment processes through connectivity (see e.g. Cowen et al, 2000). In short: a network of MPAs should capture and be able to sustain the regional elements of marine biodiversity (Roff, 2005).

The life forms in the oceans are connected both trophodynamically and spatially. Trophodynamic connections can be considered under 'ecosystem-based management' (see Chapter 13). It is the spatial connections that concern us in this chapter. Ultimately, all ecological and physical systems on our planet are connected by water currents in the oceans, atmospheric circulation and winds, and by the movements of crustal rocks and drift of the tectonic plates. However, it is the shorter time-spans of biological life-histories of marine organisms that are relevant to the vital concept of connectivity of MPA networks. We have to consider and plan for not only the *structures* of the marine environment itself and its biological communities, but also the *processes* of the environment (e.g. currents) and its organisms (e.g. their behaviours) in the process of connectivity.

Although all parts of the oceans are biologically and ecologically connected on some time and space scale, they are also isolated in time and space. Both connections and barriers in the oceans are strong, but not obvious. For the purposes of conservation of the components of biodiversity, it is the relationships between the time and space scales of connectivity and isolation that are of paramount importance. Global biogeography (both geological and contemporary – see Chapter 5) is the result of the balance between the structures and processes of isolation versus connectivity, in relation to the dispersal and migration capabilities and life cycle durations of the marine biota.

Natural isolation is desirable because it is a fundamental prerequisite to the reduction of gene flow and enhancement of the probability of speciation. Artificial (human induced) isolation, however, can lead to increased population fragmentation with the probable end result of extinction of local subpopulations and loss of genetic diversity (see Chapter 10). Connectivity is therefore essential to natural processes of gene flow and ecological integrity. Connectivity is in fact a major component of ecological integrity.

Ecological integrity is a relatively new concept, actively discussed by ecologists, and in common usage in both terrestrial and marine conservation, but a consensus has not yet emerged as to its definition (see Box 16.1). Ecological integrity is a term used to describe ecosystems that are self-sustaining and self-regulating. For example, they have complete food webs, a full complement of native species that can maintain their populations and naturally functioning ecological processes (energy flow, nutrient and water cycles, etc.). However, because the oceans are fluid and continuous, it is logical to argue that there can be no such thing as 'ecological integrity' of an isolated MPA; the viability of any single MPA depends on its connectivity. Only a network can achieve this because the integrity of any one site depends on its connectivity with other similar sites, especially in terms of resources and recruitment. This is the fundamental reason for the importance of networks of MPAs.

It is well known that ecological stability is inversely related to size and volume of the area (or marine container). Nevertheless, qualitative statements such as 'an MPA should be large enough to preserve its ecological integrity' have no real meaning and no useful value for planning purposes. We must appreciate that the oceans are interconnected. The important issue is to know on what space and time scales protected areas are connected at the regional level.

If our conservation objective is to conserve both fisheries and biodiversity, then networks of MPAs must protect all types of organisms – not just the taxonomic and 'ecospecies' categories (e.g. as in Tables 3.1 and 3.3) – but must accommodate and respect all the different types of life cycles and species movements within the oceans (see Table 17.1). The concept of networks of MPAs within each ecoregion basically supposes that protected areas are 'oases' of the various components of biodiversity within an otherwise degraded region of the oceans. However, creating regional conservation areas as refuges for species diversity will be largely ineffective if they remain hemmed-in by human-modified habitats or remain isolated without connecting corridors.

In terrestrial environments the issue of habitat fragmentation has become a major concern for several reasons. For example, remnant pieces of habitats left undisturbed may be too small to support either home-ranges of larger or minimum viable populations (MVP) of smaller species, and corridors between such fragmented areas may not exist, leading to population disruption and extinction. However, in this respect (and many others, see Chapter 3), the marine environment differs fundamentally from the terrestrial environment. Problems of minimum habitat size do not apply in the pelagic realm, but the benthic realm may be analogous to terrestrial environments either for home–range or MVP considerations.

The issue of connectivity is, however, different between pelagic and benthic realms. In the pelagic realm, most species connect among members of a population, either because they are full-time members of the pelagic environment or because they are temporarily members of the meroplankton. But in the benthic realm relatively few species encounter continuity of substrate

Table 17.1 Life cycle types and movements of marine organisms

Life cycle number	Life cycle movements (migration, drift, dispersal)	Examples of organisms
1	Entirely pelagic, pelagic reproduction, seasonal(?) migrations within the pelagic realm	Whales, dolphins, porpoises
2	Pelagic as adult, but land-referenced, migration to terrestrial/ice reproduction	Seals, turtles, some marine birds
3	Pelagic spawning followed by larval drift, adult counter-current migration	Pelagic fish species, larger nekton
4	Holoplanktonic (sexual, asexual reproduction), retained within a biogeographic region by water mass movements and vertical migrations	Most planktonic organisms, and nekton
5	Pelagic or demersal as adult, migrations, benthic egg laying (larvae suppressed) or viviparous	Cartilaginous fish, sharks, rays
6	Benthic spawning, planktonic larval drift, adult counter-current migration	Demersal fish species
7	Benthic-referenced territorial, nest rearing	Reef/rock fish species
8	Marine adult, migration to freshwater to spawn	Anadromous fish species
9	Freshwater adult, migration to marine to spawn	Catadromous fish species
10	Benthic adult invertebrates, planktonic (meroplankton) larval drift, adult/sub-adult counter-current migration (closed retention/recruitment cells)	Crabs and lobsters
11	Benthic adult invertebrates, planktonic (meroplankton) larval drift and counter-current drift (closed retention/recruitment cells)	Crabs and lobsters
12	Benthic adult invertebrates, planktonic larval dispersal (open recruitment –'invasive', colonizing)	Many invertebrate taxa, most marine macrophytic plants
13	Benthic adult invertebrates, larvae suppressed limited dispersal	Many invertebrate taxa
14	Benthic adult invertebrates, asexual reproduction	Corals
15	Symbiotic and parasitic organisms, execute life cycles in conjunction with host	Many

Note: **migration** – an active 'purposeful' movement from one location to another; **drift** – a passive directional dissemination in an ocean current; **dispersal** – a passive non-directional dissemination from a location

over broad areas. Because benthic habitats are discontinuous (at various scales of heterogeneity) their continuity and connectivity is achieved – as various kinds of propagules and larvae – within the planktonic communities of the pelagic realm.

Thus we could argue that marine populations and communities are inherently more resilient to habitat fragmentation than terrestrial ones, because continuity and connectivity is provided largely through the aquatic medium rather than by the substrate. There will of course be exceptions, but this may be a general principle. This means that it now becomes vital to understand patterns of connectivity as mediated via the pelagic realm, and this is now why water quality in coastal waters also becomes important within the medium itself.

Understanding connectivity is critical for the design of representative networks of MPAs, and for the development of conservation strategies to protect species associated with degraded and fragmented seascapes. Without knowledge about connectivity patterns, it may be impossible to interpret the cause of changes observed through time and space in open-ocean and deep-sea ecosystems beyond national jurisdiction. The dynamics of many ecological systems that are widely separated across an ocean basin are coupled in complex ways through the activities of individuals which move between them, including in areas within national jurisdiction. Improved mapping of bioregions, and associated ecosystems and habitats, will also improve our understanding of connectivity.

An understanding of the natural patterns of oceanographic connectivity within a region is therefore vital to the establishment of true networks of MPAs. Such knowledge is required to understand and evaluate: 'source-sink' dynamics of recruitment processes for all biota and especially fish populations; meta-population structures and fragmentation; ecosystem-level isolation mechanisms; the rationale for location of an individual MPA and its ecological contribution in terms of source-sink dynamics; the number of MPAs required in a region, their size and their distance apart; the patterns of connectivity among MPAs, and to judge whether a set of MPAs actually constitutes an efficient network of MPAs.

In summary, once we have determined what constitutes a coherent set of MPAs (in other words the set that includes all representative and distinctive MPAs within a region and replicates from candidate sets), then we can analyse the patterns of connectivity among them. In order to determine whether a set of MPAs actually acts as a network, we need to consider a series of complications, including: all the categories and types of organisms in a region and their migrations and dispersals; the regional flow patterns; the retention and/or dispersal of propagules by currents.

In coastal waters especially there are many topographic complications; this is why genetic studies here are so useful (and they can be conducted on the adults as well as the larvae). In offshore areas there may be fewer complications, and models of dispersal may be more reliable. However, each region of the oceans requires its own studies of connectivity patterns, based on local knowledge of currents.

Finally, the efficiency of a proposed network can be assessed from a knowledge of regional ocean currents, atmospheric winds and larval durations and dispersal. We propose that if the proportion of propagules captured in models is equal to or greater than the proportions of region protected in a candidate set of MPAs, then we could consider that we have designed an efficient network.

Types of marine organism life cycles and movements

The major taxonomic groups of marine organisms have been summarized in Table 3.1. For conservation planning purposes we can, more usefully, recognize that marine species fall into several ecological groups depending upon how they inhabit the pelagic environment or some combination of pelagic and benthic environments during their life cycles. Life cycles and movements of marine organisms are very diverse. All marine organisms live in some definable habitat type as adults, and either recruit back to it after some migrant or dispersal phase, or recruit to another habitat of the same type elsewhere. Table 17.1 summarizes the major ecological groups, according to the characteristics of their life cycles and movements.

The first two groups and some others (e.g. corals of group 14) should be adequately considered in terms of distinctive area MPAs. However, we can only protect such species temporarily or seasonally within such MPAs; other legislation and protection measures, for example mobile pelagic MPAs, are required during migration phases.

Importantly, for the organisms of groups 2, 8 and 9 in Table 17.1, it should be noted that ecological connectivity can also be from sea to land, and from sea to freshwater and vice versa. This extends the concept of MPAs to the terrestrial and freshwater environments – often forgotten linkages. The other ecological groups in Table 17.1 would be protected within various types of representative MPAs. It is primarily the groups of fish and invertebrates within an ecoregion that are benthic as adults, but that have meroplanktonic dispersal phases (larvae or other propagules) that will be emphasized in the rest of this chapter.

Although the strategies underlying several life-history types are becoming understood (e.g. Roff, 1992), such theory does not yet seriously inform the marine conservation planning process. Life-history patterns, recruitment mechanisms, use of the environment by migrant species and dispersing larval stages all need to be assessed as part of the process of selecting MPAs. By migrant species we mean those vertebrates (reptiles, birds, mammals and fish) and invertebrates which, as key components of their life cycles, either use marine resources in more than one geographical area, or use other environments (freshwater, terrestrial) for part of their life cycle, and make ('purposeful') directed movements from one area to another. Other organisms, as part of their life cycle, may be passively disseminated as larvae (or other kinds of propagules) within the planktonic realm either advected directionally by currents (drift) or non-directionally diffused (dispersal).

Larval types durations and dispersal distances

Marine fish and invertebrates have highly variable reproductive strategies, and few generalizations are to be had. The patterns of early development and larval morphology of marine invertebrates are especially varied (see e.g. Levin and Bridges, 1995; McEdward, 1995). However, along gradients of increasing depth and increasing latitude, there is a general trend for species to have some form of demersal or direct development that bypasses a planktonic phase (Figure 17.1). In polar regions and deeper waters larvae are generally suppressed in favour of some form of more direct development. In neritic temperate and tropical shallow waters, however, the majority of species have some form of propagule – predominantly meroplanktonic larvae which are dispersed within the water column. Some estimates indicate that 70 per cent or more of benthic species on continental shelves have meroplanktonic larvae of some sort.

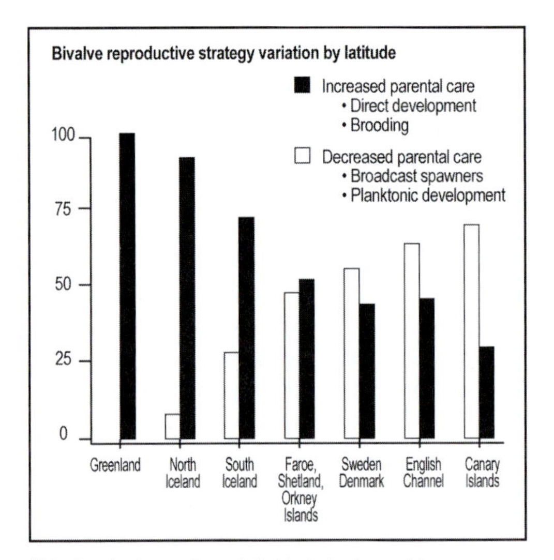

Note: Broadcast spawning and planktonic development increase towards lower latitudes

Source: After Thorson (1950)

Figure 17.1 Latitudinal variation in reproductive strategies of bivalve molluscs

The timing of release of larvae is determined by an array of factors. Like their terrestrial counterparts, marine organisms typically time their reproductive cycles according to a variety of environmental clues (see e.g. Morgan, 1995 for review), including seasonal, lunar and tidal. Cues of light-dark cycles and pressure, coupled with endogenous rhythms, enable animals to synchronize reproduction and release of gametes and larvae. The timing of release appears to include potential adaptations to periods of higher planktonic resources, reduced presence of predators or current regimes conducive to patterns of recruitment. Complexities of settlement behaviour – including availability of suitable substrate, 'choice' of substrate types and biological interactions with predators and competitors (Figure 17.2) – add further complications to already complex life cycles (see McEdwards, 1995).

Durations of the planktonic phase are highly variable (see for example compilations in Strathmann, 1987; Morgan, 1995), without apparent patterns among taxonomic groups. Some species (for example among the algae and molluscs) have extremely short larval durations measured only in hours, while others (for example among other molluscs and crustaceans) have very long durations – up to a year or more.

Fish larvae of demersal species also vary in their morphology, but to a much lesser extent than in the invertebrates; this is perhaps to be expected in a more cohesive taxonomic group. Fish larvae still vary considerably in their development times, primarily as a function of egg size (which determines hatching time and time of first feeding) and temperature – which inversely determines the time as larvae in the plankton (see e.g. www.larvalbase.org/).

As might be expected, there is a strong relationship between propagule (larval) duration and average dispersal distances. Shanks et al (2003) found a highly significant relationship among selected data ($r = 0.78$, $p < 0.001$) indicating that longer times in the plankton led to greater dispersal distance. However, the assembled data was highly bi-modal indicating strategies for dispersal of either < 1km or > 20km. From this they suggested that a series of reserves 4–6km

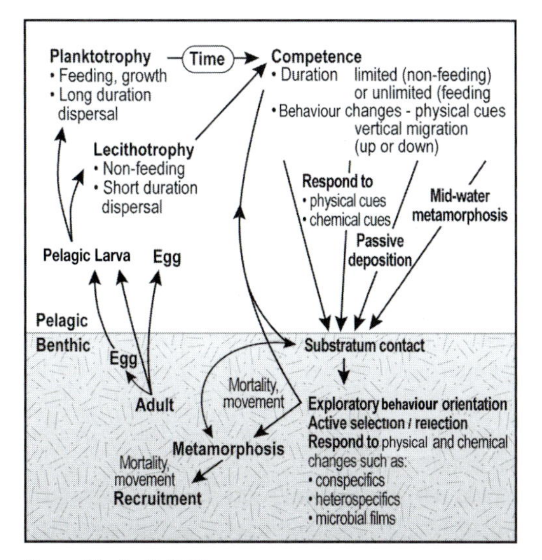

Source: After Pawlik (1992)

Figure 17.2 Generalized life histories of marine invertebrates with a planktonic larval phase

in diameter, and spaced 10–20km apart should be sufficient to constitute a network that would capture propagules among protected sites.

There is a growing number of studies on dispersal, connectivity and genetic differentiation of marine populations. An analysis of the literature by Bradbury et al (2008) indicates a historical focus on low-dispersal/low-latitude species, which are unlikely to be representative of global patterns. Dispersal within marine fishes (in other words planktonic larval duration and genetic differentiation) increased with latitude, adult body size and water depth. These global patterns represent a first step towards understanding and predicting taxon-specific and regional differences in dispersal, which are vital information for planning of MPA networks.

Though valuable as generalizations, and perhaps as starting considerations for planning, many other complications need to be considered for the overall design of regional networks. For example, Bay et al (2006) showed that pelagic larval durations for ten species of tropical reef fishes (*Pomacentridae* and *Gobiidae*) differed significantly and intraspecifically among sampling times and locations, and among regions, despite tight estimates. This indicates that the pelagic larval duration is a much more plastic trait than previously appreciated, and raises the possibility that it has become regionally adapted to dispersal and recruitment strategies.

Retention mechanisms

Because many benthic marine species have larvae that can spend up to several months in the planktonic stage, it was generally assumed that they must drift for great distances and become widely dispersed. This assumption led to the view that most coastal benthic populations are highly connected through larval transport and are demographically 'open'. Recent research – focusing on a combination of oceanographic, genetic and other biological/ecological techniques – has now indicated that benthic species can be retained in relatively local populations, by a variety of physical circulation mechanisms coupled with biological behaviours (Table 17.2). Such quantitative knowledge of connectivity contributes importantly to knowledge in an array of subjects, including design and location of MPAs, population dynamics, local processes of extinction, recolonization and source-sink dynamics, and the spread of invasive species.

The emphasis in larval studies has generally been on species' dispersal abilities, but despite their seeming continuity, the oceans exhibit many forms of barriers that cause retention of propagules within defined geographic ranges. Indeed, these retention mechanisms are the very reasons for the existence of distinct ecoregions and other biogeographic zones. They are still rather poorly researched but will engender unique features within each ecoregion.

The dispersal, and retention, of larvae in marine ecosystems is a highly complex subject, and is region-, locale and habitat-specific. Reviews by Wildish and Kristmanson (1997), Shanks (1995) and Bradbury and Snelgrove (2001) summarize many of the processes that aid in the aggregation, retention or dispersal of propagules (Table 17.2). Propagules of all kinds are subject to topographic, oceanographic, meteorological and biological effects of various kinds, and their interactions, that result in retention and recruitment cells that often involving counter-migration. Behavioural responses of larvae to a variety of environmental

Table 17.2 A listing of some aggregation and retention mechanisms in marine zooplankton, especially applicable to the meroplanktonic larvae of benthos and fish

Aggregation mechanisms
Oceanic convergences at a variety of spatial scales where water masses meet and descend
Surface slicks and convergences associated with internal waves
Patchiness generated by breaking of internal waves at a thermocline
Langmuir convergence circulation associated with wind-generated vertical vortices
Frontal regions between stratified and unstratified waters
Fronts associated with tidal circulation
Coastal boundary layer or coastal frontal zones associated with stratification
Wind-driven Ekman flow
Retention mechanisms
Oceanographic gyres at a variety of scales at water mass junctions or topographic features
Taylor columns associated with seamounts create topographic eddies
Eddies associated with islands downstream of prevailing currents
Frontal regions between stratified and unstratified waters
Eddies in coastal waters associated with bays and headlands
Tidal currents, residual tidal currents and tidal bores
Estuarine circulation – outward surface flow and entrained shoreward deeper water
Coastal changes in land-breeze and sea-breeze regimes
Entrainment in macrophytic plant canopies

Note: Some mechanisms could be included under either or both categories of effect.
Aggregation = process that accumulates individuals of a species above average background concentrations
Retention = process that retains individuals of a species within a localized area, i.e. that prevents dispersion
Sources: Various, including: Wildish and Kristmanson (1997); Shanks (1995); Bradbury and Snelgrove (2001)

cues play an under-appreciated role in the aggregation and retention of larvae (see e.g. Young, 1995). The net result of these effects is generally that dispersal is less than predicted or modelled on the basis of currents alone.

In estuaries, many species of both holoplankton and larval forms of euryhaline benthic species are adapted to be retained in the water column by taking advantage of estuarine flow patterns. In most estuaries, the net seaward flow of reduced density surface water is countered by entrainment of higher density subsurface waters from offshore flowing landward. Estuarine plankton take advantage of these opposite flows at different depths in a form of 'station-keeping' by making either diel or ontogenic vertical migrations. Such retention mechanisms are well known for a variety of crustacean species including commercially important crabs. Under normal conditions, there may therefore be relatively little dispersal of such species beyond the limits of their 'home' estuary or bay. It is likely to be the rare extreme events of tidal, wind and flushing combinations that disperse these larvae beyond their usual limits.

In nearshore coastal waters larval movements are especially complex. Here, various retention mechanisms can ensure that larvae recruit close to place of origin rather than being disseminated. Zooplanktonic organisms are subject to a variety of accumulation and retention mechanisms (Table 17.2). Thus, for example, Langmuir convergence circulation and internal wave convergences can accumulate larvae to thousands of times their mean abundance. Larvae cannot be considered simply as passive particles, because they can actively change their vertical distribution over time, thereby encountering currents of different speeds and trajectories. Neustonic forms especially are subject to accumulation or to greater than average dispersal because of meteorological events. Alternatively, diel changes in the patterns of onshore and offshore winds may lead to local coastal retention. Accumulation of zooplankton in bays as a result of topographic creation of eddies is also well known (e.g. Archambeault et al, 1998). The existence of coastal circulation cells (e.g. Carter, 1988) is also undoubtedly responsible for the local retention of many shallow water species, although the physical–biological interactions involved have been poorly investigated.

Similar retention mechanisms at generally larger scales operate in offshore waters, where water movements such as cross-shelf mixing, frontal systems between stratified and unstratified waters, oceanic convergences and Taylor caps over seamounts can all lead to retention of propagules and the creation of biogeographic boundaries either seasonally or semi-permanently.

Modelling studies based solely on estimates of currents and larval development times could therefore lead to significant errors in the estimation of dispersal distances, especially in coastal waters where physical complexities and retention mechanisms are likely to be more significant at the relevant scales of dispersal. For this reason, even comprehensive circulation models may be more suited for offshore regions in less physically complex marine environments, where they are more likely to yield reliable estimates of dispersal distributions. In complex coastal waters, other techniques such as genetic studies should yield more precise pictures of actual population dispersal and connectivity (see below).

Recruitment cells

The process of connectivity and recruitment in the marine environment is even more complex than involved simply in the dispersal of planktonic propagules. Recruitment of marine organisms is the set of processes whereby a species retains its progeny within an ecoregion conducive to its survival. This involves the balance between two opposing 'necessities': first the desirability of dispersal to colonize new available habitats; second the necessity of maintaining populations within supportive habitats. Three main strategies are apparent.

Open recruitment cells are characteristic of benthic species whose propagules are dispersed into the pelagic realm, without any obvious mechanism to return them to the habitat of origin. Presumably they rely mainly upon the kinds of mechanisms operating within the water column, described in the previous section, in order to retain at least a sustainable proportion of their propagules within suitable habits of an ecoregion. Research here has emphasized the process of settlement rather than the processes involved in retention mechanisms (see e.g. McEdward, 1995).

Closed recruitment cells comprise a period of passive larval drift followed by sub-adult or adult counter-current migration, both within the pelagic realm. This is clearly a very successful strategy, and many of the world's major marine fisheries fall into this category. A prime example of such a recruitment cell is that of the anchovy (*Engraulis* spp) in westward intensified currents (e.g. the Benguela). However, even here, counter-current migrations may be aided by a variety of complexities, including retention mechanisms, current anomalies, and navigation and orientation behaviours (e.g. Nelson and Hutchings, 1987).

Closed recruitment cells of a different type are also characteristic of many species of larger benthic invertebrates, especially the crustaceans, but now involving both a planktonic phase and a benthic counter-current migration phase. Connectivity in the marine environment is frequently conceived of as belonging only to dispersal phases (larvae or propagules) of benthos, and/or active migrations within the pelagic realm. However, in many motile benthic species there are also migrations within the benthos in corridors quite analogous to terrestrial ones. Typical of such cells is the process of recruitment in the spiny lobsters (*Palinuridae*) whose exotic phyllosoma larvae drift within the plankton. Following settlement to the benthos of a post-larval puerulus stage, the adults or sub-adults actively undertake a counter-current migration in often well-defined 'corridors' (see e.g. Booth, 1997), back towards the spawning areas where larvae were released. Similar migrations are undertaken by other lobster species (e.g. *Homarus* spp), although their patterns of connectivity may be more complex and are not as well documented (e.g. Incze et al, 2009).

Observations indicate that the patterns of recruitment processes and cells vary around the world. Even within a species, some populations undertake extensive migrations, others do not. This is to be expected as the result of natural selection of the interactions of animal behaviours (larval, juvenile and adult) and the regional patterns of currents. The recruitment cell that develops within a region is simply the set of behaviours that ensures retention of populations within a suitable ecoregion and its habitats. The species extant in a region today (including anadromous and catadromous ones) are therefore those whose larval behaviours and migrations have become well adapted to their local water movements.

However, not all oceanographic processes induce the retention of marine larvae at all times. Important processes may actually enhance the loss of larvae from their normal recruitment area. For example, warm core rings (spun off from westward-intensified ocean currents) may entrain shelf waters and reduce the recruitment of marine fish stocks through offshore transport and mass mortality of larvae. Myers and Drinkwater (1989) tested this idea using weekly satellite images to generate time-series of the positions and numbers of warm core rings from the mid-Atlantic Bight to the Grand Banks. They combined these with estimates of the timing of the spawning and the duration of larval stages to create stock-specific annual indices of ring activity and shelf-slope front variability. There was evidence that increased warm core ring activity reduced recruitment in 17 stocks of groundfish. A series of such oceanographic events, still not well understood, seem to influence the variability of recruitment success in the world's fisheries. Any single-year study of source-sink dynamics and recruitment mechanisms may therefore be inadequate to define overall patterns of connectivity.

Assessing connectivity within regions and locally

Patterns of global connectivity have been considered in Chapter 10 on genetics. As will be apparent from the foregoing, assessing connectivity at the regional and local scale is not as straightforward as some generalizations imply. However, patterns of connectivity within regions can be measured in a variety of ways. These include the historical and more conventional methods of current meters, drogues and drifters, models of ocean currents (caused by tides, winds, residual flows etc.) coupled with biological characteristics of propagules, studies of genetic relatedness of larvae or adults, and more recently tracking studies of individual organisms themselves. Finally, various combinations of these techniques are now providing insights into patterns of ocean connectivity, source-sink dynamics and recruitment processes.

There is now a growing series of publications on connectivity among MPAs, and the source-sink dynamics of local populations. Knowledge of sources and sinks is vital to plan for the proper regional sites of MPAs. For example, an MPA placed in a region that acts as a source of recruits should be more valuable than one in a sink location (e.g. Crowder et al, 2000).

Current meters, drogues, drifters

Historically, scientific understanding of global and regional current patterns has come from a combination of direct measurements by in situ current meters, and indirectly by calculations of density-driven geostrophic currents based on in situ observation of salinity, temperature and depth. Surface ocean currents (down to depths of 600–1000m) are predominantly driven by the wind and controlled by the rotation of the earth (as a function of the Coriolis parameter), whereas currents in deep waters (below about 600m) are generally considered to be driven by density differences. These conventional oceanographic techniques have described the broad scale – global to regional – patterns of ocean circulation, and defined the major surface and sub-surface ocean currents of our planet, including their seasonal variations.

As an example of a regional study, Klinger and Ebbesmeyer (2002) used drift cards to infer larval transport by surface currents in the San Juan Archipelago and Northwest Straits region of Washington State. They released 6400 cards, nearly 40 per cent of which were recovered. Drift cards tended to accumulate in restricted areas, with 70 per cent of the cards accumulating on only 15 per cent of the shoreline. The spatial distribution of recoveries suggests that certain sites are tightly linked with each other and are likely to sustain high levels of larval import (i.e. they are sink sites), indicating their potential importance in regional MPA network design.

Because not all conservation managers have access to more advanced technologies, Delgado et al (2006) suggest that low-tech methods (in this case drifters) are still valuable – despite their well-described drawbacks – to help elucidate the source of recruits. In a study designed to determine the origin of queen conch (*Strombus gigas*) larvae recruiting to the Florida Keys, they used drift vials that were released from four sites in Mexico with queen conch aggregations and at three sites in the Florida Straits. They concluded that most of the larvae found in the Keys originated from within the Keys and that the system is dependent on local recruitment, and therefore restoration efforts should target local spawning populations.

At regional to local scales these conventional techniques can illustrate circulation patterns at any given moment in time under specific conditions of residual currents, tides and winds. But regional and local currents are also extremely variable and heavily influenced by tides, basin topography and adjacent land masses. At these scales, current meter and drogue studies become major tools to identify the actual local currents. Such methods can give estimates of the actual current speed and direction at discrete depths. However, drifter studies – even when released in abundance – can only specify a pattern of connectivity at the specific time of the study and cannot be extrapolated to other times. The ability to extrapolate patterns of connectivity to cover all conditions is the major advantage of models.

Models of connectivity, sources and sinks

In order to evaluate conservation efforts, for both utilization and preservation of resources, models are increasingly being produced for a variety of scenarios and are becoming always more sophisticated. The fisheries literature is replete with early life-history models – mainly for single species – that simulate patterns of larval dispersal and mortality (e.g. Brickman and Frank, 2000). However, it is source-sink dynamics and connectivity that are emphasized here.

We need to understand source-sink dynamics in order to site MPAs efficiently (Crowder et al, 2000). Sources can be defined as areas where birth rates exceed death rates and emigration rates exceed immigration rates; conversely in sink areas death rates exceed birth rates and immigration exceeds emigration. Placement of reserves in sink habitats has the potential to harm fish populations, and if source habitat is in short supply, identifying and protecting it can be critical to fisheries conservation (Crowder et al, 2000).

Defining the scale of larval dispersal and connectivity among marine populations is therefore of prime importance for conservation, and understanding of population dynamics, genetic structure and biogeography. Biophysical models (incorporating both biological and physical parameters) have now been produced for several ecoregions (e.g. Cowen et al, 2006). However, conclusions made about dispersal distances (e.g. reef fishes' dispersal is in the order of 10–100km), and whether recruitment is predominantly local or from outside the local area (in other words the degree of retention) should not be uncritically extrapolated to other areas or to other taxonomic groups. Both dispersal distances and retention can vary greatly among taxa, and even within species among locations. For example, Cuban snapper populations exhibit levels of self-recruitment of 37 to 80 per cent, but with significant variation among areas (Paris et al, 2005).

Models can reliably reproduce all combinations of currents, tides, winds and development times of organisms. They therefore have the potential to predict dispersal patterns under a wide variety of oceanographic and meteorological conditions. However, the use of overly simple models (e.g. ignoring the interactions of advection and diffusion and biological factors – that is, non-biophysical models (Largier, 2003; Cowen et al, 2006) – can result in inaccuracies of estimates of larval concentrations by as much as nine orders of magnitude (Cowen et al, 2000)!

Additional complications in assessing source-sink dynamics and dispersal are not lacking. The optimal design of a set of MPAs requires attention to the interacting effects of larval dispersal, reserve location and reserve size (Planes et al, 2000; Stockhausen et al, 2000), as suggested in Figure 17.3. In addition, larval connectivity is an intermittent and heterogeneous process on annual time scales (Siegel et al, 2008) arising from chaotic events that range from advection to mass mortality by rings from westward intensified currents (e.g. Myers and Drinkwater, 1989).

In offshore (non-coastal) waters, where topographic effects and retention mechanisms are generally simpler, models may yield reliable

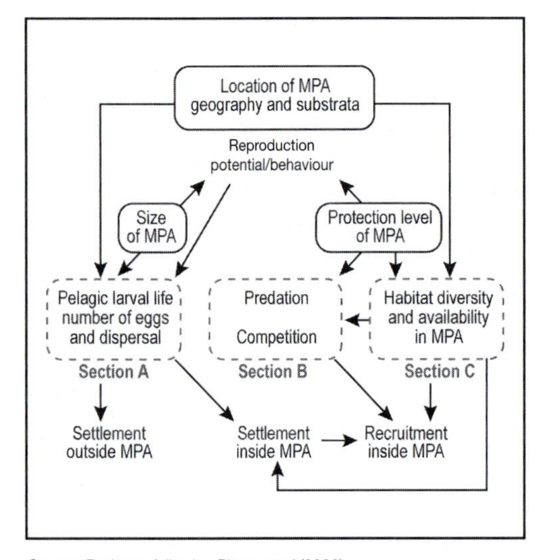

Source: Redrawn following Planes et al (2000)

Figure 17.3 The optimal design of a set of MPAs requires attention to the interacting effects of larval dispersal, reserve location and reserve size

results on patterns of connectivity. In nearshore coastal waters, spatial patterns of connectivity are more complex and dispersal kernels (probability distributions of larval settlement) become variable and distorted because of the combined effects of topography, currents and winds – among other influences (e.g. Aiken et al, 2007). Here, in coastal waters, genetic studies come into their own.

The great majority of existing models are location and species specific. Perhaps the most important models for the future will fall into two categories: those that assess connectivity among the members of entire networks of MPAs (see below); and those that can define how the potential of marine reserves for conserving biodiversity (e.g. Gerber et al, 2003) can be combined with existing expertise in the sustainable conservation of exploited fish stocks.

Genetic studies

As indicated in Chapter 10, genetic research has now become an indispensible tool for marine conservation. Genetics provides tools with broad potential application, but ones that are generally only applied to specific locations and selected species. In this section of the present chapter we can only summarize some of the concepts (often tentative) that genetics has contributed to the study of connectivity.

Genetic studies tell us nothing about water movements themselves; they allow us to infer what may or (under given conditions) must have occurred, but not how. Genetic information therefore allows us to reconstruct past events of connectivity, and their timing or periodicity. Such information is reconstructive not predictive, and is used to best advantage with knowledge of water movements and the biology and life histories of organisms. In general (and as might be expected) there appears to be high gene flow in species with high dispersal potential, and low gene flow in species with low dispersal potential (Palumbi, 1995), but there are many exceptions. A few examples must suffice.

The significance of genetics for the connectivity among MPA sites is exemplified in the study of Bell and Okamura (2005). They demonstrated genetic differentiation, isolation, inbreeding and reduced genetic diversity in populations of the dog-whelk *Nucella lapillus* in Lough Hyne Marine Nature Reserve (an isolated sea lough in southern Ireland), compared with populations on the local adjacent open coast and populations in England, Wales and France. This study clearly indicates the adverse long-term genetic consequences of selecting reserves on the basis of isolation and ease of protection.

However, in some areas that might reasonably be expected to show biological isolation and genetic differentiation it is not found. Analysis of the intra-specific genetic structure using the mitochondrial marker COI reveals that populations of two Galatheidae species (*Munida thoe* and *Munida zebra*) polymorphic for this marker, are genetically not structured, both among the Norfolk seamounts of the southwest Pacific and between the seamounts and the island slope (Samadi et al, 2006). The genetic structure of other crustaceans and a planktotrophic gastropod reveals a similar pattern. Population structure is observed only in *Nassaria problematica*, a non-planktotrophic gastropod with limited larval dispersal. Thus, the limitation of gene flow between seamounts appears to be observed only for species with limited dispersal abilities.

Conversely, even within relatively homogeneous regions genetic differentiation may still be observed. For example, Ruzzante et al (1998) provide evidence of genetic structure of cod *Gadus morhua*, at spawning-bank scales. This was consistent with variations in oceanographic features and in the spatio-temporal distribution of spawning, which may both represent barriers to gene flow among geographically contiguous populations inhabiting a highly advective environment. The differences described are consistent with post-dispersal spawning fidelity to natal areas, a behaviour that may be facilitated by topographically induced gyre-like circulations that can act as retention mechanisms. This conclusion is supported by Knutsen et al (2003). They also found that the Atlantic cod was sub-structured into genetically differentiated populations on a fine geographical scale, along a 300km region of a coastal area free of any obvious physical barriers, well within the dispersal ability of this species. They found no geographical pattern to the genetic differentiation and there was no apparent relation to distance along the coastline. These findings support the idea that low levels of genetic differentiation are due to passive transport of eggs or larvae by the ocean currents rather than to adult dispersal.

Even in species which have apparently similar larval dispersal potential, analysis of mtDNA can show quite divergent patterns. For example the temperate continental species *Strongylocentrotus purpuratus* shows large-scale genetic homogeneity, whereas a tropical insular *Echinometra* species shows substantial genetic differentiation (Uehara et al, 1986). A number of other allozyme and DNA studies (including those above) show strong genetic structure in species with high dispersal potential, while some individual species exhibit high dispersal potential and low genetic differentiation at some times and in some areas, but at other times and places have low dispersal and high genetic differentiation. These kinds of exceptions and apparent contradictions come from many taxonomic groups and many geographic areas (Palumbi, 1995), presumably as the result of the local variability of recruitment cells under biological and oceanographic influences. These apparent contradictions warrant much closer study of genetics and oceanography in combination (see below).

Among coral reef fishes, the general lack of correlation between pelagic larval duration and genetic connectivity across barriers indicates that life history and ecology can be as influential as oceanography and geography in shaping evolutionary partitions within ocean basins (see Rocha et al, 2007). Hence conservation strategies require recognition of ecological hotspots, those areas where habitat heterogeneity may promote speciation, in addition to more traditional approaches based on biogeography.

It therefore becomes vital to know the details of the life histories of organisms, the dynamics of biogeographic boundaries and their physical oceanography, and in coastal regions the details of (largely ignored) coastal cell circulations (see Chapter 11 and Carter, 1988). It appears more and more that the geographic separation of populations is at least as much a function of the mechanisms of larval dispersal, currents and recruitment cells (involving both biological and oceanographic factors), rather than just the topographic localities of adult populations (see e.g. Zhan et al, 2009).

Taxonomic and genetic information tells us that biogeographic boundaries are real – even at the sub-specific (population) level – within ecoregions. They must be the product of some combination of interactions of topography and current patterns and the biology of dispersal propagules. They can be distinguished by careful oceanographic and genetic studies, and are likely to be region specific.

It is perhaps in the immediate coastal zone, where physical circulation models often do not perform well due to the topographic and circulation complexities of coastal cells, that genetic techniques are most useful. Here larvae may periodically connect adult benthic populations by a series of 'stepping stones' (which assumes that populations only exchange propagules with adjacent populations) following enhanced dispersal by a series of extreme or variable events including floods, tides and winds. Sotka and Palumbi (2006) argue that the geographic width of a stable genetic cline is determined by a balance between the homogenizing effects of dispersal and the diversifying effects of selection. Thus, if selection and clinal width are quantified, then the average geographic distances that larvae move (their connectivity) can be inferred. Their theory of genetic clines indicates that the average dispersal distance of larvae is some fraction (generally ~ 35 per cent) of the clinal width, and the dispersal distances so inferred should be of the correct order. Such estimates of larval dispersal are extremely valuable, as they can provide important guidance to conservation efforts.

Finally, inter-annual variations in extinctions, the fraction of the population spawning at times conducive to survival (e.g. Hedgcock's (1994) 'sweepstakes' – chance matching hypothesis), mass mortalities of larvae, disturbances and recolonization from outside a region are all features of all marine systems, which lead to recruitment and natural selection within populations and the evolution of local communities (e.g. Lessios et al, 1994). There are apparently many 'rules' still to learn in terms of the genetics of connectivity.

Tracking studies

In strict contrast to dispersals, tracking studies of individual organisms (fish, marine mammals, turtles etc.) allow us to define the routes and timing of migrations of individual species. Using various kinds of tags, valuable ancillary information on animal physiology and oceanography can also be collected from larger animals, giving intimate details of their environment and biological responses to it. Such data is presently restricted to a few individuals of selected larger species, but would be invaluable for definition of pelagic mobile MPAs and the protection of animals during the migratory phase.

An international leader in the field of ocean tracking is the Ocean Tracking Network (OTN) (http://oceantrackingnetwork.org/), a cooperative global initiative. Through the OTN, thousands of commercial and endangered marine species will be tagged to help improve fishing practices and better understand the oceans. Animals are being tagged to telemeter information on where they go, what conditions they experience, how they interact and how individuals' behaviours change on time scales relevant to climate change. This is information that scientists and managers need in order to protect and restore ocean productivity. Knowing where marine animals actually travel means that it is easier to designate new MPAs, set shipping routes, approve resources exploration and deliver a picture of the complex interactions of biology and physics in the world's oceans.

The OTN is global in scale, but tracking techniques can also be used at regional and local scales. For example, Jones et al (2005) solved the mystery of the natal origin of clownfish (*Amphiprion polymnus*) juveniles by mass-marking all larvae in a population by immersion in tetracycline, while at the same time parentage of new recruits was established by DNA genotyping (see below for details). Becker et al (2007) successfully applied an in situ larval culturing technique in order to develop elemental microchemical fingerprinting as a tracking tool to determine sources of settled invertebrates. Elemental fingerprinting takes advantage of location-specific chemical signatures recorded in hard parts of marine organisms at the time of their formation, which can serve as geographic 'tags' – for example, elemental ratios of Co/Ca, Pb/Ca, Cu/Ca and Mn/Ca. Becker et al (2007) showed that coastal mussel larvae, previously thought to be highly dispersed, can be retained within 20–30km of their natal origin. In addition, they found that two closely related and co-occurring species, *Mytilus californianus* and *Mytilus galloprovincialis*, exhibited substantially different patterns of connectivity among sites.

Similar chemical tracking techniques have also been applied to other benthic invertebrates, including lobsters, in order to determine their benthic migration pathways and corridors. For example Chou et al (2002) measured five metal variables (Ag, Cd, Cu, Mn and Zn) in order to document movements of lobsters within the upper Bay of Fundy in Canada.

Combinations of techniques

Molecular tools probably perform at their best when integrated with other data and approaches (Selkoe et al, 2008), especially when the underlying genetic signal is relatively weak, as occurs in many marine species. Recent studies combining genetic, oceanographic, behavioural and modelling approaches have provided new insights into the spatial ecology of marine populations, in particular regarding larval migration, barriers to dispersal and source-sink population dynamics. Combinations of techniques have been used by Galindo et al (2006) to develop connectivity estimates from oceanographic models to predict genetic patterns resulting from larval dispersal in a Caribbean coral. Their coupled oceanographic-genetic model successfully predicts many of the patterns observed, including the isolation of the Bahamas and an east–west divergence near Puerto Rico.

As a specific example of combinations of techniques, Jones et al (2005) solved the mystery of the natal origin of clownfish (*Amphiprion polymnus*) juveniles by mass-marking all larvae in a population by immersion in tetracycline. At the same time, parentage of all potential adults and all new recruits arriving in the population was established by DNA genotyping. They established that although no individuals settled into the same anemone as their parents, many settled remarkably close to home. Even though this species has a larval duration of 912 days, remarkably one-third of settled juveniles had returned to a two-hectare natal area, with many settling < 100m from their birth site. This represents the smallest scale of dispersal known for any marine fish species with a pelagic larval phase.

Evaluation of connectivity among sets of MPAs

Multi-species networks

Although there have been many studies of source-sink dynamics in local regions, and the general principles for designing networks of MPAs have been stated (e.g. Gerber et al, 2003; Hastings and Botsford, 2003), there have been fewer attempts to investigate the features of population dynamics and regional hydrography necessary to support potential networks. Perhaps this is because we have so few real networks! There are now several examples of models of MPA networks based on recruitment patterns of individual species, but such networks are generally lacking for multi-species assemblages (see Grantham et al, 2003).

Unfortunately, even when 'networks' have been established, planned or claimed, they may fall well short of their expected purpose. In an interesting recent study, Johnson et al (2008) undertook an analysis of the proposed marine Natura 2000 sites (n = 298) in the North Atlantic between Portugal and Denmark. The median size of sites was 7.6km^2 with a median distance among neighbour sites of 21km (range 2–138km). Although the issue of habitat specificity was not addressed, based on assumptions about dispersal capacity and local retention of species, Johnson et al (2008) suggested that at least half of the proposed sites were both too small and too isolated to support regional populations within a 'network' consisting of these sites.

A study with similar objectives was carried out by Robinson et al (2005) on the north Pacific coast of Canada. The main purpose was to use a three-dimensional oceanographic simulation model to understand connectivity among the proposed Gwaii Haanas National Marine Conservation Area (GHNMCA) and ten other proposed or existing MPAs. To give the particle experiments some realism, the selection of simulation time period, larval duration and depths was determined by literature values for 44 species of fish and 22 species of invertebrates present in the proposed GHNMCA. February to April was identified as an important recruitment period, with high abundances of fish and invertebrate larvae. The simulations were conducted using passive particles placed at three depths and vertically migrating particles for 30 or 90 days in late winter. Importantly, simulated surface particle dispersion was found to be consistent with winter ocean current observations made from analysis of satellite imagery, current mooring and drifter data. The GHNMCA would contribute to a network of MPAs because it supplies and receives particles from other MPAs in northern British Columbia. Model simulations also indicated that the greatest source of particles to GHNMCA originate from 30m and not 2m flows. Finally, the simulated mean daily dispersal rate of 2.0km/day^{-1} would allow fish and invertebrates to self-seed northern portions of the GHNMCA in winter. Together, the GHNMCA and other MPAs appear to contribute a large percentage of particles to non-MPA regions in northern Hecate Strait, which may be considered a particle sink in winter.

An efficient multi-species network

Ideally in a region planned to contain several MPAs (i.e. a set as a putative network) a significant proportion of dispersing propagules should either auto-recruit (within the MPA of origin) or allo-recruit to another MPA with similar geophysical properties. But what constitutes a 'significant' proportion? Calculations such as those of Johnson et al (2008) represent only average conditions and generalities concerning size and spacing of MPAs. What is required is a sensitive model of an actual or proposed network that takes the full spectrum of regional life cycles, hydrography and meteorology into account, and actually calculates the probabilities of recruitment of propagules within the network. We clearly need some method to estimate the probabilities of recruitment of propagules which are exchanged among sites, that are (or could become) members of a network of MPAs. In short we need a measure of the entire 'efficiency' of the network as a whole – that is, whether its component MPAs as a whole act as a regional source or sink – and the ability of the entire network to retain sufficient recruits to ensure sustainability of its varied populations and communities.

Conceptually, this could be a difficult modelling challenge. There are several potential problems including: probability of fertilization of adults; mortality of propagules after release; dispersal patterns of the planktonic phase; settlement and benthic recruitment success, and so on (see e.g. McEdward, 1995). However, assuming that these and most other biological events associated with recruitment are stochastically distributed within a region, it is the patterns of dispersal that need to be evaluated, and indeed this is where most research effort (for conservation purposes) has been directed. A simple working principle and null hypothesis is now suggested, namely that the percentage efficiency of a network (as evaluated by a model) should be equal to or greater than the percentage of a region covered by a set of MPAs (see Figure 17.4).

Efficiency is defined simply as the sum of the proportions of propagules retained within any MPA of origin, plus those that recruit to another MPA with a similar array of habitats. In order to

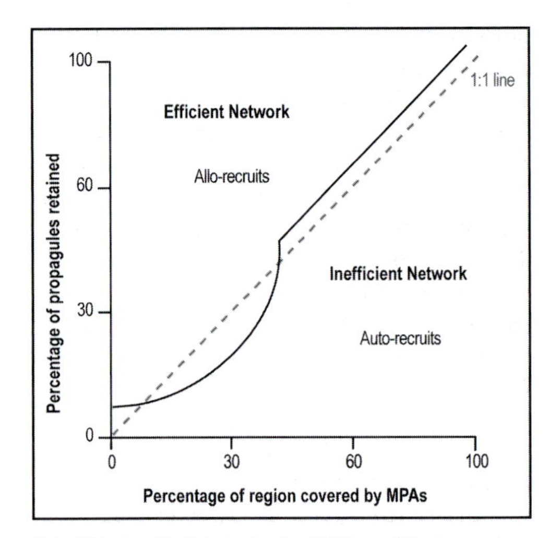

Note: Efficient and inefficient networks of MPAs would lie above and below the 1:1 line as indicated

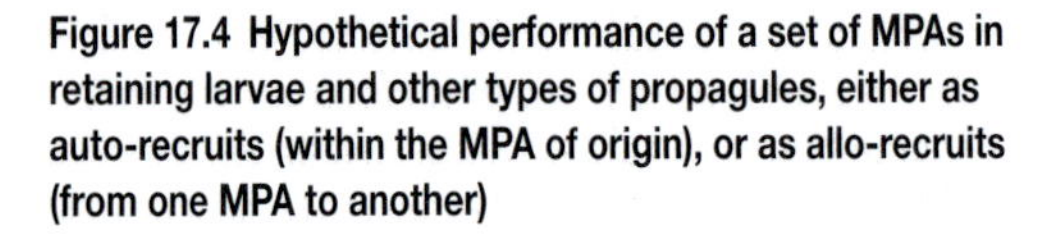

Figure 17.4 Hypothetical performance of a set of MPAs in retaining larvae and other types of propagules, either as auto-recruits (within the MPA of origin), or as allo-recruits (from one MPA to another)

illustrate how this can be calculated, we summarize the findings of an unpublished study by Roff, Toews and Bryan using the model Webdrogue (www2.mar.dfo-mpo.gc.ca/science/ocean/coastal_hydrodynamics/WebDrogue/webdrogue.html), whose properties are similar to the model used by Robinson et al (2005).

Webdrogue is a graphical user interface to a drift trajectory program. It can be used to obtain drift predictions for any point in the model domain. Drift trajectories are computed using circulation derived from the tides, the seasonal mean circulation, wind-driven circulation and a surface-wind drift. The technique for computing the wind-driven circulation and for combining all the circulation components is described in Hannah et al (2000). Velocity fields are computed as the sum of five components: the long-term seasonal mean, the responses to along-shelf and cross-shelf wind stress and sea level forcing at the upstream boundary, and the M2 tidal currents.

The fields are a realistic representation of 3-d seasonal circulation on the western and central Scotian Shelf obtained from historical observations and a combination of diagnostic and prognostic numerical models with forcing by tides, wind stress and baroclinic and barotropic pressure gradients. The major current features – the southwestward Nova Scotian and shelf-edge currents, and partial gyres around Browns and Sable Island Banks (see Figure 16.1b) – persist year-round but with significant seasonal changes. The quality of the flow fields varies across the region. As described in Hannah et al (2001), comparison with current meter observations shows good agreement for the Browns Bank, southwest Nova Scotia and inner-shelf regions, and for observations on Georges Bank is also good. The wind-driven flow has two components – the local response to wind and the response to the sea surface pressure field set up by the large-scale wind field.

The Webdrogue program has been used for a variety of biological studies in eastern Canada, including prediction of the distribution and extent of right whale feeding grounds and genetic differentiation of scallop populations. Kenchington et al (2006) showed that there was good correspondence between differentiation of scallop populations based on separation of areas from Webdrogue studies, and those indicated by independent genetic analyses. All this leads to confidence in the predictions from the model.

The Webdrogue model was applied to the set of MPAs in the Gulf of Maine and Scotian Shelf derived by the CLF/WWF (2006) study (see Chapter 16 and Figure 16.6). That study identified a set of potential priority areas for conservation, including representative and distinctive areas. The CLF/WWF report rather arbitrarily set a target of 20 per cent of conservation features for the entire study region. But is this a sufficient proportion of the region, and do the candidate MPAs actually constitute a true network or an efficient network as defined in Box 16.1? The hard question is not: is 20 per cent enough?' but rather: is enough 20 per cent?' We can only determine this from extensive genetic surveys, or by models of the region. Although the MPA set was referred to as network in the CLF/WWF report, in fact its connectivity patterns were not evaluated. This aspect of connectivity among the set of MPAs described in the CLF/WWF (2006) report is addressed here.

The null hypothesis tested on the set of MPAs in the CLF/WWF report was that the proportion of larvae that successfully recruit, either within the candidate MPA of origin or from one candidate MPA to another, will be equal to or greater than the proportion of the area of the region occupied by the set of candidate MPAs. If the proportion of a region protected is too small, or the individual MPAs are too small or too far apart to be mutually supporting for natural recruitment processes, then the proportion of propagules 'captured' by the MPAs will be less than the area covered by the MPAs – for example as discovered in the analysis of Johnson et al (2008). If the MPAs constitute a sufficient or an efficient network, then they should collectively 'capture' more propagules than the proportion of the region they occupy, or at least the same proportion (see Figure 17.4).

Of the 30 MPAs designated by the CLF/WWF report, 26 were examined (four at the extremes were dropped), covering some 19 per cent of the study region. These were subjected to repeated simulations in a modification of the Webdrogue model; this allowed for multiple particles (propagules) to be released from each of the candidate MPAs in turn, and to track their movements within the region over time. Mortality of propagules was ignored, and a particle was assumed to have 'survived' if it was either retained within the MPA of origin or if it reached another designated MPA. Thus settlement behaviour was also not assessed. Three depth intervals were chosen for tracking of particles: 2.5m – the neustonic layer representative of many species of crustacean larvae; 25m – representative of the upper mixed water layer; and 100m. Conditions during a single year (2006) were chosen for study, using the modelled hydrographic flow regimes for the periods of study and the actual records of weather conditions of that year.

Each MPA was simulated during two seasons, representing spring and summer, when a majority of benthic larvae would be released, and each simulation ran consecutively for a period of 15 days. Longer periods would of course be more appropriate for longer-lived larvae such as those of the larger benthic crustaceans. The spring simulations were released on 15 April and ended on 29 April 2006, whereas the summer simulations were released on 15 August and ended on 29 August 2006. As such, each individual MPA had six corresponding scenarios, leading to a total of 156 individual simulations representing tens of thousands of individual propagules.

For the purpose of this analysis,'retention' was defined as the maintenance of released propagules within the MPA of origin throughout a simulation. It was observed that percentage retention decreased as a function of time in all scenarios (e.g. Figure 17.5). This was an expected result and can be attributed to the ability of both wind and water currents to force material outside of the initial starting MPA over time. In addition, the rate at which retention decreased corresponded with depth of release (Figure 17.5). Propagules released closer to the surface moved out of an MPA of origin faster than those released deeper in the water column. This result can be attributed to the increased susceptibility of propagules to both wind and current forces which have greater strength closer to the surface. In total, 14 of 26 MPAs retained at least some fraction of the initial starting material released, and as a general trend larger MPAs exhibited higher overall retention.

The term 'settlement' was used to identify the proportion of propagules that was contained anywhere within the network of all 26 MPAs at some specific time. There was an overall net reduction in percentage settlement over the duration of the simulation. Again it was observed that propagules moving at shallow depths were lost from the network more quickly than those released in deeper simulations. The 100m simulations from both seasons indicated a settlement rate approximately double of those observed at the other depths. Spring simulations exhibited a greater loss of total settlement than in summer simulations. This is likely due to more severe weather conditions found during spring, including higher winds and enhanced currents due to larger freshwater influences.

Overall the proportion of settlement of propagules was 25.6 per cent in the spring simulations, and 29.1 per cent in summer, both figures exceeding the proportion of the study area covered by the 26 MPAs (~ 19 per cent). The conclusion from this study is therefore that under

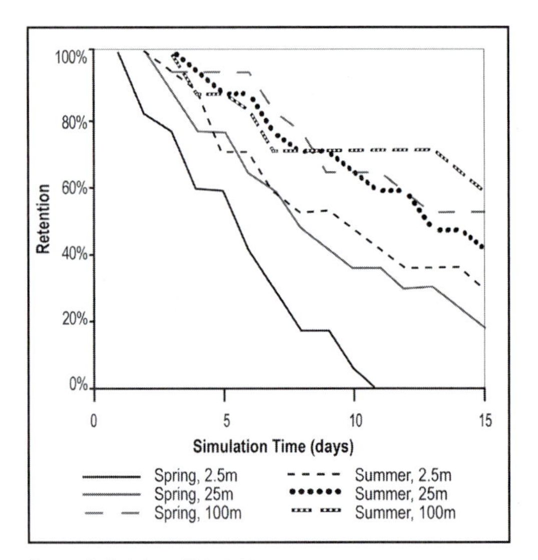

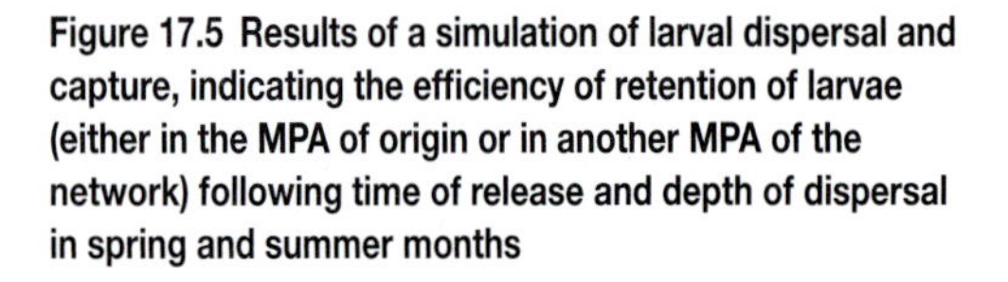

Source: Roff et al unpublished data

Figure 17.5 Results of a simulation of larval dispersal and capture, indicating the efficiency of retention of larvae (either in the MPA of origin or in another MPA of the network) following time of release and depth of dispersal in spring and summer months

the conditions tested, the set of MPAs proposed by CLF/WWF (2006) actually do comprise an efficient network. These analyses could be extended further to include propagule dispersal of longer durations, and could also include more detailed examination of source-sink dynamics and the contributions of individual MPAs to the overall network. For example, one MPA off the southwest coast of Nova Scotia appears to be a strong sink area (Figure 17.6) which may contribute disproportionately to overall network retention characteristics. Such continued analyses would permit further optimization of reserve design, and location and size of individual MPAs.

How large an area and what proportion?

In this chapter we now return to the issue raised in Chapter 16, of how large a total area or what proportion of a region needs to be protected. It is important to note that even with analyses such as that by Webdrogue above, we still do not have

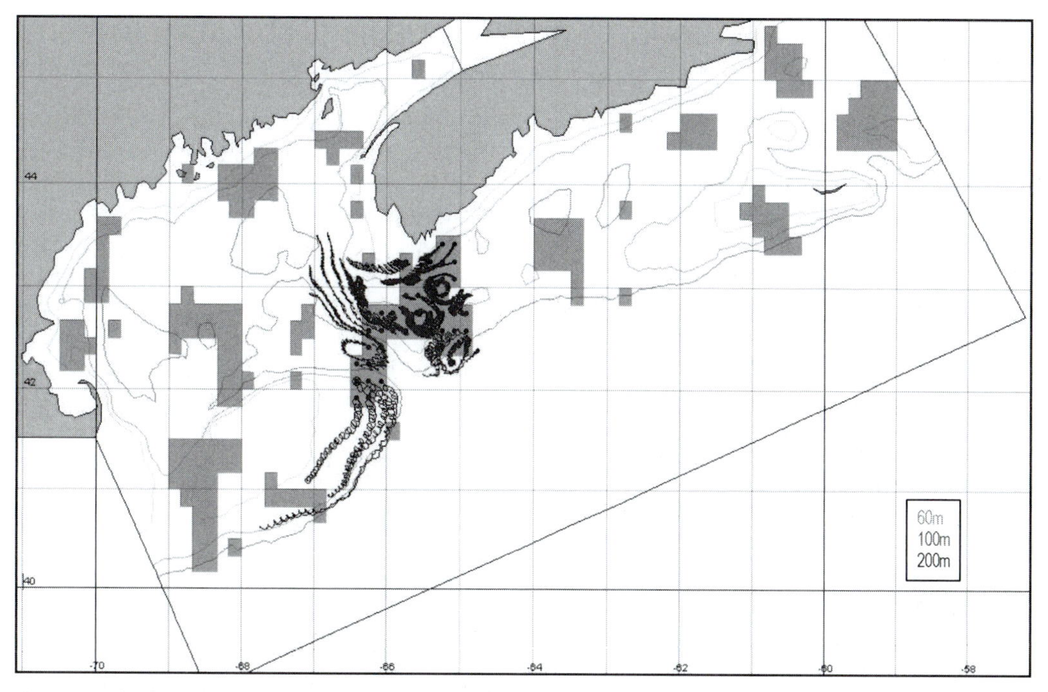

Figure 17.6 Simulation example of one MPA off the southwest coast of Nova Scotia which appears to be a strong sink and auto-recruitment area, and which may contribute disproportionately to the overall network retention characteristics

an objective measure of how large an area within a region should be protected by MPAs.

Until the theory is further developed (see Box 16.4), the best estimates we have of requirements for total area come only from models of exploited species (e.g. Guenette et al, 1998; Beattie et al, 2002). These models suggest that we need to protect between 25 and 40 per cent of an entire region and implement catch quotas in order to ensure a sustainable fishery. However, since the assemblages of commercially important demersal fish appear to be determined largely by depth and association with water masses (see Chapter 6), the distribution of this 25–40 per cent of a region assigned to MPAs could be largely discretionary among substrate types.

A second method to estimate the entire proportion of a region that should be protected comes from calculations of the search for an efficient network of MPAs. A whole series of considerations need to converge in order to design a true network of MPAs, including:

- Determination of distinctive MPAs and their size (Chapters 7 and 14).
- Determination of representative MPAs and sizes (Chapters 5, 6 and 14).
- Designation of entire area to be conserved – from fisheries models (Chapter 13).
- Determination of coherent sets of MPAs (Chapter 16).
- Determination of network characteristics and efficiency (this chapter) etc. (see Box 16.5).

The process of network design and calculation of the total proportion of a region required in MPAs in fact becomes highly iterative (Figure 17.7) in a search to incorporate all desired biodiversity components in an efficient network.

Even more sophisticated analyses of networks of MPAs are now being undertaken, for example Lockwood et al (2002), who considered a range of larval dispersal distributions from leptokurtic to platykurtic. The effect of different dispersal patterns was considered for both single isolated reserves of varying size receiving no

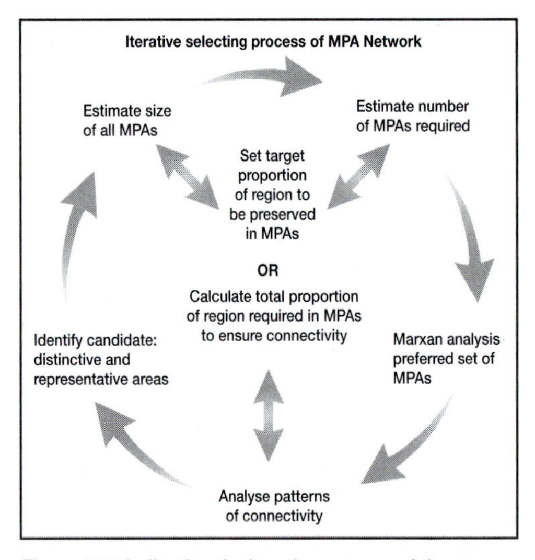

Figure 17.7 Indicating the iterative nature and the steps in the process of site selection for membership in a network of MPAs

external larvae, as well as multiple reserves with varying degrees of connectivity. Persistence in an isolated reserve required a size approximately twice the mean dispersal distance. Regardless of the dispersal pattern the population in a patch was not persistent if the reserve size was reduced to just the mean dispersal distance. With an idealized coastline structure consisting of an infinite line of equally spaced reserves, separated by regions of coastline in which reproduction is nil, the relative settlement was a function of the fraction of coastline and size of reserves over a broad range of dispersal patterns. A combination of sophisticated modelling studies and genetic information should allow a clear understanding of the complexities of connectivity patterns in coastal regions.

Lastly, we note that the subject of connectivity between networks of MPAs among ecoregions (in other words the patterns of connectivity among connectivities) has not yet been addressed. This means another level of spatial integration that would ensure that regional and national conservation efforts are globally coordinated.

Conclusions and management implications

The real significance of networks of MPAs is that they represent the only framework that has so far been conceived that can potentially ensure the preservation of the components of marine biodiversity (in other words the persistence of the ecological integrity of the oceans – their sustainability) in the face of continued human exploitation and degradation of the marine environment. Properly designed networks of MPAs, along with pollution control, are the indispensable condition to ensure survival of the ocean's biodiversity – including its fisheries.

Despite national commitments and international agreements for networks of MPAs, and despite the existence of tens of thousands of individual MPAs covering some 40 million square kilometres of the oceans (www.wdpa.org/), there are few declared networks of MPAs. Even those so declared appear to be deficient when examined for connectivity. Even the Great Barrier Reef Marine Park consists of sets of MPAs not known to comprise true networks.

More encouragingly, there are growing numbers of studies on dispersal, connectivity and genetic differentiation of marine populations that should soon lead to confidence in the design of MPA networks. However, an analysis of the literature by Bradbury et al (2008) indicates a historical focus on low-dispersal/low-latitude species, which are unlikely to be representative of global patterns. Nevertheless, these existing steps towards understanding and predicting taxon-specific and regional differences in dispersal are vital information for planning of MPA networks.

Elucidation of source-sink patterns and dynamics is vital not only to the design of MPA networks, but also to proper management of fisheries resources. Specific locations can vary by orders of magnitude in importance as both sources and sinks for larvae and recruits of all species (Roberts, 1997). It is likely that sites supplied copiously from 'upstream' areas will be more resilient to recruitment overfishing, less susceptible to species loss and less reliant on local management than other places. The mapping of connectivity patterns will enable the identification of beneficial management partnerships among nations and the design of networks of interdependent reserves.

Because of the high connectivity of the pelagic realm – with its open 'corridors' – the marine environment may be less susceptible to habitat fragmentation than the terrestrial environment. However, the subject of benthic corridors (analogues of those on land), which are vital for the successful connectivity of counter-migrating benthic species such as crabs and lobsters, has received far less attention than pelagic connectivity.

For offshore regions the required patterns of connectivity can probably be obtained from good local oceanographic models of currents and winds, coupled with knowledge of development times and behaviours for a range of dominant species, although genetic studies are also always informative. For inshore areas, especially in regions of complex topography and poorly researched coastal circulation cells, genetic studies will probably yield the best information on connectivity, where oceanographic studies of water movements are complex.

We should always be cautious in pleading 'special status' whether for individual species, habitats or 'ecosystems' unless they can truly be considered unique. However, when it comes to defining networks of MPAs, it is absolutely necessary to conduct site-specific studies, because oceanographic regimes vary according to regional characteristics of topography, tides and other current patterns.

Finally, although we have now examined the process of planning for marine conservation from the global to the local level, the remaining challenge is to integrate plans across the spatial hierarchy and the ecological hierarchy. Unfortunately, little thought has yet been directed to the spatial integration of networks across regional and ecoregional thresholds – that is, to the development of networks of networks or what they might 'look like'.

References

Aiken, C. M., Navarrete, S. A. Castillo, M. I. and Castilla, J. C. (2007) 'Along-shore larval dispersal kernels in a numerical ocean model of the central Chilean coast', *Marine Ecology Progress Series*, vol 339, pp13–24

Archambault, P., Roff, J. C. and Bourget, E. (1998) 'Nearshore abundance of zooplankton in relation to coastal topographic heterogeneity, and the mechanisms involved', *Journal of Plankton Research*, vol 20, pp671–690

Bay, L. K., Buechler, K., Gagliano, M. and Caley, M. J. (2006) 'Intraspecific variation in the pelagic larval duration of tropical reef fishes', *Journal of Fish Biology*, vol 68, pp1206–1214

Beattie, A. Sumaila, U. R., Christensen, V. and Pauly, D. (2002) 'A model for the bioeconomic evaluation of marine protected area size and placement in the North Sea', *Natural Resources Modelling*, vol 15, pp413–437

Becker, B. J., Levin, L. A., Fodrie, F. J. and McMillan, P. A. (2007) 'Complex larval connectivity patterns among marine invertebrate populations', *Proceedings of the National Academy of Sciences*, vol 104, no 9, pp3267–3272

Bell, J. J. and Okamura, B. (2005) 'Low genetic diversity in a marine nature reserve: Re-evaluating diversity criteria in reserve design', *Proceeding of the Royal Society*, vol 272, pp1067–1074

Booth, J. D. (1997) 'Long distance movements in *Jasus* spp. and their role in larval recruitment', *Bulletin of Marine Science*, vol 61, pp111–128

Bradbury, I. R., and Snelgrove, P. V. R. (2001) 'Contrasting larval transport in demersal fish and benthic invertebrates: The roles of behaviour and advective processes in determining spatial pattern', *Canadian Journal of Fisheries and Aquatic Sciences*, vol 58, pp811–823

Bradbury, I. R., Laurel, B., Snelgrove, P. V. R., Bentzen, P. and Campana, S. E. (2008) 'Global patterns in marine dispersal estimates: The influence of geography, taxonomic category and life history', *Proceeding of the Royal Society B*, vol 275, pp1803–1809

Brickman, D. and Frank, K. T. (2000) 'Modelling the dispersal and mortality of Browns Bank egg and larval haddock (*Melanogrammus aeglefinus*)', *Canadian Journal of Fisheries and Aquatic Sciences*, vol 57, pp2519–2535

Carter, R. W. G. (1988) *Coastal Environments. An Introduction to the Physical, Ecological and Cultural Systems of Coastlines*, Academic Press, London

Chou, C. L., Paon, L. A. and Moffatt, J. D. (2002) 'Metal contaminants for modelling lobster (*Homarus americanus*) migration patterns in the Inner

Bay of Fundy, Atlantic Canada', *Marine Pollution Bulletin*, vol 44, no 2, pp134–141

CLF/WWF (2006) 'Marine ecosystem conservation for New England and maritime Canada: A science-based approach to identifying priority areas for conservation', Conservation Law Foundation and World Wildlife Fund Canada, Toronto, Canada

Cowen, R. K., Lwiza, K. M. M., Kamazimz, M. M., Sponaugle, S., Paris, C. B. and Olson, D. B. (2000) 'Connectivity of marine populations: open or closed?', *Science*, vol 287, pp857–859

Cowen, R. K., Paris, C. D. and Srinivasan, H. (2006) 'Scaling of connectivity in marine populations', *Science*, vol 311, pp522–527

Crowder, L. B., Lyman, S. J., Figueira, W. F. and Priddy, J. (2000) 'Source-sink population dynamics and the problem of siting marine reserves', *Bulletin of Marine Science*, vol 66, pp799–820

Delgado, G. A., Glazer, R. A., Hawtof, D., Aranda, D. A., Rodríguez-Gil, L. A. and de Jesús-Navarrete, A. (2006) 'Do Queen Conch (*Strombus gigas*) Larvae Recruiting to the Florida Keys Originate from Upstream Sources? Evidence from plankton and drifter studies', in R. Grober-Dunsmore and B. D. Keller (eds) *Caribbean Connectivity: Implications for Marine Protected Area Management*, National Oceanic and Atmospheric Administration, Office of National Marine Sanctuaries, Silver Spring, MD

Galindo, H., Olson, D. and Palumbi, S. (2006) 'Seascape genetics: A coupled oceanographic-genetic model predicts population structure of Caribbean corals', *Current Biology*, vol 16, no 16, pp1622–1626

Gerber, L. R., Botsford, L. W. Hastings, A., Possingham, H. R., Gaines, S. D., Palumbi, R. and Andelman, S. (2003) 'Population models for marine reserve design: A retrospective and prospective synthesis', *Ecological Applications*, vol 1, pp47–64

Grantham, B. A., Eckert, G. L. and Shanks, A. L. (2003) 'Dispersal potential of marine invertebrates in diverse habitats', *Ecological Applications*, vol 13, pp108–116

Guenette, S., Lauck, T. and Clark, C. (1998) 'Marine reserves: From Beverton and Holt to the present', *Reviews in Fish Biology and Fisheries*, vol 8, pp251–272

Hannah, C. G., Shore, J. A. and Loder, J. W. (2000) 'The retention-drift dichotomy on Browns Bank: A model study of interannual variability', *Canadian Journal of Fisheries and Aquatic Sciences*, vol 57, pp2506–2518

Hannah, C. G., Shore, J., Loder, J. W. and Naimie, C. E. (2001) 'Seasonal circulation on the western and central Scotian Shelf', *Journal of Physical Oceanography*, vol 31, pp591–615

Hastings, A. and Botsford, L. W. (2003) 'Comparing designs of marine reserves for fisheries and for biodiversity', *Ecological Applications*, vol 13, ppS65–S70

Hedgecock, D. (1994) 'Does Variance in Reproductive Success Limit Effective Population Sizes of Marine Organisms?', in A. R. Beaumont (ed) *Genetics and Evolution of Aquatic Organisms*, Chapman & Hall, London

Incze, L. Xue, H. Wolff, N., Xu, D., Wilson, C., Steneck, R., Wahle, R., Lawton, P., Pettigrew, N. and Chen, Y. (2009) 'Connectivity of lobster (*Homarus americanus*) populations in the coastal Gulf of Maine: Part II. Coupled biophysical dynamics', *Fisheries Oceanography*, vol 19, pp1–20

Johnson, M. P., Crowe, T. P., Mcallen, R. and Alcock, A. L. (2008) 'Characterizing the marine NATURA 2000 network for the Atlantic region', *Aquatic Conservation: Marine and Freshwater Ecosystems*, vol 18, pp86–97

Jones, G. P., Planes, S. and Thorrold, S. R. (2005) 'Coral reef fish larvae settle close to home', *Current Biology*, vol 15, no 14, pp1314–1318

Kenchington, E. L., Patwary, M. U., Zouros, E. and Bird, C. J. (2006) 'Genetic differentiation in relation to marine landscape in a broadcast-spawning bivalve mollusc (*Placopecten magellanicus*)', *Molecular Ecology*, vol 15, pp1781–1796

Klinger, T. and Ebbesmeyer, C. (2002) 'Using Oceanographic Linkages to Guide Marine Protected Area Network Design', in T. Droscher (ed) *Proceedings of the 2001 Puget Sound Research Conference*, Puget Sound Action Team, Olympia, WA

Knutsen, H. P., Jorde, P. E., Andr, C. and Stenseth, N. C. (2003) 'Fine-scaled geographical population structuring in a highly mobile marine species: The Atlantic cod', *Molecular Ecology*, vol 12, pp385–394

Largier, J. L. (2003) 'Considerations in estimating larval dispersal distances from oceanographic data', *Ecological Applications*, vol 13, pp71–89, www.esajournals.org/doi/abs/10.1890/1051-0761(2003)013 per cent5B0071:CIELDD per cent5D2.0.CO per cent3B2 - aff01#aff01

Lessios, H. A., Weinberg, J. R. and Starczak, V. R. (1994) 'Temporal variation in populations of the marine isopod *Excirolana*: How stable are gene frequencies and morphology?', *Evolution*, vol 48, pp549–563

Levin, L. A. and Bridges, T. (1995) 'Pattern and Diversity in Reproduction and Development', in L. McEdward (ed) *Ecology of Marine Invertebrate Larvae*, CRC Press. Boca Raton, FL

Lockwood, D. R., Hastings, A. and Botsford, L. W. (2002) 'The effects of dispersal patterns on marine reserves: Does the tail wag the dog?', *Theoretical Population Biology*, vol 61, pp 297–309

McEdward, L. R. (1995) *Ecology of Marine Invertebrate Larvae*, CRC Press, Boca Raton, FL

Morgan, S. (1995) 'The Timing of Larval Release', in L. McEdward (ed) *Ecology of Marine Invertebrate Larvae*, CRC Press. Boca Raton, FL

Myers, R. A. and Drinkwater, K. (1989) 'The influence of Gulf Stream warm core rings on recruitment of fish in the northwest Atlantic', *Journal of Marine Research*, vol 47, pp635–656

Nelson, G. and Hutchings, L.(1987) 'Passive Transport of Pelagic System Components in the Southern Benguela Area', in A. I. L. Payne, J. A. Gulland and K. H. Brink (eds) *The Benguela and Comparable Ecosystems*, South African Journal of Marine Science 5, Sea Fisheries Research Institute, Cape Town, South Africa

Palumbi, S. R. (1995) 'Using Genetics as an Indirect Estimator of Larval Dispersal', in L. McEdward (ed) *Ecology of Marine Invertebrate Larvae*, CRC Press, Boca Raton, FL

Paris, C. B., Cowen, R. K., Claro, R. and Lindeman, K. C. (2005) 'Larval transport pathways from Cuban snapper (*Lutjanidae*) spawning aggregations based on biophysical modeling', *Marine Ecology Progress Series*, vol 296, pp93–106

Pawlik, J. R. (1992) 'Chemical ecology of the settlement of benthic marine invertebrates', *Oceanography and Marine Biology Review*, vol 30, pp273–335

Planes, S., Galzin, R., Garcia Rubies, A., Goni, R., Harmelin, J. G., Le Direach, L., Lenfant, P. and Quetglas, A. (2000) 'Effects of marine protected areas on recruitment processes with special reference to Mediterranean littoral ecosystems', *Environmental Conservation*, vol 27, pp126–143

Roberts, C. M. (1997) 'Connectivity and management of Caribbean coral reefs', *Science*, vol 278, pp1454–1457

Robinson, C. L. K., Morrison, M. and Foreman, M. G. G. (2005) 'Oceanographic connectivity among marine protected areas on the north coast of British Columbia, Canada', *Canadian Journal of Fisheries and Aquatic Sciences*, vol 62, pp1350–1362

Rocha, L. A., Craig, M. T. and Bowen, B. W. (2007) 'Phylogeography and the conservation of coral reef fishes', *Coral Reefs*, vol 26, pp501–512

Roff, D. A. (1992) *The Evolution of Life Histories: Theory and Analysis*, Chapman and Hall, New York

Roff, J. C. (2005) 'Conservation of marine biodiversity: too much diversity, too little co-operation', *Aquatic Conservation: Marine and Freshwater Ecosystems*, vol 15, pp1–5

Ruzzante, D. E., Taggart, C. T. and Cook, D. (1998) 'A nuclear DNA basis for shelf- and bank-scale population structure in northwest Atlantic cod (*Gadus morhua*): Labrador to Georges Bank', *Molecular Ecology*, vol 7, pp1663–1680

Samadi, S., Bottan, L., Macpherson, E., De Forges, B. R. and Boisselier, M. C. (2006) 'Seamount endemism questioned by the geographic distribution and population genetic structure of marine invertebrates', *Marine Biology*, vol 149, pp1463–1475

Selkoe, K. A. Henzler, C. M. and Gaines, S. D. (2008) 'Seascape genetics and the spatial ecology of marine populations', *Fish and Fisheries*, vol 9, no 4, pp363–377

Shanks, A. L. (1995) 'Mechanisms of Cross-shelf Dispersal of Larval Invertebrates and Fish', in L. McEdward (ed) *Ecology of Marine Invertebrate Larvae*, CRC Press, Boca Raton, FL

Shanks, A. L., Grantham, B. A. and Carr, M. H. (2003) 'Propagule dispersal distance and the sizing and spacing of marine reserves', *Ecological Applications*, vol 13, pp159–169

Siegel, D. A., Mitarai, S., Costello, C. J., Gaines, S. D., Kendall, B. E., Warner, R. R. and Winters, K. B. (2008) 'The stochastic nature of larval connectivity among nearshore marine populations', *Proceedings of the National Academy of Sciences*, vol 105, pp8974–8979

Sotka, E. E. and Palumbi, S. R. (2006) 'The use of genetic clines to estimate dispersal distances of marine larvae', *Ecology*, vol 87, pp1094–1103

Stockhausen, W. T., Lipcius, R. N. and Hickey, B. M. (2000) 'Joint effects of larval dispersal, population regulation, marine reserve design, and exploitation of production and recruitment in the Caribbean spiny lobster', *Bulletin of Marine Science*, vol 66, pp957–990

Strathmann, M. F. (1987) *Reproduction and Development of Marine Invertebrates of the Northern Pacific Coast: Data and methods for the study of eggs, embryos, and larvae*, University of Washington Press, Seattle, WA

Thorson, G. (1950) 'Reproductive and larval ecology of marine bottom invertebrates', *Biological Review*, vol 25, pp1–45

Uehara, T., Shingaki, M. and Taira, K. (1986) 'Taxonomic studies in the sea urchin, genus Echinometra, from Okinawa and Hawaii', *Zoological Science*, vol 3, no 1114

Wildish, D., and Kristmanson, D. (1997) *Benthic Suspension Feeders and Flow*, Cambridge University Press, New York

Young, C. M. (1995) 'Behaviour and Locomotion During the Dispersal Phase of Larval Life', in L. McEdward (ed) *Ecology of Marine Invertebrate Larvae*, CRC Press, Boca Raton, FL

Zhan, A., Hu, J., Hu, X., Zhou, Z., Hui, M., Wang, S., Peng, W., Wang, M. and Bao, Z. (2009) 'Fine-scale population genetic structure of Zhikong scallop (*Chlamys farreri*): Do local marine currents drive geographical differentiation?', *Marine Biotechnology*, vol 11, pp223–235

18

Approaches to the Establishment of Marine Monitoring Programmes

Stabilizing the baselines

Preserving health by too severe a rule is a worrisome malady.

Francois de La Rochefoucauld (1613–1680)

Introduction

The past two decades have seen a suite of new national and international statutes and agreements requiring the development of monitoring programmes to support, among other things, the ecosystem approach to management discussed in Chapters 4 and 13. For example, at the international level, the United Nations Convention on the Law of the Sea (UNCLOS, part XII, s. 4, a. 204) commits signatories to monitor the risks or effects of pollution, publish these monitoring results and assess the potential effects of planned activities under their jurisdiction. In addition, the collaborative United Nations Environment Programme (UNEP) – World Conservation Monitoring Centre has been developing terrestrial and marine monitoring programmes for over two decades, informed by the various conventions and decisions under these agreements.

At the regional level, marine indicators of various types are required to assess progress towards issues regulated under the European Union (EU) Water Framework Directive and EU Marine Strategy Framework Directive. The Oslo and Paris Commission Convention (OSPAR), composed of 15 European nations within the western coasts and catchments of Europe, commits signatories to the marine environment and has resulted in a framework of ecological quality objectives (EcoQQ) for the northeast Atlantic (Rogers et al, 2007). At the national level, marine monitoring is required under the US Federal Water Pollution Control Act, the UK Marine Monitoring and Assessment Strategy (UKMMAS), as well as by a number of other nations.

These overarching legal and policy directives to monitor the often nebulous concepts of ecosystem health and human well-being, however well intentioned, provide little direction on what specifically is required to be monitored, how it is to be monitored and for how long. Scientific working groups established in the wake of these agreements have been working on developing methods to assess progress towards these directives. However, the breadth and scope of what potentially could be monitored spans from global networks of ocean sensors used for weather prediction, navigation and atmospheric science to determining local effects of contaminants in the tissues of individual organisms. Marine monitoring is therefore a collective, interdisciplinary endeavour linking fields as diverse as molecular biology, oceanography and economics to identify ecologically and economically relevant objectives and indicators.

UNEP defines monitoring as a repetitive observation (for defined purposes) of one or more chemical or biological elements according to a prearranged schedule over time and space using comparable and standardized methods (van der Oost et al, 2003). While this definition is comprehensive it fails to consider the human activities that could be monitored, such as boating (e.g. leading to noise, vessel strikes), catch per unit effort (CUPE) or the degree of shoreline development. Thus marine monitoring for the purposes of evaluating and informing marine conservation and management efforts is much broader than most monitoring programmes. Unfortunately there are few, if any, systematic attemps to fully consider the suite of abiotic, biotic and human-use characteristics, features and conditions that could contribute to assessments to determine the ecological and economic status of marine systems.

While long-term oceanographic monitoring programmes have been in effect for decades, many jurisdictions have recently established, or are currently establishing, marine biological monitoring programmes to comply with international and national legislation and conventions to monitor whether MPAs are achieving their objectives, and to monitor whether various fisheries management decisions are resulting in measurable changes to fished populations (NRC, 1990; Schiff et al, 2002; Bernstein and Weisberg, 2003). In addition, other recent monitoring programmes have been designed to establish environmental baselines to assist with various tasks, which include: predicting fisheries recruitment success in a given year; monitoring changes in sea-level rise as a result of global warming; and monitoring radionuclides in coastal waters.

Broadly, all monitoring programmes can be summarized as one of the following three types: *implementation monitoring* to assess whether strategies are implemented, such as progress towards a commitment to protect some percentage of a marine area; *compliance monitoring* to determine whether strategies are being followed, such as monitoring fisheries by-catch; and *effectiveness monitoring* to assess whether strategies are meeting targets.

The purpose of this chapter is to focus on effectiveness monitoring and to broadly outline some of the characteristics of a well thought-out marine monitoring programme as well as to suggest the types of characteristics, features or conditions that can be monitored in marine environments. In keeping with the hierarchical approach to conservation used by Zacharias and Roff (2000), this chapter examines the various options for monitoring at the genetic, population, community and ecosystem levels of organization. This chapter is not meant to be a comprehensive resource on all aspects of monitoring (e.g. pollution monitoring, monitoring fisheries harvests), but is intended to provide the reader with various monitoring options which they may then choose to explore further. One of the difficulties with monitoring is that there are many potential ways to address a monitoring question and few texts provide an overview of each methodology and its strengths and weaknesses. Those interested in a more thorough treatment of marine monitoring programmes are encouraged to review Davies et al (2001).

Perspectives on marine monitoring

While marine monitoring programmes are established for many purposes, there are four general reasons for their establishment (modified from Davies et al, 2001):

1 To measure the environmental impact of human activities.
2 To give early warnings of problems.
3 For scientific interest and to understand the behavior and function of ecosystems.
4 To advise decision-making and track the success of management actions.

Measuring the environmental impact of human activities includes monitoring the effects of consumptive (e.g. fishing) and non-consumptive (e.g. whale-watching) activities as well as monitoring levels of pollutants in various marine systems. Monitoring for early warnings or problems generally includes the ambient monitoring of water quality or fish populations. Ambient monitoring is not targeted at any particular threat, but features are monitored that may indicate that threats are present. Scientific monitoring can be used to establish baselines and benchmarks (Dayton et al, 1998) as well as to understand the behaviour and functioning of marine systems. Establishing relationships such as community responses to oceanographic oscillations may provide valuable information to fisheries management and the location and operation of MPAs. Lastly, monitoring can be used to evaluate the success or failure of management decisions and can be used as an input into adaptive management. Examples of these types of programmes include evaluating the effects of fisheries closures in relation to fisheries management objectives.

Regardless of the various perspectives on marine monitoring, the types of things that can be monitored in marine environments can be broadly categorized into *characteristics*, *features* and *conditions*. While there is considerable overlap within these definitions, they are proposed here to ensure that the full range of the types of things that can be monitored is considered. Examples of characteristics that can be monitored include the size and age classes of populations and morphological characteristics. Features that can be monitored include physiographic features such as reefs, seamounts and other physiographic features. Conditions that can be monitored include water mass properties, water quality parameters or the health of a marine organism.

A rigorous and well thought-out monitoring programme will be designed to answer one or more management questions. Where possible, monitoring programmes should be viewed as analogous to scientific investigations where a hypothesis is tested to determine whether monitoring objectives (targets) are achieved. Monitoring programmes should also compare the current situation to an established standard to determine the condition of a feature. Lastly, effective monitoring requires that target conditions should be clearly defined.

Initiating a monitoring programme

Marine monitoring is a broad field that encompasses very different objectives. All monitoring programmes, however, are linked in that they attempt monitor the change(s) of a variable over some period of time. While there are many different features of the marine environment that could be monitored, monitoring can generally be separated into the following types.

Ambient monitoring programmes

Ambient monitoring programmes do not address any specific threat or specific objectives, but are established to look for changes in background/ baseline conditions. Ambient monitoring programmes are generally established over longer time periods (decades in the case of many ocean buoy measurement systems), over larger geographic areas, and often do not presuppose any particular threat in any area. The programmes examine broad-scale ecological properties determined by direct observation, automated collection or remote sensing of abiotic or biotic information. The placement of ambient monitoring stations may be based on either a systematic or random scheme, or may use ships of opportunity. Types of abiotic ambient programmes include light stations and ocean buoys, which collect information on a number of variables, including salinity, temperature, wind speed and direction, and solar irradiance.

Examples of ambient biotic monitoring programmes include 'Mussel Watch', which monitors oysters and mussels for various contaminants in a number of countries (Goldberg and Bertine, 2000). The Great Barrier Reef Marine Park Authority's Reef Plan Marine Monitoring Programme is an ambient programme to assess the long-term effectiveness of the Reef Plan in halting and reversing declines in water quality from watersheds that input into the Great Barrier Reef. The programme consists of the following five components: river mouth water quality monitoring; inshore marine water quality monitoring; marine biological monitoring; bio-accumulation monitoring; and monitoring of socio-economic factors. Lastly, the Puget Sound Ambient Monitoring Programme (PSAMP) is a multi-agency effort in Washington State, USA to monitor the health of Puget Sound, a constrained waterway supporting multiple uses with several million people living in the adjacent watersheds. Since 1989, the PSAMP has assessed the health of Puget Sound fishes and macro-invertebrates through monitoring the status and trends of the following five indicators: contaminant levels in tissue and bile; liver disease in adult English sole; endocrine disruption in male fish; spawning success; and fish abundance (Newton et al, 2000). The challenge of ambient monitoring is to distinguish between 'signal' and 'noise', where signal becomes noise and vice versa when spatial and temporal scales are changed (Osenberg et al, 1994).

Targeted monitoring programmes

Targeted monitoring programmes are directed at specific features or objectives, and are often tied with management objectives or evaluating recovery efforts. Targeted programmes are the most common type of monitoring, where the effectiveness of management decisions is assessed. Common targeted monitoring programmes include monitoring communities and habitats before and after a development project to assess the impacts of the development. Nearshore examples include monitoring marine communities before and after the establishment of an industrial site. Offshore examples include monitoring near oil and gas platforms and monitoring fish populations to ensure fisheries management decisions are achieving their objectives. More advanced targeted monitoring programmes operate under global or regional conventions and agreements and address multiple issues simultaneously.

A frequently used targeted approach is the pressure-state-response (PSR), which attempts to link human-induced pressures on the environment with changes in the state (condition) of the environment that are in turn addressed by changes in activities through legislation or policy (OECD, 1993). PSR in marine environments consists of the identification of key forcing variables (e.g. oceanographic indices, fishing mortality), measures of system state (condition) for the ecosystem and for individual components (e.g. stock status for target species), indicators of system response to the pressures identified and management actions to address these pressures.

Integrated monitoring programmes

Integrated monitoring programmes are designed to report on environmental conditions and trends and rely on the combination of a number of measures to report on trends both expected and unexpected at various spatial and temporal scales. Functionally, integrated monitoring combines both ambient and targeted approaches at multiple levels of the ecological hierarchy but no single, agreed-upon definition has yet been proposed or adopted. Integrated monitoring approaches have been primarily applied towards understanding ecosystem function to inform fisheries management. The US Environmental Protection Agency (EPA) and European Union have a number of integrated monitoring programmes under way to address multiple human activities and their impacts on marine environments and to integrate the monitoring of contaminants, biomarkers and population and community indicators of ecosystem health.

The US EPA has developed a three-tiered approach to provide monitoring information towards what the EPA terms an integrated assessment (Messer et al, 1991). The EPA intends to use monitoring to: characterize the problem; diagnose the causes; set management actions; assess the effectiveness of actions; re-evaluate the causes; and continue assurance of effectiveness of actions. The three tiers are as follows:

Characterization of the problem (Tier 1) examines broad-scale ecological response properties determined by survey, automated collection and/or remote sensing. This monitoring may be ambient or targeted depending on the objectives of the study.

Diagnosis of the causes (Tier 2) examines issue- or resource-specific surveys and observations concentrating on cause–effect interactions.

This monitoring may also be ambient, but is primarily targeted.

Diagnosis of interaction and forecasting (Tier 3) consists of intensive monitoring and research index sites with higher spatial and temporal resolution to determine specific mechanisms of interaction needed to build cause–effect models (NSTC, 1997). Information generated at each tier is designed to assist with the direction and interpretation of results from the other tiers.

Stages in a monitoring programme

Monitoring programmes generally consist of four stages. The first stage is to determine what to monitor, which is generally set in response to some problem or objective. Any variable which is expected to change over time can be monitored; however, what is monitored should either be a feature of interest or vary in some predictable manner with a feature of interest. The most common monitoring objectives relate to near-shore water quality, where there are a number of measures (e.g. chlorophyll-*a*, turbidity, dissolved oxygen, faecal coliforms etc.) which can be used as indicators of water quality.

The second stage is to determine the most appropriate monitoring technique to use. This again is determined by evaluating the monitoring objectives. There are hundreds of different types of monitoring at the genetic, population, community and ecosystem levels (Table 18.1), where each technique has its own strengths and weaknesses. Questions that should be asked when determining the appropriate monitoring technique include (modified from Davies et al, 2001):

- Will the technique damage the species or environment?
- Will the technique provide a type of measurement consistent with the target objectives?
- Will the technique measure the attribute across an appropriate range of conditions?
- Will the technique provide sufficiently precise observations to detect appropriate scales of change?
- Is the technique within the budget available?

The third stage is to organize the deployment of the technique in the field and the final step is to assess the condition of the features of interest.

The role of indicators in monitoring

Indicators are used as surrogates to assess a trend or condition when the target variable cannot be measured due to limitations in either the ability or cost of directly monitoring the variable. A typical definition of an indicator is that it '...is a measure, index, or model used to estimate the current state and future trends in physical, chemical, biological, or socioeconomic conditions of the environment, along with thresholds for management action to achieve desired ecosystem goals' (Fisher, 2001). Indicators are differentiated from other types of statistics (e.g. measurement of an event or phenomena that produces raw data) in that indicators are linked to a specific management question or problem. Indicies (e.g. water quality index, index of biotic integrity) are aggregations of indicators according to some formula to produce a single metric for communication and summarization purposes.

For example, temperature can be used as a surrogate for the reproductive success and survival of certain species, and dissolved oxygen levels may indicate the level of human disturbance in a system. Trend monitoring over time generally utilizes condition indicators (as defined by Zacharias and Roff, 2001; Chapter 9), which vary in response to the ecological state, health or integrity of a system. Indicators or surrogates for water quality frequently use concentration of chlorophyll, sediments, dissolved oxygen and total suspended solids. Indicators at the genetic level include the presence or frequency of certain alleles. At the species/population level, indicators may include condition indicators and keystone species. At the community level, indicators that could be monitored include the abundance relationships between certain species as well as guilds or functional groups. At the ecosystem level, indicators are often water quality measures such as chlorophyll, turbidity and dissolved solids.

Table 18.1 Types of biophysical and human-use features, characteristics and conditions that can be monitored at the population, community and ecosystem levels

Population level	Community level	Ecosystem level	Human-use
Density	Species richness, evenness,	Productivity	*Upland uses*
Area	abundance	Water motion	Population
Size	Succession	Entrainment	Population density
Presence	Disturbance	Water properties	Zoning density
Range	Alternate stable states	photosynthetically active	Land use
Distribution	Predation	radiation (PAR)	Land cover
Age structure	Competition	turbidity	Impervious cover
Genetic diversity	Parasitism	chlorophyll	Stormwater condition
Gene frequency	Mutualism	temperature	Sewered vs non-sewered areas
Number of alleles	Disease	salinity	Fish and shellfish closures for
Degree of linkage	Amensalism	dissolved oxygen	contaminants
Inbreeding or outbreeding	Transition areas	total organic carbon	Discharge reports and pollution
depression	Functional groups	total solids	permits
Genetic drift	Meta populations	total volatile solids	Habitat alteration
Bottlenecks	Heterogeneity	total sulfide	Wetland loss
Gene flow	Endemism	Retention	Lost urban streams
Migration	Diversity	Illumination	*Marine uses*
Recruitment	Representative and distinct	Stratification	Fish and shellfish closures for
Retention	areas	Patchiness	contaminants
Evolution	Biomass	Dissolved gasses	Catch per unit effort (CUPE)
Molecular markers		Boundaries	Number of boats
Dispersion		Water clarity	Number of boat trips
Viability		Nutrient status	Time on site (TIA)
Pathogens		Nitrogen compounds	Person-day visits
Mortality/Morbidity		• Nitrate, nitrite, and	Number of divers
Strandings		ammonium	Number of dives
Gene flow		• Phosphorous	Income redistribution (e.g. fishers
Level of contaminants/oiling		Sediment character	and tourism operators)
Disease		Sediment type	Wealth flows
Deformities		Bathymetry	Adjacent visitorship and tourism
Individual size and condition		Topography	Transportation type (powered,
index		Particle size	non-powered)
Growth rates		Recruitment	
Rates of harvest		Population viability	
Contaminants		Genetic diversity	
artificial radionuclides		Benthic cover	
petroleum hydrocarbons		Area for breeding	
chlorinated hydrocarbons		Availability of haulout sites	
metals		Area for feeding	
organotin		Sea level rise	
carcinogens		Contamination	
mutagens		Metals (Cd, Zn, Pb, Hg, Cu, Fe,	
pesticides		Mn and Co)	
endocrine disrupters			
physical debris			

Characteristics of a good indicator are that it is scientifically sound, easily understood, sensitive to the change(s) it is intended to measure, measurable to a level of accuracy available with current technology and capable of being updated regularly. Other questions that should be considered in the selection of indicators include (adapted from http://cleanwater.gov/coastalresearch/report.html, Rees et al, 2008):

- Can the proposed indicator be quantified in a simple manner?
- Does the indicator respond to a broad range of conditions?
- Is the indicator sensitive and anticipatory to problematic conditions or concerns?
- Can the indicator resolve meaningful differences in such environmental conditions?
- Can the measurement provide an integrated view of effects over various times, scales, and changes in environmental conditions?
- Are the results from the measurement reproducible and transportable?
- Is there reference information by which to judge the results obtained?
- Can the results be compared across differences in time and space?
- Is the indicator non-destructive where the measurement does not cause ecosystem damage?
- Is the indicator easy to understand and communicate to non-specialists and decision-makers?
- Is the indicator scientifically and legally defensible?
- Does the indicator involve relevant stakeholders in its formulation and use?

There have also been a number of efforts to categorize different marine indicators. The United Nations Educational, Scientific and Cultural Organization (UNESCO, 2006) identified the following three classes of ecosystem indicators (slightly modified):

- Ecological indicators to characterize and monitor changes in the state of various physical, chemical and biological aspects of the environment relative to defined quality targets with thresholds for management action.
- Socio-economic indicators to measure whether environmental quality is sufficient to maintain human health, human uses of resources and favourable public perception.
- Governance indicators to monitor the progress and effectiveness of management and enforcement practices towards meeting environmental policy targets.

A final consideration is how indicators should be selected, as there are potentially hundreds of different characteristics, features and conditions that could be utilized for monitoring and indicators developed (Table 18.1). Ultimately, indicators must survive the scrutiny of peer review and legal challenge, and must have the confidence of stakeholders in that they are objectively derived and useful for a particular purpose. Rice and Rochet (2005) propose an eight-step framework for choosing and using indicators which is summarized below:

1. Identify the users, their needs and their objectives for all potentially affected stakeholders.
2. Translate stakeholder objectives into candidate indicators that consider ecological and economic aspects.
3. Assign weights to the following nine screening criteria of three types:
   - interpretation (concreteness, public awareness, theoretical basis);
   - implementation (availability of historic data, cost, measurability);
   - application (sensitivity, specificity, responsiveness).
4. Score the indicators on the screening criteria through evaluation of the information content or quality of each indicator relative to each criterion as well as the strength of the evidence by which information content or quality is judged.
5. Summarize the results of the scoring through quantitative means.
6. Determine how many indicators are required.
7. Select the final suite of indicators to be used.
8. Report on the ecological and economic status using the indicators to inform and, if necessary, alter management direction.

In summary, indicators are a method of presenting and summarizing complex information in a straightforward and understandable manner that can also be used to gauge the general status of a system. While there is a wide range of choice of indicators at the various ecological levels that could be applied towards marine monitoring programmes, 'good' indicators follow the SMART philosophy (specific, measurable, achievable, relevant and timely). Indicators can also be applied in a sequential manner where urgent requirements for them can be addressed through the development of preliminary indicators that do not directly measure a change in a biological system but rather measure changes to human activities that may affect marine environments. This interim step provides time for more complex indicators of ecosystem state to be developed to monitor stability, diversity and 'ecosystem health'.

Challenges continue to exist in applying marine indicators. While a full discussion on the strengths and weaknesses of using marine indicators is beyond the scope of this book, some of the more serious issues associated with marine indicators include: breaks in time-series due to political and funding constraints; insufficient historical data with which to compare to the present state; a lack of systematic standards for the use of marine indicators; lack of qualified staff to work with indicators; indicator terminology that is often inconsistent between disciplines and nations; and indicators that lack sufficient sensitivity to be applicable.

Approaches to marine monitoring and the types of characteristics, features and conditions that may be monitored

The two primary kinds of features or conditions that can be monitored in marine environments include: biophysical characteristics, features and conditions; and human-use characteristics, features and conditions. The biophysical characteristics, features and conditions can be monitored at the following levels (after Zacharias and Roff, 2000):

- Genetic
- Populations/species
- Communities
- Habitats/ecosystems
- Combinations of levels (e.g. biotopes)

For the purposes of marine monitoring, the genetic level is generally incorporated into the species/population level as genetic monitoring relates predominantly with genetic diversity and the degree of distinctiveness (as expressed through isolation and connectivity) among populations. Also, certain characteristics, features and conditions (e.g. area) can be monitored at multiple levels of the ecological hierarchy.

Population/species level monitoring

Population/species level monitoring has traditionally been concerned with either monitoring changes in population abundance and structure, or monitoring levels of contaminants within a population at the organismal level. Monitoring changes in population characteristics is a vital aspect of stock assessment and may also signal adverse effects by contaminants or habitat loss in non-harvested populations. However, outside of intertidal environments (which are easily observed), changes in population structure is difficult to detect. Often, only mass mortalities and morbidities are detected, as are changes in primary sexual characteristics (imposex) in certain benthic invertebrates.

The purpose of this section is not to discuss in detail the various methods of monitoring at the population/species level, but rather to provide the various characteristics, features and conditions that can be monitored at this level, as well as some popular techniques and recent advances in population-level monitoring. At the population/species level, characteristics, features and conditions which can be monitored are shown in Table 18.1.

Biomonitoring

A frequently used population-level monitoring approach that focuses on the ability to use individual living organisms to detect or signal changes that cascade upwards to cause either population- and community-level impacts or changes to the physical environment has been termed 'biomonitoring'. Biomonitoring is predicated on the assumption that stress-induced changes at the community and ecosystem levels are to be avoided but indicators of these changes often lack the sophisitication for early detection. Therefore, changes detected at the molecular and cellular levels may provide an early warning of incipent change at the population and ecosystem levels (Moore et al, 2004).

Biomonitoring differs from chemical, human-use or ecosystem (in other words oceanographic) monitoring in that it is assumed to be a more accurate predictor of true ecological risk because the approach links direct cause and effect between a contaminant and an environmental stress. While biomonitoring could be applied to many types of environmental stresses (e.g. noise, fishing, introduced species) the approach is primarilly directed towards determining the consequences of contaminants (xenobiotic compounds) on organisms and their subsequent upward effects on the ecological hierarchy. Thus the biomonitoring approach is assumed to act as an early-warning system where 'biomarkers' signal the adverse biological responses towards certain xenobiotic compounds at the organismal level. Biomarkers are measurments in body fluids, cells or tissues that indicate changes due to the presence and magnitude of various compounds (NRC, 1989). The World Health Organization defines a biomarker to include almost any measurement reflecting an interaction between a biological system and a potential hazard, which may be chemical, physical or biological. Finally, the US National Academy of Sciences defines a biomarker as '...a xenobiotically induced variation in cellular or biochemical components or processes, structures or functions that is measurable in a biological system or sample' (NRC, 1989).

Biomarkers have been further subdivided in terms of their use at each level of the ecological hierarchy. The application of four different biomonitoring levels using biomarkers, bioassays, bioindicators and ecological indicators was proposed by Van Gastel and Van Brummelen (1994) and they are defined as follows:

- Biomarkers measure biochemical and physiological processes and deviations from the normal situation ('health') at the sub-organismal level.
- Bioassays measure survival, growth and reproduction of individuals in the face of contaminants using classic laboratory ecotoxicity tests at the population level.
- Bioindicators measure changes in genetic structure, age structure or abundance of a population.
- Ecological indicators measure changes in species composition, abundance and diversity that may be indicative of the effects of pollution on communities.

Most biomonitoring efforts to date have been directed at detecting the effects and consequences of trace metals and organic pollutants, including polychlorinated biphenyls (PCBs), organochlorine pesticides (OCPs), polycyclic aromatic hydrocarbons (PAHs), polychlorinated dibenzofurans (PCDFs) and polychlorinated dibenzodioxins (PCDDs).

Perhaps the earliest and most well-known biomonitoring programme is the Mussel Watch programme which both monitors ambient conditions as well as targets the extent and impact of specific contaminants. Initiated in 1975 to monitor radionuclides, petroleum hydrocarbons, chlorinated hydrocarbons and metals, oysters and mussels were selected as indicators because they bioaccumulated toxins to levels that were measurable using the detection equipment of the time (Goldberg, 1975). The programme was marketed as both a national and international solution to monitoring pollution in nearshore coastal waters. In 1988, the programme was reviewed by Goldberg (1988) who – among others – noted that detecting the amounts of pollution in organisms, while useful, failed to evaluate the effects of these pollutants on both the indicators as well as their environments. In addition, new toxic chemicals were continuously added to the marine environment, but insufficient budgets were available to analyse these chemicals (Goldberg and Bertine, 2000). The Mussel Watch programme has been adopted in several dozen nations.

In 1986, the US Mussel Watch programme began monitoring a suite of trace metals and organic contaminants at 145 sites that has now increased to approximately 140 different chemical compounds at nearly 300 sites. Earlier monitoring in the 1960s and 1970s for a subset of these sites extends the contaminant record to over 40 years. A 20-year review of the monitoring data concluded that, since 1986, metal trends were slightly downward depending on location but that, nationwide, trends in organic contaminants significantly decreased (Kimbrough et al, 2008). For most organic contaminants, downward trends were the result of state and federal regulation. The highest concentrations for both metal and organic contaminants are found near urban and industrial areas.

There have also been a number of recent developments in population-level monitoring which are expected to reduce costs and improve monitoring results. While readers should consult van der Oost et al (2003) for a thorough discussion of recent advances in the biomonitoring field, a partial list of the various biomarkers now in use includes biotransformation enzymes (phase I and II), oxidative stress parameters, biotransformation products, stress proteins, metallothioneins (MTs), MXR proteins, hematological parameters, immunological parameters, reproductive and endocrine parameters, genotoxic parameters, neuromuscular parameters, physiological, histological and morphological parameters (van der Oost, 2003).

Among the more interesting developments are methods to avoid analysing large numbers of pollutants in either the marine environment (e.g. water, sediments) or organisms (e.g. mussels), through assaying certain detoxifying enzymes and hormonal active substances. In this way, the overall effect of a suite of pollutants on a population or community can be evaluated in a cost effective manner. This methodology is broadly known as assessing collectives of pollutants. Detoxifying enzymes (e.g. metallothionein and cytochrome P 450) are produced when organisms are exposed to a number of compounds, including PAHs, PCBs, dioxins and furans (Roesijadi et al, 1991; Anderson et al, 1995). Measurement units for cytochrome P450 are given in units of benzo[*a*] pyrene equivalents, where elevated levels of these enzymes signal the potential for toxic, carcinogenic or mutagenic responses by the organisms. Cytochrome P450 has been extensively used in monitoring studies due to low assay costs. However, more work needs to be undertaken to relate cytochrome P450 levels with overall species and community health, or integrity (Goldberg and Bertine, 2000). Hormonal active substances regulate metals and detoxification at the cellular level, and are often induced by copper and cadmium.

Another recent field of study has been to determine the endocrine disruptor effects on marine populations caused by compounds with oestrogenic properties. Oestrogen mimics may be natural or synthetic and are believed to impair the reproductive success of certain organisms. Natural sources that have demonstrated effects on marine organisms include estrone and 17*B*-estradiol. Humans have introduced large numbers of chemicals that may function as oestrogen mimics; however, oestrogen ethinyl estradiol (found primarily in oral contraceptives) has the potential for the largest effects (Matthiessen and Law, 2002). While demonstrated linkages between these compounds and organismal reproductive success are thus far lacking, anecdotal evidence suggests that these mimics may have a significant effect on certain marine species. There are a number of different techniques to assess the effect of different mimics on hormonal activity.

Lastly, toxicity identification and evaluation (TIE) and scope-for-growth (SFG) are new methods to examine the effects of contaminants on organisms. TIE has been developed by the EPA and seeks to identify the effects of individual chemicals and chemical compounds on a single or small group of organisms, through isolation of each chemical compound. SFG is a measure of the amount of energy available to an organism for somatic growth. SFG declines in polluted organisms which are having to devote energy to detoxification and tissue repair. If SFG is zero, the organism will fail to grow. If it is negative, the organism will lose body weight and potentially die. Pollutants most commonly affecting SFG are PAHs, polar organics and tributyltin (Kroger et al, 2002).

Marine monitoring at the population level is continuing to evolve. While cost effective, accurate and reliable sensors for temperature, conductivity, depth and tubidity have been developed and deployed world-wide, recent research into the field of biosensors to monitor additional chemical and biological parameters is progressing. One of the more interesting aspects of advances in biosensors is the application of the technique towards the field of molecular taxonomy, where the measurement of the type and abundance of species (generally plankton based on characteristic nucleic acid sequences) may indicate anthropogenic disturbance in a system (Kroger et al, 2000). Another research area is molecular imprinting (MIP), which introduces recognition properties into synthetic polymers, which can then be used to recognize ions, peptides and proteins, steroids and whole cells. The process develops synthetic polymer templates which are created by mixing together the target compound (template) with an appropriate monomer (often methacrylic acid) and cross-linker (e.g. ethylene glycol dimethacrylate) in appropriate solvents. This mixture is then polymerized using UV or chemical initiation. Unlike many biological compounds such as antibodies and enzymes, which are inherently unstable, MIPs are resistant to changes in pH, pressure and temperatures, are generally inexpensive, and are compatible with existing micromachine technology (Kroger et al, 2002).

Community level monitoring

Marine communities are complex, cryptic and difficult to observe and census. Consequently, they are also difficult to monitor. The most common community-level monitoring is directed towards species compositional and diversity measures such as species richness, evenness and abundance discussed in Chapter 8 (Table 18.1). Often, assessments of the 'health' or 'integrity' of marine communities is made by monitoring at the population and ecosystem levels because of the complexity of monitoring at the community level; however, there are benefits of monitoring at the community level. A recent study by Samhouri et al (2009) tested the strength and utility of 22 potential community-level indicators through a modelling experiment that simulated increased fishing effort in seven temperate marine systems where trophic interactions are reasonably well known. Results demonstrated that six of the indicators – biomass of detrivoires, flatfish, phytoplankton, jellyfish, benthic invertebrates, and the proportion of commercial and non-commercial species – showed relatively strong relationships with at least half of the 22 ecosystem attributes (Samhouri et al, 2009).

One community-level technique which has shown promise in assessing the impacts of pollutants is the use of 'ecosystem indictors' in the context of Van Gastel and Van Brummelen (1994), where human activities are evaluated using the presence, absence, abundance or composition of assemblages of different species. Marine ecosystem indicators may be sessile or mobile species as well as micro-organisms. This approach differs from traditional use of community diversity measures discussed in Chapter 8 in that the intent is to use changes in community structure to signal the presence of a particular stressor, or at least the presence of some type of stress. Ecosystem indicators are a means of turning data into information and are a way of reducing ecosystem complexity into a form that is most informative and useful to management. Examples of community-level ecosystem indicators that have been developed include (modifed from FAO, 2009):

- relative biomass of gelatinous zooplankton, cephalopods, small pelagics, scavengers, demersal fish, piscivores, top predators (TL 4+, which synthesize over large temporal and spatial scales), and biogenic habitat (cover forming species);
- biomass ratios focusing in particular on the piscivore: planktivore (PS:ZP), pelagic: demersal (P:D), and infauna: epifauna biomass ratios where values have been published for PS:ZP and P:D that can guide reference direction;

- size spectra, which indicate perturbations in system structure (using the slope of the curve), but can also highlight changes in system productivity (via the intercept);
- maximum (or mean) length which can be used if fishing size preferences are considered;
- total fisheries removals (catch+bycatch+discards) that considers the total biomass removed from the system versus what is left cycling in the system;
- diversity (counts of species) as discussed in Chapter 8;
- size at maturity (weight and length), which can indicate changes in the system and stock structure.

Ecosystem (habitat) level monitoring

The ecosystem, or habitat level, is probably the least difficult of the hierarchies to monitor given the relative ease of collecting ecosystem-level data (Table 18.1). Beyond its ease of collection, monitoring ecosystem-level data has significant value to marine monitoring initiatives. First, information on marine biological communities is sparse and often difficult to census. Ecosystem data, however, can be readily acquired using a number of in situ and remote sensing techniques (Tables 18.2, 18.3, 18.4) and, depending on the bottom-up or top-down drivers in a particular ecosystem, may be closely coupled to biotic communities and therefore used as a surrogate for biological information (Denman and Powell, 1984). Ecosystem-level monitoring technology is relatively well developed (e.g. CTD (conductivity, temperature, depth) sensors) and many of the biological responses to ecosystem variables (e.g. temperature and growth rates) have been quantitatively demonstrated. Second, biological data have often been affected by anthropogenic activities, which, depending on the monitoring objective, may reduce their value to monitoring efforts. As one example, considerable amounts of marine data are obtained from fishery catch statistics, where the collection process itself changes community composition and biomass. Lastly, the most enduring and recurrent features in marine systems are abiotic in nature (Roff and Taylor, 2000), meaning that deviations from conditions that exhibit medium- and long-term permanence may be detected more easily than fluctuations in biological attributes at the organismal, population and community level.

Recently, there have been significant international efforts to establish seafloor observatories, which are broadly defined as unmanned collections of instruments, sensors and command modules at fixed sites connected to shore stations via fibre optic networks. As an example, the US Ocean Observatories Initiative (OOI) consists of the following three elements: a regional cabled network of interconnected sites on the seafloor spanning several oceanographic features and processes, such as the North East Pacific Time-series Undersea Networked Experiment (NEPTUNE); relocatable deep sea buoys that could be deployed in harsh environments; and new construction of enhancements of existing facilties.

Other international efforts to establish major ocean observation systems include the establishment of the Integrated Global Observing Strategy (IGOS, 2003) that consists of the global climate observing system, global ocean observing system (GOOS) and the global terrestrial observing system. The GOOS is a partnership between the Intergovernmental Oceanographic Commission (IOC) of UNESCO, the World Meteorological Organisation (WMO) and the UNEP to: develop a 'globally interlinked' system for the acquisition, storage and distribution of marine and oceanographic data; develop and implement the technologies to manipulate and work with these data; and build capacity in developing countries to aquire and use marine and oceanographic data.

Monitoring human uses

As conservation and management strategies are designed to assess and mitigate human impacts on marine environments, it becomes imperative to be able to document and assess the level of these activities over time to determine relationships with changes in marine community and habitat structure (Table 18.5). As humans are a terrestrial species and marine environments are biogeochemically downstream from the terrestrial realm, monitoring programmes must therefore

Table 18.2 A comparison of benthic inventory and monitoring techniques

Benthic sampling method	Description	Coverage (km^2 h^{-1})	Characteristic spatial resolution (m)	Comments
Satellite remote sensing	Multispectral (visible and IR) systems include IKONOS, GOES, SeaWIFs, Landsat, SPOT, Quickbird.	> 100	1–1000	Applicable only in shallow environments. Temporal resolution affected by weather and orbital repeat time.
Airborne remote sensing	May be passive (e.g. multispectral, film) or active (e.g. LIDAR).	> 10	0.1–10	Applicable only in shallow environments. Spectral signature difficult to interpret.
Side scan sonar	Provides information on sediment texture, topography, bedforms and object detection.	10	1	Does not normally produce bathymetric data. Image swaths can be mosaiced to produce photo-realistic images.
Multi beam bathymetry	Produces shaded relief topographic maps which can be used to interpret seabed geology, relief and processes.	5	1	Backscatter can be used to characterize substrata. Less useful than side scan sonar for object detection.
Acoustic ground discrimination system (AGDS)	Produces maps of seabed roughness and therefore seabed characteristics.	1.5	1	Requires significant processing prior to interpretation. Existing systems include QTC – View, RoxAnn and EchoPlus.
Sub-bottom profilers	Provide high-resolution definition of the seabed sediments to approx. 50m below the seafloor.	0.8	1	Can be used to map sediment thickness and infaunal communities.
Video camera	Can be used to identify biological communities as well as field-truth other methods.	0.2	0.1	Can be operated by divers, submersibles, ROVs, or by remote methods. Difficult to establish exact positions as video is often shot obliquely.
Bottom trawls	Various methods remove objects on and/or near the bottom.	0.2	Variable	Destructive, not applicable in certain bottom types.
Photography	Can be used to identify biological communities as well as field truth other methods.	0.1	0.01	Can be used to identify biological communities as well as field truth other methods. Difficult to establish exact positions as photographs are often shot obliquely.
Benthic grab/core sampling	Fixed volume samples are sampled from benthos using divers or remotely operated equipment.	0.003	0.01	Requires additional analysis in laboratory.
Cabled seafloor observatories	Seafloor array of nodes of scientific instruments connected to each other and a shore station through fibre optic cables.	> 1000	1–1000	Currently being established in the northeast Pacific by Canada and the US as well as in the European Union.

Source: Adapted from Kenny et al (2000)

Table 18.3 A comparison of pelagic inventory and monitoring techniques

Pelagic sampling method	Description	Coverage (km^2 h^{-1})	Characteristic spatial resolution (m)	Comments
Optical satellite remote sensing	Multispectral (visible and IR) systems include IKONOS, GOES, SeaWIFs, Landsat, SPOT, Quickbird.	> 100	1–1000	Imagery only captures the first several metres of the water column. Temporal resolution affected by weather and orbital repeat time. Provides sea surface temperature and colour.
Radar satellite remote sensing	Radar systems include Radarsat, JRS-1, SIR.	> 100	10–100	Imagery captures surface water conditions which may be used to infer pelagic structures (fronts, internal waves etc.).
Airborne remote sensing	May be multispectral or film-based.	> 10	0.1–10	Used to calculate chlorophyll and suspended solids as well as census marine animals. Spectral signatures difficult to interpret.
Vessels	Many instruments can be deployed during a cruise.	> 1	Variable	Sampling rate depends on vessel speed and rate of sampling by instruments.
Passive acoustic monitoring	Acoustic detection of primarily marine mammals and fish.	> 1	Variable	May consist of a single detector or large, regional arrays such as the US Navy's Sound Surveilance System (SOSUS).
Drifters	Provide information on ocean currents. May also provide surface temperature, wind, ocean colour, pressure or salinity.	N/A	100	Spatial resolution dependent on the number of drifters deployed in an area. Currently used to model recruitment.
Sonar	Provides information on density, distribution and abundance.	< 2	0.1	Discriminate between species based on sonar data and knowledge of species' habitat requirements. Measurements of size can be made from larger species.
Submersibles/ROVs/AUVs	Provide the ability to collect samples and take video and photographs.	1	Variable	
Divers	Knowledgeable divers can identify communities and habitats in situ.	0.2	0.1	Applicable for detailed studies of small areas at shallower depths.
Cabled seafloor observatories	Seafloor array of nodes of scientific instruments connected to each other and a shore station through fibre optic cables.	> 1000	1–1000	Currently being established in the northeast Pacific by Canada and the US as well as in the European Union.

Table 18.4 A comparison of intertidal/estuarine inventory and monitoring techniques

Intertidal sampling method	Description	Coverage ($km^2\ h^{-1}$)	Characteristic spatial resolution (m)	Comments
Satellite remote sensing	Multispectral (visible and IR) systems include IKONOS, GOES, SeaWIFs, Landsat, SPOT, Quickbird.	> 100	1–1000	Applicable only in wider intertidal environments and estuaries. Temporal resolution affected by weather and orbital repeat time.
Airborne remote sensing	May be passive (e.g. multispectral or film) or active (e.g. LIDAR).	> 10	0.1–10	Applicable only in wider intertidal environments and estuaries. Can be a very effective inventory/monitoring tool.
Quadrats	Provides estimates of abundance, density over larger areas.	0.1	0.1	Very applicable in wider intertidal environments and estuaries.
Transects	Provides estimates of abundance, density over larger areas.	0.1	0.1	Applicable in narrow intertidal zones.
Photographs	Can be used to identify biological communities as well as field-truth other methods.	0.1	0.01	Difficult to establish exact positions as photographs are often shot obliquely.

assess marine environments as well as the terrestrial activities and uses that are known to have an impact on marine environments. Human-use monitoring can be separated into programmes that monitor terrestrial and upland characteristics, features and conditions and programmes that monitor marine characteristics, features and conditions (Tables 18.1 and 18.2).

A major difficulty with monitoring human activities is that, with the exception of direct, quantifiable impacts to marine resources (e.g. fishing), linking human activities (e.g. number of divers, boat trips, changes in terrestrial land use/cover) to environmental changes is difficult.

Challenges in marine monitoring

In contrast to most terrestrial monitoring programmes, marine monitoring is made difficult by a number of characteristics of marine environments beyond the fact that marine systems are often difficult to observe and census. While an in-depth treatment of the differences between terrestrial and marine systems is provided in Chapter 3, marine monitoring programmes need to consider and account for the following unique aspects of marine systems:

Determining baselines/background states: Many marine monitoring programmes are established in systems which have a long history of human impacts; therefore there may be no true baseline or 'natural state' with which to compare monitoring results (Dayton et al, 1998).

Monitoring in non-climax communities: In many temperate and Arctic marine communities, it is unknown if these communities ever reach a climax community state or are continually within early to middle successional stages. Therefore monitoring efforts (especially at the population and community level) may be monitoring natural succession rather than changes due to anthropogenic influences. The most extreme examples of this effect are regime shifts and altered stable states, which may change the successional path for communities over time scales of decades.

Terrestrial contributions: Monitoring signals may be a result of terrestrial contributions (usually via nutrients and energy inputs), which may operate

Table 18.5 General types of human-use information collected as part of human-use inventories

Structures	Consumptive activities	Non-consumptive activities	Environmental quality	Zoning
Wharves/floating moorage	Commercial fisheries	Marine camping sites	Drinking water	Water licence locations/ amounts
Industrial facilities	Recreational fisheries	Boating areas	Residences with septic tanks	Waste permit locations/ amounts
Log dumps/storage areas	Indigenous fisheries	Other recreational sites	Ocean dump sites	Upland land use zoning
Pipelines [submerged]	Marine vegetation harvesting sites	Cruising routes	Point and non-point pollution sources	Land use capability
Present land/water use	Aquaria harvest sites	Nature appreciation sites	Ballast water exchange sites	Coastal zoning
Outfalls [permitted/ non-permitted]	Aggregate extraction areas	Finfish/shellfish mariculture sites		Marine protected areas
Storm sewers	Hydrocarbon extraction areas	Wildlife viewing sites		Marine tenures
Construction sites	Other mineral extraction areas	Recreation capability surveys		Commercial operations
Water intakes				
Airports/landing strips				
Emergency response depots and staging areas				
Dykes/levees				

independently of marine processes. Monitoring efforts in neritic areas, therefore, must consider inputs from terrestrial environments.

Decreasing spatial and temporal stability from shore: Spatial and temporal stability increase with increasing distance from shore, therefore inshore monitoring information will tend to be more temporally and spatially variable than offshore results for the same variable. This has importance for designing sampling programmes as programmes should recognize and account for increasing variability closer to the coast.

Conclusions and management implications

Marine monitoring holds significant promise for detecting natural and anthropogenic changes, evaluating the success of management strategies, meeting national and international conservation commitments and assisting with environmental impact assessments. While most efforts have been directed towards collecting baseline oceanographic data, monitoring specific contaminants and setting sustainaible fishery quotas, the field is beginning to explore how monitoring can support more cross-cutting, interdisciplinary issues such as ecosystem approaches to management, climate change and cumulative impacts. However, to date, there has been only moderate progress towards applying marine monitoring for sustainable ocean management, which has been hampered by: a lack of consistent indicator-evaluation frameworks; no institution being charged with periodic collection and assessment of data on oceans; and no regular collection and assessment of socio-economic data related to the human well-being of those communities and jurisdictions that rely on marine environments (Cicin-Sain, 2007).

Experiences with marine monitoring programmes to date have concluded that suites of indicators at multiple levels of the ecological hierarchy are necessary to adequately detect, explain and, in certain instances, predict ecosystem change. Much of the current research into monitoring and indicators is segregated into groups specializing in biomonitoring at the organismal level, applying monitoring techniques to improve fisheries management, or monitoring abiotic characteristics to inform oceanographic and atmospheric research. Only recently, and for certain ecosystems (e.g. Northeast Pacific, northern Europe) have these groups begun to cooperate on developing integrated monitoring initiatives that incorporate multiple types of monitoring.

Marine monitoring programmes will continue to face challenges in identifying indicators that can be reliably censused and interpreted, and will link specific causes and effects. Longer-term efforts to develop indicators that signal and predict changes at other levels of the ecological hierarchy are still some way off. A final challenge for all marine monitoring programmes will be to ensure that monitoring results are effectively communicated to stakeholders, decision-makers and the public and that revised management direction is in turn applied to mitigate threats and reduce anthropogenic risks to marine environments. Funding agencies must also be convinced of the need for uninterrupted times series to identify changes and evaluate management strategies.

References

Anderson, J. W., Jones, J. M., Steinert, S., Sanders, B., Means, J., McMillin, D., Vue, T. and Tukeye, R. (1995) 'Correlation of CYP1A1 induction, as measured by the P450 RGS biomarker assay, with high molecular weight PAHs in mussels deployed at various sites in San Diego Bay in 1993 and 1995', *Marine Environmental Research*, vol 48, nos 4–5, pp389–405

Bernstein, B. and Weisberg, S. B. (2003) 'Southern California's marine monitoring system ten years after the National Research Council evaluation', *Environmental Monitoring and Assessment*, vol 81, pp3–14

Cicin-Sain, B. (2007) 'Johannesburg five years on: Marine policy and the world summit on sustainable development; How well are we doing?', *RGS-IBG Conference People and the Sea*, 25 October 2007, UK National Maritime Museum, Greenwich

Davies, J., Baxter, J., Bradley, M., Connor, D., Khan, J., Murray, E., Sanderson, W., Turnbull, C. and Vincent, M. (2001) *Marine Monitoring Handbook*, Joint Nature Conservation Committee, Peterborough

Dayton, P. K., Tegner, M. J., Edwards, P. B. and Riser, K. L. (1998) 'Sliding baselines, ghosts, and reduced expectations in kelp forest communities', *Ecological Applications*, vol 8, pp309–322

Denman, K. L. and Powell, T. M. (1984) 'Effects of physical processes on planktonic ecosystems in the coastal ocean', *Oceanography and Marine Biology: An Annual Review*, vol 22, pp125–168

FAO (2009) *State of the World Fisheries and Aquaculture*, Food and Agriculture Organization, Rome

Fisher, W. S. (2001) 'Indicators for human and ecological risk assessment: A US EPA perspective', *Human and Ecological Risk Assessment*, vol 7, pp961–970

Goldberg, E. D. (1975) 'The Mussel Watch: A first step in global marine monitoring', *Marine Pollution Bulletin*, vol 6, pp111–114

Goldberg, E. D. (1988) 'Information needs for marine pollution studies', *Environmental Monitoring and Assessment*, vol 11, no 3, pp293–298

Goldberg, E. D. and Bertine, K. K. (2000) 'Beyond the Mussel Watch: New directions for monitoring marine pollution', *The Science of the Total Environment*, vol 247, pp165–174

Integrated Global Observing Strategy (2003) www.fao.org/gtos/igos/docs/Igos_brochure_Jul03v07.pdf, accessed 23 December 2010

Kenny, A. J., Andrulewicz, E., Bokuniewicz, H., Boyd, S. E., Breslin, J., Brown, C., Cato, I., Costelloe, J., Desprez, M., Dijkshoorn, C., Fader, G., Courtney, R., Freeman, S., de. Groot, B., Galtier, L., Helmig, S., Hillewaert, H., Krause, J. C., Lauwaert, B., Leuchs, H., Markwell, G., Mastowske, M., Murray, A. J., Nielsen, P. E., Ottesen, D., Pearson, R., Rendas, M.-J., Rogers, S., Schuttenhelm, R., Stolk, A., Side, J., Simpson, T., Uscinowicz, S. and Zeiler, M. (2000) 'An overview of seabed mapping technologies in the context of marine habitat classification', ICES Annual Science Conference, Bruges, Belgium

Kimbrough, K. L., Johnson, W. E., Lauenstein, G. G., Christensen, J. D. and Apeti, D. A. (2008) 'An assessment of two decades of contaminant monitoring in the nation's coastal zone', *NOAA Technical Memorandum* NOS NCCOS 74, Silver Spring, MD

Kroger, S., James, D. W. and Malcolm, S. J. (2000) 'Bio-Probe – towards a sensor-based "Taxonomist on a SmartBuoy"', The World Congress of Biosensors, 24–26 May 2000, San Diego, CA

Kroger, S., Piletsky, S. and Turner, A. P. F. (2002) 'Biosensors for marine pollution, research, monitoring and control', *Marine Pollution Bulletin*, vol 45, pp24–34

Matthiessen, P. and Law, R. J. (2002) 'Contaminants and their effects on estuarine and coastal organisms in the United Kingdom in the late twentieth century', *Environmental Pollution*, vol 120, pp739–757

Messer, J. J., Linthurst, R. A. and Overton, W. S. (1991) 'An EPA programme for monitoring ecological status and trends', *Environmental Monitoring and Assessment*, vol 17, no 1, pp67–78

Moore, M. N., Depledge, M. H., Readman, J. W. and Paul Leonard, D. R. (2004) 'An integrated biomarker-based strategy for ecotoxicological evaluation of risk in environmental management', *Mutation Research – Fundamental and Molecular Mechanisms of Mutagenesis*, vol 552, nos 1–2, pp247–268

NRC (National Research Council) (1989) *Biological Markers in Reproductive Toxicology*, National Academy Press, Washington, DC

NRC (1990) *Managing Troubled Waters: The Role of Marine Environmental Monitoring*, National Academy Press, Washington, DC

Newton, J., Mumford, T., Hohrmann, J., West, J., Llanso, R., Berry, H. and Redman, S. (2000) 'A conceptual model for environmental monitoring of a marine system', Puget Sound Ambient Monitoring Programme, Olympia, WA

National Science and Technology Council (1997) *Our Changing Planet: The FY 1998 U.S. Global Change Research Programme*, Global Change Research Information Office, Washington, DC

OECD (1993) *OECD Core Set of Indicators for Environmental Performance Reviews. A Synthesis Report by the Group on the State of the Environment*, OECD, Paris

Osenberg, C. W., Schmitt, R. J., Holbrook, S. J., Abu-Saba, K. E. and Flegal, A. R. (1994) 'Detection of environmental impacts: Natural variability, effect size, and power analysis', *Ecological Applications*, vol 4, pp16–30

Rees, H. L., Hyland, J. L., Hylland, K., Mercer Clarke, C. S. L., Roff, J. C. and Ware, S. (2008) 'Environmental indicators: Utility in meeting regulatory needs. An overview', *ICES Journal of Marine Science*, vol 65, pp1381–1386

Rice, J. C. and Rochet, M. J. (2005) 'A framework for selecting a suite of indicators for fisheries management', *ICES Journal of Marine Science*, vol 62, pp516–527

Roff, J. C. and Taylor, M. (2000) 'A geophysical classification system for marine conservation', *Journal of Aquatic Conservation: Marine and Freshwater Ecosystems*, vol 10, pp209–223

Rogers, S. I., Tasker, M. L., Earll, R. and Gubbay, S. (2007) 'Ecosystem objectives to support the UK vision for the marine environment', *Marine Pollution Bulletin*, vol 54, pp128–144

Roesijadi, G., Vestling, M. M., Murphy, C. M., Klerks, P. L. and Fenselau, C. (1991) 'Structure and time-dependent behavior of acetylated and non-acetylated forms of a molluscan metallothionein', *Biochimica et Biophysica Acta*, vol 1074, pp230–236

Samhouri, J. F., Steele, M. A. and Forrester, G. E. (2009) 'Intercohort competition drives density dependence and selective mortality in a marine fish', *Ecology*, vol 90, pp1009–1020

Schiff, K. C., Weisberg, S. B. and Raco-Rands, V. (2002) 'Inventory of ocean monitoring in the Southern California Bight', *Environmental Management*, vol 29, pp871–876

UNESCO (2006) *A Handbook for Measuring the Progress and Outcomes of Integrated Coastal and Ocean Management*, UNESCO, Paris

van der Oost, R., Beyer, J. and Vermeulen, N. P. E. (2003) 'Fish bioaccumulation and biomarkers in environmental risk assessment: A review', *Environmental Toxicology and Pharmacology*, vol 13, pp57–149

Van Gastel, C. A. and Van Brummelen, T. C. (1994) 'Incorporation of the biomarker concept in ecotoxicology calls for a redefinition of terms', *Ecotoxicology*, vol 5, pp217–225

Zacharias, M. A. and Roff, J. C. (2000) 'An ecological framework for the conservation of marine biodiversity', *Conservation Biology*, vol 14, no 5, pp1327–1334

Zacharias, M. A. and Roff, J. C. (2001) 'Use of focal species in marine conservation and management: A review and critique', *Aquatic Conservation: Marine and Freshwater Ecoystems*, vol 11, pp59–76

19

Remaining Problems in Marine Conservation

Present problems, future solutions

The significant problems we have cannot be solved at the same level of thinking with which we created them.

Albert Einstein (1879–1955)

Introduction

Our objective in this book was to present the science of marine biodiversity and marine conservation. The book began with a discussion about the need for marine conservation and then explored the theory and practice of marine conservation from the perspective of ecological principles. Because our focus in this book is on the ecological basis and principles for marine conservation, we were less concerned with other aspects of marine conservation such as international conventions, marine management, policy, legislation, enforcement, socio-economics and the human considerations in marine conservation. We also acknowledged that other subjects, including population biology, ecology and fisheries biology, are covered elsewhere in standard texts.

Our emphases were primarily on: the identity and components of marine biodiversity; potential approaches to the conservation of marine biodiversity; its relationships to environmental structure and heterogeneity; and the planning of practical marine conservation strategies at the regional and national levels. The main rationale for this approach is that although concerns for the preservation of marine biodiversity are truly global and international, nevertheless most planning and practical initiatives to conserve marine biodiversity will be undertaken at the national and regional levels. The history of science (whatever the discipline) shows how '...the development of comprehensive theoretical systems seems to be possible only after a preliminary classification has been achieved' (Nagel, 1961). The intention was not to ignore or gloss over other aspects of, or disciplines in, marine conservation; rather it was to emphasize the fundamental importance of ecological knowledge and planning.

Marine protected areas (MPAs) are a primary interest of this text not because the establishment of MPAs is the only thing we should do, but because it has been repeatedly shown that MPAs are effective in protecting 'pieces' of the marine environment and their constituent species. In addition, many jurisdictions already have the social and political licence to establish MPAs, and many MPAs have been established worldwide. Thus, much of this text is simply advocating the use of existing legal and policy tools to manage and conserve marine environments. Our perspective on the application of MPAs is that, while not alone sufficient to protect marine environments, they are a place to start and perhaps 'buy time' while other management prescriptions (e.g. fisheries conservation, regulating land-source pollution) are implemented.

Our book is fundamentally about how to address the various impacts (climate change, habitat loss, introduced species, overharvesting, pollution) of humans on marine ecosystems. The role of this book is not to elaborate on the ecological, social or geographical extent of these impacts as this has been done elsewhere but to propose approaches/techniques at the genetic, population, habitat/community and ecosystem levels to understand, quantify and mitigate these impacts. As such, this book provides only a cursory overview of global threats to marine ecosystems (e.g. climate change, pollution) as the knowledge and understanding of these threats is increasing daily and any detailed summary of these threats is likely to be dated by the time this book is published.

In this book we could not hope to address all the problems of marine conservation. To date, traditional approaches to marine conservation (fisheries management, coastal zone management, ecosystem approaches to management and MPAs), with the exception of certain localized successes, have not slowed, halted or reversed degradation of the marine environment. At best, most conservation efforts have achieved a 'managed decline' rather than maintenance or restoration of ecological components.

However, through an examination of ecological principles and a study of the nature of the marine environment, we can present the issues that need to be addressed in a systematic way. What we were attempting then is a 'codification' of the undertaking of marine conservation. By this we mean that we examined the various problems that arise in attempting to systematically deal with the practicalities of marine conservation, to show what they are and how they can be dealt with from available or obtainable data. The main aim of this book therefore was to present the major ecological concepts of, and approaches to, marine conservation, with an emphasis on protected areas – MPAs.

There are a number of chapters that we would have liked to have written but this will need to wait for a second edition. We conducted a straw poll among peers asking the question: 'What are the remaining problems in marine conservation?' These remaining problems presented here are not an exhaustive list and are not mutually exclusive but are the issues that, through the course of writing this book, we believe are necessary to consider in order to move the marine conservation discipline forwards. These 'remaining problems' also assume that fundamental research into understanding the structure and function of marine ecosystems (knowledge gaps) is an overarching problem and a prerequisite for further improvement in the management and conservation of marine environments. Furthermore, this chapter stipulates that improved and expanded inventories at the genetic through to ecosystem levels are also necessary for improved conservation and management outcomes. As such, we have focused the remainder of this chapter on what we term 'ecological management problems' so as to differentiate from other types of 'remaining problems' (e.g. shared jurisdiction, lack of inventories) that are outside the scope of this book.

'Remaining problems' in the various ecological hierarchies

There are hundreds, if not thousands, of 'remaining problems' in marine conservation and the significance of these problems depends largely on who you ask and where they live. At the genetic level it is still relatively rare for molecular data to be directly applied towards natural resource conservation (Latch and Ivy, 2009). Molecular ecology is still a relatively new discipline where techniques continue to be refined to be able to estimate effective population sizes, determine population structure, detect hybridization and bottlenecks, resolve taxonomic uncertainties (e.g. cryptic species), identify dispersal patterns, and identify molecular markers that may be used to indicate environmental change and understand how species respond to change (Primmer, 2009). Indeed, the journal *Conservation Genetics* was not established until 2000 and the first textbook in the field (*Introduction to Conservation Genetics*) only published in 2002 (Frankham et al, 2002).

At the species/population level, many fundamental ecological questions that have conservation implications are yet to be answered. While new tools such as trace elements, isotopes and satellite tracking are improving knowledge at the species/population level, the 'remaining problems' at this level of the hierarchy are fundamentally a result of two issues: first, knowledge of the diversity and abundance in most of the world's oceans is undersampled. Second, most species are only observed during part of their life cycles (e.g. Burton, 2009). While an exhaustive list of remaining problems at the species/population level is beyond the scope of this book some conservation-related examples include: the geographic range of individual populations; the role of behaviour in the development of management strategies; whether aquaculture/mariculture is a problem or a solution for marine management and conservation; the role of population augmentation/enhancement in marine conservation; determining what types of fisheries recover from collapse and why.

At the community level, 'remaining problems' are primarily related to the lack of knowledge of the interactions between species and populations and how human activities influence these interactions. Fundamental questions such as the degree to which competition structures marine communities, the relationships between the classical and microbial food webs, and the role(s) of top predators in structuring marine food webs have yet to be answered. Many outstanding community-level questions have imminent real-world significance, such as whether whale populations should be controlled to assist the recovery/management of commercial fish stocks (Gerber et al, 2009).

At the habitat/ecosystem level, 'remaining problems' are primarily related to the lack of understanding of biophysical processes within the ocean and the broader relationships between the hydrosphere–atmosphere and hydrosphere–lithosphere. From a conservation perspective the more immediate issues at the habitat/ecosystem level are related to: further biogeographic characterization particularly in offshore pelagic and benthic realms (e.g. Agardy, 2010); modelling the fate and persistence of pollutants (land- and marine-based) in marine environments; the impacts of climate change on marine habitat composition and quality; global-scale prioritization of geographic areas or habitats for conservation attention.

'Remaining problems' across the various ecological hierarchies requiring an integrative approach

While the previous section outlines some of the broader issues that confound attempts to conserve marine environments, these issues, or 'remaining problems', are primarily related to a lack of understanding of the composition, function and structure of marine systems. However, there are additional remaining problems that cross ecological hierarchies but, while related to a lack of understanding of marine systems, also highlight problems associated with theories of marine conservation and how these theories are applied in marine decision-making.

What is the goal of marine conservation?

The current ecological state of much of the marine environment leaves no doubt that marine conservation is necessary. Many vertebrate species (fish, whales) have been harvested to the brink of commercial, if not ecological, extinction. Vast areas of the ocean (e.g. polar regions) are expected to be significantly altered as a result of increasing greenhouse gas concentrations. Many important habitats (e.g. corals, mangroves, seagrasses) have been lost or degraded as a result of human activities. While organizations and jurisdictions throughout the world are currently occupied with addressing impacts to marine environments, very few individuals stop to ask the fundamental question: 'what is the goal of marine conservation?' This is far from a rhetorical query or naval-gazing exercise, for without a goal to work towards, conservation efforts become open-ended endeavours where scarce resources are diluted to accommodate an increasing number of semi-permanent conservation programmes. In addition, without an overall goal for marine conservation, whether globally, regionally or subregionally, incremental gains (e.g. recovery of single species) cannot be evaluated in the larger context of what success looks like. Furthermore, in the absence of an overarching conservation goal, disparate groups may be working at cross-purposes. Perhaps the most egregious example of this type of disconnect is the debate about whether to reduce marine mammal populations in order to help restore fish stocks (e.g. Gerber et al, 2009).

Most answers to the question: what is the goal of marine conservation? involve one or more of the following: dodging the question by reiterating the reasons for protecting marine biodiversity (see Chapter 1; Thorne-Miller and Catena, 1991); focusing conservation efforts on certain geographic areas or functional groups (e.g. Kelleher et al, 2004); stating what humans value in marine systems (see Chapter 15); and invoking the need to maintain structures and processes at the genetic, species/population, community and ecosystem levels (see Chapter 4; Convention on Biological Diversity, 2010). This book is guilty of these failings, not because we lack the fortitude to address this topic but because very little discussion has occurred regarding what, exactly, the purpose of our conservation efforts is and what it is that humanity is trying to achieve.

The answers to the question of: what is the goal of marine conservation?' potentially range from the very focused (e.g. protect rare alleles in marine populations), to the all-encompassing (e.g. protect ecosystem structure and function), to the nearly tautological (e.g. the goal of marine conservation is ecosystem-based management), to the utilitarian (e.g. maintain the sustained provision of ecological goods and services to humans), to the focus on process (e.g. establish MPAs). Outside of certain exceptions, such as the Great Barrier Reef, a common consensus on the goal of conservation efforts in most habitats and ecosystems is lacking, which potentially leads to working at cross-purposes, or potentially confusing activity with achievement.

While many ocean 'action plans' and 'blueprints' have been and are being developed, the broader questions related to the overall objectives of marine conservation remain unanswered. These questions are becoming particularly pertinent as marine environments face significant impacts as a result of climate change (see below).

Prioritizing conservation actions

Marine conservation has yet to properly consider the full array of prioritization options and how these options should be incorporated into conservation efforts. Prioritization, broadly defined, is the establishment of priorities, including the sequence and timing of actions such as the establishment of regulations or protected areas. The scope of prioritization is broad and could encompass: allocating conservation resources (funding, staff, protected area budgets) among regions; designing MPAs; operating species conservation programmes; designing biodiversity surveys and monitoring programmes; managing threatened, migratory or invasive species and investing in greenhouse gas mitigation schemes (McCarthy et al, 2010).

'Prioritizing' in the marine conservation lexicon is primarily interpreted as the selection of areas (e.g. MPAs, priority conservation areas) to be conserved or managed differently from surrounding areas, or the selection of which populations or species should receive preferential management attention (e.g. prioritizing based on stock status). Examples of prioritization issues in marine systems include questions related to determining when MPAs are a better management tool than fishery closures or whether conservation investments should be proactive (e.g. prohibiting damaging impacts) or reactive (e.g. mitigating damaging impacts).

At the population/species level marine prioritization is deficient primarily due to a lack of information regarding population/species conservation status, which is particularly important for the conservation of socio-economically important species (e.g. fisheries, Brehm et al, 2010). While it is not surprising – given the difficulties in observing and censusing the marine environment – that less than 5 per cent of species on the IUCN red list are marine species, at the population/species level what is required are better methods to evaluate and prioritize conservation threats across taxa and functional groups (IUCN,

2010). While different population/species prioritization systems have been developed for terrestrial conservation purposes, marine conservation continues to lag behind terrestrial efforts in this area.

Marine conservation needs to review the rule-based, scoring and ranking systems used in terrestrial environments for their applicability in marine systems (Brehm et al, 2010). In particular, many terrestrial prioritization systems use 'traditional' ranking criteria (e.g. threat, endemicity, rarity, population decline, quality of habitat, intrinsic biology vulnerability, human impact, species abundance in relation to their geographical range size, recovery potential and estimated budget for conservation, taxonomic uniqueness, phylogenetic criteria, cultural values, economic criteria, level of species knowledge, state of present research) with modifiers such as the number of countries in which the taxa occur as well as 'use' categories (e.g. food crop, industrial, ornamental, cultural value) (Ford-Lloyd et al, 2008).

At the community level, additional work is required on prioritizing community-level considerations. Currently, research is under way on a number of community-level prioritization topics, including: developing methods to assess the cumulative impacts on marine communities (e.g. Halpern et al, 2008; Ban et al, 2010); identifying and using focal species for marine conservation and management (e.g. Caro, 2010); and continuing efforts to model how energy moves through marine food webs (e.g. Christensen et al, 2009). However, methods to balance conservation objectives related to, for example, linking species diversity, richness and abundance with other concerns such as endemism, succession and regime shifts has been recognized as necessary but not yet operationalized into a consistent methodology.

At the ecosystem/global level, there are currently nine different terrestrial conservation prioritization systems based on various criteria that recommend where conservation efforts should be made (Brooks et al, 2006). While these disparate systems are based on varied criteria (e.g. 'hotspots', intact forests, centres of endemism) and often recommend opposing areas (e.g. intact boreal forests are not centres of endemism), these efforts have few marine corollaries. Perhaps the nearest marine effort of this kind is the work of Halpern et al (2008) who synthesized 17 global datasets of anthropogenic drivers to develop a global map of human impact on marine ecosystems. While an important exercise, the subsequent task is to determine whether to invest in those areas with relatively low human disturbance (e.g. polar regions) or those areas subject to significant human degradation (e.g. most coastal tropical regions).

Lastly, marine conservation needs to develop methods to link ecological considerations to return on investment (ROI) approaches that provide direction on the optimal allocation of funds across a spectrum of potential projects and actions (Murdoch et al, 2007). The ROI approaches differ from existing cost-based approaches used by conservation planners (e.g. MARXAN, SITES) in that, unlike existing approaches, the ROI outputs recommend the conservation tool (e.g. protected areas) as well as the optimal spatial configuration to maximize conservation objectives (Murdoch et al, 2007).

'Thresholds' and 'baselines' and their efficacy in marine conservation and management

Until recently, setting conservation objectives based on historical knowledge of the past composition, structure and function of an ecosystem was established practice (Chapter 13, Christensen et al, 2009). Reconstructing pre-impact food webs through modelling, molecular techniques and historical information is now common and most marine areas of commercial fisheries importance have been described in this manner. This information is used to set backward-looking conservation objectives (e.g. restore top predator populations to some percentage of historical level, maintain the ecological function of certain functional groups) that attempt to restore a system, or components of a system, to some historical state.

However, in an era of climate change and other human impacts we need to acknowledge the extent to which the past is becoming a less and less reliable predictor of the future. In some instances our knowledge of the past and inherent desire to predict the future handicaps decision-making as we cling to models predicting how much of a population can be removed from a system while still keeping the system 'natural' within some ill-defined criteria. This reconstructive approach continues to be used even when we fully acknowledge that the system under management bears a diminishing resemblance to some past historical condition given centuries of past human impacts combined with rapid greenhouse gas-induced ecological changes.

Thus, marine conservation needs to revisit the question: what are the acceptable limits of change in those marine ecosystems that cannot be restored to some past state? While many managers and scientists are struggling with this issue, the emerging discipline of 'threshold dynamics' perhaps has the best chance of identifying where the proverbial ecological 'lines in the sand' – or rather in the water – actually exist.

Ecological thresholds can be defined as a state or condition beyond which rapid state or regime shifts occur. Threshold dynamics is the study of these tipping points and broadly encompasses the concepts of resilience, regime shifts, trophic shifts, altered stable states and, to a lesser extent, ecological integrity (Osman et al, 2010), which have independently been the subject of many studies over the past several decades. While these concepts appear throughout this book in various guises, only recently have these previously partially-linked concepts been amalgamated under the aegis of 'threshold dynamics', which is an important step forward for marine conservation and management and therefore worthy of discussion here.

If ecological 'thresholds' are upper boundaries of acceptable change that are not to be exceeded without ecological, often cascading consequences, then 'baselines' are the initial, pre-industrialized ecological conditions targeted by management efforts. Baselines are loosely defined as the pre-existing condition or state of a population, habitat or ecosystem prior to human impacts. Baselines can be set or identified through a number of means, including: direct or indirect (e.g. passed down over time) traditional knowledge of aspects of a system (e.g. past average catches or presence of certain species); past written or visual records (e.g. landed catches, inventories); or use of molecular techniques to determine past population estimates. The immediately apparent problem with the use of baselines as a conservation construct is that 'true' baselines may not have existed in many nearshore marine systems for millennia given the long and heavy use of marine environments by pre- and post-Western societies.

The use of traditional knowledge of a system may lead to 'sliding baselines' where conservation and management outcomes are based on a false understanding of the characteristics of a pre-impact system (Dayton et al, 1998). Sliding baselines inevitably lead to weakened conservation targets, which in turn may lead to further degradation of a system. A further twist on the sliding baselines problem is the existence in certain marine systems of 'alternate stable states', which are long periods of what appear to be stable, often healthy climax communities free of human impacts but are actually communities that have been changed to such a degree that the dominant biota have completely changed from one climax community (e.g. invertebrates) to another (e.g. macroalgae).

As discussed above, setting conservation targets becomes even more difficult in the face of anthropogenic-induced climate change. Direct physical and oceanographic consequences of human-induced rapid increases in greenhouse gas concentrations are currently understood to result in: increasing global ocean temperatures; acidification; rising sea levels; altered ocean circulation; and changing weather patterns (Brierley and Kingsford, 2009). Marine biological responses to increased greenhouse gas concentrations are still not fully understood but currently include: decreased ocean productivity; altered food web dynamics; reduced abundance of habitat forming species; shifting species distributions; and a greater incidence of disease (Hoegh-Guldberg and Bruno, 2010). Consensus has been reached that the rate of change in greenhouse gas concentrations is unprecedented in the earth's history; however, the impact on human and ecological communities is only recently beginning to be understood. The broader question related to climate change is whether conservation targets have any meaning in a system where impacts are as yet unknown and currently cannot be reliably predicted.

Therefore, if past baselines are unknown and it may not be feasible to restore a system to a previous condition due to climate change, how should conservation outcomes be identified and conservation targets set? The fundamental premise of this book is that identifying and protecting representative and distinctive marine areas is the foundation for all conservation efforts, which is a departure from terrestrial species/population management that focuses on the recovery of individual species (single-species management, often driven by legislative requirements).

Our suggested approaches to maintaining ecological structure, process and function may not lead to a desired outcome (e.g. abundance of an identified species and its ecological role) but will, in the absence of a past baseline or threshold, continue to allow evolutionary processes to unfold. If human impacts are properly managed, this approach may result in the emergence of a long-term stable state, however different from past historical configurations, that can be understood and managed for the long-term production of goods and services.

Integrating human and ecological considerations in marine decision-making

This book, for the most part, intentionally excluded human considerations as this topic has been considered elsewhere and because integrating human and ecological considerations is a recent development in marine conservation. Terrestrial conservation is rife with examples of how human activities, choices and behaviours can be used as inputs into models that predict ecosystem status. For example, the impacts of snowmobiles on wildlife (either through triggering avoidance or providing corridors for predator movement as a result of compacting the snowpack) can be modelled, as can the ecological impacts with varying levels of snowmobile activity at different times, elevations and geographic locations (Seip et al, 2007).

Marine ecology has lacked a formal framework for incorporating human activities into decision-making. At best, certain reserve design algorithms and approaches (Chapters 16 and 17) allow human activities to be captured spatially where they are used to modify the conservation 'score' or desirability of a particular geographic location. Certain conservation programmes are beginning to use human-use information such as commercial shipping routes, commercial fishing areas, marine infrastructure (e.g. oil rigs) and recreational uses as inputs into conservation decision-making; however, most of these applications are geared towards reserve design and cumulative impacts (Ban et al, 2010; Halpern et al, 2008).

What continues to be missing from marine conservation efforts are models of human behaviour. For example, a relationship exists between high-seas commercial fishing and fuel prices; however, this is not simply a relationship based on the costs of production and landed value but a more complex calculus involving human perceptions and emotion that could be modelled and input into marine conservation decision-making. With a greater understanding of how human behaviour is modified by fuel prices, the effects of fuel taxes on marine biodiversity objectives can be explored.

Understanding the minimum amount of information needed to make informed management decisions

The discipline of conservation biology has several characteristics distinguishing it from ecology or other forms of biological science. Foremost, however, is how conservation biology as a discipline addresses risk. Conservation biology has been formulated as a 'crisis discipline' where the traditional, adversarial, time-consuming model of developing and testing abstract ecological models is replaced with a precautionary philosophy of acting sooner even when decisions are not fully informed by the best available information (Frankel and Soule, 1981).

Marine conservation must thus balance the need to make informed decisions using the best available science with the need to make timely decisions on biological values under threat. Population growth, extinction risk and reserve design are examples of model outputs used in decision-making that can vary widely given small changes or uncertainties in input data (McDonald-Madden et al, 2008). Given that many marine environments lack full inventories of their species, communities and habitats, combined with a dearth of monitoring over time, the question of how much data is needed to make conservation decisions is germane. One of the most effective routes in blocking or delaying conservation decisions is to question the scientific inputs into the decision-making process.

Conversely, even if information is available it may not be used to inform decision-making. Cook et al (2010) reviewed the information used to manage over 1000 protected areas in Australia and found that approximately 60 per cent of conservation management decisions relied on experience-based, rather than evidence-based, information. This study suggests that, regardless of the amount of information available, decisions will still be made based on personal experience, 'gut' feeling and political considerations. This work supports the conclusions by Gerber and DeMaster (1999) and Norris (2002) that, although data may accumulate in a linear fashion, the power of data to address policy issues does not.

Concerns regarding the scientific basis of decision-making generally fall into the following categories:

- Biological inventories are insufficient to accurately estimate population status.
- Models are insufficiently accurate/detailed/robust to provide credible results.
- Existing management measures have not been given sufficient time to see if they will work.
- The effects of climate change nullify any efforts to predict future condition or success of management efforts.

Marine conservation is particularly susceptible to delaying decisions due to a real or perceived lack of information, particularly on stock status for commercially harvested species, combined with the difficulty of engaging the public on often cryptic, unknown species. However, outside of discussions related to the information required to de-list endangered species (Gerber and Hatch, 2002) there have been few discussions on what 'burden of proof' is required before embarking on a particular conservation action. The Food and Agricultural Organization (FAO) states that: 'Conservation and management decisions for fisheries should be based on the best scientific evidence available, also taking into account traditional knowledge of the resources and their habitat, as well as relevant environmental, economic and social factors' (Cochrane, 2002). FAO guidelines further state that this requirement involves three steps:

1. The collection of suitable data and information on the fisheries, including on the resources and on the environmental, economic and social factors.
2. Appropriate analysis of these data and information so that they may be used to address the decisions that need to be made by the fisheries managers.
3. The consideration and use of the analysed data and information in actually making the decisions. (Cochrane, 2002)

However, this guidance is vague at best and can easily be used as an excuse to delay potentially controversial decisions until further information is collected. This phenomenon has been observed in terrestrial conservation biology, which was initially theoretically informed and guided by the principles of island biogeography theory and genetics to guide reserve design (Linquist, 2008). This approach, challenged due to its seemingly simplistic view of the function of ecological systems, shifted towards a more rigorous autoecological (case-by-case) approach in the hopes to bring more credibility to the discipline. The onerous requirements of the autoecological approach often exceeded the financial capacity of the organizations responsible for decision-making, which led to the development of reserve selection algorithms (e.g. MARXAN) based on maximizing the number of objectives in the minimum area. These algorithms have received criticism for their apparent lack of consideration for certain genetic (e.g. genetic consequences of small population), population (e.g. minimum viable populations), community (e.g. species–area relationships) and ecosystem (e.g. retention mechanisms) processes.

An interesting marine exception is the current global moratorium on the harvest of large whales. In this instance, a very cautious and precautionary harvest regime (the Revised Management Procedure – RMP) sits idly while species and stocks that would not be allowed to be harvested under the RMP are currently harvested under scientific permit (Zacharias et al, 2006). In this instance, a robust decision-making framework that considers the quality and quantity of input information exists but has not been applied due to a lack of social licence.

There continues to be a need to address the question: how much information is enough in marine systems? While techniques such as the info-gap decision theory (Ben-Haim, 2006) have shown promise in assessing the effects of incomplete information on model accuracy, many important conservation decisions are held up pending more accurate information, and a broader discussion needs to occur on how decisions should be made in the absence of complete information.

Conclusions

The above remaining problems are only a sample of the work that needs to be completed to both achieve parity between terrestrial and marine conservation as well as to move the discipline to a place where 'managed decline' is no longer the best achievable outcome.

References

Agardy, T. (2010) *Ocean Zoning: Making Marine Management More Effective*, Earthscan, London

Ban, N. C., Alidina, H. M and Ardron, J. A. (2010) 'Cumulative impact mapping: Advances, relevance and limitations to marine management and conservation, using Canada's Pacific waters as a case study', *Marine Policy*, vol 34, pp876–886

Ben-Haim, Y. (2006) *Info-gap Decision Theory: Decisions under Severe Uncertainty*, Academic Press, Sydney, Australia

Brehm, J. M., Maxted, N., Martins-Loução, M. A. and Ford-Lloyd, B.V. (2010) 'New approaches for establishing conservation priorities for socio-economically important plant species', *Biodiversity Conservation*, vol 19, pp2715–2740

Brierley, A. S. and Kingsford, M. J. (2009) 'Impacts of climate change on marine organisms and ecosystems', *Current Biology*, vol 19, pp602–614

Brooks. T. M., Mittermeier, R. A., da Fonseca, G. A. B., Gerlach, J., Hoffmann, M. Lamoreux, J. F., Mittermeier, C. G., Pilgrim, J. D. and Rodrigues, A. S. L. (2006) 'Global biodiversity conservation priorities', *Science*, vol 313, pp58–61

Burton, R. S. (2009) 'Molecular markers, natural history, and conservation of marine animals', *BioScience*, vol 59, no 10, pp831–840

Caro, T. (2010) *Conservation by Proxy: Indicator, Umbrella, Keystone, Flagship, and Other Surrogate Species*, Island Press, Washington, DC

Christensen, V., Walters, C. J., Ahrens, R., Alder, J., Buszowski, J., Christensen, L. B., Cheung, W. W. L., Dunne, J., Froese, R., Karpouzi, V., Kastner, K., Kearney, K., Lai, S., Lam, V., Palomares, M. L. D., Peters-Mason, A., Piroddi, C., Sarmiento, J. L., Steenbeek, J., Sumaila, R., Watson, R., Zeller, D. and Pauly, D. (2009) 'Database-driven models of the world's large marine ecosystems', *Ecological Modelling*, vol 220, pp1984–1996

Cochrane, K. L. (2002) *A Fishery Manager's Guidebook: Management measures and their application*, FAO Fisheries Technical Paper, No. 424, FAO, Rome

Convention of Biological Diversity (2010) 'A new era of living in harmony with Nature is born at the Nagoya Biodiversity Summit', www.cbd.int/doc/press/2010/pr-2010-10-29-cop-10-en.pdf, accessed 3 November 2010

Cook, C. N., Hockings, M. and Carter, R. W. (2010) 'Conservation in the dark? The information used to support management decisions', *Frontiers in Ecology and the Environment*, vol 8, pp181–186

Dayton, P. K., Tegner, M. J., Edwards, P. B. and Riser, K. L. (1998) 'Sliding baselines, ghosts and reduced expectations in kelp forest communities', *Ecological Applications*, vol 8, pp309–322

Ford-Lloyd, B.V., Brar, D., Khush, G. S., Jackson, M.T. and Virk, P. S. (2008) 'Genetic erosion over time of rice landrace agrobiodiversity', *Plant Genetic Resources*, vol 7, no 2, pp163–168

Frankel, O. H. and Soulé, M. E. (1981) *Conservation and Evolution*, Cambridge University Press, Cambridge

Frankham, R., Ballou, J. D. and Briscoe, D. A. (2002) *Introduction to Conservation Genetics*, Cambridge University Press, Cambridge

Gerber, L. R., and Hatch, L.T. (2002) 'Are we recovering? An evaluation of recovery criteria under the U.S. Endangered Species Act', *Ecological Applications*, vol 12, pp668–673

Gerber, L. R., and DeMaster, D. P. (1999) 'A quantitative approach to Endangered Species Act classification of long-lived vertebrates: Application to the North Pacific humpback whale', *Conservation Biology*, vol 13, no 5, pp1203–1214

Gerber, L. R., Morissette, L., Kaschner, K. and Pauly, D. (2009) 'Should whales be culled to increase fishery yield?', *Science*, vol 323, pp880–881

Halpern, B. S., Walbridge, S., Selkoe, K. A., Kappel, C.V., Micheli, F., D'Agrosa, C., Bruno, J. F., Casey, K. S., Ebert, C., Fox, H. E., Fujita, R., Heinemann, D., Lenihan, H. S., Madin, E. M., Perry, M. T., Selig, E. R., Spalding, M., Steneck, R. and Watson, R. (2008) 'A global map of human impact on marine ecosystems', *Science*, vol 319, pp948–952

Hoegh-Guldberg, O. and Bruno, J. F. (2010) 'The impact of climate change on the world's marine ecosystems', *Science*, vol 328, pp1523–1528

IUCN (2010) 'Assessment process', www.iucnredlist.org/technical-documents/assessment-process, accessed 05 November 2010

Kelleher, G., Glover, L. and Earle, S. (2004) *Defying Ocean's End: An Agenda for Action*, Island Press, Washington, DC

Latch, E. K. and Ivy, J. A. (2009) 'Meshing molecules and management: A new era for natural resource conservation', *Biology Letters*, vol 5, pp3–4

Linquist, S. (2008) 'But is it progress? On the alleged advances of conservation biology over ecology', *Biology and Philosophy*, vol 23, pp529–544

McCarthy, M. A., Thompson, C. J., Hauser, C., Burgman, M. A., Possingham, H. P., Moir, M. L., Tiensin, T. and Gilbert, M. (2010) 'Resource allocation for efficient environmental management', *Ecology Letters*, vol 13, pp1280–1289

McDonald-Madden, E., Baxter, P.W. and Possingham, H. P. (2008) 'Making robust decisions for conservation with restricted money and knowledge', *Journal of Applied Ecology*, vol 45, pp1630–1638

Murdoch, W. Polasky, S., Wilson, K. A., Possingham, H. P., Kareiva, P. and Shaw, R. (2007) 'Maximizing return on investment in conservation', *Biological Conservation*, vol 139, pp375–388

Nagel, E. (1961) *The Structure of Science*, Hackett, Cambridge

Norris, S. (2002) 'How much data is enough? Lessons on quantifying risk and measuring recovery from the California Gray Whale', *Conservation Magazine*, vol 3, no 1

Osman, R. W., Munguia, P. and Zajac, R. N. (2010) 'Ecological thresholds in marine communities: Theory, experiments and management', *Marine Ecology Progress Series*, vol 413, pp185–187

Primmer, C. R. (2009) 'From conservation genetics to conservation genomics', *The Year in Ecology and Conservation Biology*, vol 1162, pp357–368

Seip, D. R., Johnson, C. J. and Watts, G. S. (2007) 'Displacement of mountain caribou from winter habitat by snowmobiles', *Journal of Wildlife Management*, vol 71, no 5, pp539–544

Thorne-Miller, B. and Catena, J. (1991) *The Living Ocean: Understanding and Protecting Marine Biodiversity*, Island Press, Washington, DC

Zacharias, M. A., Gerber, L. R. Hyrenbach, K. D. (2006) 'Review of the Southern Ocean Sanctuary: Marine Protected Areas in the context of the International Whaling Commission Sanctuary Programme', *Journal of Cetacean Resource Management*, vol 8, no 1, pp1–12

INDEX